INTERSCIENCE MONOGRAPHS AND TEXTS IN PHYSICS AND ASTRONOMY

Edited by **R. E. MARSHAK**

Volume I: E. R. Cohen, K. M. Crowe, and J. W. M. DuMond, **The Fundamental Constants of Physics**

Volume II: G. J. Dienes and G. H. Vineyard, **Radiation Effects in Solids**

Volume III: N. N. Bogolubov and D. V. Shirkov, **Introduction to the Theory of Quantized Fields**

Volume IV: J. B. Marion and J. L. Fowler, *Editors*, **Fast Neutron Physics** *in two parts* **Part I: Techniques** **Part II: Experiments and Theory**

Volume V: D. M. Ritson, *Editor*, **Techniques of High Energy Physics**

Volume VI: R. N. Thomas and R. G. Athay, **Physics of the Solar Chromosphere**

Volume VII: Lawrence H. Aller, **The Abundance of the Elements**

Volume VIII: E. N. Parker, **Interplanetary Dynamical Processes**

Volume IX: Conrad L. Longmire, **Elementary Plasma Physics**

Volume X: R. Brout and P. Carruthers, **Lectures on the Many-Electron Problem**

Volume XI: A. I. Akhiezer and V. B. Berestetskii, **Quantum Electrodynamics**

Volume XII: John L. Lumley and Hans A. Panofsky. **The Structure of Atmospheric Turbulence**

Volume XIII: Robert D. Heidenreich, **Fundamentals of Transmission Electron Microscopy**

Volume XIV: G. Rickayzen, **Theory of Superconductivity**

Volume XV: Raymond J. Seeger and G. Temple, *Editors*, **Research Frontiers in Fluid Dynamics**

Volume XVI: C. S. Wu and S. A. Moszkowski, **Beta Decay**

Volume XVII: Klaus G. Steffen, **High Energy Beam Optics**

Volume XVIII: V. M. Agranovich and V. L. Ginzburg, **Spacial Dispersion in Crystal Optics and the Theory of Excitons**

Volume XIX: A. B. Migdal, **Theory of Finite Fermi Systems and Applications to Atomic Nuclei**

Volume XX: B. L. Moiseiwitsch, **Variational Principles**

Volume XXI: L. S. Shklovsky, **Supernovae**

Volume XXII: S. Hayakawa, **Cosmic Ray Physics**

Volume XXIII: J. H. Piddington, **Cosmic Electrodynamics**

Volume XXIV: R. E. Marshak, Riazuddin, and C. P. Ryan, **Theory of Weak Interactions in Particle Physics**

Volume XXV: Norman Austern, **Direct Nuclear Reaction Theories**

Volume XXVI: Gérard G. Emch, **Algebraic Methods in Statistical Mechanics and Quantum Field Theory**

INTERSCIENCE MONOGRAPHS AND TEXTS IN PHYSICS AND ASTRONOMY

Edited by R. E. MARSHAK

University of Rochester, Rochester, New York

VOLUME XXVI

Algebraic Methods in Statistical Mechanics and Quantum Field Theory

GERARD G. EMCH

UNIVERSITY OF ROCHESTER

WILEY–INTERSCIENCE, a division of

John Wiley & Sons, Inc. New York · London · Sydney · Toronto

Library of Congress Catalog Card Number: 70-175792

ISBN 0-471-23900-3

Printed in the United States of America.

10 9 8 7 6 5 4 3 2 1

Preface

La nature contient les éléments, en couleur et en forme, de toute peinture, comme le clavier contient les notes de toute musique.

Mais l'artiste est né pour en sortir, et choisir, et grouper avec science, les éléments, afin que le résultat en soit beau—comme le musicien assemble ses notes et forme des accords—jusqu'à ce qu'il éveille du chaos la glorieuse harmonie.

Dire au peintre qu'il faut prendre la nature comme elle est, vaut de dire au virtuose qu'il peut s'asseoir sur le piano.

Mallarmé [1885]

Contemporary scientific literature offers ample evidence that the algebraic methods which have revolutionized pure mathematics are now in the process of having a similar impact in the physical sciences. The algebraic approach to statistical mechanics and quantum field theory is an example of this new orientation.

The motivation of this approach was a dissatisfaction with the discord too often perceived in the sounds produced from the keyboard of theoretical physics. These flaws were felt to be due to a technique that was too undiscriminating and too restricted in its range. The idea that algebraic methods could help improve the situation is almost as old as quantum mechanics itself. The hope was that these methods would capture the elements that should provide the physical foundations of a mathematically consistent formalism. This program required a long time to mature. A level of understanding in which we do not need to cast the theory in the rigid mold of one Hilbert space representation has now been reached. Different Hilbert space representations are necessary to describe different physical situations and we know how to construct the proper representation, given the physical situation to be considered. This flexibility provides an essential extension of the traditional Fock-space formalism of quantum mechanics. It allows us to work out, with some justifiable confidence, the properties of systems in which an infinite number of degrees of freedom are present. These systems are encountered in statistical mechanics when the thermodynamical limit is considered and in quantum field theory when a fully relativistic theory of interacting local fields is approached.

The algebraic approach to physical problems also presents an interesting aspect when viewed in the light of the history of mathematics and physics. Its development has served as an effective catalyst in the pursuit of the dialogue between physicists and mathematicians. This is so true that this area of research is now actually an interdisciplinary enterprise. The advantages that this kind of situation can offer to both disciplines are well known.

The present book is an attempt to offer a systematic introduction to the ideas and techniques that form the core of this new area of mathematical physics. It is intended to be readable either as a monograph or a textbook. The level of the presentation has been determined by the criterion that it should be explicit enough to be within reach of graduate students in mathematics or physics and advanced enough to sustain their respective interests.

To achieve these aims we chose to put the accent on the general aspects of the theory. Specialized results relative to particular models are touched on only occasionally for the purpose of illustrating the general theory. This choice of material is also motivated by the fact that a newcomer in the field often faces unnecessary frustrations when trying to find his way through the extensive cross-referencing present in the original literature, especially when it comes to understanding the physical and mathematical foundations of the theory.

The dynamics of the book is provided by the interplay of two principles. The first is that the physical motivation should be formulated carefully enough for the reader to know at each step what the theory contains and what is expected to come out of it. The second is that the general mathematical setting should be presented in a self-contained manner to enable the reader to gain fluency in the language without having to learn its vocabulary and grammar from a variety of other sources. These two principles are constantly woven together to produce a theory, the principal results of which are singled out for rapid identification and retrieval as theorems. Proofs have been chosen on the basis of their didactic value, and those given are as much a part of the text as the theorems themselves.

The book is divided into four chapters. Each is in turn divided into sections and subsections and contains an introduction that describes in general terms what is going to happen in that chapter. These indications are then made more specific in an outline which opens each section. When necessary, these outlines are further elaborated on in the form of preliminary remarks, which can be found at the beginning of some subsections. Theorems are numbered consecutively in each section (e.g., Theorem II.1.5 appears in Chapter 2 Section 1, and the table of contents indicates that it is to be found on p. 99). Original works referred to throughout the text are listed in a single bibliography. Finally, an index lists the terms we use and gives the page on which they are introduced.

In Chapter 1 we study the general motivation for the algebraic approach to certain physical problems. This motivation has two components. The first stems from the failure of Fock-space techniques; after a review of the usual formalism of quantum mechanics we demonstrate this failure in the case of the van Hove model. This is done in the first section. The second component of the motivation is the desire to locate the "first principles" on which quantum mechanics operates; this "axiomatic" or "epistemological" vein is followed in Section 2. The reader who is not too curious about general motivations may want to skip this chapter or perhaps its second section; he should, however, at least read the last two subsections carefully, and in particular the proof of Theorem I.2.14, which provides the key to many of the subsequent developments.

The main tools of the algebraic approach are described in Chapter 2. In the first section we present the aspects of the mathematical theory of C^*-algebras and von Neumann algebras which are most useful for physical applications. The second section is devoted to the study of symmetries and symmetry breaking. Subsections II,1.a,b,c and II,2.a,b,d,e,f are most essential; a reader in a hurry might then be temporarily satisfied by merely skimming through Subsections II,1.d,e,f and II,2.c and by passing Subsection II,1.g.

Chapters 3 and 4 cover the applications of the algebraic methods to two types of problem: the representations of the CCR/CAR and quasi-local theories. The existence of nonequivalent representations of the CCR has been one of the most efficient motivations for the algebraic approach. An algebraic account of this part of the theory is to be found in Subsections III,1.c and d. The roles played by the thermodynamical limit and by the requirement of space-time covariance are discussed in Chapter 4. This chapter has been condensed intentionally and even a somewhat superficial reader should go through it all. Suggestions for further readings, dispersed throughout the book, are to be found in the highest concentration in this chapter.

GÉRARD G. EMCH

Rochester, New York
October 1971

Acknowledgments

I want to record here my gratitude to those who contributed in various ways to the preparation of this book. First among them is Professor R. E. Marshak without whose initiative and periodic insistence this book would never have been started, let alone finished in any reasonable amount of time. The material contained in it was taught as a special topics course at the University of Rochester during the academic year 1968–1969. To the students and faculty members who attended it I am indebted for the many comments that enabled me to determine the form that the book's presentation should take to fulfill its function. Professor E. J. Verboven of the University of Nijmegen read an earlier version in the summer of 1969; I wish to thank him for some valuable remarks. Dr. Charles Radin went through the first and final drafts, and made numerous suggestions, several of which have been worked into the book. With Mr. Jeffrey Friedman he also undertook the formidable task of making up for the many deficiencies of my style in a language that was not whispered to me in the cradle. Mrs. Shirley McDonnel and Mrs. Katherine Woods produced the typescript with much expertise. Finally, I wish to express my deepest gratitude to my wife for her understanding, patience, and help in preparing the index and bibliography.

The final version of this book was worked out while I was lecturer at the NSF Workshop in Statistical Mechanics held in Bozeman, Montana, during the summer of 1970.

G. G. E.

Contents

Algebraic Methods in Statistical Mechanics and Quantum Field Theory

CHAPTER 1

General Motivation

In spite of the successes registered by quantum mechanics in the solution of physical problems involving only a finite number of degrees of freedom, it should be recognized that most of the extensions of this theory to more general situations have been plagued with difficulties, such as divergences, which physicists have had to learn to live with, short of having a better way to handle them. To lay claims to a physicist's interest any new approach should at least demonstrate the following three points:

First, it should isolate some of the difficulties just alluded to and establish that they are indeed due to some intrinsic limitations of the usual framework. Second, the new approach should contain ordinary quantum and classical theories to the extent of being able to reproduce the successes of earlier approaches. Third, it should provide a framework leading to a proper solution of some of the traditional difficulties. The algebraic approach claims to satisfy these three requirements, the first of which is analyzed in Section 1 with the help of two illustrative models; this section concludes with some anticipations related to the third claim. The second claim is discussed in Section 2 from an axiomatic point of view.

SECTION 1. WHY NOT STAY IN FOCK SPACE?

OUTLINE. The aim of this section is to present some evidence of the necessity for an algebraic approach to physical systems involving an infinite number of degrees of freedom; these systems are encountered in both the quantum theory of fields and in the statistical mechanics of the many-body systems considered in the thermodynamical limit. We shall proceed mainly under the assumption that we can work within the familiar textbook setting; in doing so, however, we shall try to alert the reader's attention to some of the difficulties linked to what must now be regarded as a too rigid frame of thought. The form of this section therefore is that of an analysis, leading to a diagnosis of some of the ills of the traditional theory; a cure is then suggested, the value of which is discussed in the following chapters.

We discuss first some mathematical aspects of usual quantum field theories, without which no further discussion could possibly make sense. We shall then study one representative example—namely the van Hove model—for which we can illustrate the working of the usual formalism and delineate its mathematical limitations and their physical implications. The relevance of that discussion for physics is broadened by some remarks about the BCS-model, a typical problem of statistical mechanics. This section closes with a short list of those fields in physics in which the new approach—as suggested these by two typical examples—might turn out to be useful.

a. Quantum Mechanics

Dirac's [1930] masterful exposition of the principles of quantum mechanics is, in the sequel, our prototype for what we call the traditional framework of quantum mechanics. In a different mathematical style the same ideas also form the core of von Neumann's [1932a] account. The structure of this theory is briefly recalled in this subsection.

The *observables* on a physical system are identified with the self-adjoint linear operators acting on some *Hilbert space* $\mathcal{H}$. In this theory $\mathcal{H}$ is a finite- or infinite-dimensional complex vector space (the vectors of which we denote as $\Phi, \Psi, \ldots$), equipped with a scalar product (Φ, Ψ) linear in Ψ and antilinear in Φ; furthermore, $\mathcal{H}$ is complete with respect to the norm $|\Phi| \equiv (\Phi, \Phi)^{1/2}$, i.e., every sequence $\{\Phi_n\}$ of vectors in $\mathcal{H}$ which is Cauchy with respect to the norm converges in $\mathcal{H}$.

The *states* of the physical system considered are identified with the *density matrices* on $\mathcal{H}$, i.e., with the self-adjoint, positive, linear operators ρ of unit trace. The *expectation value* of an observable A on a state ρ is then computed with the formula

$$\langle \rho; A \rangle = \operatorname{Tr} \rho A$$

In particular we notice that to every vector Φ in $\mathcal{H}$ such that $|\Phi| = 1$ there corresponds a state ϕ defined by

$$\langle \phi; A \rangle = \operatorname{Tr} P_\phi A = (\Phi, A\Phi),$$

where P_ϕ is the projector on the one-dimensional subspace $\{\lambda\Phi \mid \lambda \in \mathbb{C}\}$ generated by Φ; P_ϕ is then the operator defined by

$$P_\phi \Psi \equiv (\Phi, \Psi)\Phi \quad \text{for all } \Psi \text{ in } \mathcal{H},$$

or, in Dirac's notation,

$$P_\phi = |\Phi\rangle\langle\Phi|.$$

These states play a central role in the formulation of the theory: the addition $\Phi + \Psi$ in $\mathcal{H}$ transcribes mathematically the fundamental *superposition principle.* Furthermore, for every such pure state ϕ the projector P_ϕ can be interpreted as the observable corresponding to the statement, "the system is in the state ϕ," so that the *transition probability* between two states ϕ and ψ (i.e., the probability that the system will be found in the state ϕ when we know that it is in the state ψ) is given by

$$\text{pr}\,\{\phi \mid \psi\} = (\Psi, P_\phi \Psi) = |(\Psi, \Phi)|^2,$$

which confirms that Φ and $\omega\Phi$ (where ω is any complex number of modulus 1) describe the same physical state. When $\mathcal{H}$ is realized as the space of all square-integrable functions on $\mathbb{R}^n$, Φ is referred to as the *wave function* corresponding to the state ϕ; Φ is clearly determined by ϕ only up to a phase ω. Finally, any density matrix can be expressed as a "statistical mixture" of these states:

$$\rho = \sum_{n=0}^{\infty} p_n P_{\phi_n},$$

with $0 \leqslant p_n \leqslant 1$ and

$$\sum_{n=0}^{\infty} p_n = 1.$$

The time evolution is described by a continuous one-parameter group of unitary† operators U_t, acting either as a transformation of the states (Schrödinger picture)

$$\Phi_t^{(S)} = U_t \Phi,$$

hence

$$\rho_t^{(S)} = U_t \rho U_{-t} \equiv \nu_t[\rho],$$

or as a transformation of the observables (Heisenberg picture)

$$A_t^{(H)} = U_{-t} A U_t \equiv \alpha_t[A].$$

These two pictures are equivalent in the sense that the evolution of the expectation values—the only measurable numbers of the theory—is given by

$$\begin{aligned} \langle \phi; A \rangle_t &= (\Phi_t^{(S)}, A\Phi_t^{(S)}) \\ &= (\Phi, A_t^{(H)}\Phi), \end{aligned}$$

and, in general,

$$\begin{aligned} \langle \rho; A \rangle_t &= \text{Tr}\,(\nu_t[\rho] A) = \langle \nu_t[\rho]; A \rangle \\ &= \text{Tr}\,(\rho \alpha_t[A]) = \langle \rho; \alpha_t[A] \rangle. \end{aligned}$$

† This is a consequence of Wigner's theorem which is discussed in Section 2.2a; the literature on this theorem is quite extensive and we wish only to mention here Wigner [1931] pp. 251–254 ([1959] pp. 233–236), Bargman [1964], Uhlhorn [1963], and Emch and Piron [1963].

The differential form of the evolution laws is then (Stone theorem [Stone, 1932b])

$$\frac{d}{dt}\Phi_t^{(S)} = -iH\Phi_t^{(S)},$$

$$\frac{d}{dt}\rho_t^{(S)} = -i[H, \rho_t^{(S)}] = -i\mathfrak{L}\rho_t^{(S)},$$

$$\frac{d}{dt}A_t^{(H)} = +i[H, A_t^{(H)}] = +i\mathfrak{L}A_t^{(H)}.$$

The generator H of the time evolution of the vectors (the "wave functions") is identified with the *Hamiltonian* of the theory; in analogy with classical mechanics the generator $\mathfrak{L}$ of the evolution of the states is called the *Liouville operator*.

In a similar manner we associate the *momentum* P with the space translations U_a, the *angular momentum* L, with the rotations U_θ, etc.

Finally, we mention that the observable C defined for any pair $\{A, B\}$ of observables by the relation

$$C = -i[A, B]$$

has the following interpretation: for any state ρ the quantity $\langle\rho; C\rangle^2$ provides the actual lower bound to the simultaneous observability of A and B in the sense that

$$\langle\rho; (A - \langle\rho; A\rangle)^2\rangle\langle\rho; (B - \langle\rho; B\rangle)^2\rangle \geqslant \langle\rho; C\rangle^2;$$

this is the *uncertainty principle*.

This brief account of the traditional framework of quantum mechanics will suffice for our purpose in this section. The epistemological problems linked to this formalism are discussed in detail in Section 2 with the aim of helping to separate the most essential aspects of this theory from the more accidental ones.

b. Scattering Theory

The typical scattering situation can be schematized as follows. In the presence of a *scattering center* (whatever the latter might be) the evolution U_t is generated by a total Hamiltonian H, whereas, if the scattering center were not present, the ("unperturbed" or "free") evolution U_t^0 would be generated by a free Hamiltonian H^0. It seems intuitively correct to assume that "far" from the scattering center the evolution is "almost free," and this is actually what the experimentalist sees: he prepares (respectively, detects) in the remote past (respectively, in the distant future) states ϕ^{in}, ψ^{in}, . . . (respectively, ϕ^{out}, ψ^{out}, . . .) which evolve freely. The central assumption of scattering theory is that there exists, in the mathematical description at hand, interpolating states ϕ, ψ, . . . , which evolve according to the full equation of evolution (i.e., according to the prescriptions imposed by H) and which approximate ϕ^{in}, ψ^{in} (respectively, ϕ^{out}, ψ^{out}, . . .) in the remote past (respectively, in the distant future) in the sense that in the limit these states cannot be distinguished by means of the set $\{A\}_0$ of observables that the experimentalist has at his disposal. Mathematically this condition is as follows:

Asymptotic condition. For each incoming state ϕ^{in} (respectively, outgoing state ϕ^{out}) there exists an interpolating state ϕ such that

$$\left.\begin{aligned}\lim_{t\to-\infty} |\langle \nu_t^0[\phi^{\text{in}}]; A\rangle - \langle \nu_t[\phi]; A\rangle| = 0\\ \lim_{t\to+\infty} |\langle \nu_t^0[\phi^{\text{out}}]; A\rangle - \langle \nu_t[\phi]; A\rangle| = 0\end{aligned}\right\} \quad \text{for all } A.$$

REMARKS. This condition can be written equivalently as

$$\lim_{t\to\mp\infty} |\langle \phi^{\substack{\text{in}\\ \text{out}}}; \alpha_t^0[A]\rangle - \langle \phi; \alpha_t[A]\rangle| = 0;$$

furthermore, in actual physical situations we might wish to work with mixtures and not only with pure states; the asymptotic condition is then modified accordingly by substituting ρ for ϕ.

In the sense prescribed by the asymptotic condition the interpolating states ϕ can be considered as the limit, as $t \to -\infty$ (respectively, $t \to +\infty$) of $\nu_{-t}[\nu_t^0[\phi^{\text{in}}]]$ (respectively, $\nu_{-t}[\nu_t^0[\phi^{\text{out}}]]$. This limiting procedure formally† defines the two operators called Møller matrices:

$$\Omega_{\pm} = \lim_{t\to\pm\infty} U_{-t} U_t^0.$$

The natural question the experimentalists want to see answered is, "What is the probability pr $\{\psi^{\text{out}} \leftarrow \phi^{\text{in}}\}$ that the detected state ψ^{out} will come from

† Several attempts have been made to give a precise meaning to these operators and, in particular, to prove that they are partial isometries. This has been achieved in potential scattering, when more or less stronger asymptotic conditions are assumed, essentially by requiring the limits to exist with respect to other topologies on the set of states; for this aspect of scattering theory the reader is referred to Jauch [1958] and Jauch, Misra, and Gibson [1968].

the prepared state ϕ^{in}?". We have

$$\begin{aligned}\text{pr}\,\{\psi^{\text{out}} \leftarrow \phi^{\text{in}}\} &= \text{pr}\,\{\psi \mid \phi\}\\ &= |(\psi, \phi)|^2\\ &= |(\Omega_+\psi^{\text{out}}, \Omega_-\phi^{\text{in}})|^2\\ &= |(\psi^{\text{out}}, \Omega_+^*\Omega_-\phi^{\text{in}})|^2\\ &= |(\psi^{\text{out}}, S\phi^{\text{in}})|^2.\end{aligned}$$

The operator S so "defined" is called the collision operator, or *S-matrix.* The task of scattering theory is then (a) to devise a scheme for calculating the matrix elements of S, knowing H and H^0, and (b) to find a model H that would provide agreement between the predicted values of the matrix elements of S and measurable quantities, such as, for instance, the cross section. None of these steps is trivial, especially for relativistic scattering: the quantum theory of fields was essentially engineered for the latter problem. In field theory the translation of the asymptotic condition is the following: given the free fields $F^{\text{in}}(x, t) = \alpha_t^0[F^{\text{in}}(x, 0)]$ (respectively, $F^{\text{out}}(x, t) = \alpha_t^0[F^{\text{out}}(x,0)]$) which describe the theory in the remote past (respectively, the distant future), there exists an interpolating field $F(x, t)$ which satisfies the full equation of motion, i.e., $F(x, t) = \alpha_t[F(x, 0)]$, and approximates $F^{\text{in}}(x, t)$ in the remote past (respectively, $F^{\text{out}}(x, t)$ in the distant future) in the following sense:

$$\lim_{t\to\mp\infty} |\langle\phi; F^{\substack{\text{in}\\\text{out}}}(x, t) - F(x, t)\rangle| = 0$$

for all (physically attainable) states ϕ on the system.

Aside from the above asymptotic condition, the standard assumption is that the theory could be entirely carried out in Fock space. This standard assumption was, however, demonstrated as untenable, from a physical standpoint, by Haag [1955], who in doing so opened physicists' eyes to the necessity for a radical departure from the traditional approach; we shall come back later (see, in particular, Section 3.1.d) to the general evidence available to support this statement. To gain some intuitive insight into the problem we nevertheless ignore, on a temporary basis, of course, Haag's objection, proceed with the theory, apply it to a model, and watch the gathering storm.

c. Fock Space

The so-called Fock space was introduced in the quantum theory of fields by Fock [1932] and has since become the standard tool of this theory (see, for instance, Schweber [1961]). Its mathematical structure has been settled in a definitive paper by Cook [1953]. In this subsection we present briefly the essentials of the so-called Fock-space techniques in the spirit of Cook's approach.

Let $\mathcal{H}^{(1)}_{(s_i)}$ be the Hilbert space used in the quantum mechanical description of a single particle of species s_i. We first define the Hilbert space

$$\overline{\mathcal{H}}^{(N)}_{(s_1,s_2,\ldots,s_N)} = \mathcal{H}^{(1)}_{(s_1)} \otimes \mathcal{H}^{(1)}_{(s_2)} \otimes \cdots \otimes \mathcal{H}^{(1)}_{(s_N)}$$

to be used for the description of a system of N particles of species $s_1, s_2, \ldots, s_N$; this space is obtained by considering all sequences

$$\begin{aligned}\Phi^{(N)}_{(s_1,s_2,\ldots,s_N)} &= \{\Phi^{(1)}_{(s_1)}, \Phi^{(1)}_{(s_2)}, \ldots, \Phi^{(1)}_{(s_N)}\}\\ &\equiv \Phi^{(1)}_{(s_1)} \otimes \Phi^{(1)}_{(s_2)} \otimes \cdots \otimes \Phi^{(1)}_{(s_N)},\end{aligned}$$

with $\Phi^{(1)}_{(s_k)}$ running over $\mathcal{H}^{(1)}_{(s_k)}$ $(k = 1, 2, \ldots, N)$; we then define a scalar product between two such sequences as

$$(\Phi^{(N)}_{(s_1,s_2,\ldots,s_N)}, \Psi^{(N)}_{(s_1,s_2,\ldots,s_N)}) = \prod_{k=1}^{N} (\Phi^{(1)}_{(s_k)}, \Psi^{(1)}_{(s_k)});$$

we consider next the vector space $\mathcal{H}^{(N)}_{(s_1,s_2,\ldots,s_N)}$ of all finite linear combinations of these sequences and extend by linearity the scalar product to this space; $\overline{\mathcal{H}}^{(N)}_{(s_1,s_2,\ldots,s_N)}$ is finally the completion of this pre-Hilbert space.

In the case in which some of the particles are of the same species we have to take into account quantum statistics. In particular, if the N particles of the system considered are *identical bosons*, the wave function $\Phi^{(N)}$ in $\mathcal{H}^{(N)}$ is admissible only if it is totally symmetric. We therefore define the symmetric part ${}^{S}\overline{\mathcal{H}}^{(N)}$ of $\overline{\mathcal{H}}^{(N)}$, i.e., the Hilbert space of all admissible wave functions as follows: first, for the elements $\Phi^{(N)}$ of $\mathcal{H}^{(N)}$ which are of the form $\Phi^{(N)}_{(1,2,\ldots,N)} = \Phi^{(1)}_1 \otimes \Phi^{(1)}_2 \otimes \cdots \otimes \Phi^{(1)}_N$, with $\Phi^{(1)}_k$ in $\mathcal{H}^{(1)}$, we define

$${}^{S}\Phi^{(N)}_{(1,2,\ldots,N)} \equiv \frac{1}{\sqrt{N!}} \sum_{P\in\mathcal{T}(N)} \Phi^{(N)}_{P(1,2,\ldots,N)},$$

where the sum is carried over all permutations of the N indices $1, 2, \ldots, N$. We then extend by linearity the mapping $\Phi^{(N)} \to {}^{S}\Phi^{(N)}$ to $\mathcal{H}^{(N)}$; this mapping is bounded and can therefore be extended to $\overline{\mathcal{H}}^{(N)}$ by continuity. The range of this mapping is by definition ${}^{S}\overline{\mathcal{H}}^{(N)}$, the symmetrized tensor product of N identical copies of $\mathcal{H}^{(1)}$. An elementary example might be in order at this point: if $\mathcal{H}^{(1)}$ is the space $\mathcal{L}^2(\mathbb{R}^3)$ of all complex-valued, square-integrable functions from the three-dimensional euclidian space $\mathbb{R}^3$, $\overline{\mathcal{H}}^{(N)}$ is the space of all complex-valued, square-integrable functions $f(\mathbf{x}_1, \mathbf{x}_2, \ldots, \mathbf{x}_N)$ (with $\mathbf{x}_k$ in $\mathbb{R}^3$ for $k = 1, 2, \ldots, N$) which are totally symmetric in their arguments.

We have then, up to here, constructed the Hilbert space for the description of a system of N identical bosons: ${}^{S}\overline{\mathcal{H}}^{(N)}$. The Hilbert space ${}^{A}\overline{\mathcal{H}}^{(N)}$ for the description of a system of N *identical fermions* can be obtained in a similar fashion, replacing the condition that $\Phi^{(N)}$ be totally symmetric with the condition that it must be totally antisymmetric.

If we then wanted to describe a process in the course of which particles are produced, absorbed, or exchanged in one way or another, the space just constructed, accommodating only a fixed number of particles, is clearly not adequate. Fock then suggested that the Hilbert space appropriate for the *second quantization* is the following, which we shall refer to in the sequel as the *Fock space:*

$$^{\varepsilon}\mathcal{H} = \sum_{N=0}^{\infty} \oplus\, {}^{\varepsilon}\overline{\mathcal{H}}^{(N)}.$$

In this sum ε stands for either S or A, depending on whether we are considering a system of bosons or a system of fermions; $\mathcal{H}^{(0)}$ is identified with the set $\mathbb{C}$ of all complex numbers.

To construct $^{\varepsilon}\mathcal{H}$ in a practical way we consider the set $^{\varepsilon}\mathcal{F}$ of all sequences

$$\Phi = \{\Phi^{(0)}, \Phi^{(1)}, \ldots, \Phi^{(N)}, \ldots\}$$

with a finite (but otherwise arbitrary) number of nonzero entries; $\Phi^{(N)}$ (vector in $^{\varepsilon}\overline{\mathcal{H}}^{(N)}$) is called the *N-particle component* of Φ; we then equip $^{\varepsilon}\mathcal{F}$ with the structure of a pre-Hilbert space by the composition laws

$$(\lambda\Phi + \mu\Psi)^{(N)} = \lambda\Phi^{(N)} + \mu\Psi^{(N)}$$

$$(\Phi, \Psi) = \sum_{N=0}^{\infty} (\Phi^{(N)}, \Psi^{(N)});$$

$^{\varepsilon}\mathcal{H}$ is obtained as the completion of $^{\varepsilon}\mathcal{F}$ with respect to the metric induced by this scalar product.

The main characteristic of Fock space is that it accommodates the wave function of any system with a finite (but otherwise arbitrary) number of particles and that operators can be defined in it which express in a natural fashion the creation and annihilation of a particle with a definite wave-function. Let us first consider our previous example in which the one-particle space $\mathcal{H}^{(1)}$ is $\mathcal{L}^2(\mathbb{R}^3)$ and the statistics is that of the *bosons*. For all f in $\mathcal{L}^2(\mathbb{R}^3)$ we define the two operators $a(f)$ and $a^*(f)$ from $^{S}\mathcal{F}$ to $^{S}\mathcal{F}$ by

$$(a(f)\Phi)^{(N)}(\mathbf{x}_1, \ldots, \mathbf{x}_N) = \sqrt{N+1}\int d\mathbf{x}\, f^*(\mathbf{x})\, \Phi^{(N+1)}(\mathbf{x}, \mathbf{x}_1, \ldots, \mathbf{x}_N)$$

$$(a^*(f)\Phi)^{(N)}(\mathbf{x}_1, \ldots, \mathbf{x}_N) = \frac{1}{\sqrt{N}} \sum_{k=1}^{N} f(\mathbf{x}_k)\Phi^{(N-1)}(\mathbf{x}_1, \ldots, \not{\mathbf{x}}_k, \ldots, \mathbf{x}_N).$$

We verify that for all $N < \infty$ and all f in $\mathcal{H}^{(1)}$ $a(f)$ [resp. $a^*(f)$] is a linear operator from $^{S}\overline{\mathcal{H}}^{N}$ to $^{S}\overline{\mathcal{H}}^{N-1}$ (resp. to $^{S}\overline{\mathcal{H}}^{N+1}$), bounded by $N\,\|f\|$ [resp. by

$(N + 1) \|f\|]$. Consequently $a(f)$ and $a^*(f)$ are well-defined linear (but unbounded) operators from ${}^S\mathcal{F}$ to ${}^S\mathcal{F}$; since ${}^S\mathcal{F}$ is a dense linear manifold in ${}^S\mathcal{H}$, we conclude that $a(f)$ and $a^*(f)$ are defined on a common dense domain (namely, ${}^S\mathcal{F}$), which is invariant under these mappings, so that all polynomials in $a^*(f_1) \cdots a^*(f_N) \cdots a(g_1) \cdots a(g_M)$, with $f_1 \cdots f_N g_1 \cdots g_M$ in $\mathcal{H}^{(1)}$, are properly defined on ${}^S\mathcal{F}$. In particular, the following identities are true *on* ${}^S\mathcal{F}$:

$$
\begin{aligned}
a(f)^* &= a^*(f), \\
a^*(\lambda f + \mu g) &= \lambda a^*(f) + \mu a^*(g), \\
[a(f), a(g)] &= 0 = [a^*(f), a^*(g)], \\
[a(f), a^*(g)] &= (f, g),
\end{aligned}
$$

for all f and g in $\mathcal{H}^{(1)}$; the last two relations are the so-called *canonical commutation relations* (CCR).

This is the first place in this book where we cannot avoid dealing explicitly with unbounded operators; we therefore collect a few elementary facts pertinent to their study. Let A be a linear operator defined on the elements of a linear manifold $\mathfrak{D}(A)$ of a Hilbert space $\mathcal{H}$ and taking its values in $\mathcal{H}$; $\mathfrak{D}(A)$ is said to be the *domain* of A and the linear manifold $\mathfrak{R}(A) = \{A\Phi \mid \Phi \in \mathfrak{D}(A)\}$ is said to be the *range* of A. Let B be similarly defined. If $\mathfrak{D}(A)$ is contained in $\mathfrak{D}(B)$ and $A\Phi = B\Phi$ for all Φ in $\mathfrak{D}(A)$, A is said to be the *restriction* of B to $\mathfrak{D}(A)$ and B is said to be an *extension* of A to $\mathfrak{D}(B)$. Now let A and A' be two linear operators respectively defined on $\mathfrak{D}(A)$ and $\mathfrak{D}(A')$ in $\mathcal{H}$; they are said to be adjoint to each other if $(\Psi, A\Phi) = (A'\Psi, \Phi)$ for all Φ in $\mathfrak{D}(A)$ and all Ψ in $\mathfrak{D}(A')$. A linear operator A is said to be *densely defined* if the norm-closure of $\mathfrak{D}(A)$ in $\mathcal{H}$ coincides with $\mathcal{H}$ itself. If A is densely defined, there exists a unique linear operator A^*, called *the adjoint* of A, such that every operator A' adjoint to A is the restriction of A^* to some linear manifold $\mathfrak{D}(A')$ contained in $\mathfrak{D}(A^*)$. A linear operator B is said to be *closed* if for every sequence $\{\Phi_n\}$ in $\mathfrak{D}(B)$ such that Φ_n converges (in the norm) to some vector Φ in $\mathcal{H}$ and $B\Phi_n$ converges (in the norm) to some vector Ψ in $\mathcal{H}$ we have necessarily Φ in $\mathfrak{D}(B)$ and $B\Phi = \Psi$. If A is densely defined, A^* is closed. A densely defined operator A is said to be *symmetric* if $\mathfrak{D}(A)$ is contained in $\mathfrak{D}(A^*)$ and A is the restriction of A^* to $\mathfrak{D}(A)$. In this case A^{**} is a closed symmetric extension of A and is called the *closure* of A. A linear operator A is said to be *self-adjoint* if in addition to being symmetric it satisfies $\mathfrak{D}(A) = \mathfrak{D}(A^*)$. In general, a symmetric operator might have several, none, or exactly one self-adjoint extension. In particular, a symmetric operator is said to be *essentially self-adjoint* if its closure is self-adjoint; in this case A admits only one self-adjoint extension, namely A^{**}. A linear operator A is said to be bounded (on its domain) if there exists a finite positive number M such that $|A\Phi| \leqslant M$ for all Φ in $\mathfrak{D}(A)$. When this condition is not satisfied, A is said to be *unbounded.* A linear operator A admits a (unique) bounded extension to the subspace $\overline{\mathfrak{D}(A)}$ [the norm-closure of $\mathfrak{D}(A)$] if and only if it is bounded on its domain. If A is densely defined and bounded on its domain, we implicitly extend it by continuity to $\mathcal{H}$ and simply speak of a *bounded* operator. In this case the above definitions reduce to their usual meaning; in particular, the adjoint of a bounded operator is bounded and a symmetric bounded operator is also self-adjoint.

We now come back to our study of the Fock formalism. The physical interpretation of the operators $a(f)$ and $a^*(f)$ is checked from their very definition: if the many-body system considered is in a state represented by the wave function Φ, then $a^*(f)\Phi$ is the wave function of the same system in a state that differs from the state described by Φ only in that it has one more particle, the new particle being in a state described by the one-particle wave function f; in this sense $a^*(f)$ creates a particle in wave function f and in a similar way $a(f)$ annihilates this particle. The expectation value $\langle \phi; a^*(f)\, a(f)\rangle$ calculated on any element Φ in ${}^S\mathcal{F}$ of the symmetric operator $N(f) \equiv a^*(f)\, a(f)$ is equal to the expectation of the number of particles of wave function f present in the state described by Φ; hence the name *number operator* for the (self-adjoint) closure N of the symmetric operator $\sum_{i=1}^{\infty} a^*(f_i)\, a(f_i)$, where $\{f_i \mid i = 1, 2, \ldots\}$ is any orthonormal basis in $\mathcal{H}^{(1)}$; it is plain that the expectation value of N calculated on any Φ in ${}^S\mathcal{F}$ is finite. Let us now denote by Φ_0 the vector of ${}^S\mathcal{F}$, the Nth components of which are $\Phi_0^{(N)} = \delta_{0,N}$. We have $\langle \phi_0; N(f)\rangle = 0$ for all f in $\mathcal{H}^{(1)}$; hence the state Φ_0 contains no particle and is therefore interpreted as *the vacuum* of the theory. It is unique and satisfies the condition $a(f)\Phi_0 = 0$ for all f in $\mathcal{H}^{(1)}$ (which is referred to as the condition of *stability* of the vacuum). Moreover, every vector Φ in ${}^S\mathcal{H}$ can be approximated as closely as we wish by a vector obtained from Φ_0 by acting on the vacuum with an appropriate polynomial in the creation operators; we refer to this property by saying that the vacuum is *cyclic.* Finally, if $f = \{f_i \mid i = 1, 2, \ldots\}$ is any arbitrary orthonormal basis in $\mathcal{H}^{(1)}$, the only subspaces of ${}^S\mathcal{H}$ left invariant by all $a^*(f_i)$ are 0 and ${}^S\mathcal{H}$ itself; this situation is described by saying that $\{a^*(f_i) \mid f_i \in f\}$ is *irreducible.*

Proceeding now from the particular to the general, we define the operators $a(f)$ and $a^*(f)$, whatever the one-particle space $\mathcal{H}^{(1)}$ might be, by going through the following steps: for every vector Φ in ${}^S\overline{\mathcal{H}}{}^{(N)}$ of the form

$$\Phi = \frac{1}{\sqrt{N!}} \sum_{\mathcal{T}(N)} \Phi^{(N)}_{P(1,2,\ldots,N)},$$

where

$$\Phi^{(N)}_{P(1,2,\ldots,N)} = \Phi^{(1)}_{p(1)} \otimes \Phi^{(1)}_{p(2)} \otimes \cdots \otimes \Phi^{(1)}_{p(N)},$$

and any vector f in $\mathcal{H}^{(1)}$, we define the vectors $a(f)$ and $a^*(f)$ by

$$a(f)\Phi = \frac{1}{\sqrt{(N-1)!}} \sum_{\mathcal{P}(N)} (f, \Phi^{(1)}_{p(1)})\, \Phi^{(1)}_{p(2)} \otimes \cdots \otimes \Phi^{(1)}_{p(N)}$$

$$a^*(f)\Phi = \frac{1}{\sqrt{(N+1)!}} \sum_{\mathcal{P}(N+1)} \Phi^{(1)}_{p(0)} \otimes \Phi^{(1)}_{p(1)} \otimes \cdots \otimes \Phi^{(1)}_{p(N)}$$

(where, for notational convenience, $\Phi_0^{(1)} \equiv f$); then by linearity we extend $a(f)$ and $a^*(f)$ to linear mappings from ${}^S\mathcal{H}^N$ to ${}^S\mathcal{H}^{N-1}$ and ${}^S\mathcal{H}^{N+1}$, respectively; these mappings, denoted by the same symbols $a(f)$ and $a^*(f)$, are bounded by $\sqrt{N}\,|f|$ and $\sqrt{(N+1)}\,|f|$, respectively, and can therefore be extended by continuity as mappings from ${}^S\overline{\mathcal{H}}^N$ to ${}^S\overline{\mathcal{H}}^{N-1}$ and ${}^S\overline{\mathcal{H}}^{N+1}$, respectively; finally, by linearity, we extend these mappings to linear operators from ${}^S\mathcal{F}$ to ${}^S\mathcal{F}$; these operators are unbounded and can therefore not be extended by continuity to the entire Fock space ${}^S\mathcal{H}$.

We easily verify that in the case in which $\mathcal{H}^{(1)}$ is $\mathcal{L}^2(\mathbb{R}^3)$ the operators $a(f)$ and $a^*(f)$ just constructed coincide exactly with the operators previously defined in this particular case; furthermore† the properties of these operators, previously stated in this particular case, *all* carry over in the general case just constructed for bosons statistics.

In Fermi statistics the same theory can be built, to parallel step by step the boson case, if we replace the symmetrizer S with the antisymmetrizer A. The only significant difference is that in the Fermi case the creation and annihilation operators satisfy the *canonical anticommutation relations* (CAR):

$$[a(f), a^*(g)]_+ = (f, g),$$
$$[a(f), a(g)]_+ = 0 = [a^*(f), a^*(g)]_+;$$

as a result of these relations the operators $a(f)$ and $a^*(f)$ are bounded on ${}^A\mathcal{F}$ and can therefore be extended by continuity to the *entire* Fock space ${}^A\mathcal{H}$.

These remarks end our general discussion of the Fock space ${}^\varepsilon\mathcal{H}$ itself.

We shall need in the sequel a definition of the *second-quantized form* $\Omega(A)$ of an arbitrary one-particle observable A, which will give a mathematically precise meaning to the heuristic expression

$$\sum_{\mathbf{k},\mathbf{k}'} A_{\mathbf{k}\mathbf{k}'} a^*_{\mathbf{k}} a_{\mathbf{k}'},$$

where

$$A_{\mathbf{k}\mathbf{k}'} = (f_{\mathbf{k}}, A f_{\mathbf{k}'}),$$
$$a^{(*)}_{\mathbf{k}} = a^{(*)}(f_{\mathbf{k}})$$

† Cook [1953].

for an orthonormal basis $\{f_k\}$ in $\mathcal{H}^{(1)}$. Such a definition has been formulated in a convenient way by Cook [1953] and we shall follow his method. Keeping in mind that we shall have to consider self-adjoint operators which are not necessarily bounded, nor have in general a completely discrete spectrum, we have first of all to define the N-fold direct product $A_1 \otimes A_2 \otimes \cdots \otimes A_N$ in a way general enough to accommodate the situations we shall have to deal with later. Let $\{\mathcal{H}_i\}$ be a finite sequence of Hilbert spaces and $\{A_i\}$, a similarly indexed sequence of linear, densely defined closed operators acting in $\mathcal{H}_i$; let us denote by $\mathfrak{D}(A_i)$ the domain of A_i. Since† A_i is closed and densely defined, its adjoint A_i^* has for domain $\mathfrak{D}(A_i^*)$ a dense linear manifold in $\mathcal{H}_i$. Let $\mathfrak{D}_*$ be the linear manifold generated in $\overline{\mathcal{H}} = \mathcal{H}_1 \otimes \mathcal{H}_2 \otimes \cdots \otimes \mathcal{H}_N$ by the cartesian product $\mathfrak{D}(A_1^*) \times \mathfrak{D}(A_2^*) \times \cdots \times \mathfrak{D}(A_N^*)$; $\mathfrak{D}_*$ is dense in $\overline{\mathcal{H}}$, since all the $\mathfrak{D}(A_i^*)$ are dense in their respective $\mathcal{H}_i$'s. On $\mathfrak{D}_*$ we define the linear operator A_* by

$$A_*(\Phi_1 \otimes \Phi_2 \otimes \cdots \otimes \Phi_N) = A_1^*\Phi_1 \otimes A_2^*\Phi_2 \otimes \cdots \otimes A_N^*\Phi_N \qquad \text{for all } \{\Phi_i \in \mathfrak{D}(A_i^*)\};$$

we now define $A_1 \otimes A_2 \otimes \cdots A_{\otimes N}$ as the adjoint $(A_*)^*$ of A_*; its domain obviously contains the linear manifold generated in $\overline{\mathcal{H}}$ by the cartesian product $\mathfrak{D}(A_1) \times \mathfrak{D}(A_2) \times \cdots \times \mathfrak{D}(A_N)$, hence is dense in $\overline{\mathcal{H}}$. $A_1 \otimes A_2 \otimes \cdots \otimes A_N$ is therefore a linear, densely defined closed operator in $\overline{\mathcal{H}}$. To come back to our main purpose we now consider the case in which all the $\overline{\mathcal{H}}_i$'s of the above definition are copies of the one-particle space $\mathcal{H}^{(1)}$ and in which consequently the A_i's are one-particle operators. In particular, for each linear, densely defined, and closed operator $A^{(1)}$ in $\mathcal{H}^{(1)}$ we define the following operators in $\overline{\mathcal{H}}^{(N)}$:

$$A_1^{(N)} = A^{(1)} \otimes I \otimes I \otimes \cdots \otimes I$$

$$A_2^{(N)} = I \otimes A^{(1)} \otimes I \otimes \cdots \otimes I$$

$$\vdots$$

$$A_N^{(N)} = I \otimes I \otimes \cdots \otimes A^{(1)}$$

and their sum $\sum_{j=1}^{N} A_j^{(N)}$, which we define on the intersection of their respective domains; the latter contains at least the linear manifold generated by $\mathfrak{D}(A^{(1)}) \times \mathfrak{D}(A^{(1)}) \times \cdots \times \mathfrak{D}(A^{(1)})$, hence is dense in $\overline{\mathcal{H}}^{(N)}$. Let us denote by $A^{(N)}$ the minimal closed extension of this sum and by $\mathfrak{D}(A^{(N)})$, its domain.

† See, for instance, p. 302 in Riesz and Sz.-Nagy [1955].

We finally consider the linear manifold $\mathfrak{D}(\Omega(A))$ in $\mathcal{H} = \sum_{N=0}^{\infty} \overline{\mathcal{H}}^{(N)}$ defined as the set of all Ψ in $\mathcal{H}$ such that $\Psi^{(N)}$ is in $\mathfrak{D}(A^{(N)})$ and $\sum_{N=0}^{\infty} \|A^{(N)}\Psi^{(N)}\|^2 < \infty$; $\mathfrak{D}(\Omega(A))$ is dense in $\mathcal{H}$. We then define the linear operator $\Omega(A)$ on this linear manifold by $(\Omega(A)\Phi)^{(N)} = A^{(N)}\Phi^{(N)}$; i.e., $\Omega(A)\Phi = \sum_{N=0}^{\infty} A^{(N)}\,\Phi^{(N)}$. We verify that since each $A^{(N)}$ is closed so is $\Omega(A)$. Furthermore, if A is a one-particle observable (hence self-adjoint with dense domain $\mathfrak{D}(A)$ in $\mathcal{H}^{(1)}$), then $\Omega(A)$ is a self-adjoint, densely defined operator in $\mathcal{H}$. The physical interpretation of this operator is now straightforward; suppose, indeed, that $\Phi^{(N)}$ in $\overline{\mathcal{H}}^{(N)}$ is of the form $\Phi_1 \otimes \Phi_2 \otimes \cdots \otimes \Phi_N$ with Φ_i in $\mathfrak{D}(A)$; then $\langle \phi^{(N)}; A^{(N)}\rangle = \sum_{i=1}^{N} \langle \phi_i; A\rangle$. This extends by linearity to $\mathfrak{D}(A^{(N)})$, so that $A^{(N)}$ is the N-particle observable corresponding to the one-particle observable A; consequently for any Φ in $\mathfrak{D}(\Omega(A))$ we have $\langle \phi; \Omega(A)\rangle = \sum_{N=0}^{\infty} \langle \phi^{(N)}; A^{(N)}\rangle$ and $\Omega(A)$ appears as the proper definition of the "second quantized form" in $\mathcal{H}$ of the one-particle observable A. The symmetry requirements to be satisfied when dealing with a system of identical particles are then taken care of trivially as follows: we first notice that $\Omega(A)$ maps ${}^{\varepsilon}\mathfrak{D} = \mathfrak{D}(\Omega(A)) \cap {}^{\varepsilon}\mathcal{H}$ into ${}^{\varepsilon}\mathcal{H}$; since ${}^{\varepsilon}\mathcal{H}$ is a closed subspace of $\mathcal{H}$, $\Omega(A)$, restricted to ${}^{\varepsilon}\mathcal{H}$, is self-adjoint on this domain (dense in ${}^{\varepsilon}\mathcal{H}$) and is then the "second quantized form" in ${}^{\varepsilon}\mathcal{H}$ of the one-particle observable A. The point of the above discussion was, as announced, to give a precise mathematical meaning to $\Omega(A)$ as a self-adjoint operator in ${}^{\varepsilon}\mathcal{H}$ and, in particular, to specify its domain ${}^{\varepsilon}\mathfrak{D}$.

As an exercise left to the reader we mention the following theorem proved by Cook [1953]:

Theorem 1. *Let* H *and* A *be linear, densely defined operators in* $\mathcal{H}^{(1)}$ *and* $\mathrm{a}^{(*)}(\mathrm{f})$ *the annihilation (respectively creation) operators already defined; suppose, further, that* H *is self-adjoint and* A *is closed. Then*

$$\Omega(\mathrm{e}^{\mathrm{iHt}}\mathrm{A}\mathrm{e}^{-\mathrm{iHt}}) = \mathrm{e}^{\mathrm{i}\Omega(\mathrm{H})\mathrm{t}}\Omega(\mathrm{A})\mathrm{e}^{-\mathrm{i}\Omega(\mathrm{H})\mathrm{t}}$$

$$\mathrm{a}^{(*)}(\mathrm{e}^{-\mathrm{iHt}}\mathrm{f}) = \mathrm{e}^{-\mathrm{i}\Omega(\mathrm{H})\mathrm{t}}\mathrm{a}^{(*)}(\mathrm{f})\mathrm{e}^{\mathrm{i}\Omega(\mathrm{H})\mathrm{t}}.$$

Again this theorem is expected to hold for physical reasons; it expresses, for instance, the fact that if H is the one-particle Hamiltonian then $\Omega(H)$ is the free Hamiltonian for the system described in Fock space; another consequence of this theorem is to give the explicit form of the relativistic transformation laws for the creation and annihilation operators when, for instance, $\mathcal{H}^{(1)}$ is chosen to be the space of an (irreducible) unitary representation of the Lorentz group.

With this theorem we conclude our review of the basic aspects of the Fock-space formalism as we intend to use it.

d. The Relativistic, Free, Scalar-Meson Field

One of the main reasons for the early success of Fock-space techniques is that the space constructed in the manner described in the preceding subsection accommodates the relativistic free fields so well. For the sake of completeness and definiteness we now consider briefly the prototype provided by the scalar meson.

In a relativistic theory we choose for the one-particle space $\mathcal{H}^{(1)}$ the Hilbert space that accommodates the irreducible representation of the Lorentz group associated with the particle we want to consider. These representations have been discussed repeatedly in the literature; in his original paper Wigner [1939] solved completely the problem of the classification of all irreducible representations of this group; the generality of his method can now be understood best from the general theory established by Mackey [1949, 1952, or 1955], who extended to a wide class of continuous groups the theory of induced representations devised by Frobenius in the case of discrete groups; abbreviated accounts of the theory, as applied to the Lorentz group, can be found in Wightman [1959, 1960, or 1962], Wigner [1962], Bargmann, Wightman, and Wigner [undated], Michel and Wightman [undated], Jauch [1959], Emch [1961 and 1963, II, Appendix], etc.; for the connection between the representations and the relativistic wave equations see Bargmann and Wigner [1948]. In view of the wide availability of these many references, we recall here only that the one-particle space for a particle of mass m and spin zero (i.e., our *scalar, neutral meson*) can be realized as the Hilbert space of all measurable square integrable functions on the real Minkowski space $\mathfrak{M}^4$ with respect to the measure $\delta(\mathbf{p}^2 - m^2)(1/|p^4|)\, dp^1\, dp^2\, dp^3$. This measure being concentrated on the hyperboloid $(\mathbf{p}^2 = m^2)$, we shall consider in the sequel $\mathcal{H}^{(1)}$ as being $\mathfrak{L}^2_\mu(\mathbb{R}^3)$; i.e., to every Φ in $\mathcal{H}^{(1)}$ there corresponds $\tilde{\Phi}:\mathbf{k} \in \mathbb{R}^3 \to \tilde{\Phi}(\mathbf{k}) \in \mathbb{C}$ with

$$|\Phi|^2_\mu \equiv \int d^3\mathbf{k}\,\omega_k^{-1}|\tilde{\Phi}(\mathbf{k})|^2 < \infty,$$

and accordingly we write the "relativistic scalar product" as

$$(\Phi, \Psi)_\mu = \int d^3\mathbf{k}\,\omega_k^{-1}\tilde{\Phi}^*(\mathbf{k})\,\tilde{\Psi}(\mathbf{k}),$$

with

$$\omega_k = (m^2 + \mathbf{k}^2)^{1/2}.$$

The one-particle Hamiltonian $H_0^{(1)}$ is then the self-adjoint operator defined by

$$(H_0^{(1)}\tilde{\Phi})(\mathbf{k}) = \omega_k\tilde{\Phi}(\mathbf{k})$$

on the set $\mathfrak{D}(H_0^{(1)})$ of all Φ in $\mathfrak{L}^2_\mu(\mathbb{R}^3)$ such that

$$|H_0^{(1)}\Phi|^2_\mu \equiv \int d^3\mathbf{k}\omega_k\, \tilde{\Phi}^*(\mathbf{k})\tilde{\Phi}(\mathbf{k}) < \infty.$$

The Fock space relative to our scalar neutral meson will then be the space ${}^S\mathcal{H} = \sum_{N=0}^{\infty} {}^S\mathcal{H}^{(N)}$ constructed from $\mathfrak{L}^2_\mu(\mathbb{R}^3)$ in the manner described in detail in the preceding subsection. On this space we define the operators $a(f)$, $a^*(f)$, and $H_0 \equiv \Omega(H_0^{(1)})$; we further construct the self-adjoint operator

$$F(f) = \frac{1}{\sqrt{2}}\,[a^*(f) + a(f)]$$

and its time-development

$$F_t(f) \equiv e^{iH_0t}F(f)e^{-iH_0t}.$$

We can now use Theorem 1 to conclude that

$$F_t(f) = F(e^{iH_0^{(1)}t}f)$$

and (denoting the time-derivative by a dot)

$$P_t(f) \equiv \dot{F}_t(f) = F_t(iH_0^{(1)}f),$$

with, in particular at $t = 0$,

$$P(f) = i\frac{1}{\sqrt{2}}\,[a^*(H_0^{(1)}f) - a(H_0^{(1)}f)]$$

and furthermore

$$\dot{P}_t(f) = \ddot{F}_t(f) = F_t(-(H_0^{(1)})^2f)$$

From the canonical commutation rules we now have

$$\begin{aligned}
[F(f), F(g)] &= \tfrac{1}{2}\{(f, g)_\mu - (g, f)_\mu\}I,\\
[P(f), P(g)] &= -\tfrac{1}{2}\{(H_0^{(1)}f, H_0^{(1)}g)_\mu - (H_0^{(1)}g, H_0^{(1)}f)_\mu\}I,\\
[F(f), P(g)] &= i\tfrac{1}{2}\{(f, H_0^{(1)}g)_\mu + (H_0^{(1)}g, f)_\mu\}I\\
&= i\tfrac{1}{2}\{(f, g) + (g, f)\}I.
\end{aligned}$$

Four brief comments on the notation used in these relations are in order: first, these relations have to be understood as expressing the equality of the left-hand side and the operator-closure of the right-hand side: second, (f, g) denotes the scalar product with respect to the Lebesgue measure $d^3\mathbf{k}$; third, these "equal-time" commutation relations can easily be translated to the general ones by the use of the explicit time-development of $F_t(f)$ and $P_t(f)$ obtained above (the validity of this last remark depends in an essential way on the fact that we are dealing here with a *free* field); fourth, when f and g

are real, these relations reduce to

$$[F(f), F(g)] = 0,$$
$$[P(f), P(g)] = 0,$$
$$[F(f), P(g)] = i(f, g)I.$$

These results conclude our summary of the mathematical formalism attached to the theory of the *free*-scalar neutral-meson field.

The connection between this development and the usual heuristic formulation can be formally exhibited. We first enclose the system in a cubic box of volume V and impose periodic boundary conditions; the functions $[f_{\mathbf{k}}(\mathbf{x}) = (1/\sqrt{V})e^{i\mathbf{k}\cdot\mathbf{x}}]$, with the familiar discrete values of $\mathbf{k}$, form an orthonormal basis of eigenfunctions of the one-particle Hamiltonian. We then consider

$$a_{\mathbf{k}}^{(*)} = a^{(*)}(\sqrt{\omega_k} f_{\mathbf{k}})$$

and

$$F(\mathbf{x}) \equiv \frac{1}{\sqrt{2V}} \sum_{\mathbf{k}} \frac{1}{\sqrt{\omega_{\mathbf{k}}}} (a_{\mathbf{k}} + a_{-\mathbf{k}}^{*}) e^{i\mathbf{k}\mathbf{x}}.$$

We can verify formally that for real f

$$F(f) \equiv \int d\mathbf{x}\, F(\mathbf{x}) f(\mathbf{x}),$$

which is actually the only mathematically meaningful definition of $F(\mathbf{x})$; the formal commutation relations

$$[F_t(\mathbf{x}), F_{t'}(\mathbf{x}')], \quad \text{etc.},$$

involving the Jordan-Pauli invariant functions can be "derived" formally from the commutation relations established above for the "smeared" fields $F(f)$ and $P(f)$, the latter being the only physically sensible quantities to appear in the theory. In the same formal sense we obtain

$$H_0 = \sum_{\mathbf{k}} \omega_{\mathbf{k}} a_{\mathbf{k}}^{*} a_{\mathbf{k}},$$

$$P(\mathbf{x}) = -\frac{i}{\sqrt{2V}} \sum_{\mathbf{k}} \sqrt{\omega_{\mathbf{k}}} (a_{\mathbf{k}} - a_{-\mathbf{k}}^{*}) e^{i\mathbf{k}\mathbf{x}}$$

and the Klein-Gordon equation

$$\ddot{F}(\mathbf{x}) = (\nabla^2 - m^2) F(\mathbf{x}).$$

Our intent in the sequel is to carry over the analysis without any reference to these formal quantities, and to use only the well-defined operators $a(f)$, $a^*(f)$, $F(f)$, $P(f)$, and $H_0 = \Omega(H_0^{(1)})$.

e. A Prototype for Quantum Field Theory: The van Hove Model

We have now gained enough knowledge about the Fock-space formalism for *free* fields to understand why this formalism is actually not sufficient for a *general* description of *interacting* fields. One of the most striking—and simplest—counterexamples to have been proposed as concrete evidence of the latter statement is the *van Hove model.*

To various degrees of mathematical rigor and physical insight, the properties of this model have been discussed by van Hove [1951, 1952], Friedrichs [1953, Part III], Schweber [1961, Section 12a], Kato [1961], Cook [1961], Segal [1963, Chapter V; see also other references cited there], Greenberg and Schweber [1958], and Guenin and Velo [1968]. Actually, the role of this model as an archetype can be traced back to the textbook by Wentzel [1943, §7] and its physical motivation, to the *Yukawa* theory of nuclear forces [1935]; it might also be mentioned that the method used by van Hove presents strong analogies with that used by Bloch and Nordsieck [1937] in their discussion of the "infrared" divergences in quantum electrodynamics.

The model is a caricature of the nuclear interaction, drawn in a way that emphasizes the influence of the nucleons on the meson field. Specifically, we are considering a neutral scalar meson field in interaction with "classical sources," the latter mimicking recoilless nucleons. This downgrading of the role of the nucleons—the fact that their energy is momentum independent—is one of the essential simplifying features of the model responsible for its exact solutility.

The classical wave equation for a field $F(\mathbf{x})$ interacting with a source distribution $\rho(\mathbf{x})$,

$$-\ddot{F}(\mathbf{x}) = -(\nabla^2 - m^2)\, F(\mathbf{x}) + \rho(\mathbf{x}),$$

suggests that the time evolution of its quantum analog F is generated by the "Hamiltonian"

$$H = H_0 + F(\rho),$$

where H_0 is the free Hamiltonian of the meson field $F(f)$ already discussed. Formally, this Hamiltonian corresponds to the heuristic form

$$H = \sum_{\mathbf{k}} \omega_{\mathbf{k}} a_{\mathbf{k}}^* a_{\mathbf{k}} + \frac{1}{\sqrt{V}} \sum_{\mathbf{k}} \frac{\tilde{\rho}(\mathbf{k})}{\sqrt{2\omega_{\mathbf{k}}}} (a_{\mathbf{k}} + a_{-\mathbf{k}}^*),$$

where $\tilde{\rho}(\mathbf{k})$ is the Fourier transform of $\rho(\mathbf{x})$.

Our first preliminary task is to give a precise meaning to this Hamiltonian as a self-adjoint operator. Let us denote by $\mathfrak{D}$ the domain (in Fock space) of the free Hamiltonian $H_0 = \Omega(H_0^{(1)})$ and verify that $F(\rho)$, hence H, is defined on this domain. To do this we notice that for every Ψ in $\mathfrak{D}$ we have

$$\infty > |H_0\Psi|^2 = \sum_N |H_0^{(N)}\Psi^{(N)}|^2 \geqslant m^2 \sum_N N^2 |\Psi^{(N)}|^2,$$

since $H_0^{(1)}$ is bounded *below* by m. Therefore for any Ψ in $\mathfrak{D}$ we have

$$\sum_N N^2 |\Psi^{(N)}|^2 < \infty,$$

$$\sum_N N |\Psi^{(N)}|^2 < \infty,$$

$$\sum_N (N + 1) |\Psi^{(N)}|^2 < \infty.$$

Consequently $a(\rho)$ and $a^*(\rho)$ are defined on $\mathfrak{D}$, and so, therefore, is $F(\rho)$; furthermore we have

$$|F(\rho)\Psi| = \frac{1}{\sqrt{2}} |a(\rho)\Psi + a^*(\rho)\Psi| \leqslant \frac{1}{\sqrt{2}}(|a(\rho)\Psi| + |a^*(\rho)\Psi|)$$

hence

$$|F(\rho)\Psi|^2 \leqslant |a(\rho)\Psi|^2 + |a^*(\rho)\Psi|^2 \leqslant |\rho|_\mu^2 \sum_N (2N + 1) |\Psi^{(N)}|^2 < \infty.$$

This inequality enables us to use the following theorem† to conclude that H is self-adjoint on $\mathfrak{D}$.

Theorem 2. *Let* A *be self-adjoint on the linear manifold* $\mathfrak{D}$ *of a Hilbert space* $\mathcal{H}$ *and let* B *be symmetric on* $\mathfrak{D}$ *such that*

$$|\mathrm{B}\Psi| \leqslant \mathrm{a}\, |\mathrm{A}\Psi| + \mathrm{b}\, |\Psi|$$

for all Ψ *in* $\mathfrak{D}$ *and some constants* a *and* b *with* $0 < \mathrm{a} < 1$ *and* $0 \leqslant b < \infty$. *Then* (A + B) *is self-adjoint on* $\mathfrak{D}$.

Hence H will be self-adjoint on $\mathfrak{D}$ if we can find two constants a and b that satisfy the conditions of the theorem and such that

$$|F(\rho)\Psi|^2 \leqslant a^2 \, |H_0\Psi|^2 + b^2 \, |\Psi|^2.$$

† This theorem is actually a particular case of the very general results obtained by Kato [1951, 1966], Trotter [1958, 1959], and Nelson [1964] and used repeatedly ever since in the study of Hamiltonian operators appearing in quantum theories.

This inequality will be satisfied if we can find a and b such that

$$a^2m^2 \sum_N N^2 |\Psi^{(N)}|^2 + b^2 \sum_N |\Psi^{(N)}|^2 - |\rho|_\mu^2 \sum_N (2N+1) |\Psi^{(N)}|^2 \geqslant 0$$

which is satisfied in turn if for all N

$$a^2m^2N^2 - 2 |\rho|_\mu^2 N + (b^2 - |\rho|_\mu^2) \geqslant 0.$$

We can always satisfy this last inequality with $a < 1$ by taking

$$a = \frac{|\rho|_\mu}{\lambda m} \quad \text{with} \quad \lambda > \frac{|\rho|_\mu}{m}$$

$$b > |\rho|_\mu (\lambda + 1).$$

Consequently H is actually self-adjoint on $\mathfrak{D}$.

Our next task is to establish the relation between the spectrum of H_0 and the spectrum of H to determine how the energy of the free field F has been modified by the introduction of its interaction with the source distribution ρ. To achieve this aim we notice that straightforward application of techniques similar to those just used allows us to show that $P(f)$ is the self-adjoint closure of its restriction to $\mathfrak{D}$, hence generates a continuous one-parameter group of unitary operators $V_\lambda(f) \equiv \exp[-iP(f)\lambda]$ which maps $\mathfrak{D}$ into itself. On writing

$$V = V_1((H_0^{(1)})^{-2}\rho)$$

we further see that

$$\begin{aligned} V^{-1}H_0V &= H_0 + i[P((H_0^{(1)})^{-2}\rho), H_0] - \tfrac{1}{2}[P((H_0^{(1)})^{-2}\rho), [P((H_0^{(1)})^{-2}\rho), H_0]] \\ &= H_0 + F(\rho) + \tfrac{1}{2} |(H_0^{(1)})^{-1}\rho|^2 I, \end{aligned}$$

i.e.,

$$VHV^{-1} = H_0 + W \cdot I$$

with

$$W = -\tfrac{1}{2} |(H_0^{(1)})^{-1}\rho|^2.$$

We mention in passing that to derive this result we used, in an essential way, the fact that the one-particle Hamiltonian is bounded below (i.e., by $m > 0$); Cook [1961], using a more sophisticated technique, obtained the following modification of this result, valid even if our particular assumptions were not satisfied: For any f in the domain $\mathfrak{D}(H)$ of the one-particle observable H the *closure* of $[\Omega(H) + F(Hf) + \tfrac{1}{2}(Hf, f)_\mu]$ is unitarily equivalent to $\Omega(H)$, with the similarity operator $\exp[ip(f)]$, where $p(f)$ is $i(a^*(f) - a(f))/\sqrt{2}$.

We then have

$$W = \tfrac{1}{2} \int d^3\mathbf{k} \, \frac{1}{\omega_\mathbf{k}^2} \, \tilde{\rho}^*(\mathbf{k}) \, \tilde{\rho}(\mathbf{k})$$

and, since $(-1)/\omega_{\mathbf{k}}^2$ is the Fourier transform of the Yukawa potential

$$Y(\mathbf{x}) \equiv -\frac{e^{-m|\mathbf{x}|}}{4\pi\,|\mathbf{x}|}$$

we have

$$VHV^{-1} = H_0 + W \cdot I,$$

where

$$W = \tfrac{1}{2}\iint d\mathbf{x}\, d\mathbf{y}\; Y(\mathbf{x} - \mathbf{y})\rho^*(\mathbf{x})\,\rho(\mathbf{y}).$$

Since the spectrum of a self-adjoint operator is invariant under unitary transformation of this operator, we conclude from the last two relations that the result of the interaction between the meson field F and the source distribution ρ is to shift the energy (of the field) by a finite constant W; furthermore, this constant is equal to the contribution that we would obtain from a model in which the sources interacted among themselves via a Yukawa potential. Physically, this result is well known; it expresses the old idea that the nuclear forces are mediated by the meson field, and the above derivation only shows, for the moment, that this statement can be made mathematically rigorous.

We notice further that subtracting the constant W from the total Hamiltonian would not affect the time-evolution of the expectation value of any observable. We can, therefore, if we so wish, replace in any of the equations of motion relevant to our purpose the total Hamiltonian H with the "renormalized Hamiltonian" $\hat{H}$ defined by

$$\hat{H} = H - W \cdot I,$$

which is then unitarily equivalent to the free Hamiltonian H_0. Whenever the latter circumstance is encountered in scattering theory we conclude immediately that the S-matrix is I. Before doing so here we have to see whether we can actually speak of a scattering situation in our model. Specifically, we have to determine whether our asymptotic condition, as we formulated it in the beginning of this section, can be satisfied. The candidates for the asymptotic free fields F^{in} and F^{out} are obviously the free field

$$F_t(f) = e^{iH_0t}F(f)e^{-iH_0t} = F(e^{iH_0^{(1)}t}f),$$

whereas we submit that

$$\hat{F}_t(f) \equiv V^{-1}\, F_t(f)V$$

is the correct interpolating field. Since $F_t(f) = F(f_t)$, where f_t stands for $e^{iH_0^{(1)}t}f$, we have

$$\hat{F}_t(f) = V^{-1}\, F(f_t)V = \hat{F}(f_t),$$

a relation characteristic of a *quasi-free field*. Furthermore, since $V\hat{H}V^{-1} = H_0$,

$$\begin{aligned}\hat{F}_t(f) &= V^{-1}e^{iH_0t}F(f)e^{-iH_0t}V\\ &= e^{i\hat{H}t}V^{-1}F(f)Ve^{-i\hat{H}t}\\ &= e^{i\hat{H}t}\hat{F}(f)e^{-i\hat{H}t},\end{aligned}$$

so that the time evolution of the field $\hat{F}(f)$ is governed, as required in the formulation of the asymptotic condition, by the total (renormalized) Hamiltonian. Finally, from the commutation relations between $F(f)$ and $P(g)$ we conclude that

$$\hat{F}(f) = V^{-1}F(f)V = F(f) + c(f, \rho)I$$

where

$$c(f, \rho) = \tfrac{1}{2}[((H_0^{(1)})^{-2}f, \rho) + \text{c.c.}];$$

hence, by replacing f with f_t in the above expression we have

$$\hat{F}_t(f) = V^{-1}F(f_t)V = F(f_t) + c(f_t, \rho)I.$$

The asymptotic condition

$$\lim_{t\to\pm\infty}\langle\phi;\, \hat{F}_t(f) - F_t(f)\rangle = 0$$

then reduces to the condition

$$\lim_{t\to\pm\infty} c(f_t, \rho) = 0,$$

which is always satisfied for f and ρ in $\mathcal{H}^{(1)}$. Mathematically the latter conclusion is easily reached by writing this scalar product in k-space and then using the Riemann-Lebesgue lemma (we recall that this lemma asserts that the Fourier transform maps $\mathcal{L}^1(\mathbb{R}^1)$ into $\mathcal{C}^1(\mathbb{R}^1)$, the space of all functions on $\mathbb{R}^1$ vanishing at infinity and continuous in the sup-norm). Physically, the fact that this limit vanishes is related to the well-known spreading of the free wave packet.

Hence we actually have a scattering situation that can be solved exactly: the total Hamiltonian has been "diagonalized," the interpolating field $\hat{F}$ has been obtained, and the asymptotic fields F^{in} and F^{out} are identical so that the S-matrix is actually I for all physical purposes; incidentally, we also saw explicitly that the interpolating field is quasi-free (in the sense specified above).

Let us now denote by Ψ the *bare vacuum* for the free-field, i.e., the vector in Fock space satisfying

$$\begin{aligned}a(f)\Psi_0 &= 0 &&\text{for all } f \text{ in } \mathcal{H}^{(1)},\\ U_t^0\Psi_0 &= \Psi_0 &&\text{for all } t;\ U_t^0 = e^{iH_0t}.\end{aligned}$$

From our preceding remarks it is clear that the vector $\hat{\Psi}_0$ (which we refer to as the *physical* or *dressed vacuum*) defined by

$$\hat{\Psi}_0 = V^{-1}\Psi_0$$

satisfies the relations

$$\hat{a}(f)\hat{\Psi}_0 = 0 \qquad \text{for all } f \text{ in } \mathcal{H}^{(1)},$$
$$\hat{U}_t\hat{\Psi}_0 = \hat{\Psi}_0 \quad \text{for all } t;\ \hat{U}_t = e^{i\hat{H}t},$$

where $\hat{a}(f)$ are the annihilation operators for the "physical" or "dressed" meson corresponding to the interpolating field $\hat{F}$. We can now introduce the explicit form of V in the definition of $\hat{\Psi}_0$ to get

$$\hat{\Psi}_0 = e^{-\frac{1}{4}|(H_0{}^{(1)})^{-1}\rho|_\mu{}^2} \sum_{N=0}^{\infty} \frac{(-1)^N}{2^{N/2}N!} [(H_0^{(1)})^{-1}\rho]^{(N)},$$

where

$$[(H_0^{(1)})^{-1}\rho]^{(N)} \equiv [a^*((H_0^{(1)})^{-1}\rho)]^N\Psi_0 \in \mathcal{H}^{(N)}.$$

We could then reconstruct the Fock space for the "dressed" field by the usual application on $\hat{\Psi}_0$ of all polynomials in the $\hat{a}$ and $\hat{a}^*$, followed by the norm closure. This space would coincide with our original Fock space, since the dressed field is unitarily equivalent to the bare field.

Up to this point of the analysis all the results we have obtained have been established rigorously within the Fock-space formalism; we saw that whenever $\rho \in \mathcal{H}^{(1)}$ no difficulty such as improperly defined transformations, infinities, and the usual paraphernalia of quantum field theories arises at all. In the standard language of the physicist this is due to our introduction of a proper "cutoff" function $\tilde{\rho}$ which decreases fast enough in k to cut off the contribution of large momentum values, hence to obliterate the usual effects linked to the presence of an infinite number of degrees of freedom associated with the field; we have, then, in effect, circumvented in the traditional manner (with perhaps a somewhat unusual insistence on the mathematical details of the proof) the "ultraviolet catastrophe."

Suppose now that we want to remove this cutoff and treat the case in which the source distribution is concentrated at the origin, letting $\tilde{\rho}(\mathbf{k}) \to 1$ (any *finite* real *constant* will do, for that matter). The traditional formalism, as discussed up to this point, cannot be used to handle the limiting situation: to start with, the total Hamiltonian, written as above in the case in which $\tilde{\rho}(\mathbf{k}) = 1$, loses its meaning as an operator acting in the Fock space of the bare mesons; other difficulties would appear if we were to try unduly to force this problem into the old formalism: for instance, the renormalization constant W would become infinite (one of the symptoms of the ultraviolet catastrophe). Still the physicist would like to have a way to treat this problem and others of the same type.

To get a feeling for what should be attempted in a case like this let us try to separate the essential aspects of what we did in the Fock-space formalism from those that are more accidental and appear to be connected in some special way to the method of calculation used.

We first have to agree that the primary objects of the theory are the fields and their expectation values. In Fock space the $F(f)$ appear as well-defined operators; the linear manifold $\mathcal{F}$ (defined in our construction of the Fock space), dense in $\mathcal{H}$, is contained in the domain of these $F(f)$ and stable under their action; let us concentrate our attention on the $F(f)$ as restricted to $\mathcal{F}$. We can then form all finite linear combinations and products of them; in mathematical terms this means exactly that they generate, together with the identity I, an algebra which we call $\mathfrak{A}$; we specify the states as normalized linear functionals $\langle\phi;\cdot\rangle$ on $\mathfrak{A}$. We finally remark that the problem at hand is so simple that we do not have to worry now about the topology to be placed on $\mathfrak{A}$.

The next thing we are interested in is the time evolution, which is entirely defined, as far as we are concerned, if we prescribe the dynamical law $F \to F_t$. From our analysis in Fock space we see that the free time evolution (resp. the time evolution in the presence of the source distribution ρ) induces a linear mapping α_t^0 (resp. α_t) from $\mathfrak{A}$ into itself such that

$$F(f) \to F_t^0(f) = F(f_t)$$

resp.

$$F(f) \to F_t(f) = F(f_t) + c(f_t - f, \rho)I,$$

where f_t and $c(g, \rho)$ were previously defined.

In line with our agreement to concentrate on those aspects of the theory that are directly connected with the fields we take these relations, together with the requirement that α_t^0 (resp. α_t) acts linearly on $\mathfrak{A}$, as our *definition* of the free evolution (resp. the dynamical law in the presence of the source distribution ρ). We finally conclude the algebraic summary of our previous analysis by proposing a candidate in $\mathfrak{A}$ for the interpolating field $\hat{F}(t)$; we define to this effect:

$$\hat{F}(f) = F(f) + c(f, \rho)I;$$

from our algebraic specification of the dynamical laws we get

$$\begin{aligned}\hat{F}_t(f) &= F_t(f) + c(f, \rho)I \\ &= F_t^0(f) + c(f_t, \rho)I,\end{aligned}$$

so that the asymptotic condition can be written, using the fact that $\langle\phi; I\rangle = 1$,

$$\lim_{t\to\pm\infty} |c(f_t, \rho)| = 0.$$

The reader may wonder, at this point, about what was to be gained by this algebraic reformulation of the analysis carried out originally in Fock space. To answer this we remark first that this way to define the free evolution holds quite generally and moreover is of immediate physical significance. In real-life problems, however, it is not so easy to determine from the start what form the mapping α_t should indeed take; this problem is similar to that of integrating the Schrödinger equation in the usual framework of quantum mechanics and should therefore not be dismissed as a trivial technicality. The hope nevertheless is that a scheme of some sort (perturbation technique, limiting procedure, or whatever) could be devised on $\mathfrak{A}$ which would not necessarily be plagued by the difficulties of the usual calculation scheme, since the latter we maintain is too strongly linked to some of the inessential aspects of the theory (e.g., the unitary implementability of the time-evolution and the existence of Møller matrices).

The van Hove model is an example in point. In the limit $\tilde{\rho}(\mathbf{k}) \to 1$ the total Hamiltonian $H(\tilde{\rho}(\mathbf{k}) = 1)$ cannot be given a meaning as an operator acting in Fock space. Still, and this is precisely the place in which an algebraic formulation of the model presents an advantage, the mapping $\alpha_t(\rho)$, as defined above, admits a limit as $\tilde{\rho}(\mathbf{k}) \to 1$ in the sense that

$$F_t(f) = F_t^0(f) + c(f_t - f, \rho_1)I, \quad \text{with } \tilde{\rho}_1(\mathbf{k}) = 1,$$

is a well-defined element of $\mathfrak{A}$, since this condition reduces to the requirement

$$c(g, \rho_1) = \tfrac{1}{2}\left[\int d^3\mathbf{k}\,\omega_{\mathbf{k}}^{-2}\tilde{g}^*(\mathbf{k}) + \text{c.c.}\right] < \infty.$$

For the same reason the interpolating field exists in this limit† as an element of $\mathfrak{A}$ and the asymptotic condition is satisfied.

To get the above results we had to abandon the idea of calculating directly in Fock space the evolution $F \to F_t$ from the ill-defined Hamiltonian $H(\rho(k) = 1)$. It is still clear that we can nevertheless represent F_t and $\hat{F}_t$ as operators acting in the Fock space of the bare mesons; some traditional aspects of quantum field theory, however, are lost in the process. Let us now devote some time to substantiating this statement. It is usually assumed that the space of the theory hosts a unitary representation of the Lorentz group (and, in particular, the time-evolution) and that a vacuum $\hat{\Psi}_0$ exists such that

$$\hat{a}(f)\hat{\Psi}_0 = 0,$$
$$\hat{U}_t\hat{\Psi}_0 = \hat{\Psi}_0.$$

Let us now show that $\hat{\Psi}_0 \neq 0$ cannot exist in the Fock space of the bare

† Whenever a limit is taken it presupposes a topology; in the present case we satisfy ourselves with the statement that the expectation values of the fields do converge.

mesons and satisfy even the first of these two conditions. Suppose, on the contrary, that such a vector exists and write its decomposition in $\mathcal{H} = \sum_N \overline{\mathcal{H}}^{(N)}$ as

$$\hat{\Psi}_0 = \sum_N \hat{\Psi}_0^{(N)}.$$

We notice that

$$\hat{a}(f) = a(f) + d(f, \rho),$$

where

$$d(f, \rho) = \int d^3k \omega_k^{-2} \tilde{f}^*(k)\, \tilde{\rho}(k).$$

We then have

$$\begin{aligned} 0 = \hat{a}(f)\hat{\Psi}_0 &= \sum_N \hat{a}(f)\hat{\Psi}_0^{(N)} \\ &= \sum_N a(f)\hat{\Psi}_0^{(N)} + \sum_N d(f, \rho)\hat{\Psi}_0^{(N)}. \end{aligned}$$

From the fact that

$$a(f)\hat{\Psi}_0^{(N)} \in \overline{\mathcal{H}}^{(N-1)} \quad \text{for all } N > 0$$

and

$$\overline{\mathcal{H}}^{(N)} \perp \overline{\mathcal{H}}^{(N')} \quad \text{for } N \neq N'$$

we get

$$-d(f, \rho)\hat{\Psi}_0^{(N)} = a(f)\hat{\Psi}_0^{(N+1)};$$

hence

$$|d(f, \rho)|\,|\hat{\Psi}_0^{(N)}| = |a(f)\hat{\Psi}_0^{(N+1)}| \leqslant (N+1)^{1/2}\, |f|_\mu\, |\hat{\Psi}_0^{(N+1)}| \quad \text{for all } N$$

and therefore either $\hat{\Psi}_0^{(N)} = 0$ or $|d(f, \rho)| \leqslant C\,|f|_\mu$. The second term of this alternative would, however, imply, by Riesz's theorem, that an element η exists in $\mathcal{H}^{(1)}$ such that

$$d(f, \rho) = (f, \eta) \quad \text{for all } f,$$

which cannot be the case when $(H_0^{(1)})^{-1}\rho \notin \mathcal{H}^{(1)}$. Hence in our case, in which $\tilde{\rho}(\mathbf{k}) = 1$, we must have $\hat{\Psi}_0^{(N)} = 0$ for all N. We therefore conclude that there is no vector $\hat{\Psi}_0 \neq 0$ in the Fock space of the bare mesons that could serve as a vacuum for the dressed meson field $\hat{F}(f)$. This result confirms the suspicion we may already have by looking at the explicit form given above for $\hat{\Psi}_0$ when $\rho \in \mathcal{H}^{(1)}$ and passing formally to the limit $\tilde{\rho}(\mathbf{k}) = 1$. Incidentally, the absence of a dressed (or physical) vacuum in the Fock space of the bare mesons has an interesting corollary; i.e., that the representations $\{a(f), a^*(g)\}$ and $\{\hat{a}(f), \hat{a}^*(g)\}$ of the canonical commutation relations are *unitarily inequivalent;* suppose, on the contrary, that there existed a unitary operator V such that

$$\hat{a}(f) = V^{-1}a(f)V;$$

then $V^{-1}\Psi_0$ (where Ψ_0 is the bare vacuum) would be a vacuum for the dressed meson field and its existence would then contradict the result just obtained. This fact† by itself already indicates that *the Möller matrices could not possibly exist* in the usual sense when the limit $\tilde{\rho}(\mathbf{k}) = 1$ is taken; this statement is confirmed by the actual form of the "renormalized" Møller operators calculated by Cook [1961] in the case in which ρ belongs to $\mathcal{H}^{(1)}$. Some general reasons for the nonexistence of Møller matrices are discussed later on as Haag's theorem(s) (see Chapter 3, Section 1.d).

The reasoning that led us to conclude that the interpolating field and the free field are unitarily inequivalent can be used to show that F_t (with $t \neq 0$) is not unitarily equivalent to F. Suppose, on the contrary, that there exists a unitary operator $\bar{U}_t$ such that

$$F_t(f) = \bar{U}_t^{-1} F(f) \bar{U}_t;$$

then $\bar{U}_t^{-1}\Psi_0 \equiv \Psi_{0,t}$ (where Ψ_0 is the bare vacuum) would be a vacuum for F_t; comparing this form of $F_t(f)$ with that given by the dynamical law α_t, namely

$$F_t(f) = F_t^0(f) + c(f_t - f, \rho)I,$$

we can use the same path as before to conclude that either $|d(f_{(t)} - f, \rho)| \leqslant C\, |f_{(t)} - f|_\mu$ or $\Psi_{0,t}^{(N)} = 0$; furthermore, since $H_0^{(1)}$ has a completely continuous spectrum,

$$\{f_t - f \,|\, f \in \mathcal{H}^{(1)}\} = \mathcal{H}^{(1)},$$

and therefore we get the same contradiction. This proves *ad absurdo* that $\bar{U}_t$ does not exist. Hence the automorphism α_t, which gives the time evolution in the presence of the singular source distribution $\tilde{\rho}(\mathbf{k}) = 1$, is not unitarily implementable. This shows a fortiori that the limiting "Hamiltonian" $H(\tilde{\rho}(\mathbf{k}) = 1)$ is not only *ill*-defined but actually *cannot* be defined at all: there is no self-adjoint operator $\bar{H}$ acting on our *original* Fock space that could possibly generate the time evolution we are concerned with.

It might be interesting to note that the argument presented above would have miscarried if we had restricted our test function space space from $\mathcal{H}^{(1)}$ to, for instance, the domain of any positive power of $H_0^{(1)}$; this remark might also be regarded as one more reason for abandoning the Fock-space formalism which is too strongly linked to one specific particle interpretation, in favor of a more general field theory which would postpone the correct particle interpretation to a later stage.

An extension of the argument presented above would show that F_t and $F_{t'}$ are unitarily inequivalent whenever $t \neq t'$. The van Hove model, then, leads

† This is sometimes alluded to in the literature (e.g., see, Barton [1963]) by the heuristic statement that the unitary transformation $V(\rho)$ becomes an "improper unitary transformation" in the limit when the cutoff is removed.

to the conclusion that a wealth of inequivalent representations of the canonical commutation relations might occur in a given physical problem. This is a new phenomenon, proper to systems involving an infinite number of degrees of freedom. It is in sharp contradistinction to the results of von Neumann [1931] (see also Jordan and Wigner [1928]), who proved that, up to a unitary equivalence, there exists only *one* irreducible representation of CAR and CCR, whenever the test function space is *finite*-dimensional. For the time being we shall content ourselves with the mere mention of this fact of life and postpone until Chapter 3 a more detailed discussion of this fundamental aspect of the theory.

To conclude our study of the van Hove model let us give, without proofs, some more or less immediate consequences of our analysis as well as a glimpse of what is coming up.

First, the analysis carried out above in the case of a source distribution concentrated at the origin can be extended without essential modifications to the case of a discrete source distribution.

Second, we can see that with the same source distribution the interpolating fields corresponding to different values of the coupling constant are also unitarily inequivalent. This ruins any attempt at a perturbative approach to the evolution following the original Dyson scheme, for what one is trying to do there is to give a series expansion of quantities, such as $\bar{U}_t$ or Møller matrices, which do not properly exist; hence we should not be surprised to see all kinds of myriotic divergences occur at some or all orders of perturbation when such a scheme is forced on a theory to which it does not belong. For an intuitive and instructive discussion of this aspect of the problem the reader is referred to van Hove's original papers [1951] and [1952]. This does not preclude using a perturbative technique to compute α_t *itself*, even when it is not so readily available as it is in the van Hove model. For that, however, some topology has to be put on the algebra $\mathfrak{A}$; therefore we have to postpone this question until we have a clearer view of what the topological algebra $\mathfrak{A}$ should be and what kind of automorphisms it does admit. We could still speculate that a proper development along these lines might produce a rational explanation for the success of quantum electrodynamics, in which a lot of meaningful results have been obtained in spite of the various renormalizations of infinities and of the infrared and ultraviolet catastrophies that plague this theory.

The appearance of inequivalent representations of CCR happens to be quite a general feature and actually is intimately connected to the spatial invariance of the theory; this statement, known as Haag's theorem, is proved and discussed in the appropriate context later on.

The question whether the theory could be represented at all in a Hilbert space has been the subject of many speculations, starting with van Hove's proposal of a "universal receptacle"†; the properties and limitations of this formalism are discussed later.

It should at least be clear by now that the Fock-space formalism of quantum field theory, in spite of its original usefulness and the seemingly natural way in which it imposed itself, presents some definite defects that an algebraic reformulation of the problem might be able to cure.

f. A Prototype for Statistical Mechanics: the BCS Model

The occurrence of "strange representations" of CCR and CAR—one of the principal reasons for advocating a formulation of quantum field theory that would go beyond Fock space—should also be expected to be encountered in other theories involving systems with an infinite number of degrees of freedom. The algebraic approach formalism, originally developed with quantum field theories in mind, might therefore be relevant to classical as well as quantum statistical mechanics insofar as the thermodynamical limit plays an essential role in these theories. This is actually the case, and we now want to review *briefly* some of the problems in which this approach has been tried and proved to be useful.

One of the nicest illustrations of this situation is the BCS model; we recall that the BCS Hamiltonian describes an interaction between electrons in a solid and provides a microscopic interpretation of the phenomenon of superconductivity.

This model was suggested by Bardeen, Cooper, and Schrieffer [1957], and a textbook account of the physics involved in it can be found in Schrieffer [1964]; its rather peculiar mathematical properties, especially in connection with the algebraic approach, were first pointed out by Haag [1962]; his paper proved to be essentially right in its conclusions and extremely stimulating in its outlook, as witnessed by the numerous refinements that followed. Among them we mention, as an orientation with no claim to exhaustiveness, the following contributions directly related to the mathematical structure of the BCS model: Ezawa [1964], Emch and Guenin [1966], Thirring and Wehrl [1967], Thirring [1968], Jelinek [1968], and Dubin and Sewell [1970].

The model is nonrelativistic and we formulate it in a box, say a cubic one, centered at the origin and of volume $V = L^3$; we assume periodic boundary conditions to make life simpler but they are by no means necessary. The one-particle space is then $\mathfrak{L}^2(V)$ and an orthogonal basis can be chosen, in agreement with our boundary conditions, as

$$f_{\mathbf{p}}(\mathbf{x}) = V^{-1/2} e^{i\mathbf{p}\cdot\mathbf{x}}$$

† The coining of this vocable is obscure.

with

$$\mathbf{p} = \frac{\mathbf{n}\pi}{L} \quad \text{and} \quad \mathbf{n} \in \mathbb{Z}^3.$$

We can then construct the antisymmetrized Fock space ${}^{A}\mathcal{H}$ on $\mathfrak{L}^2(V)$ and the Fermi operators $F^*_{\pm}(f)$, $F_{\pm}(g)$, as indicated in Subsection c. In particular, we define

$$a_i^*(\mathbf{p}) = F_i^*(f_{\mathbf{p}}) \quad \text{and its adjoint } a_i(p) \qquad (i = +, -),$$

which create and annihilate electrons with momentum p and spin, respectively, "up" or "down." The Hamiltonian of the system is then written formally as

$$H(V) = \sum_{\mathbf{p},i} \varepsilon(p)\, a_i^*(\mathbf{p})\, a_i(\mathbf{p}) + \sum_{\mathbf{p},\mathbf{q}} a_+^*(\mathbf{p})\, a_-^*(-\mathbf{p})\, v(\mathbf{p}, \mathbf{q})\, a_-(-\mathbf{q})\, a_+(\mathbf{q}),$$

where $\varepsilon(p)$ and $v(\mathbf{p}, \mathbf{q})$ are scalar functions defined by

$$\varepsilon(p) = \frac{p^2}{2m}$$

$$v(\mathbf{p}, \mathbf{q}) = \iint_V d\mathbf{x}\, d\mathbf{y}\, f_{\mathbf{p}}(\mathbf{x})\, v(\mathbf{x}, \mathbf{y})\, f^*_{\mathbf{p}}(y);$$

$v(\mathbf{x}, \mathbf{y})$ expresses the nonlocal character of the interaction between Cooper pairs as mediated by the phonons. The following properties of this interaction

(i) $$v(\mathbf{x}, \mathbf{y}) = v(\mathbf{y}, \mathbf{x})^*$$

(ii) $$v \equiv \iint_{\mathbb{R}^6} d\mathbf{x}\, d\mathbf{y}\, |v(\mathbf{x}, \mathbf{y})| < \infty,$$

(iii) $$\sum_{\mathbf{q}} |v(\mathbf{p}, \mathbf{q})| < \infty$$

are used to arrive at the conclusion that the following approximation of the Hamiltonian becomes *exact* in the thermodynamical limit $V \to \infty$:

$$H_{\text{eff}}(V) = \sum_{\mathbf{p},i} E(p)\, \gamma_i^*(\mathbf{p})\, \gamma_i(\mathbf{p}),$$

where the Fermion operators $\gamma_i(p)$ are defined by

$$\gamma_+(\mathbf{p}) = u(p)\, a_+(\mathbf{p}) + v(p)\, a_-^*(-\mathbf{p})$$

and

$$\gamma_-(\mathbf{p}) = -v(-p)\, a_+^*(-\mathbf{p}) + u(-p)\, a_-(\mathbf{p})$$

with

$$u(p) = \frac{\Delta^*(p)}{D(p)},$$

$$v(p) = \frac{E(p) - \varepsilon(p)}{D(p)},$$

$$E(p) = \{\varepsilon(p)^2 + \Delta^*(p)\,\Delta(p)\}^{1/2},$$

$$D(p) = \{[E(p) - \varepsilon(p)]^2 + \Delta^*(p)\,\Delta(p)\}^{1/2},$$

where $\Delta(p)$ appears as a solution to the self-consistency equation

$$\Delta(p) = -\lim_{V\to\infty}\sum_{\mathbf{q}} v(\mathbf{p}, \mathbf{q})\,\frac{\Delta(q)}{2E(q)}\tanh\left\{\beta\,\frac{E(q)}{2}\right\}.$$

This equation always admits the solution $\Delta(p) = 0$; however, below a certain critical value T_c of the temperature ($\beta = (kT)^{-1}$), depending on $v(\mathbf{x}, \mathbf{y})$, another solution $\Delta(p) \neq 0$ occurs, in particular with the feature that $\Delta(0)$ is strictly positive (the famous "energy gap" in the one-particle excitation spectrum); we should also notice that if $\Delta(p)$ is a solution of this equation so is $e^{i\alpha}\,\Delta(p)$, with α real.

Our purpose now is not to present a derivation of these results nor a proof of the fact that they are thermodynamically exact, although this can be done (see the bibliography at the beginning of this subsection). Rather we shall take them for granted and concentrate instead on the inescapable paradoxes linked to these results when the need for a complete analysis which is allowed to go beyond the usual Fock-space formulation is not recognized.

We first observe that we will run into trouble if we naïvely assume—as we should in the usual framework of, say, solid-state physics—that there exists a bonafide self-adjoint operator H such that

(i) H acts on the Fock space constructed from the vacuum Ψ_0 defined by

$$F_{\pm}(f)\Psi_0 = 0 \quad \text{for all } f \text{ in } \mathfrak{L}^2(\mathbb{R}^3)$$

(ii) H represents the energy of the system in the thermodynamical limit, and can be obtained as the limit (in some sense) as $V \to \infty$ of either $H(V)$ or $H_{\text{eff}}(V)$.

The trouble is that the one-particle excitation spectrum of H is given by $E(p)$ *which depends on the temperature* via $\Delta(p)$; then so will the spectrum of H itself. This is an intolerable situation in the traditional framework, since the spectrum of an operator is an invariant property of this operator and should not depend on the "diagonalization procedure" used to compute it.

We further remark that the original Hamiltonian is invariant under the gauge transformations

$$F^{(*)}(f) \to e^{\pm i\alpha} F^{(*)}(f),$$

whereas H_{eff} is not invariant under this transformation; this is true independently of V, with which this symmetry has nothing to do; we should then wonder how taking the thermodynamical limit could restore this symmetry.

Finally the vacuum of the γ-particles, which is the ground state of H_{eff} and is defined by

$$\gamma(f)\hat{\Psi}_0 = 0 \quad \text{for all } f \text{ in } \mathfrak{L}^2(\mathbb{R}^3),$$

can be calculated in terms of the vacuum of the a-particles and turns out to be given by the expression

$$\hat{\Psi}_0 = \prod_{\mathbf{p}} (u(p) + v(p)\, b^*(\mathbf{p}))\Psi_0$$

where $b^*(\mathbf{p})$ creates the Cooper pair $a_+^*(\mathbf{p})\, a^*(-\mathbf{p})$. Hence $\hat{\Psi}_0$ depends on the temperature and leads, in the thermodynamical limit, to the same kind of troubles as those encountered with the vacuum-candidates of the van Hove model; we have here the further complication that $\hat{\Psi}_0$, the ground state of the Hamiltonian, is not gauge invariant; a quieter way to express this somewhat distressing fact can be found in the literature under the assertion that the vacuum, or the ground state of the BCS Hamiltonian for any given temperature $T \leqslant T_c$, is infinitely (and continuously) degenerate.

For all these reasons (as well as for some others that are not so immediately obvious) it seems impossible to understand properly, within the traditional approach, the fact that the proposed diagonalization becomes exact in the thermodynamical limit; yet this diagonalization must be exact in some sense to account for the sharp transition between the normal and the superconductive phases.

As mentioned earlier, the mathematical structure of the BCS model has attracted the attention of several authors; from their works the fact emerges that the distinctive features of the model can actually be understood if and when we work in a Hilbert space large enough to host a continuous infinity of irreducible representations of the canonical anticommutation rules; this evidently implies that we should go much beyond the usual Fock-space formalism. The problem—solved in part in the papers mentioned—is to construct the appropriate representation space in which to conduct the study of the model.

A satisfactory answer to the questions raised is possible within the algebraic formalism we shall study in the sequel; in particular, the gauge-invariance puzzle is solved when we notice that the representation of the CAR to be associated with the partition function corresponding to a given temperature, calculated in the thermodynamical limit, is not irreducible, in contrast to the

Fock representation, and can be decomposed as a direct integral of "primary"† representations, the latter being mapped into one another by the gauge transformations in such a manner that, in spite of the fact that none of them is gauge-invariant, the resulting integral is itself gauge-invariant, as it should be, since it is defined from a Hamiltonian that is invariant under this symmetry.

g. Outlook

The models presented in the preceding two subsections are clearly extreme oversimplifications of the physical reality; we could rightly point out that the van Hove model does not describe a genuine scattering situation ($S = I$ can be understood physically from the fact that the sources are recoilless), whereas the BCS model is nothing but a case in which the molecular field method (Weiss-type theory now carried in k- rather than x-space) works. Thus these models at best can be discussed as prototypes for some of the intrinsic difficulties encountered in quantum field theory and statistical mechanics. Their value in this respect comes from the fact that we know for sure how things go wrong in the usual formalism applied to these models, since they are *exactly soluble* when treated within the proper formalism. Finding a method to cure the ills presented by these models is, however, no proof by itself of the universality of the method. We can, nevertheless, and this is what we do in the next section, defend the method on better grounds if we can show in addition that it involves some basic principles and that a solution of the above models is just an illustration of how the method works when no "hard analysis" is involved.

The basic principle of the algebraic approach is to avoid starting with a specific Hilbert space scheme and rather to emphasize that the *primary objects* of the theory are the fields (or the observables) considered as purely algebraic quantities, together with their linear combinations, products, and limits in the appropriate topology. This forms the bulk of Section 1.2. We then define symmetries, such as time evolution and gauge transformations, as automorphisms that preserve the structure just established. We postpone this aspect of the problem to Section 2.2. The representations of these algebraic objects as operators acting on a Hilbert space are then introduced in a way that depends essentially on the states to the study of which the investigation is directed. This last step is known as the GNS construction; it is first described toward the end of Section 1.2 and its properties are analyzed in more detail in Section 2.1; for the time being we just want to mention that it

† These "primary" representations turn out actually to be "irreducible" in the extreme case in which $T = 0$ (i.e., $\beta = \infty$).

actually is the extension to the general quantum case of the well-known Koopman formalism for classical statistical mechanics.

The value of the algebraic method is further supported by some of the successes it has already registered and which extend significantly the generality of some of the remarks made about the particular models studied in the preceding two subsections. We noticed, for instance, in Subsection e that the "physical vacuum" escapes the Fock space of the free field when the cutoff in the interaction is removed and can then provide a basis from which to construct a new representation of the interacting fields; this fact is not proper to the van Hove model and is encountered again in the *constructive field theories* of Glimm and Jaffe. In Subsection f we saw that in the BCS model the degeneracy of the ground state is linked to the spontaneous breaking of gauge symmetry; this suggests that the algebraic approach should be tried against the general problem of *spontaneous symmetry breaking*. Some progress has been made along these lines. A related problem has also received some light from this kind of approach, namely, the understanding of the mechanism responsible for *phase transitions*. These algebraic methods have also been exploited with some success in several problems of classical and quantum statistical mechanics, ranging in generality from ergodic theory to the study of Bose-Einstein condensation, the interpretation of the computations of the spontaneous magnetization in the Ising model, and the understanding of the approach to equilibrium. In a somewhat different domain of physics we might also mention that the study of optical coherence, using Bargmann space formalism, actually goes farther than the ordinary Fock space formalism in a direction that can also be understood from an algebraic point of view.

It appears, then, that there is plenty of motivation from the study of various physical situations, for *not* staying within Fock space, and for turning to a more algebraic approach.

SECTION 2 THE EMERGENCE OF THE ALGEBRAIC APPROACH

OUTLINE. Since the early days of quantum mechanics, pioneers like Jordan, von Neumann, and Wigner seemingly perceived the suitability of an algebraic formulation of the new theory. The aim of this approach was to obtain a more phenomenological justification of some of the tools already in use and subsequently to allow us to disregard those aspects of the previous formalisms that were either physically unjustifiable or mathematically inappropriate.

These efforts, however, had little influence—if any—on the actual formal developments that were rapidly taking place and that culminated in the standard formulation of quantum electrodynamics. The latter theory—in spite of its successes, which indicate that it obviously contains at least some element of truth—is generally recognized as being plagued by conceptual difficulties which have been circumvented only by the use of formal artifices emphasized, for instance, by van Hove's model (see Section 1). The mathematical difficulties of relativistic quantum field theory were later carefully analyzed in Wightman's axiomatics. For many years, however, this attempt to straighten out the mathematical language has led to more and more difficult problems to such an extent that it, in turn, fell into some sort of disrepute among too many physicists who could not refrain from a feeling of frustration toward a mathematical construction that fell short of producing a theory of genuinely relativistic interacting quantum fields.

With this in the background, a more radical return to the sources has lately been advocated by Haag and his school. The purpose of this section is to review the earlier contributions, up to the work of I. E. Segal, directly connected to this line of thought; it culminates in the representation theorem known as the *GNS construction*.

a. The Jordan Algebra of Observables in Traditional Quantum Mechanics

The idea of an algebraic approach is already present in the initial matrix formulation of quantum mechanics to which the names of Heisenberg, Born, Jordan, Dirac, . . . are attached. At this early stage of the development of the theory von Neumann [1927a, b] made two contributions relevant to our purpose. First he formulated the quantum theory as an eigenvalue problem in Hilbert space and indicated in an unambiguous way the rules of the operator calculus pertinent to the game. Second, he analyzed the concept of state from the point of view of the theory of probability. Since his ideas in this connection will constantly recur in the sequel, let us now examine briefly the structure of quantum mechanics as it was recognized at that time.

Postulate 1. *To each observable* A *on a given physical system there corresponds a linear self-adjoint operator* π(A) *acting on a Hilbert space* $\mathcal{H}_\pi$ *and conversely.*

Some remarks are immediately in order. First, we should notice that the "converse" part of the postulate is now known to be untenable (existence of "superselection rules"; see below in this subsection); since, however, von Neumann made a rather mild use of the second part of the postulate, we shall keep it on a temporary basis and naturally exclude it in a more definitive axiomatization. Second, if we denote by $\mathfrak{A}$ the set of all observables on the physical system considered and by $\pi(\mathfrak{A})$, its image through π, we can already observe that

(i) for "any" sequence $\{A_i\}$ of elements in $\mathfrak{A}$ and "any" sequence $\{\lambda_i\}$ of real numbers $\Sigma_i \lambda_i \, \pi(A_i)$ also belongs to $\pi(\mathfrak{A})$;

(ii) for any observable A and "any" real function f of one real variable $f(\pi(A))$ belongs to $\pi(\mathfrak{A})$;
(iii) if, however, A and B are two arbitrary elements of $\mathfrak{A}$, $\pi(A)\,\pi(B)$ in general *does not* belong to $\pi(\mathfrak{A})$, whereas $\pi(A)\,\pi(B) + \pi(B)\,\pi(A)$ *does*;
(iv) The symmetrized product

$$\{\pi(A), \pi(B)\} \equiv \tfrac{1}{2}(\pi(A)\,\pi(B) + \pi(B)\,\pi(A))$$

satisfies a number of properties, some of which are worth noticing [Jordan, 1932, 1933a, b; Jordan, von Neumann, and Wigner, 1934; von Neumann, 1936; Segal, 1947].

First, as mentioned in (iii), it belongs to $\pi(\mathfrak{A})$ for any two observables A and B. *Second*, it is commutative and bilinear. *Third*, its introduction does not require, as we did above, the knowledge of the ordinary product of two noncompatible observables (i.e., two observables such that the corresponding operators do not commute in the ordinary sense). We have, indeed,

$$\{\pi(A), \pi(B)\} = \tfrac{1}{4}((\pi(A) + \pi(B))^2 - (\pi(A) - \pi(B))^2),$$

the right-hand side of this expression involving only those operations that we have already mentioned under remarks (i) and (ii). There is a possibility, therefore of defining it without going out of $\pi(\mathfrak{A})$. *Fourth*, this symmetrized product is *not* associative in general; i.e.,

$$\{\pi(A), \pi(B), \pi(C)\} \equiv \{\{\pi(A), \pi(B)\}, \pi(C)\} - \{\pi(A), \{\pi(B), \pi(C)\}\}$$

can be different from zero. This lack of associativity, which we see by simple inspection, will allow us to recover a notion that we apparently lose if we discard the ordinary product, namely, the notion of compatibility between two observables. Writing

$$[\pi(A), \pi(B)] = \pi(A)\,\pi(B) - \pi(B)\,\pi(A),$$

we readily have

$$\{\pi(A), \pi(B), \pi(C)\} = \tfrac{1}{4}[\pi(B), [\pi(A), \pi(C)]],$$

from which we conclude that a sufficient condition for the left-hand side to vanish is that A and C are compatible. It is a nontrivial property of *bounded* operators that the four conditions

(α) $\quad \{\pi(a), \pi(B), \pi(C)\} = 0, \quad$ for all B in $\mathfrak{A}$,

(β) $\quad \{\pi(A), \pi(A), \pi(C)\} = 0,$

(β') $\quad \{\pi(A), \pi(C), \pi(C)\} = 0,$

(γ) $\quad [\pi(A), \pi(C)] = 0,$

are equivalent.

Indeed $(\alpha) \to (\beta)$ and (β'), $(\gamma) \to (\alpha)$ are trivial implications. One way to see that (β) or $(\beta') \to (\gamma)$ is to prove first that this implication is true in the particular case in which $\pi(A)$ is a projector; the general case is then obtained by the successive use of the spectral decomposition theorem, the Jacobi identity, and the spectral theorem again. This method of proof has the advantage of suggesting the proper criterion for compatibility in the case of unbounded observables. Notice, in particular, that the implication $(\alpha) \to (\gamma)$ is wrong in the case of unbounded operators a counterexample being provided by the canonical commutation relations.

As a result of this discussion we see that the important concept of compatibility can be expressed without using the ordinary product between operators, but only the weakened notion of symmetrized product. As a particular case of this property we mention the following mathematically important fact, which we want to list separately as a *fifth* property of $\pi(\mathfrak{A})$, namely,

$$\{\pi(A)^2, \pi(B), \pi(A)\} = 0 \quad \text{for all } B \in \mathfrak{A}.$$

The physical meaning of this relation is to express that $\pi(A)$ and $\pi(A)^2$ commute, provided that they are bounded operators.

The properties we have outlined so far equip $\pi(\mathfrak{A})$ with a structure known in mathematics as that of a *real, commutative Jordan algebra*,† the composition laws of which are realized here by addition, multiplication by real numbers, and symmetrized product. As we have seen, the notion of compatibility between observables can be expressed completely within this structure. The natural question to ask now is whether this structure can be transferred back to the set $\mathfrak{A}$ itself. If this were the case, we could then claim to have achieved an algebraic axiomatization of quantum mechanics that would not require the postulate of an underlying Hilbert space. Incidentally, the set of all observables on a quantum system with "superselection rules"‡ also satisfies these axioms *except* for the fact that the "converse" part of Postulate 1 is not assumed to hold in general; the presence of a superselection rule is expressed by the fact that

$$\{\pi(A), \pi(B), \pi(C)\} = 0 \quad \text{for all } B \text{ and } C \text{ in } \mathfrak{A}$$

does not imply that $\pi(A)$ is a multiple of the identity operator. Furthermore, we would also have the possibility of characterizing classical systems as particular cases of the structure obtained so far, in which $\mathfrak{A}$, in addition to

† For a general mathematical discussion of the properties of such algebras see, for instance, Braun and Koecher [1966].

‡ The occurrence of superselection rules was first discovered on a particular case by Wick, Wightman, and Wigner [1952]; a brief axiomatic discussion can be found in Emch and Piron [1963]; for a didactic account see Jauch [1968a].

being a real commutative Jordan algebra, would also be *associative*, i.e.,

$$\{\pi(A), \pi(B), \pi(C)\} = 0 \quad \text{for all } A, B, \text{ and } C \text{ in } \mathfrak{A}.$$

To achieve this aim, namely, to get a completely algebraic characterization of physical theories and to justify on a phenomenological basis the very use of Jordan algebras, we need a further concept to which we now want to turn our attention.

The state of a physical system is understood intuitively as a way to express the simultaneous knowledge of the expectation values of all observables on the physical system considered.

Postulate 2. *To each state ϕ of the physical system considered corresponds a "density matrix" $\rho = \pi(\phi)$ acting on the Hilbert space $\mathcal{H}_\pi$ of Postulate 1, and such that the expectation values $\langle\phi; \mathrm{A}\rangle$ can be computed by the rule $\langle\phi; \mathrm{A}\rangle = Tr\, \pi(\phi)\, \pi(\mathrm{A})$.*

As with Postulate 1, our knowledge has progressed a little in the meantime and we know (precisely as a result of the algebraic approach we are to discuss) that we cannot always represent any state in the above way. For the time being, however, we want to play down the latter fact and to analyze the structure already emerging from our postulate as it stands; in particular, three properties of states are worth noticing; they follow immediately from Postulate 2 and seem to be so linked to the very concept of state that we shall later take them as state axioms:

(i) for "any" sequence $\{A_i\}$ of elements in $\mathfrak{A}$ and "any" sequence $\{\lambda_i\}$ of real numbers

$$\begin{aligned}\mathrm{Tr}\, \pi(\phi)\Sigma_i\lambda_i\pi(A_i) &= \Sigma_i\lambda_i\, \mathrm{Tr}\, \pi(\phi)\, \pi(A_i)\\ &= \Sigma_i\lambda_i\langle\phi; A_i\rangle;\end{aligned}$$

(ii) for any observable A $\mathrm{Tr}\, \pi(\phi)\pi(A)^2 \geqslant 0$;

(iii) $\mathrm{Tr}\, \pi(\phi)I = 1$.

Moreover we notice that

(iv) for any sequence $\{\phi_i\}$ of states and any sequence $\{\lambda_i\}$ of positive real numbers with $\Sigma_i\lambda_i = 1$

$$\langle\phi; A\rangle \equiv \Sigma_i\lambda_i\langle\phi_i; A\rangle$$

defines a state ϕ with all three of the above properties. A state ϕ is said to be pure if it cannot be properly decomposed as above. A well-known example of a pure state (when the converse part of Postulate 1 is assumed) is provided by any vector Φ_π in $\mathcal{H}_\pi$:

$$\langle\phi; A\rangle = (\Phi_\pi, \pi(A)\Phi_\pi).$$

We denote by $\mathfrak{S}$ the set of all states satisfying properties (i) to (iii).

We are now ready for an attempt to motivate on a phenomenological basis the axioms of an algebraic approach. Our task is to provide the couple $(\mathfrak{A}, \mathfrak{S})$ with a structure rich enough to allow existing mathematics to help and still not so restrictive that it will rule out applications of physical interest. We learned from Section 1, for instance, that Postulates 1 and 2 above are probably too restrictive. From the analysis, as carried up to here, emerges the feeling that the structure of $\mathfrak{A}$ and of $\mathfrak{S}$ can be summarized as far as physics is concerned by assuming that $\mathfrak{A}$ is a real commutative Jordan algebra. Our first problem, then, is to try to justify this choice of structure.

b. Structure Axioms 1 to 5 (*composition laws of observables*)

Structure Axiom 1. *To each physical system Σ we can associate the triple $(\mathfrak{A}, \mathfrak{S}, \langle;\rangle)$ formed by the set $\mathfrak{A}$ of all its observables, the set $\mathfrak{S}$ of all its states, and a mapping $\langle;\rangle: (\mathfrak{A}, \mathfrak{S}) \to \mathbb{R}$ which associates with each pair* (A, ϕ) *in* $(\mathfrak{A}, \mathfrak{S})$ *a real number* $\langle \phi; \mathrm{A} \rangle$ *we interpret as the expectation value of the observable* A *when the system is in the state* ϕ.

Hence the elements $\phi \in \mathfrak{S}$ can be considered as mappings from $\mathfrak{A}$ to $\mathbb{R}$, and conversely we can regard each element $A \in \mathfrak{A}$ as a mapping $\langle \phi; A \rangle : \mathfrak{S} \to \mathbb{R}$. Accordingly, if $\mathfrak{T}$ is a subset of $\mathfrak{S}$, we denote by $A|_{\mathfrak{T}}$ the restriction of this mapping to $\mathfrak{T}$. We say that $A|_{\mathfrak{T}} \leqslant B|_{\mathfrak{T}}$ whenever $\langle \phi; A \rangle \leqslant \langle \phi; B \rangle$ for all ϕ in $\mathfrak{T}$; in particular, we simply write $A \leqslant B$ if the above inequality holds for all states on $\mathfrak{A}$, and $A \geqslant 0$ if and only if $\langle \phi; A \rangle \geqslant 0$ for all $\phi \in \mathfrak{S}$ so that by definition states are *positive* functions on $\mathfrak{A}$.

Definition. *A subset $\mathfrak{T} \subseteq \mathfrak{S}$ is said to be* full *with respect to a subset $\mathfrak{B} \subseteq \mathfrak{A}$ if the inequality* $\mathrm{A}|_{\mathfrak{T}} \leqslant \mathrm{B}|_{\mathfrak{T}}$ *between any two elements* A *and* B *of* $\mathfrak{B}$ *implies* $\mathrm{A} \leqslant \mathrm{B}$.

REMARK. In the literature a subset $\mathfrak{T} \subseteq \mathfrak{S}$ is often said to be full with respect to $\mathfrak{B} \subseteq \mathfrak{A}$ if the *equality* $A|_{\mathfrak{T}} = B|_{\mathfrak{T}}$ between two elements A and B in $\mathfrak{B}$ implies that A and B are identical.

Structure Axiom 2. *The relation $\leqslant$ defined above is a partial ordering relation on $\mathfrak{A}$; in particular* $\mathrm{A} \leqslant \mathrm{B}$ *and* $\mathrm{A} \geqslant \mathrm{B}$ *imply* $\mathrm{A} \equiv \mathrm{B}$.

The fact that $\mathfrak{A}$ can be equipped with a structure of partially ordered set plays an important role in some of Segal's [1947] proofs and has been extensively used by Sherman [1951 and 1956] and Lowdenslager [1957] in their discussion of Segal's postulates. At this point in our formulation of the postulates we insist rather on the fact that $\mathfrak{S}$ is such that $\langle \phi; A \rangle = \langle \phi; B \rangle$ for all $\phi \in \mathfrak{S}$ implies $A \equiv B$; hence we assume in particular that there are enough states in $\mathfrak{S}$ so that we can distinguish between two elements of $\mathfrak{A}$

(i.e., two observables) by measuring their expectation values, or, in other words, we identify two observables whose expectation values coincide on all states.

We notice that if A and B are any two observables $\{\langle\phi; A\rangle + \langle\phi; B\rangle \mid \phi \in \mathfrak{S}\}$ defines a measurable quantity on Σ, i.e., an observable. Formally this is an assumption on the structure of $(\mathfrak{A}; \mathfrak{S})$ which we strengthen (on similar physical grounds) as our next axiom.

Structure Axiom 3.

(i) *There exists in* $\mathfrak{A}$ *two elements* O *and* I *such that, for all* $\phi \in \mathfrak{S}$ *we have* $\langle\phi; \mathrm{O}\rangle = 0$ *and* $\langle\phi; \mathrm{I}\rangle = 1$.
(ii) *For each observable* A *in* $\mathfrak{A}$ *and any real number* λ *there exists an element* $(\lambda\mathrm{A})$ *in* $\mathfrak{A}$ *such that* $\langle\phi; \lambda\mathrm{A}\rangle = \lambda\langle\phi; \mathrm{A}\rangle$ *for all* $\phi \in \mathfrak{S}$.
(iii) *For any pair of observables* A *and* B *in* $\mathfrak{A}$ *there exists an element* $(\mathrm{A} + \mathrm{B})$ *in* $\mathfrak{A}$ *such that* $\langle\phi; \mathrm{A} + \mathrm{B}\rangle = \langle\phi; \mathrm{A}\rangle + \langle\phi; \mathrm{B}\rangle$ *for all* ϕ *in* $\mathfrak{S}$.

REMARK. By virtue of our second axiom we have that $O, I, \lambda A$, and $A + B$ are *uniquely* defined. For the same reason we conclude that the sum is distributive with respect to the multiplication by real numbers and is both commutative and associative. We then see that this axiom equips $\mathfrak{A}$ with the structure of a *real vector space*. Moreover, by virtue of the very definition of "sum" and "multiplication by scalars," the elements of $\mathfrak{S}$ (i.e., the states) are real *linear* functionals on $\mathfrak{A}$.

It is not our intention here to enter into details of the controversial subject of the measuring process theory, although we can assume from the recent literature that it has maintained a vivid philosophical interest since von Neumann's investigations. We shall content ourselves with the perhaps naïve view that the state of a system is a way to characterize the method used for its preparation. The state then manifests itself to the observer when for each observable A he performs a sequence (in principle infinite, in practice big enough to reach a reasonable degree of confidence) of independent trial measurements on systems prepared in an identical way. What the observer receives is a distribution of real numbers, whose "proper" average he calls the expectation value $\langle\phi; A\rangle$ of the observable A in the state ϕ. We say that a state ϕ is *dispersion-free on the observable* A when the above distribution is concentrated on a single number, namely $\langle\phi; A\rangle$. Let us now denote by $\mathfrak{S}_A$ the set of all dispersion-free states on A and by σ_A, the set of all the values that A assumes on its dispersion-free states; in short, we shall refer to σ_A as the *spectrum* of A and omit the proof that this concept actually coincides with what is meant by the spectrum of an observable in the usual formulation of quantum theory.

The notion of dispersion-free states is closely related to that of simultaneous observability. We note first that two observables A and B can be simultaneously measured with arbitrary precision whenever the system is in a state ϕ in $\mathfrak{S}_A \cap \mathfrak{S}_B$. The latter subset of $\mathfrak{S}$ can actually be empty, in which case A and B can never be measured simultaneously with arbitrary precision. At the other extreme lies the notion of *compatibility*, which we want to approach in the following way.

Definition. *A subset* $\mathfrak{T} \subseteq \mathfrak{S}$ *is said to be* complete *if it is full with respect to the subset* $\mathfrak{A}_{\mathfrak{T}} \subseteq \mathfrak{A}$ *defined by* $\mathfrak{A}_{\mathfrak{T}} \equiv \{A \in \mathfrak{A} \mid \mathfrak{S}_A \supseteq \mathfrak{T}\}$. *A complete subset* $\mathfrak{T} \subseteq \mathfrak{S}$ *is said to be* deterministic *for a subset* $\mathfrak{B} \subseteq \mathfrak{A}$ *whenever* $\mathfrak{B} \subseteq \mathfrak{A}_{\mathfrak{T}}$. *A subset* $\mathfrak{B} \subseteq \mathfrak{A}$ *is said to be* compatible *if the set* $\mathfrak{S}_{\mathfrak{B}} \equiv \bigcap_{B \in \mathfrak{B}} \mathfrak{S}_{\mathfrak{B}}$ *is complete.*

Lemma. $\mathfrak{B}$ *is compatible if and only if it admits a deterministic set of states.*

Proof. If $\mathfrak{B}$ is compatible, $\mathfrak{S}_{\mathfrak{B}}$ is complete by definition and $\mathfrak{B} \subseteq \mathfrak{A}(\mathfrak{B}) \equiv \mathfrak{A}_{\mathfrak{S}_{\mathfrak{B}}}$, so that $\mathfrak{S}_{\mathfrak{B}}$ is actually deterministic for $\mathfrak{B}$. Conversely, suppose that there exists a set $\mathfrak{T}$ that is deterministic for $\mathfrak{B}$; by definition this means that $\mathfrak{T}$ is complete and $\mathfrak{B} \subseteq \mathfrak{A}_{\mathfrak{T}}$. The latter condition means, in turn, that $\mathfrak{T} \subseteq \mathfrak{S}_B$ for all B in $\mathfrak{B}$, hence $\mathfrak{T} \subseteq \mathfrak{S}_{\mathfrak{B}}$ and $\mathfrak{A}(\mathfrak{B}) \subseteq \mathfrak{A}_{\mathfrak{T}}$. $\mathfrak{T}$ complete means that it is full with respect to $\mathfrak{A}_{\mathfrak{T}}$. This implies a fortiori $\mathfrak{S}_{\mathfrak{B}}$ full with respect to $\mathfrak{A}_{\mathfrak{T}}$, hence with respect to $\mathfrak{A}(\mathfrak{B})$. Consequently $\mathfrak{S}_{\mathfrak{B}}$ is complete. ∎

REMARK. In loose terms our notion of compatibility states that all the observables of a compatible set are simultaneously measurable with arbitrary precision on a set of states that is "large enough" (how "large" is precisely what the exact definition actually says). We notice in this respect that it would *not* have been sufficient, to cover at least the usual concept of compatibility, to demand the existence of a set $\mathfrak{T} \subseteq \bigcap_{B \in \mathfrak{B}} \mathfrak{S}_B$ that is full with respect to $\mathfrak{B}$; it is actually necessary to enlarge $\mathfrak{B}$ considerably to be able to write a definition of compatibility along these lines. Our definition, in a sense, shows how $\mathfrak{B}$ should be enlarged: it should be enlarged to the set $\mathfrak{A}_{\mathfrak{T}}$ of all observables which admit at least the states of some complete $\mathfrak{T} \subseteq \bigcap_{B \in \mathfrak{B}} \mathfrak{S}_B$ as dispersion-free states. Using an algebraic and topological structure on $\mathfrak{A}$ which we have not yet obtained, Segal [1947] gave an alternative procedure to achieve the necessary enlargement of $\mathfrak{B}$.

We now want to say that an observable A is completely determined by its set $\mathfrak{S}_A$ of dispersion-free states, its spectrum σ_A, *and* the mapping $A: \mathfrak{S}_A \to \sigma_A$.

Structure Axiom 4. *For every observable* A *the set* $\mathfrak{S}_{\mathrm{A}}$ *is deterministic for the one-dimensional subspace of* $\mathfrak{A}$ *generated by* A; *for any two observables* A *and* B: $\mathfrak{S}_{A+B} \supseteq \mathfrak{S}_A \cap \mathfrak{S}_B$; *and* $\mathfrak{S}_I = \mathfrak{S}$.

REMARKS. We notice first of all that for any theory that is not completely trivial this axiom implies that σ_A and $\mathfrak{S}_A$ are nonempty. Actually this axiom ensures a certain maximality property to $\mathfrak{S}_A$. We see, indeed, as a particular consequence of this axiom that for any observable A there is no observable $B \neq A$ with $\mathfrak{S}_B \supseteq \mathfrak{S}_A$ and $A|_{\mathfrak{S}_A} = B|_{\mathfrak{S}_A}$. Hence two observables that have the same dispersion-free states and coincide on them are necessarily identical. This expresses the intuitive fact that motivated this axiom. Finally, we point out that this axiom requires that $\mathfrak{S}_{\mu A+\lambda I} \supseteq \mathfrak{S}_A$ for all λ, μ in $\mathbb{R}$. Hence $\mathfrak{S}_A$ is actually deterministic on the subspace of $\mathfrak{A}$ generated by A and I.

Aside from its interest with respect to the self-consistency and completeness of the theory, Structure Axiom 4 enables us to formulate in a concise way the postulates related to the existence of powers of observables and their properties.

Structure Axiom 5. *For any element* A *in* $\mathfrak{A}$ *and any non-negative integer* n *there exists at least one element, denoted* A^{n}, *in* $\mathfrak{A}$ *such that*

(i) $\mathfrak{S}_{\mathrm{A}^\mathrm{n}} \supseteq \mathfrak{S}_{\mathrm{A}}$
(ii) $\langle \phi; \mathrm{A}^{\mathrm{n}} \rangle = \langle \phi; \mathrm{A} \rangle^{\mathrm{n}}$ *for all* ϕ *in* $\mathfrak{S}_{\mathrm{A}}$.

REMARKS

(i) Using Axiom 4, we see that A^n, as just defined, is unique in $\mathfrak{A}$; we can then call it unambiguously the *nth power of A*.
(ii) We conclude from the already existing structure of $\mathfrak{A}$ that the operation of raising an observable to a power is associative (i.e., $(A^n)^m = A^{nm}$) and distributive with respect to scalar multiplication (i.e., $(\lambda A)^n = \lambda^n A^n$). The usual rules (namely, $A^0 = I$, $A^1 = A$, $I^n = I$ and $0^n = 0$ for $n > 1$) are easily proved to hold true.
(iii) Since $\langle \phi; A^2 \rangle \geqslant 0$ for all ϕ in $\mathfrak{S}_A$, which is deterministic for the set formed by A and A^2, and $\mathfrak{S}_0 \supseteq \mathfrak{S}_A$, we see that $A^2 \geqslant 0$, the equality holding only when A itself is 0. Using the linearity of any ϕ on $\mathfrak{A}$ and the fact that $\mathfrak{S}$ is full for $\mathfrak{A}$, we conclude that $A^2 + B^2 + C^2 + \cdots = 0$ implies $A = B = C = \cdots = 0$.
(iv) One important property, however, is missing, namely the distributivity of power raising with respect to addition for an arbitrary pair A, B. We are not able to prove it at this point, neither do we want to postulate it; nevertheless we are in the position to derive it in some particular important cases.
(v) Let us denote by $\mathfrak{P}(A)$ the subset of $\mathfrak{A}$ obtained from A by addition, multiplication by scalars, and power raising; in other words $\mathfrak{P}(A)$ is the subspace of $\mathfrak{A}$ generated from the powers of A. We notice that for any element B of $\mathfrak{P}(A)$ we have $\mathfrak{S}_B \supseteq \mathfrak{S}_A$. Using Axiom 4, we see that $\mathfrak{S}_A$ is deterministic on $\mathfrak{P}(A)$, hence that $\mathfrak{P}(A)$ is a compatible set of observables. This reasoning actually generalizes without difficulty to the following statement:

Theorem 1. *Let $\mathfrak{B}$ be a compatible set of observables and $\mathfrak{P}(\mathfrak{B})$, the subset of $\mathfrak{U}$ generated from the elements of $\mathfrak{B}$ by addition, multiplication by scalars, and power raising. Then $\mathfrak{P}(\mathfrak{B})$ is a compatible set of observables.*

REMARK. In particular, $\mathfrak{P}(\mathfrak{B})$ has a full set of states all dispersion-free for every observable in $\mathfrak{P}(\mathfrak{B})$. Incidentally, we might remark that the result of this theorem is so thoroughly expected from an intuitive point of view that it is sometimes (e.g., see Segal, [1947]) accepted as part of the definition of compatibility of a set $\mathfrak{B}$ of observables. Our formulation differs only slightly from that of Segal in that we wanted to arrive naturally at the power structure of $\mathfrak{U}$ (instead of postulating from the outset the whole algebraic structure) and that the concept of compatibility seemed to us to be logically a priori independent of that structure. Theorem 1 shows that the two alternative definitions of compatibility are related.

A final remark should be made on the route we are trying to follow; our aim is to write the postulates in such a manner that their content will be expressed in terms that can be tested directly on the phenomenological level, even if this operation involves some idealizations, thus making their critical analysis easier. An example in point has recently been provided by Jordan himself,† who doubted the power-associative structure ‡ of $\mathfrak{U}$; he actually constructed an example for which

$$[Q, Q, Q^2] \equiv (Q \circ Q) \circ Q^2 - Q \circ (Q \circ Q^2) = Q^2 \circ Q^2 - Q^4 = 2eI \neq 0,$$

where Q is interpreted as the position coordinate and e is an "elementary length." This is interpreted as meaning that, in addition to the principle of complementarity which imposes a limit to the precision with which position and momentum can be measured *simultaneously*, there is a "supplementary impossibility to measure the position with an arbitrary precision." Expressed in our language, this amounts to denying the existence of enough (and, indeed, of any) dispersion-free states on Q; hence it is not only our Axiom 5 but also our axiom 4 which is brought into question in this theory. However, the basic assumption underlying Jordan's proposal, namely that there exists an intrinsic "elementary length," is still open to experimental verification. In any case, it should be pointed out that this unusual theory is *not* covered by our axioms.

c. Structure Axiom 6 (*Jordan algebraic structure of* $\mathfrak{A}$)

The subset $\mathfrak{P}(\mathfrak{B})$ of $\mathfrak{A}$ formed from a compatible set $\mathfrak{B}$ enjoys an interesting property, namely that within $\mathfrak{P}(\mathfrak{B})$ the power-raising operation is distributive with respect to the addition. We actually prove this fact in a somewhat stronger form in the immediate sequel.

† Jordan [1968, 1969] and references quoted therein.

‡ See (ii) following Axiom 5 and the first definition of the next subsection.

As of now $\mathfrak{A}$ is equipped with a structure of real vector space on which an associative power structure is superimposed; the link between these two structures is still rather tenuous and in particular $\mathfrak{A}$ is not yet an algebra. An important step in providing $\mathfrak{A}$ with a richer structure is provided by the following preliminary observation:

Definition. *With each pair* A, B *of elements in* $\mathfrak{A}$ *we associate an element* $\mathrm{A} \circ \mathrm{B}$ *of* $\mathfrak{A}$, *called the* symmetrized product of A and B, *defined by* $\mathrm{A} \circ \mathrm{B} \equiv \frac{1}{2}\{(\mathrm{A} + \mathrm{B})^2 - \mathrm{A}^2 - \mathrm{B}^2\}$.

REMARKS. This product enjoys some immediate properties worth mentioning.

(i) We have obviously $A \circ B = B \circ A$.
(ii) Using Axioms 4 and 5, we see furthermore that

$$\langle \phi; A \circ B \rangle = \langle \phi; A \rangle \langle \phi; B \rangle \quad \text{for all } \phi \in \mathfrak{S}_A \cap \mathfrak{S}_B;$$

we conclude from this that $A^n \circ A^m = A^{n+m}$, since $\mathfrak{S}_A$ is deterministic on $\mathfrak{P}(A)$; we then have, in particular, $A \circ I = A$. Hence the power structure of $\mathfrak{A}$ can be viewed from a mathematical standpoint as being generated from the symmetrized product defined above, which admits I as its neutral element.

Our task now is to determine whether (or when) this product is associative and distributive. For any triple A, B, C of elements of $\mathfrak{A}$ we define the *associator* $\{A, B, C\} = (A \circ B) \circ C - A \circ (B \circ C)$. We consider first a particular case. Let $\mathfrak{B}$ be any subset of $\mathfrak{A}$, $\mathfrak{S}_{\mathfrak{B}}$ be defined as $\bigcap_{B \in \mathfrak{B}} \mathfrak{S}_B$, and $\mathfrak{A}_{\mathfrak{S}_{\mathfrak{B}}}$ be denoted $\mathfrak{A}(\mathfrak{B})$; for every triple A, B, C of elements in $\mathfrak{A}(\mathfrak{B})$ we conclude from the second remark following the definition of the symmetrized product that $\langle \phi; \{A, B, C\} \rangle = 0$ for all ϕ in $\mathfrak{S}_{\mathfrak{B}}$. *If* $\mathfrak{B}$ is now compatible, $\mathfrak{S}_{\mathfrak{B}}$ is complete, i.e., full with respect to $\mathfrak{A}(\mathfrak{B})$, which contains in particular $\{A, B, C\}$. Hence $\{A, B, C\} = 0$. We see from an analogous argument that $A \circ (\lambda B + \mu C) = \lambda(A \circ B) + \mu(A \circ C)$; this completes the proof of the following results:

Theorem 2. *The symmetrized product is associative and distributive over* $\mathfrak{A}(\mathfrak{B})$ *for any compatible set* $\mathfrak{B}$ *of observables.*

Corollary. $\mathfrak{B}$ *compatible implies* $\mathfrak{P}(\mathfrak{B})$ *associative.*

REMARK. We can paraphrase Theorem 2 by saying that for $\mathfrak{B}$ compatible, $\mathfrak{A}(\mathfrak{B})$, hence $\mathfrak{P}(\mathfrak{B})$, are equipped as a result of our axioms with the structure of abelian, distributive, and associative algebras. If $\mathfrak{A}(\mathfrak{B})$ were already $\mathfrak{A}$ itself, we would have a situation characteristic of *classical* mechanics.

We now want to extend our discussion beyond classical theories. In the early days of the algebraic approach to physical theories it had already been noted† that "an algebraic discussion is scarely possible" if the distributive law does not hold for the symmetrized product. Whereas all other axioms for the structure of $\mathfrak{A}$ have a rather intuitive physical justification, it seems, however, extremely hard to find any such reason for postulating distributivity in general (i.e., with the exception of the particular case of a classical theory in which all observables are mutually compatible). It is perhaps of interest to mention here that in contrast to the assumptions of the papers just mentioned Segal [1947] did *not* assume distributivity in his axioms and that Sherman [1956] was actually able to exhibit a mathematical example which satisfies all of Segal's fundamental axioms and still leads to a nondistributive product. No physical application, however, has yet been found for the more general objects constructed by Sherman.

It seems fair, therefore, to say that although no compelling reason has yet been found for assuming that the distributivity of the product will hold neither has any been proposed in favor of dropping this postulate from the realm of physically relevant theories. In view of this situation we do not assume distributivity here in its most explicit form but rather proceed inductively.

We first recall that in any classical theory in which all observables are compatible with one another the "associator" $\{A, B, C\}$ among any three of them vanishes. We propose to consider only those physical theories that are generalizations of classical theories in the sense that they satisfy the following condition:

Structure Axiom 6. *For any triple* A, B, C *of observables in which* A *and* C *are compatible the associator* {A, B, C} *vanishes.*

REMARKS

(i) Classical theories satisfy this axiom, as seen in Theorem 2. Furthermore, none of the quantum theories known so far violates this axiom.

(ii) We indicate in anticipation that for bounded observables von Neumann [1936] defines the compatibility of two observables A and C precisely by the relation $\{A, B, C\} = 0 \ \forall\ B \in \mathfrak{A}$.

(iii) We point out here, essentially for mathematical reasons, that this axiom could be derived as a consequence of the following alternate postulate:

For any triple A, B, C of observables in which A and C are compatible the associator $\{A, B, C\}$ is compatible with both A and C. Furthermore, for any observable A and any ϕ in $\mathfrak{S}_A$: $\langle\phi; A \circ B\rangle = \langle\phi; A\rangle\langle\phi; B\rangle$.

With this alternate formulaton we consider the set containing the three elements A, C and $\{A, B, C\}$; this set is compatible by virtue of the first

† Jordan, von Neumann, and Wigner [1934]; von Neumann [1936].

part of the postulate; with the help of the second part we easily adapt the proof of Theorem 2 to the present case to conclude that $\{A, B, C\} = 0$.

(iv) In the immediate sequel we actually use only a weaker form of our Axiom 6:

For any pair of observables A and B the associator $\{A, B, \lambda I\}$ vanishes.

This weakened form of the axiom is exactly equivalent to the following statement:

For any pair of observables A and B the symmetrized product is homogeneous in both of its factors, i.e.,

$$\lambda(A \circ B) = (\lambda A) \circ B = A \circ (\lambda B).$$

hence

$$A \circ B = \tfrac{1}{4}[(A + B)^2 - (A - B)^2].$$

We notice that this is the form in which we encountered the symmetrized product in our discussion of the usual formulation of quantum mechanics (see Subsection a). Hence our postulates are also satisfied for this case.

(v) By comparing this form of the symmetrized product with the form used for its definition we get

$$(A + B)^2 + (A - B)^2 = 2A^2 + 2B^2.$$

The importance of this equality with respect to the distributivity of the symmetrized product has long been recognized.† Let us substitute $(A + C)$ and $(A - C)$ successively for A in this equality and then subtract the resulting equalities from one another; we get

$$(A + B) \circ C + (A - B) \circ C = 2(A \circ C).$$

Replacing $A + B$ and $A - B$ with A and B, respectively, we get

$$(A + B) \circ C = A \circ C + B \circ C.$$

By combining this result with the fact that the symmetrized product is commutative and homogeneous we get the distributive law

$$A \circ (\lambda B + \mu C) = \lambda A \circ B + \mu A \circ C$$

for any triple of observables and any real numbers.

(vi) Looking back at our structure Axiom 6, we notice finally that since all powers of an observable are compatible we have $\{A^n, B, A^m\} = 0$ for all $A, B \in \mathfrak{A}$ and all non-negative integers n and m. In particular, $A \circ (B \circ A^2) = (A \circ B) \circ A^2$.

We can now collect all these remarks into the following statement:

Theorem 3. *The set $\mathfrak{A}$ of all observables on a physical system is an abelian (distributive but not necessarily associative) real Jordan algebra.*

† See, for instance, von Neumann [1936].

We now have equipped $\mathfrak{A}$ with a mathematical structure that seems rich enough possibly to allow a classification of all its realizations. In particular, the problem arises whether this structure will lead to a more general and possibly more tractable theory than the usual classical or quantum theories. Jordan, von Neuman, and Wigner [1934] attacked this problem in their fundamental paper and made a considerable dent in it. It might be of interest here to quote their axioms:

In the algebra $\mathfrak{A}$ with elements $A, B, C, \ldots,$ one can form from the elements A and B the elements $A \pm B$, λA (λ a real number), and A^m ($m = 1, 2, \ldots, A^1 = A$); all the usual rules of calculation hold for the operations $A \pm B$ and λA; and $\mathfrak{A}$ has a finite linear basis. A^m is a continuous function of A (that is, if we represent all the elements as linear aggregates of the basis elements, then the coefficients of A^m depend continuously on those of A). $\mathfrak{A}$ is "formally real," that is,

(I) if $A^2 + B^2 + C^2 + \cdots = 0$, then $A = B = C = \cdots = 0$.

It is always the case that

(II′) $(A^m + A^n)^2 = A^{2m} + A^{2n} + 2A^{m+n}$.

(II′) is equivalent to

(II) $A^m A^n = A^{m+n}$, $m, n = 1, 2, \ldots,$ but it has the definition $A^1 = A$, $A^{m+1} = AA^m$ as a consequence. Finally,

(*) $(A + B + C)^2 + A^2 + B^2 + C^2 = (A + B)^2 + (A + C)^2 + (B + C)^2$.

(*) is equivalent to

$$(A + B) \circ C = A \circ C + B \circ C,$$

but, because of the continuity of A^2 [and, therefore, of $A \circ B$ (which is defined as our symmetrized product)], it has the condition $(\lambda A) \circ B = \lambda(A \circ B)$ as a consequence. This completes the list of the desired rules of calculation. (*) could be replaced by the simpler rule

$$(A + B)^2 + (A - B)^2 = 2A^2 + 2B^2.$$

All of these postulates, except for (*), necessarily arise from the assumed physical conditions.

These authors called the mathematical object just defined an *r-number algebra* and they proved as a consequence of their axioms that $(A \circ B) \circ A^2 = A \circ (B \circ A^2)$; an algebra satisfying all these properties is now called a *commutative real Jordan algebra*, a name apparently introduced by A. A. Albert [1946], and several mathematical generalizations of this concept have been worked out by mathematicians. The study of Jordan algebras has become a vigorous field of mathematics, as witnessed, for instance, by Braun and Koecher [1966]. A brief review of the role and place of Jordan algebras in the general mathematical study of algebras has been written by L. J. Paige [1963] and could serve as an introduction to this field. The physical inception of this notion is to be traced back to Jordan [1933a, b] in his pioneering study of possible generalizations and improved formulations of quantum mechanics.

By comparing the Jordan-von Neumann-Wigner axioms with ours we notice two significant (and actually related) differences on which we now want to comment.

First, the notion of state does not appear explicitly in the paper by Jordan, von Neumann, and Wigner, although it was certainly lying in the background of their investigation, as could be guessed, for instance, from the papers by Jordan [1933a, b] and von Neumann [1927b, 1936]; in particular, von Neumann's discussion contains the germs of the symmetric postulation we are presenting. We indeed introduced the postulates in an inductive sequence, devised to emphasize the complementary roles played by the observables and the states. In doing so our desire was to take into account most seriously the fact that physicists perceive the "physical world" only through expectation values, a concept involving both observables *and* states. The relative emphasis given to observables over states, or conversely, varies from one extreme to the other in the various axiomatic schemes to be found in the literature; the scheme chosen here lies more or less in the middle of the road for reasons that appear natural to my "physical intuition"; so does the approach followed by Mackey [1963], who starts with an axiomatization of the probability measure p which associates with the triples (A, ϕ, M) (formed by an observable A, a state ϕ, and a Borel subset M of $\mathbb{R}$), the probability $p(A, \phi, M)$ that the observable A will take a value in M when the system is in the state ϕ. Piron [1964] is primarily concerned with the structure of a certain class of observables,† whereas states appear later; so is Segal [1947] who, however, considers from the outset a much larger class of observables; at the opposite extreme we mention a paper by Edwards [1970] who starts his account with an axiom on the structure of the set of all states on a physical system.

The second difference between the axioms of Jordan, von Neumann, and Wigner and ours is that we do not restrict $\mathfrak{A}$ to have a *finite* linear basis; this restriction is obviously a severe one and was to be dropped‡ from any general theory, since it excludes any ordinary quantum theory formulated on an infinite-dimensional Hilbert space (e.g., the description of a quantum particle moving on a real line!). The mathematical simplification introduced by this restriction is that it allows the analysis of the properties of $\mathfrak{A}$ without having to use any explicit topological notion. In particular, Jordan, von Neumann, and Wigner were able to prove from their postulates two important results that already indicate the power of this line of approach. The first of these results is that a spectral theory can be completely worked out, and the second is that a complete classification of all the realizations of their axioms can be given.

† The "propositions" in Subsection e.

‡ As Jordan, von Neumann, and Wigner naturally remarked.

Jordan, von Neumann, and Wigner's results can be summarized in the following two theorems:

Theorem 4 (JNW I). *Let $\mathfrak{A}$ be an* r-*number algebra. Then there exists in $\mathfrak{A}$ a unit element* I *such that* $I \circ A = A$ *for all* A *in $\mathfrak{A}$. Furthermore, each element* A *of $\mathfrak{A}$ can be written in the form* $A = \Sigma a_i P_i$, *where* $P_i^2 = P_i \in \mathfrak{A}$; $P_i \circ P_j = 0$ *for* $i \neq j$; $\sum_i P_i = I$, *and where* a_i *are the "proper values" of* A, *defined as the* μ *distinct simple real roots of the characteristic equation* $f(a) = 0$, *where*

$$f(A) = A^\mu + \lambda_{\mu-1}A^{\mu-1} + \cdots + \lambda_1 A + \lambda_0$$

is the polynomial of the lowest degree in A *which vanishes.*

REMARKS

(i) The proof, as given in the original paper, depends in an essential way on the existence of a *finite* linear basis for $\mathfrak{A}$, hence is not reproduced here, since we eventually want to dispose of this restriction.

(ii) Restriction (i), however, should not overshadow the interest of the theorem which points in the right direction. In particular, it emphasizes the role of the elements of $\mathfrak{A}$ of the form $P = P^2$; we refer to these special elements as "propositions" for reasons made clear in Subsection e.

The following consequence of this theorem is proved separately in Subsection e:

Corollary to JNW I. *Under the assumptions of Theorem* 4 *the spectrum* σ_A *of* A *in $\mathfrak{A}$ coincides exactly with the set of proper values of* A; *furthermore,* $\mathfrak{S}_A = \bigcap_i \mathfrak{S}_{P_i}$.

The second of the Jordan, von Neumann, and Wigner results, which we want to mention here, is the classification of all the realizations of their axioms. To present this classification we need some preliminary definitions. Two subspaces $\mathfrak{B}$ and $\mathfrak{C}$ of $\mathfrak{A}$ are said to be *orthogonal* if $\mathfrak{B} \circ \mathfrak{C} = 0$; $\mathfrak{A}$ is said to be the *direct sum* of two of its orthogonal subspaces (say $\mathfrak{B}$ and $\mathfrak{C}$) if every element A of $\mathfrak{A}$ can be written uniquely in the form $A = B + C$ with B in $\mathfrak{B}$ and C in $\mathfrak{C}$. $\mathfrak{A}$ is said to be *simple* if it has *no proper ideal*, i.e., if it has no proper subspace $\mathfrak{T}$ such that $\mathfrak{A} \circ \mathfrak{T} \subseteq \mathfrak{T}$. Using in an essential way the fact that their *r*-number algebras have finite linear basis, Jordan, von Neumann, and Wigner proved the following:

Lemma. *Every* r-*number algebra can be written as the finite direct sum of simple* r-*number algebras.*

Consequently, it is sufficient for the classification of the realizations of $\mathfrak{A}$ to know all the simple *r*-number algebras. A classification of these algebras was also obtained by Jordan, von Neumann, and Wigner. If $\mathfrak{X}$ is an associative (but not necessarily commutative) algebra, we can define on $\mathfrak{X}$ a symmetrized (or antisymmetrized) product $A \circ B \equiv \frac{1}{2}(AB + BA)$ (or $[A, B] \equiv AB - BA$), where AB denotes the ordinary product of A and B in $\mathfrak{X}$. We denote by $\mathfrak{X}^+$ (or $\mathfrak{X}^-$) the vector space $\mathfrak{X}$ equipped with the symmetrized (or antisymmetrized) product just described. We verify immediately that $\mathfrak{X}^+$ is a commutative Jordan algebra (i.e., the symmetrized product is distributive and commutative but not necessarily associative and it satisfies $A \circ (B \circ A^2) = (A \circ B) \circ A^2$). Similarly, $\mathfrak{X}^-$ is a Lie algebra. A Jordan algebra $\mathfrak{A}$, isomorphic to a subalgebra of $\mathfrak{X}^+$ for some $\mathfrak{X}$, is called *special*; the Jordan algebras that are not special are called *exceptional*. In contrast to the theory of Lie algebras, in which the Poincaré-Birkhoff-Witt theorem asserts that every Lie algebra is isomorphic to a subalgebra of $\mathfrak{X}^-$ for some $\mathfrak{X}$, exceptional Jordan algebras *do* exist and have been studied (see Albert [1958]). An early example was provided by Albert [1934],† namely the algebra $\mathfrak{M}_3^8$ of all hermitian matrices of order 3 over the quasi-quaternions (or Cayley numbers). In this connection Jordan, von Neumann, and Wigner proved the following result:

Theorem 5 (JNW II). *Every simple* r*-number algebra is a special Jordan algebra except one, namely* $\mathfrak{M}_3^8$.

The role of the exceptional algebra $\mathfrak{M}_3^8$ in quantum mechanics has not yet been settled within the algebraic approach. As a matter of curiosity we mention in anticipation that Sherman [1956] proved that $\mathfrak{M}_3^8$ can be normed to become a system of observables according to Segal [1947].

The following special Jordan algebras are of interest in the context of quantum mechanics, as Theorem 6 shows.

1. $\mathbb{R}$, the algebra of the real numbers in which $A + B$, λA, and $A \circ B$ are defined in the usual way.

2. $\mathfrak{S}_N$ ($N = 1, 2, \ldots$), the algebra with linear basis $I, X_1, \ldots, X_{N-1}$ in which $A + B$ and λA are defined in the usual way but in which $A \circ B$ is defined by $I \circ I = I$, $I \circ X_i = X_i$ and $X_i \circ X_j = \delta_{ij} I$ (where δ_{ij} is the Kronecker symbol).

3. $\mathfrak{M}_\gamma^\chi$ ($\chi = 1, 2, 4$; $\gamma = 1, 2, \ldots$), the algebra of hermitian matrices of order γ whose entries are real numbers ($\chi = 1$), complex numbers ($\chi = 2$), or real quaternions ($\chi = 4$); $A + B$ and λA are defined in the usual way, but $A \circ B$ is defined by $A \circ B = \frac{1}{2}(AB + BA)$, where AB is the usual matrix product of A by B.

A resolution of the identity I in $\mathfrak{A}$ is a set $\{P_i\}$ of pairwise orthogonal propositions adding up to I (e.g., the set $\{P_i\}$ appearing in Theorem 4 is such

† Incidentally, at the suggestion of Jordan, von Neumann, and Wigner.

a resolution of the identity). In an r-number algebra $\mathfrak{A}$ a proposition is said to be *unresolvable* if it cannot be properly represented as the sum of pairwise orthogonal propositions. We can show that a proposition P is unresolvable exactly when the subspace $N_1(P) = \{X \in \mathfrak{A} \mid X \circ P = X\}$ has linear dimension 1. The *order* γ of $\mathfrak{A}$ is the (well-defined) number of terms in a resolution of the identity into pairwise orthogonal *unresolvable* propositions. With these notations we have the following classification:

Theorem 6. *A simple special* r*-number algebra of order* γ *and linear dimension* N *is isomorphic to*:

$$\mathbb{R} \quad \textit{if } \gamma = 1$$
$$\mathfrak{S}_N \quad (\textit{with } \mathrm{N} = 3, 4, \ldots) \textit{ if } \gamma = 2$$
$$\mathfrak{M}_\gamma^\chi \quad (\textit{with } \chi = 2(\mathrm{N} - \gamma)/\gamma(\gamma - 1) = 1, 2, 4) \textit{ if } \gamma \geqslant 3.$$

REMARKS

(i) $\mathfrak{S}_1$ is $\mathbb{R}$; $\mathfrak{S}_2$ is not simple; all $\mathfrak{M}_1^\chi$ are $\mathbb{R}$; $\mathfrak{M}_2^\chi$ with $\chi = 1, 2, 4$ are, respectively, $\mathfrak{S}_3$, $\mathfrak{S}_4$, and $\mathfrak{S}_5$.

(ii) With Theorem 5 and the preceding lemma this theorem completes the classification of all realizations of the axioms of Jordan, von Neumann, and Wigner.

(iii) The appearance of real numbers and quaternions as matrix elements when $\gamma \geqslant 3$ has been confirmed in the proposition-calculus approach, as already indicated by Birkhoff and von Neumann [1936]. This suggested that possible generalizations of the usual complex Hilbert space formalism for quantum mechanics could be obtained by considering real or quaternionic Hilbert spaces. The case of real Hilbert spaces has been extensively considered by Stueckelberg et al. [1960, 1961a, b, 1962] with special emphasis on the formulation of the uncertainty principle; an exact correspondence between their approach and the usual one has emerged from their work. Quantum mechanics on quaternionic Hilbert spaces has been investigated by Finkelstein et al.† The functional analysis pertinent to quantum mechanical purposes has been developed for both of these unconventional approaches‡, and the theory of group representations has been worked out to some extent§; in particular, a complete classification of all physically relevant irreducible representations of the Poincaré group has been obtained‖ with

† Finkelstein et al. [1959, 1962, 1963]; Emch [1963a, b]; see also Tavel [1964].

‡ Aside from the references already given, we could mention Stone [1932a], von Teichmuller [1935], and Goldstine and Horwitz [1962; 1964].

§ Dyson [1962]; Emch [1963a].

‖ Emch [1963b, 1965]; these results have recently been rederived by Viswanath [1968].

the result that a one-to-one correspondence has been established between these representations and those obtained in the complex case. For completeness we might mention that some questions of interpretation remain in the related case of the Galilei group. Some further generalizations along these lines have been proposed by Goldstine and Horwitz [1962, 1964], Horwitz and Biederharn [1965], and Horwitz [1966], who considered possible physical applications (in particular to theories of the Yang and Mills type) of Hilbert spaces constructed over arbitrary finite algebras, thus investigating generalizations of the algebra $\mathfrak{M}_3^8$ mentioned above.

d. Structure Axioms 7 and 8 (*topological structure of* $\mathfrak{A}$)

With the wealth of information contained in their paper, Jordan, von Neumann, and Wigner demonstrated the power of a purely algebraic approach to quantum theories. As we remarked earlier, however, there is a major weakness in this pioneering work, namely that they have to assume that $\mathfrak{A}$ has a *finite* linear basis. This had to be corrected by the introduction of an appropriate *topological* structure before the claim could be made that the theory provides a formalism general enough for the need of quantum problems. The first attempt in this direction was made by von Neumann [1936]. His declared intention was to imitate the *weak topology* of operators acting on Hilbert spaces in apparent connection with the theory of rings of operators he was developing at the same time in collaboration with Murray. Some functional-analysis tools were made extensively available within this axiomatization; in particular, a spectral theory which generalized Theorem 4 to this extended framework was obtained and the connection with the proposition-calculus approach was pointed out. Unfortunately, although the paper was identified as "Part I" of some series to appear, "Part II" has never been made available in print and we are short of a classification†—and a representation theory—comparable to that obtained in the case in which $\mathfrak{A}$ has a finite linear basis. Moreover, aside from its mathematically appealing features, the topology introduced by von Neumann was not presented in a way that gave it a plausible phenomenological motivation. Actually, in spite of the allusions made to the notion of trace and to the necessity of a "statistical interpretation," the concept of state does not appear to play any significant role in this paper. Since this concept is nevertheless of primary importance to our physical motivation, we now want to return to a

† See, however, the classification obtained by Størmer [1966] for the so-called JW-algebras of type I, i.e., the weakly closed Jordan algebras of self-adjoint operators (acting on a real, complex, or quaternionic Hilbert space), which are equipped with the natural symmetrized product and contain minimal projectors (see also Størmer (1968b]).

closer examination of it with the purpose of using it to equip $\mathfrak{A}$ with a natural topology.

For any observable A in $\mathfrak{A}$ and any *complete* set $\mathfrak{T} \subseteq \mathfrak{S}_A$ we define the following subset of the positive real axis $\tau(A, \mathfrak{T}) \equiv \{\lambda \in \mathbb{R}^+ \mid |\langle\phi; A\rangle| < \lambda \ \forall\ \phi \in \mathfrak{T}\}$ and the nonnegative number $\|A\|_{\mathfrak{T}} \equiv$ g.l.b. $\{\tau(A, \mathfrak{T})\}$.

Lemma. $\|A\|_{\mathfrak{T}}$ *is independent of* $\mathfrak{T} \subseteq \mathfrak{S}_A$ *and is equal to* $\|A\| \equiv sup_{\phi\in\mathfrak{S}} |\langle\phi; A\rangle|$.

Proof. For all $\lambda \in \tau(A, \mathfrak{T})$ we have $-\lambda I|_{\mathfrak{T}} < A|_{\mathfrak{T}} < \lambda I|_{\mathfrak{T}}$. Since $\mathfrak{T}$ is complete, it is full with respect to the set $\mathfrak{A}_{\mathfrak{T}}$ of all elements $B \in \mathfrak{A}$ with $\mathfrak{S}_B \subseteq \mathfrak{T}$. Hence we have $-\lambda I < A < \lambda I$ and then $|\langle\phi; A\rangle| < \lambda$ for all $\lambda \in \tau(A, \mathfrak{T})$ and *all* $\phi \in \mathfrak{S}$. Consequently, $\sup_{\phi\in\mathfrak{S}} |\langle\phi; A\rangle|$ is a lower bound for $\tau(A, \mathfrak{T})$ and, since $\|A\|_{\mathfrak{T}}$ is the g.l.b. of $\tau(A, \mathfrak{T})$, we get $\|A\| \equiv \sup_{\phi\in\mathfrak{S}} |\langle\phi; A\rangle| \leqslant \|A\|_{\mathfrak{T}}$. The proof is completed by noticing that $\|A\|_{\mathfrak{T}} = \sup_{\phi\in\mathfrak{T}} |\langle\phi; A\rangle| \leqslant \sup_{\phi\in\mathfrak{S}} |\langle\phi; A\rangle| \equiv \|A\|$. ∎

It follows immediately from this lemma that for all $\lambda \in \mathbb{R}$, all A and B in $\mathfrak{A}$, $\|\lambda A\| = |\lambda|\ \|A\|$, $\|A + B\| \leqslant \|A\| + \|B\|$ and that the vanishing of $\|A\|$ occurs only when $A = 0$. Therefore we have the following:

Corollary. $\|\cdot\cdot\cdot\|$ *is a* norm *for* $\mathfrak{A}$ *and* $|\langle\phi; A\rangle| \leqslant \|A\|$ *for all* (A, ϕ) *in* $(\mathfrak{A}; \mathfrak{S})$.

From a phenomenological point of view we might remark at this point that one actually never deals in the laboratory with any observable A for which $\langle\phi; A\rangle$ is not finite; it is current practice nevertheless to consider in the theory "idealized observables" that are unbounded although well-defined on at least a full set of states. It turns out, however, that at this stage of the development of our axiomatization this idealization generates more troubles than it carries advantages and we consequently elect to defer its introduction, reserving for the future the possibility of having recourse to it as the need might occur.† In the same pragmatic vein it is convenient to assume here that any Cauchy-sequence in the norm admits a limit element in $\mathfrak{A}$.

As a result of these considerations, $\mathfrak{A}$ is now equipped with the structure of a *real Banach space* relative to the natural norm introduced above, and the states ϕ in $\mathfrak{S}$ are *continuous* (positive linear) functionals on $\mathfrak{A}$ with respect to the topology induced by this norm (see preceding corollary). We further extend $\mathfrak{S}$ to the set of *all* such functionals on $\mathfrak{A}$. Hence, essentially for the sake of mathematical convenience but also because no outrageous restriction seems to be involved, we restrict our attention to those systems of observables that satisfy the following postulate:

† Some extensions of this general approach have recently been worked out to include unbounded observables and fields; we mention, only as an indication, the work of Sewell [1970], Wyss [1969], and Powers [1971].

Structure Axiom 7. *The norm of any element* A *in* $\mathfrak{A}$ *is finite and* $\mathfrak{A}$ *is topologically complete when regarded as a metric space with the distance between any two elements* A *and* B *of* $\mathfrak{A}$ *defined by* $\|A - B\|$. $\mathfrak{S}$ *is then identified with the set of all continuous positive linear functionals* ϕ *on* $\mathfrak{A}$ *satisfying* $\langle\phi; I\rangle = 1$.

We now recall that $\mathfrak{A}$ is (Theorem 3) a commutative real Jordan algebra; the question of what relations might exist between this algebraic structure and the topology just introduced then naturally arises. It is this problem we now want to consider.

Lemma. *For all* A *and* B *in* $\mathfrak{A}$, $\|A^2\| = \|A\|^2$ *and*

$$\|A^2 - B^2\| \leqslant \max\{\|A\|^2, \|B\|^2\}.$$

Proof. (i) We can assume without loss of generality that $A \neq 0$. Let λ be in $\tau(A, \mathfrak{T})$ for some complete $\mathfrak{T}$ in $\mathfrak{S}_A$; we have $-\lambda I|_{\mathfrak{T}} < A|_{\mathfrak{T}} < \lambda I|_{\mathfrak{T}}$, hence $0|_{\mathfrak{T}} < A^2|_{\mathfrak{T}} < \lambda^2 I|_{\mathfrak{T}}$. Since $\mathfrak{T}$ is full with respect to $\mathfrak{A}_{\mathfrak{T}} \supseteq \mathfrak{P}(A)$, we conclude that $0 < A^2 < \lambda^2 I$, hence $\lambda^2 \in \tau(A^2, \mathfrak{T})$. With the same reasoning we get that λ^2 in $\tau(A^2, \mathfrak{T})$ implies λ in $\tau(A, \mathfrak{T})$. Consequently $\|A^2\|_{\mathfrak{T}} = \|A\|_{\mathfrak{T}}^2$ and therefore $\|A^2\| = \|A\|^2$.

(ii) The proof of the second part of the lemma follows directly from the definition of the norm. ∎

Lemma. (i) *For any state* ϕ *in* $\mathfrak{S}$ *and any pair* (A, B) *of observables in* $\mathfrak{A}$ *the Schwartz inequality holds true, i.e.,*

$$\langle\phi; A \circ B\rangle^2 \leqslant \langle\phi; A^2\rangle\langle\phi; B^2\rangle$$

(ii) *Let* A *be an arbitrary element in* $\mathfrak{A}$ *and* ϕ *be in* $\mathfrak{S}_A$*; then* $\langle\phi; A \circ B\rangle = \langle\phi; A\rangle\langle\phi; B\rangle$ *for all* B *in* $\mathfrak{A}$.

Proof. Since ϕ is positive, $\langle\phi; (A - \lambda B)^2\rangle \geqslant 0$. From the *distributivity* of the symmetrized product (with respect to the addition), its *homogeneity* (with respect to the scalar multiplication) (both of which we have as consequences of our Structure Axiom 6), and the linearity of ϕ, we get

$$\langle\phi; A^2\rangle + \lambda^2\langle\phi; B^2\rangle - 2\lambda\langle\phi; A \circ B\rangle \geqslant 0.$$

If $\langle\phi; B^2\rangle = 0$, we get a contradiction for a convenient value of the otherwise arbitrary real number λ unless $\langle\phi; A \circ B\rangle = 0$; this proves the Schwartz inequality in this particular case. If $\langle\phi; B^2\rangle \neq 0$, we write $\lambda = \langle\phi; A \circ B\rangle/\langle\phi; B^2\rangle$ from which the Schwartz inequality follows. From this inequality we get, in particular,

$$\langle\phi; (A - \langle\phi; A\rangle I) \circ B\rangle^2 \leqslant \langle\phi; (A - \langle\phi; A\rangle I)^2\rangle\langle\phi; B^2\rangle;$$

ϕ in $\mathfrak{S}_A$ implies that the right-hand side is zero, from which we conclude to the validity of the second assertion of the lemma. ∎

Theorem 7. *For any state* ϕ *in* $\mathfrak{S}$ *and any pair* A, B *of observables in* $\mathfrak{A}$ *we have*

$$\text{(i)} \qquad |\langle\phi; \mathrm{A} \circ \mathrm{B}\rangle| \leqslant \|\mathrm{A}\| \cdot \|\mathrm{B}\|,$$

$$\text{(ii)} \qquad \|\mathrm{A} \circ \mathrm{B}\| \leqslant \|\mathrm{A}\| \cdot \|\mathrm{B}\|,$$

$$\text{(iii)} \qquad \|\mathrm{A}^2 - \mathrm{B}^2\| \leqslant \|\mathrm{A} - \mathrm{B}\| \cdot \|\mathrm{A} + \mathrm{B}\|.$$

Proof. From Schwartz inequality and our previous results

$$\langle\phi; A \circ B\rangle^2 \leqslant \langle\phi; A^2\rangle\langle\phi; B^2\rangle \leqslant \|A^2\| \cdot \|B^2\| = \|A\|^2 \cdot \|B\|^2,$$

which proves (i); (ii) follows immediately from (i), and (iii) follows from (ii) on noticing that $A^2 - B^2 = (A - B) \circ (A + B)$ follows from the distributivity of the symmetrized product. ∎

REMARK. (ii) expresses the continuity of the symmetrized product in each of its factors and (iii) expresses the continuity of A^2 in A.

The idea of introducing the norm topology on $\mathfrak{A}$, instead of trying to mimic the weak-operator topology, was strongly advocated by Segal [1947] and it might be worthwhile to reproduce here his very concise postulates:

I. Algebraic Postulates.
  1. $\mathfrak{A}$ is a real linear space.
  2. There exists in $\mathfrak{A}$ an identity element I and for every $A \in \mathfrak{A}$ and positive integer n an element A^n of $\mathfrak{A}$, these being such that the usual rules for operating with polynomials in a single variable are valid: if f, g, h are polynomials with real coefficients, and if $f(g(\mu)) = h(\mu)$ for all real μ, then $f(g(A)) = h(A)$; here $f(A) = I + \sum_{k=1}^{m} \lambda_k A^k$ if $f(\mu) = \sum_{k=0}^{m} \lambda_k \mu^k$.

II. Metric Postulates. There is defined for each observable A a nonnegative number $\|A\|$ such that
  1. If λ is an arbitrary real number and A and B are arbitrary elements of $\mathfrak{A}$, then $\|\lambda A\| = |\lambda| \cdot \|A\|$, $\|A + B\| \leqslant \|A\| + \|B\|$. The vanishing of $\|A\|$ implies $A = 0$, and $\mathfrak{A}$ is topologically complete when regarded as a metric space with the distance between A and B defined as $\|A - B\|$. (In other words, $\mathfrak{A}$ is a real Banach space relative to $\|A\|$ as a norm.)
  2. $\|A^2 - B^2\| \leqslant \max\,[\|A^2\|, \|B^2\|]$.
  3. $\|A^2\| = \|A\|^2$.
  4. $\|\sum_{B\in\mathfrak{B}} B^2\| \leqslant \|\sum_{B\in\mathfrak{C}} B^2\|$ if $\mathfrak{B} \subseteq \mathfrak{C}$, $\mathfrak{B}$ and $\mathfrak{C}$ being finite subsets of $\mathfrak{A}$.
  5. A^2 is a continuous function of A.

From an axiomatic point of view we might notice that Segal's postulate II.4 (but not II.5!) is redundant; on the other hand, the postulates are consistent in the sense that we already have some realizations of them; in particular, a norm can be introduced on each r-number algebra (including† the exceptional $\mathfrak{M}_3^8$) to make it satisfy all of Segal's postulates; moreover, these postulates offer a genuine generalization of those of Jordan, von

† Sherman [1956].

Neumann, and Wigner [1934]; consider, indeed, a uniformly closed* subalgebra $\mathfrak{X}$ (where $\mathfrak{X} \ni I$) of $\mathfrak{B}(\mathcal{H})$, the set of all bounded operators acting on some infinite-dimensional Hilbert space $\mathcal{H}$; the special (infinite-dimensional) Jordan algebra $\mathfrak{A} = \{A = A^* \mid A \in \mathfrak{X}\}$ satisfies all of Segal's postulates.

The mathematical object $\mathfrak{A}$ defined by Segal's postulates is referred to in the sequel as a *Segal algebra*.

An inspection of our results up to this point shows that the theory presented above, i.e., as specified by our Structure Axioms 1 to 7, equips the set $\mathfrak{A}$ of all observables with the structure of a Segal algebra. Some differences between Segal's axiomatization and ours are worth pointing out. First of all, our approach stresses the emphasis we want placed on the role of states in the formulation of both the algebraic and the topological structure of the theory; in this respect, however, it should be realized that the notion of states is actually implicitly present in most of Segal's motivation; we differ mainly in paying more attention to the concept of dispersion-free states at an earlier stage of the formulation. This was necessary for a proper motivation of the power structure of $\mathfrak{A}$ (see Axiom 5); moreover, it allowed us to introduce rather early the concept of compatibility between observables. The latter concept in turn was used in our Structure Axiom 6 which was designed to prescribe the generalization of classical mechanics we wanted to consider; the main consequence of this axiom was that the symmetrized product $A \circ B$ became distributive (with respect to the addition) and homogeneous (with respect to scalar multiplication). Segal also has a "formal product" which he defines in a way similar to our symmetrized product and which actually coincides with it when $\mathfrak{A}$ is distributive. Segal, however, does *not* postulate distributivity in general and Sherman [1956] actually constructed a class of "universal counterexamples" of Segal algebras which are not distributive.

We now want to devote some attention to the study of compatible sets of observables.

Lemma. *For any subset $\mathfrak{T}$ in $\mathfrak{S}$ the subset $\mathfrak{A}_{\mathfrak{T}}$ of $\mathfrak{A}$ defined by $\mathfrak{A}_{\mathfrak{T}} \equiv \{A \in \mathfrak{A} \mid \mathfrak{S}_A \supseteq \mathfrak{T}\}$ is closed.*

Proof. Let $A_n \to A$ with $\{A_n\} \subseteq \mathfrak{A}_{\mathfrak{T}}$; for any ϕ in $\mathfrak{T}$ we have

$$
\begin{aligned}
|\langle\phi; A^2\rangle - \langle\phi; A\rangle^2| &= |\langle\phi; A^2 - A_n^2\rangle - \langle\phi; A - A_n\rangle\langle\phi; A + A_n\rangle| \\
&\leqslant |\langle\phi; A^2 - A_n^2\rangle| + |\langle\phi; A - A_n\rangle\langle\phi; A + A_n\rangle| \\
&\leqslant \|A^2 - A_n^2\| + \|A - A_n\| \, \|A + A_n\| \\
&\leqslant 2\,\|A - A_n\| \,(\|A - A_n\| + 2\,\|A\|),
\end{aligned}
$$

which can be made arbitrarily small. Hence every ϕ in $\mathfrak{T}$ is dispersion-free on A and therefore A belongs to $\mathfrak{A}_{\mathfrak{T}}$. ∎

Theorem 8. *Let* $\mathfrak{B}$ *be a compatible set of observables and* $\mathfrak{P}(\mathfrak{B})$, *the set of all polynomials in the elements* B *of* $\mathfrak{B}$. *Then the closure* $\overline{\mathfrak{P}(\mathfrak{B})}$ *of* $\mathfrak{P}(\mathfrak{B})$ *is an* associative *Segal subalgebra of* $\mathfrak{A}$.

Proof. Let $\mathfrak{S}_{\mathfrak{B}} = \bigcap_{B\in\mathfrak{B}} \mathfrak{S}_B$ and $\mathfrak{A}(\mathfrak{B}) = \mathfrak{A}_{\mathfrak{S}_{\mathfrak{B}}}$. We have already seen (Theorem 1) that $\mathfrak{P}(\mathfrak{B}) \subseteq \mathfrak{A}(\mathfrak{B})$ which is associative (Theorem 2) and closed (preceding lemma); hence $\mathfrak{A}(\mathfrak{B})$ contains $\overline{\mathfrak{P}(\mathfrak{B})}$ which is then also associative; the other properties of a Segal algebra are easily verified to hold true for $\overline{\mathfrak{P}(\mathfrak{B})}$. ∎

The following representation theorem, proved by Segal [1947], shows that physical systems, all observables of which are compatible, are exactly "classical systems":

Theorem 9. *Every associative Segal algebra* $\mathfrak{A}$ *is isomorphic (algebraically and metrically) with the algebra* $\mathfrak{C}(\Gamma)$ *of all real-valued, continuous functions* A *on a compact Hausdorff space* Γ, *where the addition, scalar multiplication, and powers of a function are defined in the usual manner and where the norm is defined by* $\|\mathrm{A}\| = sup_{\gamma\in\Gamma} |\mathrm{A}(\gamma)|$. *Furthermore, every state* ϕ *in* $\mathfrak{S}$ *has the form*

$$\langle\phi; \mathrm{A}\rangle = \int_\Gamma \mathrm{A}(\gamma)\, \mathrm{d}\mu(\gamma),$$

where μ *is a regular Borel measure on* Γ *such that* $\mu(\Gamma) = 1$. *Conversely, every such measure generates through the above integral a state* ϕ *on* $\mathfrak{A}$.

An understanding of this result requires some rudimentary knowledge of topology and measure theory, which we now review briefly.† Let $\mathcal{X}$ be a set and $\mathcal{T}$, a collection of subsets of $\mathcal{X}$ such that $\mathcal{T}$ contains (a) the empty set and $\mathcal{X}$ itself, (b) the intersection of any finite family $\{\mathcal{O}_i\}$ of members of $\mathcal{T}$, and (c) the union of any family $\{\mathcal{O}_k\}$ of members of $\mathcal{T}$. These conditions define a *topological space* and the members of $\mathcal{T}$ are said to be the *open* subsets of $\mathcal{X}$. A neighborhood of a point x in $\mathcal{X}$ is a subset $\mathcal{N}$ of $\mathcal{X}$ such that (a) $x \in \mathcal{N}$ and (b) $x \in \mathcal{O} \subseteq \mathcal{N}$ for some member $\mathcal{O}$ of $\mathcal{T}$. A subset $\mathcal{C}$ is said to be *closed* if $\mathcal{X} - \mathcal{C}$ is open. The *closure* of an arbitrary subset $\mathcal{A}$ of $\mathcal{X}$ is the smallest closed subset of $\mathcal{X}$ containing $\mathcal{A}$. Let $\mathcal{X}$ be a set and τ_i ($i = 1, 2$), a topology defined on $\mathcal{X}$ by a family $\mathcal{T}_i$ of open subsets. We say that τ_1 is *weaker* than τ_2 if $\mathcal{T}_1$ is contained in $\mathcal{T}_2$; we denote this by $\tau_1 \leqslant \tau_2$. This relation implies that the τ_2-closure of an arbitrary subset of $\mathcal{X}$ is contained in

† For classic presentations see, for instance, in general topology, Bourbaki [1940], Kuratowski [1948] or Kelley [1955]; in measure theory, Halmos [1950] or Berberian [1965]; Choquet [1969] provides an introduction to both.

the τ_1-closure of this subset. A function $f: \mathfrak{X}_1 \to \mathfrak{X}_2$ is *continuous* if and only if, for every open (resp. closed) subset $\mathcal{A}$ in $\mathfrak{X}_2$, $f^{-1}(\mathcal{A})$ is open (resp. closed) in $\mathfrak{X}_1$. A function is said to be bicontinuous if both f and f^{-1} are continuous. A topological space is said to be Hausdorff if, for any x, y in it, x and y have disjoint neighborhoods. A topological space $\mathfrak{X}$ is said to be *compact* if every covering of $\mathfrak{X}$ by open subsets admits a finite subcovering. A subset $\mathcal{A}$ of $\mathfrak{X}$ is said to be *compact* if it is a compact topological space when equipped with the topology it naturally inherits from $\mathfrak{X}$. A subset $\mathcal{A}$ of $\mathfrak{X}$ is said to be *relatively compact* if its closure in $\mathfrak{X}$ is compact. A topological space is said to be *locally compact* if each of its points has an open neighborhood that is relatively compact. The measure theory required for an understanding of the statement in Theorem 9 can be summarized as follows: we recall that a *σ-ring* $\mathcal{S}$ is a nonempty class of subsets of a set $\mathfrak{X}$ such that (a) S and T in $\mathcal{S}$ implies $S - T$ in $\mathcal{S}$ and (b) $\{S_i \in \mathcal{S} \mid i = 1, 2, \ldots\}$ implies that $\bigcup_{i=1}^{\infty} S_i \equiv S \in \mathcal{S}$. If $\mathcal{E}$ is any family of subsets of $\mathfrak{X}$, there always exists a unique smallest σ-ring $\mathcal{S}(\mathcal{E})$ of subsets of $\mathfrak{X}$ containing $\mathcal{E}$; $\mathcal{S}(\mathcal{E})$ is said to be the σ-ring generated by $\mathcal{E}$. Now let $\mathfrak{X}$ be a locally compact Hausdorff space; the *Borel* subsets of $\mathfrak{X}$ are defined as the elements of the σ-ring $\mathcal{S}(\mathcal{E})$ generated by the family $\mathcal{E}$ of all compact subsets of $\mathfrak{X}$. If $\mathfrak{X}$ is compact, and not only locally compact, every closed subset of $\mathfrak{X}$ is compact and we can replace "compact" with "closed" in the above definition of Borel subsets. A *Borel measure* on $\mathfrak{X}$ is a function μ which attributes to every Borel set S a nonnegative number $\mu(S)$, called the measure of S, such that (a) $\mu(\phi) = 0$, (b) for every sequence $\{S_i \mid i = 1, 2, \ldots\}$ of disjoint Borel sets $\mu(\bigcup_{i=1}^{\infty} S_i) = \sum_{i=1}^{\infty} \mu(S_i)$, and (c) $\mu(E) < \infty$ for every compact subset E of $\mathfrak{X}$. A Borel measure is said to be *regular* if for every Borel set S of finite measure

$$\mu(S) = lub\{\mu(E) \mid E \subset S; E \in \mathcal{E}\}.$$

A real-valued function f defined on the locally compact space $\mathfrak{X}$ is said to have *compact support* if there exists a compact subset E of $\mathfrak{X}$ such that $f = 0$ on $\mathfrak{X} - E$. We denote by $\mathfrak{C}_0$ the class of all continuous functions $f : \mathfrak{X} \to \mathbb{R}$ with compact support. From the first part of the theorem we see that ϕ is a positive *Radon measure* on Γ. The second part of the theorem is an application of the Riesz-Markoff theorem which asserts that to every positive linear functional ϕ on $\mathfrak{C}_0$ corresponds a unique regular Borel measure μ such that

$$\langle \phi; f \rangle = \int_{\mathfrak{X}} f(x)\, d\mu(x)$$

for all f in $\mathfrak{C}_0$ and conversely. The integral in the right-hand side has to be understood quite generally as follows: we first decompose f into its positive and negative parts (which are measurable); $f = f^+ - f^-$ and we write

$$\int f(x)\, d\mu(x) = \int f^+(x)\, d\mu(x) - \int f^-(x)\, d\mu(x),$$

the right-hand side of which will make sense as soon as the integral of a μ-measurable positive function is defined; let f be such a function and S_f, the set of all μ-measurable functions ψ which (a) assume only a finite number of values (i.e., $\psi = \sum_{i=1}^n a_i \chi_{A_i}$, where χ_A denotes the characteristic function of the subset A of Γ) and (b) satisfy $0 \leqslant \psi \leqslant f$. By definition we have in this case

$$\int f(x)\, d\mu(x) = \sup_{\psi \in S_f} \left\{ \int \psi(x)\, d\mu(x) \right\},$$

where

$$\int \psi(x)\, d\mu(x) = \sum_{i=1}^{n} a_i\, \mu(A_i).$$

As already mentioned, Theorem 9 provides a faithful concrete realization of our algebraic approach to classical (statistical) mechanics. Following Koopman [1931], a less faithful but concrete Hilbert space realization can be obtained which is nevertheless useful† in analyzing properties of a classical system in a given state ϕ. Let $\mathfrak{L}^2(\Gamma, \mu)$ be the Hilbert space of all functions $\Psi: \Gamma \to \mathbb{C}$, which are square-integrable with respect to the measure μ associated with the given state ϕ by Theorem 9. For every A in $\mathfrak{A}$ we define the bounded self-adjoint operator

$$\pi_\phi(A): \mathfrak{L}^2(\Gamma, \mu) \to \mathfrak{L}^2(\Gamma, \mu)$$

by

$$(\pi_\phi(A)\Psi)(\gamma) = A(\gamma)\Psi(\gamma) \quad \text{for all } \Psi \text{ in } \mathfrak{L}^2(\Gamma, \mu).$$

We notice, in anticipation of the general case, that the vector Φ in $\mathfrak{L}^2(\Gamma, \mu)$, defined by $\Phi(\gamma) = 1$ for all γ in Γ, enjoys two remarkable properties: first

$$\langle \phi; A \rangle = (\Phi, \pi_\phi(A)\Phi) \quad \text{for all } A \text{ in } \mathfrak{A};$$

second, the complex linear manifold generated by $\{\pi_\phi(A)\Phi \mid A \in \mathfrak{A}\}$ is *dense* in $\mathfrak{L}^2(\Gamma, \mu)$. Finally, we might mention that $\{\pi_\phi(A) \mid A \in \mathfrak{A}\}$ is a Segal algebra when equipped with (a) the Banach space structure inherited from $\mathfrak{B}(\mathfrak{L}^2(\Gamma, \mu))$ and (b) the symmetrized product $X \circ Y = \frac{1}{2}(XY + YX)$, where XY is the ordinary composition law of the linear operators X, Y acting on $\mathfrak{L}^2(\Gamma, \mu)$; since $[\pi_\phi(A), \pi_\phi(B)] = 0$ for all A and B in $\mathfrak{A}$, the symmetrized product reduces in this case to the usual product.

A *flow* (Γ, μ, T_t) is defined in classical mechanics‡ as a one-parameter

† For instance, in ergodic theory; see Halmos [1958], Arnold and Avez [1968].
‡ See, for instance, Arnold and Avez [1968].

group $\{T_t\}$ of automorphisms (mod 0) of (Γ, μ) depending measurably on t. To this corresponds a continuous one-parameter group $\{U_\phi(t)\}$ of unitary operators defined by

$$(U_\phi(t)\Psi)(\gamma) = \Psi(T_t[\gamma]) \quad \text{for all } \Psi \text{ in } \mathfrak{L}^2(\Gamma, \mu)$$

and satisfying the following properties:

1. $U_\phi(t)\Phi = \Phi$ and

$$(U_\phi(t)\pi_\phi(A)U_\phi(-t)\Psi)(\gamma) = A(T_t[\gamma])\Psi(\gamma) \equiv \alpha_t[A](\gamma)\Psi(\gamma)$$

for all A in $\mathfrak{A}$, all t in $\mathbb{R}$, all Ψ in $\mathfrak{L}^2(\Gamma, \mu)$, and all γ in Γ. This means that

2. $U_\phi(t)\,\pi_\phi(A)U_\phi(-t) = \pi_\phi(\alpha_t[A])$.

We establish later that this realization can be generalized to cover the quantum case as well, with naturally the essential exception that $\pi_\phi(A)$ and $\pi_\phi(B)$ will in general no longer commute.

We now continue our discussion of the consequences of Theorem 9 and notice that to each point γ in Γ we can associate the state $\langle\gamma; A\rangle = A(\gamma)$, which is clearly dispersion-free on all A in $\mathfrak{C}(\Gamma) \sim \mathfrak{A}$. The set $\{\gamma\}$ of all states γ so obtained is full with respect to $\mathfrak{A}$. Hence an *associative* Segal algebra admits a full set of states which are dispersion-free on all its elements; this already strong result is improved in the sequel and extended to any *associative* Segal subalgebra of an arbitrary Segal algebra. We already know an example of this extension, namely the associative subalgebra $\overline{\mathfrak{P}(\mathfrak{B})}$ generated by a compatible set of observables.

Our axioms can be used to sharpen somewhat Sherman's conclusion [1951] that in a Segal algebra $\mathfrak{A}$ the sum of the squares of elements of $\mathfrak{A}$ can itself be written as the square of an element of $\mathfrak{A}$; this result in turn will allow us to obtain some more information about the set $\mathfrak{S}$.

Lemma. *Every positive observable admits a positive square root.*

Proof. Let A be an arbitrary positive observable and $\overline{\mathfrak{P}(A)}$, the Segal subalgebra generated by A and I; this algebra is associative (since A and I are compatible). From Theorem 9 we know that $\overline{\mathfrak{P}(A)}$ is isomorphic to some $\mathfrak{C}(\Gamma)$; let π denote this isomorphism. "A positive" means that $\langle\phi; A\rangle \geqslant 0$ for all ϕ in $\mathfrak{S}$ and, in particular, for those states corresponding to points in Γ. Consequently the function $\pi(A)$ is positive. Let f be its positive square root; f belongs to $\mathfrak{C}(\Gamma)$. Hence there is an element B in $\overline{\mathfrak{P}(A)}$ such that $\pi(B) = f$, and, since π is an isomorphism, $B^2 = A$; f positive implies $\langle\gamma; B\rangle \geqslant 0$ for all γ in Γ. Since $\{\gamma\}$ is full for $\overline{\mathfrak{P}(A)}$, $\langle\phi; B\rangle \geqslant 0$ for all ϕ in $\mathfrak{S}$; i.e., $B \geqslant 0$. ∎

We notice that since the sum of the squares of observables is itself a positive observable it is itself the square of a positive observable.

We now want to show that the states on $\mathfrak{A}$ form a large class of the set of all the linear functions that might be defined on $\mathfrak{A}$. We say that a functional χ on $\mathfrak{A}$ is *positive* if $\langle \chi; A^2 \rangle$ is a positive number for all A in $\mathfrak{A}$; because of the preceding lemma, this requirement is equivalent to $\langle \chi; A \rangle \geqslant 0$ for all positive A in $\mathfrak{A}$. The *norm* $\|\chi\|$ of a functional χ on $\mathfrak{A}$ is defined by $\sup_{\substack{A \in \mathfrak{A} \\ \|A\| \leqslant 1}} |\langle \chi; A \rangle| \equiv \|\chi\|$.

Lemma. *Every positive linear functional χ on $\mathfrak{A}$ with $\langle \chi; \mathrm{I} \rangle < \infty$ is continuous and its norm is given by the value it takes on the element $I \in \mathfrak{A}$.*

Proof. For every X in $\mathfrak{A}$ such that $\|X\| \leqslant 1$ we have

$$\langle \phi; I - X \rangle = 1 - \langle \phi; X \rangle \geqslant 1 - \sup_{\phi \in \mathfrak{S}} |\langle \phi; X \rangle| = 1 - \|X\| \geqslant 0$$

for all ϕ in $\mathfrak{S}$ and then $(I - X) \geqslant 0$; since χ is positive by assumption, $\langle \chi; X \rangle \leqslant \langle \chi; I \rangle$. We obtain $(I + X) \geqslant 0$ in the same way; hence $-\langle \chi; X \rangle \leqslant \langle \chi; I \rangle$. Consequently for every X with $\|X\| \leqslant 1$, $|\langle \chi; X \rangle| \leqslant \langle \chi; I \rangle$. For an arbitrary A in $\mathfrak{A}$ we write $X = A/\|A\|$ and get from the preceding result $|\langle \chi; A \rangle| \leqslant \|A\| \langle \chi; I \rangle$. Therefore χ is continuous. Furthermore, $\langle \chi; I \rangle \leqslant \sup_{\substack{A \in \mathfrak{A} \\ \|A\| \leqslant 1}} |\chi; A \rangle| \leqslant \langle \chi; I \rangle$ which concludes the proof of the lemma. ∎

From this lemma we readily prove the following result which makes redundant the condition that $\phi \in \mathfrak{S}$ be continuous:

Corollary. *Every positive linear functional ϕ on $\mathfrak{A}$ such that $\langle \phi; \mathrm{I} \rangle = 1$ (or equivalently with $\|\phi\| = 1$) is a state on $\mathfrak{A}$.*

We now want to discuss the notion of pure states which plays an important role in the sequel. We first notice that if $\{\phi_i\}$ is a sequence of states and $\{\lambda_i\}$, a sequence of positive numbers adding up to 1, then $\Sigma_i \lambda_i \phi_i$ is also a state, called the mixture of the states ϕ_i with coefficients λ_i. Mathematically, this fact is expressed by saying that $\mathfrak{S}$ is a *convex set.*

Definition. *A state ϕ is said to be* pure *if it cannot be written as a convex linear combination of two other states.*

In other words, ϕ pure is an *extreme point* of the convex set $\mathfrak{S}$.

A state ϕ is said to *dominate* a state ψ if there exists a real number $\lambda > 1$ such that $(\lambda\phi - \psi)$ is a positive real functional on $\mathfrak{A}$. We notice that if ϕ is a mixture it dominates every one of its components. Conversely, if, given a state ϕ, there exists another state ψ such that $(\lambda\phi - \psi) \geqslant 0$ for some $\lambda > 1$, we can form $\chi = (\lambda\phi - \psi)/(\lambda - 1)$; χ is positive and $\langle \chi; I \rangle = 1$. Hence χ is a state (see preceding corollary) and we can write $\phi = (1/\lambda)\psi + [1 - (1/\lambda)]\chi$ so that ϕ is a mixture of the states ψ and χ. This remark proves the following:

Lemma. *A state ϕ is pure if and only if it dominates no other state.*

This necessary and sufficient condition is sometimes taken in the mathematical literature as the definition of a pure state.

The question then naturally arises whether there exist pure states on $\mathfrak{A}$. In the particular case in which $\mathfrak{A}$ is associative, the answer is already given as an immediate consequence of Theorem 9: every one of the states $\langle\gamma; A\rangle$ associated with a point γ in the space Γ is *pure*. In fact, as a result of the following lemma (Segal [1947]), they are the only pure states on an associative Segal algebra:

Lemma. *Let μ be a regular measure on a compact Hausdorff space Γ such that $\mu(\Gamma) = 1$. Suppose that μ is not a convex linear combination of two other such measures. Then there is a point γ of Γ whose complement has μ-measure zero.*

In the general case we can conclude to the existence of pure states as follows. We equip the set $\mathfrak{A}^*$ of all continuous linear functionals on $\mathfrak{A}$ with the so-called *weak*-topology* (denoted also in the sequel *w*-topology* and also called the $\mathfrak{A}$-topology on $\mathfrak{A}^*$), i.e., the weakest topology relative to which the mappings $\chi \to \langle\chi; A\rangle$ are continuous for each A in $\mathfrak{A}$; for completeness we mention that this topology is obtained by taking as a basis of neighborhoods all sets of the form

$$N(\chi; \mathfrak{B}, \varepsilon) = \{\omega \in \mathfrak{A}^* \mid |\langle\chi; B\rangle - \langle\omega; B\rangle| < \varepsilon \,\forall\, B \in \mathfrak{B}\},$$

where χ is in $\mathfrak{A}^*$, $\mathfrak{B}$ is a *finite* subset of $\mathfrak{A}$, and $\varepsilon > 0$. This topology presents two advantages. First, it makes sense from the physical point of view when restricted to the subset $\mathfrak{S} \subseteq \mathfrak{A}^*$ of all states on $\mathfrak{A}$. Second, it is technically very powerful from a mathematical point of view; the reader is referred to the fifth chapter of Dunford and Schwartz [1957] for a systematic mathematical account of this and related concepts; in particular, he will find there the proofs of the theorems used in this section. Since $\mathfrak{S}$ is a subset of $\mathfrak{A}^*$, closed in the w^*-topology and bounded in the norm-topology, it is compact† in the w^*-topology. We further consider the linear space $\mathfrak{A}$ as a subset of its double dual (i.e., the elements of $\mathfrak{A}$ are considered as linear functionals on $\mathfrak{A}^*$); $\mathfrak{A}$ is *total* for $\mathfrak{A}^*$ (i.e. $\langle\chi; A\rangle = 0$ for all A in $\mathfrak{A}$ obviously implies that the element χ of $\mathfrak{A}^*$ is zero); hence‡ $\mathfrak{A}^*$ is a locally convex topological linear space when equipped with its w^*-topology. In the w^*-topology $\mathfrak{S}$ appears as a compact subset of a locally convex topological space; it therefore contains some extremal points§. This answers our previous question about the existence

† See Dunford and Schwartz [1957], V.4.3.
‡ See Dunford and Schwartz [1957], V.3.3.
§ See Dunford and Schwartz [1957], V.8.2.

of pure states. Moreover, $\mathfrak{S}$ being convex, by the Krein-Milman theorem† it is the closed convex hull of its extreme points (i.e., it is the intersection of all the closed convex subsets of $\mathfrak{A}^*$ containing the extreme points of $\mathfrak{S}$). Let us denote by $\mathfrak{S}^P$ the set of all pure states on $\mathfrak{A}$ (i.e., $\mathfrak{S}^P$ is the set of all extreme points of $\mathfrak{S}$); now suppose that for some pair (A, B) of elements of $\mathfrak{A}$ we have $\langle\phi; A\rangle \leqslant \langle\phi; B\rangle$ for all ϕ in $\mathfrak{S}^P$; since $\mathfrak{S}$ is the closed convex hull of $\mathfrak{S}^P$, we then have that $\langle\phi; A\rangle \leqslant \langle\phi; B\rangle$ for all ϕ in $\mathfrak{S}$; i.e., $A \leqslant B$. We therefore have proved the following:

Theorem 10. *The set $\mathfrak{S}^P$ of all pure states on $\mathfrak{A}$ is full with respect to $\mathfrak{A}$.*

This result was extended by Segal [1947] to an arbitrary Segal algebra $\mathfrak{A}$; the difficulty in his proof was to show that $\mathfrak{S}$, defined as the set of all (real-valued) linear functionals ϕ on $\mathfrak{A}$ (the Segal algebra considered) with $\langle\phi; A^2\rangle \geqslant 0$ for all A in $\mathfrak{A}$ and $\langle\phi; I\rangle = 1$, is such that $\langle\phi; A\rangle = 0$ for all ϕ in $\mathfrak{S}$ implies $A = 0$, i.e., by linearity that two observables which coincide on all states are identical. This result is physically so natural that we actually postulated it as our second structure axiom. We subsequently needed this axiom in our phenomenological justification of the power structure of $\mathfrak{A}$. Hence our Structure Axiom 2 is mathematically redundant once we have that $\mathfrak{A}$ is a Segal algebra, although it is physically necessary to arrive at this structure of $\mathfrak{A}$ from a phenomenological point of view.

We now want to introduce the notion of a partial state on a physical system. In practice we often obtain (either experimentally or theoretically) only partial information about a given physical system. When this information can be expressed by a positive linear functional f on a linear subspace $\mathcal{M}$ of $\mathfrak{A}$ such that $\mathcal{M} \ni I$ and $\langle f; I\rangle = 1$, f is called a *partial state* on $\mathcal{M}$. An example of a partial state is provided by the n-body correlation functions appearing in almost every problem in statistical mechanics. The study of such partial information is interesting both from a practical point of view‡ and from the point of view of checking the internal consistency of a given theory (see below). The fundamental result in this direction has already been obtained by Segal [1947]:

Lemma. *Let $\mathcal{M}$ be a linear subspace of $\mathfrak{A}$, with $\mathcal{M} \ni$ I, and f be a partial state on $\mathcal{M}$. Then there exists a state ϕ on $\mathfrak{A}$ such that f is the restriction of ϕ to $\mathcal{M}$.*

Proof. We have already proved that $|\langle\chi; A\rangle| \leqslant \|A\|$ for any normalized positive linear functional on $\mathfrak{A}$; the proof carries over for f on $\mathcal{M}$ and we therefore have $|\langle f; M\rangle| \leqslant \|M\|$ for every M in $\mathcal{M}$ and every partial state f

† See Dunford and Schwartz [1957], V.8.4.
‡ See, for instance, Emch, Knops, and Verboven [1968a, b].

on $\mathcal{M}$. For every element A in $\mathfrak{A}$ but not in $\mathcal{M}$ we form the sets $\mathcal{M}^+(A) \equiv \{M \in \mathcal{M} \mid M \geqslant A\}$ and $\mathcal{M}^-(A) \equiv \{M \in \mathcal{M} \mid M \leqslant A\}$; these sets are nonempty, since they contain $\|M\| \cdot I$ and $-\|M\| \cdot I$, respectively. We then define

$$\delta^+ = \underset{M \in \mathcal{M}^+(A)}{\text{g.l.b.}} \langle f; M\rangle$$

and

$$\delta^- = \underset{M \in \mathcal{M}^-(A)}{\text{l.u.b.}} \langle f; M\rangle$$

and notice that $\delta^- \leqslant \delta^+$. For every real number δ such that $\delta^- \leqslant \delta \leqslant \delta^+$ we form the linear functional δ defined by $\langle \delta; \lambda A + M\rangle = \lambda\delta + \langle f; M\rangle$ on the linear subspace $\mathcal{M}'$ spanned by $\mathcal{M}$ and A; we check separately for λ positive and negative that δ is a positive linear functional on $\mathcal{M}' \supset \mathcal{M}$. Since trivially $\langle \delta; I\rangle = 1$, δ is a partial state on $\mathcal{M}'$; by our initial remark we still have $\langle \delta; M'\rangle \leqslant \|M'\|$ for all M' in $\mathcal{M}'$; it is clear, however, that f is the restriction of δ to $\mathcal{M}$; we can then proceed by transfinite induction as in the proof of the usual† Hahn-Banach theorem, the only difference being that we can actually work with *positive* linear functionals, thanks to Segal's ingenious construction of the sets $\mathcal{M}^{\pm}(A)$ and the positive functional δ. This completes the proof of the lemma. ∎

The extension of f to ϕ is, in general, not unique. Two states ϕ_1 and ϕ_2 on $\mathfrak{A}$ which coincide on $\mathcal{M}$ are said to be $\mathcal{M}$-*equivalent*. This concept has been used in equilibrium statistical mechanics (e.g., Emch, Knops, and Verboven [1968a, b], in nonequilibrium statistical mechanics (e.g., Emch [1966b]), and for the theory of the measuring process (e.g., Jauch [1970] and references quoted there).

In analogy with our definition of the set $\mathfrak{S}$ of the states on the entire $\mathfrak{A}$ we say that a partial state on the subspace $\mathcal{M} \subseteq \mathfrak{A}$ is *pure* if it cannot be written as a convex linear combination of two other partial states on the same $\mathcal{M}$.

Lemma. *Let $\mathcal{M}$ be a linear subspace of $\mathfrak{A}$, with $\mathcal{M} \ni$ I, and f, a* pure *partial state on $\mathcal{M}$. There then exists a* pure *state ϕ on $\mathfrak{A}$ such that f is the restriction of ϕ to $\mathcal{M}$.*

Proof. Let $\{f\}$ be the set of all states ψ on $\mathfrak{A}$ such that $\psi|_{\mathcal{M}} = f$; this set is not empty, as proved by the preceding lemma. It is closed in the w^*-topology and bounded in the norm-topology; it is therefore compact for the w^*-topology and obviously convex; we can then use the Krein-Milman theorem again. Let ϕ be an extremal point of $\{f\}$; we want to show that ϕ is pure. Suppose, on the contrary, that there exists $0 < \lambda < 1$ and ψ_1, ψ_2 in $\mathfrak{S}$ such that $\phi = \lambda\psi_1 + (1 - \lambda)\psi_2$; by restricting this equality to $\mathcal{M}$ we

† See, for instance, Dunford and Schwartz [1957], II.3.10.

conclude from f pure that ψ_1 and ψ_2 coincide with f on $\mathcal{M}$, hence belong to $\{f\}$; this contradicts ϕ extremal in $\{f\}$ unless $\phi = \psi_1 = \psi_2$; therefore ϕ is pure. ▮

We now use this lemma to prove the following:

Theorem 11. *To any point σ in the spectrum σ_{A} of an* arbitrary *observable* A *in $\mathfrak{A}$ corresponds at least one pure state ϕ on $\mathfrak{A}$ such that $\langle\phi; \mathrm{A}\rangle = \sigma$ and ϕ dispersion-free on the Segal subalgebra $\overline{\mathfrak{P}(\mathrm{A})}$ generated by* A *and* I.

Proof. $\sigma \in \sigma_A$ means that there exists some $\psi \in \mathfrak{S}_A$ such that $\langle\psi; A\rangle = \sigma$. For all continuous functions g of one real variable we have $\langle\psi; g(A)\rangle = g(\sigma)$ and then $\langle\psi; g(A)^2\rangle = \langle\psi; g(A)\rangle^2$; hence ψ is dispersion-free on $\overline{\mathfrak{P}(A)}$. Since $\overline{\mathfrak{P}(A)}$ is associative, it is isomorphic to some $\mathfrak{C}(\Gamma)$ (see Theorem 9). The restriction of ψ to $\overline{\mathfrak{P}(A)}$ then corresponds to a point γ of Γ and ψ is pure on $\overline{\mathfrak{P}(A)}$. From the preceding lemma we conclude that there exists a state ϕ, pure on the whole $\mathfrak{A}$, coinciding with ψ on $\overline{\mathfrak{P}(A)}$, hence satisfying all the requirements of the conclusion of the theorem. ▮

Incidentally, this theorem (see also the way in which the proof is conducted) suggests that our Structure Axioms 4 and 5 are not mathematically independent of Segal's postulates; our argument in this connection is the same as that expounded in the discussion following Theorem 10.

As a consequence of this theorem, Segal pointed out the following:

Corollary. *Let $\mathfrak{A}(\mathcal{H})$ be the set of all bounded self-adjoint operators on an infinite-dimensional Hilbert space $\mathcal{H}$. There exist pure states on $\mathfrak{A}(\mathcal{H})$ which are* not *of the form $\langle\phi; \mathrm{A}\rangle = (\Phi, \mathrm{A}\Phi)$ for all* A *in $\mathfrak{A}(\mathcal{H})$ and some fixed Φ in $\mathcal{H}$.*

Proof. $\mathfrak{A}(\mathcal{H})$ contains some self-adjoint elements A with a nonempty continuous part σ_A^c in their spectrum. For such an A let $\sigma \in \sigma_A^c$; from the theorem there exists a state ϕ pure on $\mathfrak{A}(\mathcal{H})$ such that $\langle\phi; A\rangle = \sigma$ and ϕ dispersion-free on $\overline{\mathfrak{P}(A)}$. If ϕ were a vector-state, i.e., if there were a Φ in $\mathcal{H}$ such that $\langle\phi; A\rangle = (\Phi, A\Phi)$ for all A in $\mathfrak{A}(\mathcal{H})$ and, in particular, for all A in $\overline{\mathfrak{P}(A)}$, this Φ would be an eigenvector of A for the eigenvalue σ, which contradicts $\sigma \in \sigma_A^c$. ▮

REMARK. We cannot attribute this fact (apparently not so well known among physicists) to some peculiarity of our assumptions (in particular, to a too restrictive formulation of our Structure Axiom 4). To reject states of the type just mentioned would imply some mathematically drastic departure from the present formalism, since Segal [1947] proved this corollary within *his* set of postulates (including, to be complete, the assumption that the sum of the squares of observables is the square of an observable, a fact which anyhow, as we pointed out earlier, was proved by Sherman [1951] to be true for any Segal algebra). From the physical point of view, moreover, it seems that there is no serious objection to these states (which are obviously not "normal" in the usual sense of the term in the theory of C^*-algebras); we shall see later that some nonnormal states actually appear naturally in the thermodynamical limit in statistical mechanics and scattering theory.

As another interesting consequence of Theorem 9, suppose that we consider a set $\mathfrak{B}$ of observables such that $\overline{\mathfrak{P}(\mathfrak{B})}$ is associative; for all intents and purposes these oservables form a classical theory. It seems, therefore, *physically natural* to say that these observables, and in fact all observables in $\overline{\mathfrak{P}(\mathfrak{B})}$, are compatible. This statement however, is mathematically non-trivial; we consider it as our next postulate:

Structure Axiom 8. *A sufficient condition for a set $\mathfrak{B}$ of observables to be compatible is that $\overline{\mathfrak{P}(\mathfrak{B})}$ is associative.*

REMARKS

(i) We already know (Theorem 8) that this condition is necessary.

(ii) For completeness we mention that this postulate is implicit in Segal's semantic. Actually Segal [1947] defines a collection $\mathfrak{B}$ of "simultaneous observables" as one for which $\bigcap_{C\in\overline{\mathfrak{P}(\mathfrak{B})}} \mathfrak{S}_C$ is contained in some set $\mathfrak{T}$, full with respect to $\overline{\mathfrak{P}(\mathfrak{B})}$; we showed that $\mathfrak{S}_{\mathfrak{B}} = \bigcap_{B\in\mathfrak{B}} \mathfrak{S}_B$ is such a set if $\mathfrak{B}$ is compatible, so that a collection of observables, compatible in our sense, is "simultaneously observable" in Segal's. This fact is confirmed by Segal's proof that a necessary and sufficient condition for $\mathfrak{B}$ to be simultaneously observable is that $\overline{\mathfrak{P}(\mathfrak{B})}$ be associative. For compatibility we originally required more, namely that $\mathfrak{S}_{\mathfrak{B}}$ be full, not only with respect to $\overline{\mathfrak{P}(\mathfrak{B})}$ but also to $\mathfrak{A}(\mathfrak{B}) = \{A \in \mathfrak{A} \mid \mathfrak{S}_A \supseteq \mathfrak{S}_{\mathfrak{B}}\} \supseteq \overline{\mathfrak{P}(\mathfrak{B})}$. With this information we see that the above axiom requires that $\mathfrak{B}$ simultaneously observable imply $\mathfrak{B}$ compatible and that this requirement is then a restriction on the possible schemes which we want to consider for physical theories.

e. Proposition-Calculus

As we pointed out in Subsection c, the elements of $\mathfrak{A}$ that satisfy the relation $P^2 = P$, and which we called "propositions,"† are so simple that we might think of starting an axiomatic approach to the general theory by considering them as the building blocks of the formalism. This line of attack was initiated by Birkhoff and von Neumann [1936] and has been followed by Mackey

† The words "questions" or "events" are also used for these particular observables.

[1963] and Jauch [1968a] in their textbook presentations of the mathematical foundations of quantum mechanics; the reader will also find interesting discussions of this approach in Piron [1964] and Varadarajan [1962, 1968, 1970]. Plymen [1968a, b] showed how the algebraic approach can be made to satisfy the essentials in the axioms of Mackey and Piron. Pool [1968a, b] voiced some criticism of the operational meaning of some of the usual lattice-theoretical axioms in the proposition-calculus approach and suggested that the latter's generality could be extended by the exploitation of the theory of Baer*-semigroups. Aside from its natural elegance, this proposition-calculus approach can be credited with having contributed to a better understanding of some of the physical concepts, such as the definition of symmetries and their representations,† the nonoccurrence of hidden variables in quantum theories,‡ and the nature of superselection rules.§

Our intent in this subsection is not to review, even in abbreviated form, the epistemological aspects of this approach; rather we shall illustrate with propositions the meaning of some of our axioms. In particular, the notion of compatibility is specialized to these particular observables.

We first establish some elementary properties of those elements P of $\mathfrak{A}$ that satisfy $P^2 = P$ in order to give some feeling for the terminology. For a start we notice that our Structure Axiom 5 implies that for every state ϕ, dispersion-free on P, we have

$$\langle \phi; P \rangle = \langle \phi; P^2 \rangle = \langle \phi; P \rangle^2,$$

so that the spectrum of P (when $P \neq 0$ and $P \neq I$) consists exactly of the two points 0 and 1; this implies

$$0 \,|_{\mathfrak{S}_P} \leqslant P \,|_{\mathfrak{S}_P} \leqslant I \,|_{\mathfrak{S}_P},$$

and, since $\mathfrak{S}_P$ is full with respect to $\{0, P, I\}$,

$$0 \leqslant P \leqslant I$$

or, equivalently,

$$0 \leqslant \langle \phi; P \rangle \leqslant 1 \quad \text{for all } \phi \in \mathfrak{S}.$$

The term "proposition" for these elements is justified by the fact that in their dispersion-free states these observables take the values zero or one (an alternative that we can interpret figuratively as "no" or "yes," "false" or "true," etc.); $\langle \phi; P \rangle$ is then the probability that P will take the value 1 (or that the "event" P will occur) when the system is prepared in the state ϕ.

† Emch and Piron [1963] and Emch [1963a, b]; the material in these papers has been consolidated in book form by Jauch [1968b]; some generalization of EP [1963] has recently been proposed by Gudder [1970].

‡ Jauch and Piron [1963].

§ Piron [1969] and references quoted therein.

For any two propositions P and Q such that $P \leqslant Q$ we see immediately that $\langle\phi; P\rangle = 1$ can occur only if $\langle\phi; Q\rangle = 1$ and $\langle\phi; Q\rangle = 0$ can occur only if $\langle\phi; P\rangle = 0$; hence the relation $P \leqslant Q$ can be interpreted as meaning "P implies Q." We can similarly interpret $P \circ Q = 0$ as meaning "P and Q are disjoint"; indeed, $P \circ Q = 0$ implies that P and Q are smaller or equal to the proposition $P + Q$, hence $\langle\phi; P\rangle = 1$ (resp. $\langle\phi; Q\rangle = 1$) implies that $\langle\phi; Q\rangle = 0$ (resp. $\langle\phi; P\rangle = 0$).

We can now use these simple remarks to prove the corollary to Theorem 4. Suppose, indeed (using the notation of the theorem), that ϕ belongs to $\bigcap_i \mathfrak{S}_{P_i}$; since the P_i's add up to I, we have $\langle\phi; P_i\rangle \neq 0$ for at least one of them; hence $\langle\phi; P_i\rangle = 1$. Since, however, the P_i's are mutually disjoint, this implies that $\langle\phi; P_i\rangle = 1$ for exactly one P_i and $\langle\phi; P_j\rangle = 0$ for the other P_j. From this we conclude that $\langle\phi; A^n\rangle = a^n$, hence that ϕ belongs to $\mathfrak{S}_A$, so that $\bigcap_i \mathfrak{S}_{P_i} \subseteq \mathfrak{S}_A$. Suppose, conversely, that ϕ belongs to $\mathfrak{S}_A$. It is then easy to see that $\langle\phi; P_i\rangle$ vanishes on all but one of the P_i and takes the value 1 on the remaining P_i; ϕ is then dispersion-free on all P_i, i.e., $\mathfrak{S}_A \subseteq \bigcap_i \mathfrak{S}_{P_i}$. With the preceding result, this shows the second part of the corollary; the first part follows from the previous remark.

We now turn to the concept of compatibility and establish the following preliminary result:

Lemma. *A sequence $\{P_i\}$ of pairwise disjoint propositions forms a compatible set of observables.*

Proof. By distributivity we see that the subalgebra $\overline{\mathfrak{P}(\{P_i\})}$ generated by these propositions is identical to the set of all observables of the form $\lambda I + \Sigma_i \lambda_i P_i$, which is clearly associative. Hence the $\{P_i\}$ form a "simultaneously observable" set and by Structure Axiom 8 this set is compatible. ∎

We now prove the following result:

Theorem 12. *A necessary and sufficient condition for two propositions* P *and* Q *to be compatible is that there exist three pairwise disjoint propositions* R, P_1, *and* Q_1 *such that* $P = R + P_1$ *and* $Q = R + Q_1$.

Proof. The *sufficiency* follows from the preceding lemma; suppose, indeed, that such a decomposition exists; since $P \circ Q = R$, $\overline{\mathfrak{P}(P, Q)}$ is identical to $\overline{\mathfrak{P}(R, P_1, Q_1)}$, which, as we have just learned, is associative; hence P and Q are simultaneously observable and by Structure Axiom 8, compatible. The *necessity* follows by construction; suppose, indeed, that P and Q are compatible; we know (Theorem 8) that $\overline{\mathfrak{P}(P, Q)}$ is associative; we then form $R = P \circ Q$, $P_1 = P - R$, and $Q_1 = Q - R$, which all belong to $\overline{\mathfrak{P}(P, Q)}$; using the associativity of $\overline{\mathfrak{P}(P, Q)}$, we see that these are actually disjoint propositions and therefore satisfy the condition of the theorem. ∎

REMARK. The condition of the theorem is so natural that it is sometimes taken as the definition of compatibility between two propositions.† This theorem, furthermore, has the following consequence:

Corollary. $\mathrm{P} \leqslant \mathrm{Q}$ *implies that* P *and* Q *are compatible.*

This assertion, in turn, is physically expected. Mathematically, however, it is intimately linked to the fact that the lattice of all propositions is *weakly modular*; recognition of this fact has been the key to the development‡ of the proposition-calculus approach to physical theories and has allowed it to escape the pitfalls encountered by Birkhoff and von Neumann [1936] when they assumed the much stronger condition of modularity.

An important tool in the study of Jordan algebras is the Pierce decomposition.§ In the case of interest to us the existence and properties of this decomposition have been proved by von Neumann [1936] and already used by Jordan, Wigner, and von Neumann [1934]. We now want to consider this decomposition in connection with the notion of compatibility; we further notice that von Neumann's topology is not essential here and that its special features can be completely dispensed with in what follows.

For an arbitrary proposition P and an arbitrary real number λ we define $N^\lambda(P) = \{A \in \mathfrak{A} \mid (P - \lambda I) \circ A = 0\}$, which is clearly a closed subspace of $\mathfrak{A}$. We now polarize the power-associative identity $\{A, A, A^2\} = 0$ by replacing A by $P + \mu A$ with μ arbitrary real. From the linear term (in μ) in the resulting expression we get

$$2P \circ (P \circ (P \circ A)) - 3P \circ (P \circ A) + P \circ A = 0. \tag{*}$$

With $A \in N^\lambda(P)$ we get $(2\lambda^3 - 3\lambda^2 + 1)A = 0$, from which we conclude that $N^\lambda(P) = 0$ unless $\lambda = 0, \frac{1}{2}, 1$. For each A in $\mathfrak{A}$ we then form the three elements

$$A^0 = 2P \circ (P \circ A) - 3P \circ A + A,$$
$$A^{1/2} = -4P \circ (P \circ A) + 4P \circ A,$$
$$A^1 = 2P \circ (P \circ A) - P \circ A,$$

adding to A. We then notice that $(P - \lambda I) \circ A^\lambda$ reproduces for $\lambda = 0, \frac{1}{2}, 1$ exactly the relation (*), hence is 0, i.e., $A^\lambda \in N^\lambda(P)$. We then have the following:

Lemma (*Pierce decomposition*). *An arbitrary proposition* P *in* $\mathfrak{A}$ *generates a decomposition of* $\mathfrak{A}$ *into a direct sum* $\oplus\, \mathrm{N}^\lambda(\mathrm{P})$ *of the three closed subspaces* $\mathrm{N}^\lambda(\mathrm{P}) = \{\mathrm{A} \in \mathfrak{A} \mid (\mathrm{P} - \lambda \mathrm{I}) \circ \mathrm{A} = 0\}$, *with* $\lambda = 0, \frac{1}{2}, 1$.

† See, for instance, Varadarajan [1962].
‡ See Piron [1964].
§ See, for instance, Paige [1963] or Braun and Koecher [1966].

By an adequate manipulation of the properties of the symmetrized product, we get (von Neumann [1936])

$$N^0(P) \circ N^0(P) \subseteq N^0(P),$$

$$N^0(P) \circ N^1(P) = 0,$$

$$Q: N^{1/2}(P) \to N^{1/2}(P) \quad \text{for any proposition } Q,$$

$$N^1(P) \circ N^1(P) \subseteq N^1(P).$$

Coming back to the notion of compatible propositions, we now prove directly the following result:

Theorem 13. *A necessary and sufficient condition for two propositions* P *and* Q *to be compatible is that* {P, A, Q} = 0 *for all* A *in* $\mathfrak{A}$.

Proof. The necessity is already implied in Structure Axiom 6. It might, however, be interesting to note that it actually follows—granted the distributivity of the symmetrized product in $\mathfrak{A}$—as a consequence of the Pierce decomposition. The condition of the theorem can be written as $P \circ (Q \circ A) = Q \circ (P \circ A)$. From the lemma it is sufficient to prove it for $A \in N^\lambda(P)$. We first prove the result in the particular case in which $P \circ Q = 0$, i.e., $Q \in N^0(P)$; from the first three properties of the Pierce decomposition mentioned after the lemma, we get, successively, that both sides of the relation $P \circ (Q \circ A) = Q \circ (P \circ A)$ are respectively 0, 0, $\frac{1}{2}Q \circ A$ for A contained respectively in $N^0(P)$, $N^1(P)$, $N^{1/2}(P)$. This proves that the condition of the theorem is satisfied when $P \circ Q = 0$. The general case is reduced to this particular case by the use of Theorem 12. The *sufficiency* is obtained as follows: suppose that $\{P, A, Q\} = 0$ for all A in $\mathfrak{A}$; using the lemma, we can write $P = P^0 + P^{1/2} + P^1$ with

$$P^{1/2} = -4Q \circ (Q \circ P) + 4Q \circ P = 4\{Q, P, P\} - 4(Q \circ Q) \circ P + 4Q \circ P;$$

using the fact that Q is an idempotent and $\{Q, P, P\} = 0$, we get $P^{1/2} = 0$. Hence $P = P^0 + P^1$. From the fact that P is an idempotent we have, using $N^0(P) \circ N^1(P) = 0$, $(P^0)^2 - P^0 = (P^1)^2 - P^1$; since

$$N^0(P)^2 \subseteq N^0(P), N^1(P)^2 \subseteq N^1(P),$$

and $N^0(P) \circ N^1(P) = 0$, we get that P^0 and P^1 are idempotent; moreover, $P^1 = 2Q \circ (Q \circ P) - Q \circ P = 2Q \circ P - Q \circ P = Q \circ P$; we can then write $R = Q \circ P$, $P_1 = P - Q \circ P$, and $Q_1 = Q - Q \circ P$ which provide a decomposition of P and Q that satisfies the conditions of Theorem 12; we therefore conclude that P and Q are compatible. This completes the proof of the theorem. ∎

REMARK. For any Segal algebra in which the symmetrized product $A \circ B$ is homogeneous and distributive in each of its factors we proved, along with Theorems 12 and 13, that the following three conditions on a pair (P, Q) of propositions are equivalent:

(i) $\overline{\mathfrak{P}(P, Q)}$ is associative.
(ii) There exists a triple of pairwise disjoint propositions R, P_1, and Q_1 such that $P = R + P_1$ and $Q = R + Q_1$.
(iii) $\{P, A, Q\} = 0$ for all A in $\mathfrak{A}$.

This is a further plausibility argument in favor of Structure Axioms 6 and 8.

f. Structure Axiom 9 and GNS Construction

The aim of this subsection is to show under which circumstances the Koopman formalism for classical (statistical) mechanics discussed in Subsection d can be extended to cover more general Segal algebras. In this subsection we define the GNS construction, one of the most important tools of the algebraic approach to physical theories.

With this intent we first recall some standard mathematical definitions. An *involution* on a complex (or real) algebra $\mathfrak{R}$ is defined as an involutive antiautomorphism, denoted by *, of $\mathfrak{R}$, i.e., as a mapping of $\mathfrak{R}$ onto itself satisfying the following properties: $(R^*)^* = R$, $(R + S)^* = R^* + S^*$, $(\lambda R)^* = \lambda^* R^*$ and $(RS)^* = S^* R^*$ for all R, S in $\mathfrak{R}$ and all complex λ with λ^* denoting the complex conjugate of λ; when $\mathfrak{R}$ is real, we evidently have $\lambda^* = \lambda$.

A complex (or real) *involutive algebra* is a complex (or real) algebra equipped with an involution; an element A of an involutive algebra is said to be *self-adjoint* when $A^* = A$. An involutive *Banach algebra* is a complex (or real) associative, involutive, normed algebra, complete with respect to the topology induced by its norm *and* such that $\|R^*\| = \|R\|$ for all R in $\mathfrak{R}$; a *C^*-algebra* (or *R^*-algebra*) is an involutive complex (or real) Banach algebra such that $\|R\|^2 = \|R^*R\|$ for all R. We remark that *except* for the associativity our algebra $\mathfrak{A}$ of all observables satisfies all the properties of an R^*-algebra, the involution being taken as the identity mapping, which is obviously an antiautomorphism, since $\mathfrak{A}$ is abelian ($A \circ B = B \circ A$). We further notice that the set $\mathfrak{A}$ of all self-adjoint elements of an (associative!) C^*-algebra (or R^*-algebra) $\mathfrak{R}$ is a Segal algebra and that in this case the symmetrized product $A \circ B = \frac{1}{2}\{(A + B)^2 - A^2 - B^2\}$ takes the simple form $A \circ B = \frac{1}{2}(AB + BA)$, where RS denotes the product in $\mathfrak{R}$. A Segal algebra is said to be *special* (complex or real) or *exceptional*, depending on whether or not it is isomorphic to the set of all self-adjoint elements of a C^*- or R^*-algebra. As for Jordan algebras, no abstract characterization of a special Segal algebra is known,† nor is it known whether in general the algebra $\mathfrak{A}$ of all observables on a physical system is special in the sense just mentioned.

Since, on the one hand, the mathematical theory of special Segal algebras is fairly well developed and, on the other, a concrete physical system involving a description in terms of an exceptional Segal algebra has not yet been encountered, we shall concentrate our attention in the sequel exclusively on special Segal algebras. Realizing, however, that this is a mathematically severe restriction, we want to impose it in the weakest possible form.

Structure Axiom 9. *$\mathfrak{A}$ can be identified with the set of all self-adjoint elements of a real or complex, associative, and involutive algebra $\mathfrak{R}$ satisfying the following two conditions:*

(i) *For each* R *in* $\mathfrak{R}$ *there exists an element* A *in* $\mathfrak{A}$ *such that* $\mathrm{R}^*\mathrm{R} = \mathrm{A}^2$;
(ii) $\mathrm{R}^*\mathrm{R} = 0$ *implies* $\mathrm{R} = 0$.

We notice that since we know that A^2 is a positive observable and that every such observable has a unique positive square root we can reformulate condition (i) as follows: for each element R in $\mathfrak{R}$ there exists a *unique positive* element A in $\mathfrak{A}$ such that $R^*R = A^2$.

The assumption that $\mathfrak{A}$ can be identified with the set of *all* self-adjoint elements of $\mathfrak{R}$ may seem severe; we might note, from a mathematical point of view, that this assumption can be replaced by the requirement that $\mathfrak{A}$ be reversible in $\mathfrak{R}$, i.e., that the following generalization of the symmetrized product is possible in $\mathfrak{A}$: for any finite sequence $\{A_i\}$ of elements of $\mathfrak{A}$ the element

$$\prod_{i=1}^{n} A_i + \prod_{i=n}^{1} A_i$$

also belongs to $\mathfrak{A}$; in this case it is easy to show that $\mathfrak{A}$ is identical to the set of all self-adjoint elements in the real subalgebra $\mathfrak{R}(\mathfrak{A})$ of $\mathfrak{R}$ generated by $\mathfrak{A}$; we can then‡ replace $\mathfrak{R}$ with $\mathfrak{R}(\mathfrak{A})$ in the statement of Structure Axiom 9.

We emphasize that we did not impose any explicit topological conditions on $\mathfrak{R}$ aside from those already present in the preceding structure-axioms on $\mathfrak{A}$; nevertheless, the requirements of the above axiom impose actually very severe restrictions on the algebra $\mathfrak{R}$, which we now want to explore. We first notice that if $\mathfrak{R}$ is real then every element R in $\mathfrak{R}$ can be decomposed in a unique way as $R = R_+ + R_-$ with $R_\pm = \pm(R_\pm)^*$; we define, indeed, $R_\pm = \frac{1}{2}(R \pm R^*)$; the uniqueness follows *ad absurdo*. Similarly, if $\mathfrak{R}$ is complex, we

† It is even doubtful whether any abstract characterization of this "special" property is possible at all; for Jordan algebras see Cohn [1954].
‡ This remark can actually be sharpened and, with the appropriate conditions, generalized to the complex case; see Størmer [1965].

define $R_+ = \frac{1}{2}(R + R^*)$ and $R_- = (R - R^*)/2i$, so that $R = R_+ + iR_-$ with R_+ *and* R_- self-adjoint.

We next extend every state ϕ in $\mathfrak{S}$, from $\mathfrak{A}$ to $\mathfrak{R}$, by writing $\langle\phi; R\rangle = \langle\phi; R_+\rangle$ when $\mathfrak{R}$ is real and $\langle\phi; R\rangle = \langle\phi; R_+\rangle + i\langle\phi; R_-\rangle$ when $\mathfrak{R}$ is complex; in both cases ϕ, extended to $\mathfrak{R}$, is real (i.e., $\langle\phi; R^*\rangle = \langle\phi; R\rangle^*$) and linear; furthermore, from $\langle\phi; R^*R\rangle = \langle\phi; A^2\rangle \geqslant 0$ we conclude by an argument similar to that used in the lemma on page 53 that the sesquilinear form $\langle\phi; R^*S\rangle$ satisfies the generalized Schwartz inequality $|\langle\phi; R^*S\rangle|^2 \leqslant \langle\phi; R^*R\rangle\langle\phi; S^*S\rangle$. In analogy with the procedure followed in the study of $\mathfrak{A}$ we *define* the norm $\|R\|$ of an arbitrary element R in $\mathfrak{R}$ as the positive square root of

$$\|R\|^2 = \sup_{\phi\in\mathfrak{S}} \langle\phi; R^*R\rangle;$$

to verify that $\|\cdots\|$ is actually a norm on $\mathfrak{R}$ we proceed as follows:

$$\|R\| = 0 \quad \text{implies} \quad \langle\phi; R^*R\rangle = \langle\phi; A^2\rangle = 0 \quad \text{for all } \phi \text{ in } \mathfrak{S};$$

hence $A = 0$, so that by (ii) $R = 0$. For all ϕ in $\mathfrak{S}$ we have

$$\langle\phi; (R + S)^*(R + S)\rangle = \langle\phi; R^*R\rangle + \langle\phi; S^*S\rangle + \langle\phi; R^*S + S^*R\rangle,$$

from which we conclude, with the help of the Schwartz inequality, that $\|R + S\| \leqslant \|R\| + \|S\|$. Finally, from the linearity of ϕ on $\mathfrak{R}$ we see that $\|\lambda R\| = |\lambda| \cdot \|R\|$. This norm, restricted to $\mathfrak{A}$, coincides with the original norm on $\mathfrak{A}$. From the fact that R^*R is a positive element of $\mathfrak{A}$ we have

$$\|R^*R\| = \sup_{\phi\in\mathfrak{S}} \langle\phi; (R^*R)^2\rangle = \sup_{\phi\in\mathfrak{S}_{R^*R}} \langle\phi; (R^*R)^2\rangle$$

$$= \sup_{\phi\in\mathfrak{S}_{R^*R}} \langle\phi; R^*R\rangle^2 = \|R\|^2.$$

To show that $\|R^*\| = \|R\|$, we first notice that since R^*R is positive there exists an element B in $\mathfrak{A}$ such that

$$\|R\|^2I - R^*R = B^2;$$

for any S in $\mathfrak{R}$ we then have $S^*B^2S = (BS)^*(BS) \geqslant 0$. Hence for every ϕ in $\mathfrak{S}$

$$\langle\phi; S^*(\|R\|^2I - R^*R)S\rangle \geqslant 0;$$

i.e.,

$$\langle\phi; S^*R^*RS\rangle \leqslant \|R\|^2\langle\phi; S^*S\rangle,$$

which is then valid for all ϕ in $\mathfrak{S}$ and all R, S in $\mathfrak{R}$. If $S = R^*$, this inequality reduces to

$$\langle\phi; (RR^*)^2\rangle \leqslant \|R\|^2\langle\phi; RR^*\rangle$$

which reduces further, when ϕ is a dispersion-free state on RR^*, to $\langle\phi; RR^*\rangle \leqslant \|R\|^2$. On taking the sup over $\mathfrak{S}_{RR^*}$, which amounts to taking the sup over $\mathfrak{S}$ itself, we get $\|R^*\| \leqslant \|R\|$; interchanging the roles of R and R^*, we conclude indeed that $\|R^*\| = \|R\|$. Finally, if $\{R_n\}$ is a Cauchy sequence in $\mathfrak{R}$ with respect to the topology induced by the norm, so is $\{R_n^*\}$, and $\{(R_n + R_n^*)/2\}$ and $\{(R_n - R_n^*)/2i\}$ are Cauchy sequences in $\mathfrak{A}$; hence they admit limit points in $\mathfrak{A}$, so that $\{R_n\}$ itself converges to an element *in* $\mathfrak{R}$; i.e., $\mathfrak{R}$ is complete with respect to its norm-topology. We can then collect these results in the following lemma.

Lemma. *The algebra* $\mathfrak{R}$ *associated with* $\mathfrak{A}$ *by Structure Axiom* 9 *can be equipped without further restrictions with the structure of a* C*-*algebra or an* R*-*algebra, depending on whether* $\mathfrak{R}$ *is complex or real.*

The interest of Structure Axiom 9 and the immediately preceding lemma resides in the following representation theorem, known as the *GNS construction*†:

Theorem 14. *Let* $\mathfrak{A}$ *be the algebra of all observables on a physical system, and* $\mathfrak{R}$, *the algebra associated with* $\mathfrak{A}$ *by Structure Axiom* 9; *then each state* ϕ *in* $\mathfrak{S}$ *generates canonically a representation* π_ϕ *of* $\mathfrak{A}$ *by self-adjoint operators acting on a complex or real Hilbert space, depending on whether* $\mathfrak{R}$ *is a complex or real.*

Proof. From the preceding lemma we learned that $\mathfrak{R}$ can be equipped with the structure of either a C^*-algebra or an R^*-algebra; it is the norm on $\mathfrak{R}$ obtained there that we consider in the following discussion. Let ϕ be any arbitrary state on $\mathfrak{A}$; we denote by the same symbol its extension to $\mathfrak{R}$ as described above; let $\mathfrak{K}_\phi \equiv \{K \in \mathfrak{R} \mid \langle\phi; R^*K\rangle = 0 \ \forall\, R \in \mathfrak{R}\}$; $\mathfrak{K}_\phi$ is clearly a subspace of $\mathfrak{R}$. Furthermore, for any R and S in $\mathfrak{R}$ and any K in $\mathfrak{K}_\phi$ we have, using the associativity in $\mathfrak{R}$, $\langle\phi; R^*(SK)\rangle = \langle\phi; (S^*R)^*K\rangle = 0$; hence $\mathfrak{K}_\phi$ is a *left ideal* in $\mathfrak{R}$. We say that two elements R and S of $\mathfrak{R}$ are ϕ-equivalent whenever $(R - S)$ belongs to $\mathfrak{K}_\phi$; since $\mathfrak{K}_\phi$ is a linear subspace of $\mathfrak{R}$, this is clearly an equivalence relation. For every R in $\mathfrak{R}$ we denote by $\Phi(R)$ its equivalence class and consider the set $\mathfrak{R}_\phi$ of all equivalence classes in $\mathfrak{R}$. We then equip $\mathfrak{R}_\phi$ with a vector space structure by defining $\lambda\Phi(R) + \mu\Phi(S) = \Phi(\lambda R + \mu S)$. This definition is unambiguous, since $\mathfrak{K}_\phi$ is a linear subspace of $\mathfrak{R}$; suppose, indeed, that R' belongs to $\Phi(R)$ and S' belongs to $\Phi(S)$. There then exist K and L in $\mathfrak{K}_\phi$ such that $R - R' = K$ and $S - S' = L$ so that $\lambda R + \mu S - (\lambda R' + \mu S') = \lambda K + \mu L \in \mathfrak{K}_\phi$. We notice that the neutral element in the vector space $\mathfrak{R}_\phi$ is the equivalence class of $0 \in \mathfrak{R}$; i.e., $\Phi(0) = 0$. Hence, in particular, $\Phi(R) = 0$ whenever R in $\mathfrak{R}$ satisfies $\langle\phi; R^*R\rangle = 0$, since in this case we have, by Schwartz inequality,

† GNS stands for Gelfand and Naimark [1943] and Segal [1947].

$|\langle\phi; S^*R\rangle|^2 \leqslant \langle\phi; S^*S\rangle\langle\phi; R^*R\rangle = 0$ for all S in $\mathfrak{R}$; i.e., $R \in \mathfrak{K}_\phi$. Furthermore, we can equip $\mathfrak{R}_\phi$ with the structure of a pre-Hilbert space by defining $(\Phi(R), \Phi(S)) = \langle\phi; R^*S\rangle$. To check that this definition is unambiguous suppose again that R' and S' belong, respectively, to $\Phi(R)$ and $\Phi(S)$. Then $\langle\phi; R^*S - R'^*S'\rangle = \langle\phi; R'^*L + S^*K\rangle$, which is zero, since K and L belong to $\mathfrak{K}_\phi$; $(\Phi(R), \Phi(S))$ is a sesquilinear form on $\mathfrak{R}_\phi$ and generates a norm (not only a seminorm!) on $\mathfrak{R}_\phi$ defined by $|\Phi(R)|^2 = \langle\phi; R^*R\rangle$. Hence $\mathfrak{R}_\phi$ is a real (or complex) pre-Hilbert space. Let $\mathcal{H}_\phi$ be its completion with respect to this norm. We then define a representation π_ϕ of $\mathfrak{R}$ into the set of all bounded operators acting on $\mathcal{H}_\phi$ in the following manner. For each R in $\mathfrak{R}$ we define $\pi_\phi(R): \mathfrak{R}_\phi \to \mathfrak{R}_\phi$ by $\pi_\phi(R)\,\Phi(S) = \Phi(RS)$. To make sure that this definition is unambiguous we have to verify that $\Phi(RS')$ is independent of S' in $\Phi(S)$; to this effect we form

$$\begin{aligned}|\Phi(RS) - \Phi(RS')|^2 &= |\Phi(R(S - S'))|^2\\ &= \langle\phi; (R(S - S'))^*(R(S - S'))\rangle\\ &= \langle\phi; (S - S')^*R^*R(S - S')\rangle \leqslant \|R\|^2 \cdot |\Phi(S - S')|^2,\end{aligned}$$

which is zero, since $S - S'$ belongs to $\mathfrak{K}_\phi$. The next thing to do is to verify that $\pi_\phi(R)$ is a representation of $\mathfrak{R}$; we have $\pi_\phi(\lambda R_1 + \mu R_2)\,\Phi(S) = \Phi((\lambda R_1 + \mu R_2)S) = \lambda\Phi(R_1S) + \mu\Phi(R_2S) = (\lambda\pi_\phi(R_1) + \mu\pi_\phi(R_2))\Phi(S)$. Moreover, $\pi_\phi(R_1R_2)\,\Phi(S) = \Phi((R_1R_2)S) = \Phi(R_1(R_2S)) = \pi_\phi(R_1)\pi_\phi(R_2)\Phi(S)$. Finally, $(\pi_\phi(R)^*\,\Phi(S_1), \Phi(S_2)) = (\Phi(S_1), \pi_\phi(R)\Phi(S_2)) = \langle\phi; S_1^*RS_2\rangle = \langle\phi; (R^*S_1)^*S_2\rangle = (\Phi(R^*S_1), \Phi(S_2)) = (\pi_\phi(R^*)\Phi(S_1), \Phi(S_2))$. Hence π_ϕ is an homomorphism for the structure of involutive algebra of $\mathfrak{R}$ to the set of operators acting on the pre-Hilbert space $\mathfrak{R}_\phi$. The last step is to show that the operator $\pi_\phi(R)$ can be extended in a unique way to a bounded operator on the Hilbert space $\mathcal{H}_\phi$. We first show that $\pi_\phi(R)$ is bounded on $\mathfrak{R}_\phi$;

$$\begin{aligned}\|\pi_\phi(R)\|^2 &= \sup_{S\in\mathfrak{R}} \frac{|\pi_\phi(R)\,\Phi(S)|^2}{|\Phi(S)|^2}\\ &= \sup_{S\in\mathfrak{R}} \frac{\langle\phi; (RS)^*(RS)\rangle}{\langle\phi; S^*S\rangle}\\ &\leqslant \frac{\|R\|^2\,\langle\phi; S^*S\rangle}{\langle\phi; S^*S\rangle} = \|R\|^2.\end{aligned}$$

Hence $\pi_\phi(R)$ is bounded on $\mathfrak{R}_\phi$ which is dense in $\mathcal{H}_\phi$; we can therefore extend $\pi_\phi(R)$ to a bounded operator on $\mathcal{H}_\phi$ in a unique manner. Everything we said about the mapping $\pi_\phi(R): \mathfrak{R}_\phi \to \mathfrak{R}_\phi$ remains true for its extension to $\mathcal{H}_\phi$, and we have obtained a representation of the R^*-algebra (or C^*-algebra) $\mathfrak{R}$ as bounded operators on a real (or complex) Hilbert space, in particular, since $\pi_\phi(R)^* = \pi_\phi(R^*)$, $\pi_\phi(A)$ is self-adjoint for every A in $\mathfrak{A}$. This achieves the proof of the theorem. ∎

For completeness we might mention that Segal proved the following corollary:

Corollary. *Under the assumption of the theorem $\mathfrak{A}$ is isomorphic, algebraically and metrically, with the algebra of all self-adjoint operators in a uniformly closed involutive algebra of bounded operators acting on a real (or complex) Hilbert space.*

We merely sketch the proof. Since $\mathfrak{S}$ is full on $\mathfrak{A}$, there exists for each R in $\mathfrak{R}$ at least one state in $\mathfrak{S}$ (which we denote by ϕ_R) such that $\langle \phi_R; R^*R\rangle$ is strictly positive. Let π_R be the representation obtained from ϕ_R by the Segal construction described in the proof of the theorem. The Hilbert direct sum π of the representations so obtained when R runs over $\mathfrak{R}$ is obviously injective (i.e., $\pi(R) = 0$ occurs only for $R = 0$). The corollary is then proved by noticing that $\pi(R)$, being injective, implies that it is isometric, i.e., $\|\pi(R)\| = \|R\|$. ∎

g. Structure Axiom 10 (*uncertainty principle*)

We first want to illustrate Theorem 14 with an elementary example and then link the uncertainty principle to the C^*-algebraic realization of our preceding axioms.

Consider the *real* associative algebra $\mathfrak{R}$ generated by the four symbols I, x, y, z with the following multiplication table,

	I	x	y	z
I	I	x	y	z
x	x	I	z	y
y	y	$-z$	$-I$	x
z	z	$-y$	$-x$	I

and equipped with the following involution: $I^* = I$, $x^* = x$, $y^* = -y$, and $z^* = z$. The algebra $\mathfrak{A}$ of all its symmetric elements is generated by I, x, z and is equipped with the following symmetrized product inherited from the ordinary product in $\mathfrak{R}$:

	I	x	z
I	I	x	z
x	x	I	0
z	z	0	I

We then consider the state ϕ on $\mathfrak{A}$ defined by

$$\langle \phi; \alpha I + \beta x + \delta z \rangle = \alpha + \delta,$$

whose natural extension to $\mathfrak{R}$ is

$$\langle \phi; \alpha I + \beta x + \gamma y + \delta z \rangle = \alpha + \delta.$$

We find that $\mathfrak{K}_\phi = \{\lambda(I - z) + \mu(x - y)\}$ so that

$$\mathfrak{R}_\phi = \nu\Phi(I) + \rho\Phi(x) = \mathcal{H}_\phi.$$

We notice that the vectors $\Phi(I)$ and $\Phi(x)$ are orthogonal, and hence provide a basis for the Hilbert space $\mathcal{H}_\phi$. On this basis the GNS representation can be written

$$\pi_\phi(I) = \begin{pmatrix} 1 & 0 \\ 0 & 1 \end{pmatrix}, \qquad \pi_\phi(x) = \begin{pmatrix} 0 & 1 \\ 1 & 0 \end{pmatrix},$$

$$\pi_\phi(y) = \begin{pmatrix} 0 & -1 \\ 1 & 0 \end{pmatrix}, \qquad \pi_\phi(z) = \begin{pmatrix} 1 & 0 \\ 0 & -1 \end{pmatrix};$$

i.e.,

$$\pi_\phi(\alpha I + \beta x + \gamma y + \delta z) = \begin{pmatrix} \alpha + \delta & \beta - \gamma \\ \beta + \gamma & \alpha - \delta \end{pmatrix}.$$

Hence $\mathfrak{R}$ is represented as the algebra of all real two-by-two matrices and $\mathfrak{A}$ is represented as the algebra of all symmetric real two-by-two matrices. Incidentally, in the notation of Jordan, von Neumann, and Wigner [1934] the GNS construction provides the isomorphism between the Jordan algebra $\mathfrak{A} = \mathfrak{S}_3$ and the Jordan algebra $\pi_\phi(\mathfrak{A}) = \mathfrak{M}_2^1$. Physically we recognize the representation of the Pauli matrices I, σ_x, and σ_z, defined abstractly by their Jordan products and realized as 2×2 matrices.

The preceding example also has the advantage of making possible a choice between the two alternatives left by the lemma preceding Theorem 14 and of exhibiting one main difference between embedding $\mathfrak{A}$ in a *real* or in a *complex* associative algebra $\mathfrak{R}$. In the case of a general R^*-algebra, we have, indeed, no meaningful uncertainty principle, the latter being understood as follows:

Structure Axiom 10. *To each pair of observables* A *and* B *in* $\mathfrak{A}$ *corresponds an observable* C *in* $\mathfrak{A}$ *which provides the actual lower bound to the simultaneous observability of* A *and* B *in the sense that*

$$\langle \phi; (\mathrm{A} - \langle \phi; \mathrm{A} \rangle)^2 \rangle \langle \phi; (\mathrm{B} - \langle \phi; \mathrm{B} \rangle)^2 \rangle \geqslant \langle \phi; \mathrm{C} \rangle^2, \qquad \forall \phi \in \mathfrak{S}.$$

We notice that this axiom is automatically satisfied if we assume that $\mathfrak{A}$ is the set of all self-adjoint elements of a *C^*-algebra*. In this case the uncertainty principle takes its usual form:

$$\langle \phi; (A - \langle \phi; A \rangle)^2 \rangle \langle \phi; (B - \langle \phi; B \rangle)^2 \rangle \geqslant \langle \phi; i[A, B] \rangle^2.$$

The mathematical theory of C^*-algebras and their representations has been developed to a great degree in recent years; its mathematical richness on the one hand and, on the other, its almost immediate relation to some of the less restrictive postulates on the set of observables on a physical system makes the study of its applications to physics extremely promising; the remaining portion of this book is mainly an elaboration of this statement.

In closing this section we should like to emphasize that the Gelfand-Naimark-Segal construction brings to the physical theory an entirely new concept that the traditional formalism for quantum theory—say à la Dirac—was lacking. This is the fact that the Hilbert space on which the observables act as operators is not a perennial feature of the theory, nor of the model to be constructed, but is dependent on the state or preparation of the system considered, as in the Koopman formalism for classical theory. Hence, to join these considerations with those developed in the first section of this chapter, the Hilbert space appropriate to the description of the fields in interation has no general reason to be the same as that of the description of the free fields; the same remark applies to the Hilbert space corresponding to different equilibrium situations in statistical mechanics.

CHAPTER 2

Global Theories

The net result of the axiomatization carried out in Section I.2 is to provide, via the GNS construction, the flexibility that is lacking in the Fock-space formalism criticized in Section I.1. Indeed, the Hilbert space in which we represent the observables (and, by extension, the fields), is not fixed; it depends actually on the problem to be considered and specifically on the preparation of the system to be studied, i.e., in the last analysis on the states to be considered. In this formalism we can envisage, for instance, a situation in which the space constructed on the vacuum of the asymptotic fields is not the same as the space constructed on the vacuum of the interpolating fields. Similarly, the space on which we represent a physical system in thermodynamical equilibrium may depend on the temperature.

Consequently, if we want to take full advantage of the freedom provided by the algebraic approach, we must gain some familiarity with the theory of representations of C^*-algebras. This chapter is devoted to that end.

The basic facts about representations are expounded in Section 1. The presentation is limited to those aspects that are most relevant for physical applications; many interesting mathematical properties had, however, to be omitted from this review which the reader is strongly advised to complete, as the need occurs, by consulting Dixmier [1957, 1964], for instance.

In Section 2 the concept of symmetry, which plays such an important role in modern physics, is analyzed in the context of the C^*-algebraic approach; many of its applications are discussed, including continuous and discrete spacetime symmetries, the dynamical definition of equilibrium states and the process of spontaneous symmetry breaking.

SECTION 1 BASIC FACTS ABOUT REPRESENTATIONS

OUTLINE. The fundamental concepts of representation theory are defined and the corresponding properties of the GNS representations are given in detail in view of the central role played by this construction in physical applications; these concepts are then illustrated by an analysis of the free Bose gas considered in the thermodynamical limit. The concept of physical equivalence, as distinguished from that of spatial equivalence, played some role in stimulating the formulation of the C^*-algebraic approach; we discuss it in Subsection d. Finally we present a collection of results concerning von Neumann algebras and Σ^*-algebras; these properties are necessary to make available to the C^*-algebraic approach several of the powerful tools of functional analysis such as the spectral theorem and the theory of types.

a. Definition of a Representation

A mapping π of a C^*-algebra $\mathfrak{R}$ into the set $\mathfrak{B}(\mathcal{H})$ of all bounded operators on a Hilbert space $\mathcal{H}$ is said to be a *representation* of $\mathfrak{R}$ if

(i) $$\pi(\lambda R + \mu S) = \lambda\pi(R) + \mu\pi(S),$$

(ii) $$\pi(RS) = \pi(R)\,\pi(S),$$

(iii) $$\pi(R^*) = \pi(R)^*$$

for all R and S in $\mathfrak{R}$ and all complex numbers λ and μ; i.e., $\pi: \mathfrak{R} \to \mathfrak{B}(\mathcal{H})$ is a morphism.

The reader will notice that in this definition we did no require that the mapping π preserve the whole structure of $\mathfrak{R}$ as a C^*-algebra; the reason for this omission is that this requirement is automatically satisfied when the three conditions imposed in the definition are fulfilled; in particular, the continuity of the mapping π follows from the fact that a morphism π from an involutive Banach algebra (here $\mathfrak{R}$) into a C^*-algebra [here $\mathfrak{B}(\mathcal{H})$] always† satisfies the condition $\|\pi(R)\| \leqslant \|R\|$.

The *kernel* (Kerπ) of a representation $\pi: \mathfrak{R} \to \mathfrak{B}(\mathcal{H})$ of a C^*-algebra $\mathfrak{R}$ is defined as $\text{Ker}\pi \equiv \{R \in \mathfrak{R} \mid \pi(R) = 0\}$; a representation is said to be *faithful* if $\pi(R) = 0$ implies $R = 0$ (i.e., $\text{Ker}\pi = 0$). In this case π is a bijective mapping from $\mathfrak{R}$ to $\pi(\mathfrak{R})$; hence we can use the above argument for π^{-1} and get $\|\pi^{-1}\pi(R)\| \leqslant \|\pi(R)\|$, which, combined with our previous remark, gives $\|\pi(R)\| = \|R\|$ for any faithful representation π.

In general, the kernel of a representation π of a C^*-algebra $\mathfrak{R}$ is a closed, two-sided ideal of $\mathfrak{R}$, i.e.,

(i) $\|R_n - R\| \to 0$ with $R_n \in \text{Ker}\pi$ implies $R \in \text{Ker}\pi$,

(ii) $RK \in \text{Ker}\pi$ } for all $K \in \text{Ker}\pi$ and $R \in \mathfrak{R}$.

(iii) $KR \in \text{Ker}\pi$ }

† For a proof see, for instance, Dixmier [1964], Proposition 1.3.7.

In this book we identify† $\pi(\mathfrak{R})$ with the quotient $\mathfrak{R}/\text{Ker}\pi$. We notice further that if $\mathfrak{R}$ is simple (i.e., admits no closed two-sided proper ideals), then every nonzero [i.e., $\pi(R) \neq 0$ for some R in $\mathfrak{R}$] representation π of $\mathfrak{R}$ is faithful. The notion of kernel of a representation plays a central role in the concept of physical equivalence discussed in Subsection d.

A representation $\pi: \mathfrak{R} \to \mathfrak{B}(\mathcal{H})$ of a C^*-algebra is said to be *cyclic* if there exists a vector Ψ in $\mathcal{H}$ such that the linear manifold $\pi(\mathfrak{R})\Psi \equiv \{\pi(R)\Psi \mid R \in \mathfrak{R}\}$ is dense in $\mathcal{H}$ (with respect to the strong topology on $\mathcal{H}$, i.e., that induced by the norm); this vector Ψ is said to be *cyclic* for the representation π. Cyclic representations play a privileged role in the mathematical theory of representations; in particular, we might notice‡ that any representation is the direct sum of cyclic representations. There are also serious reasons to consider cyclic representations in connection with the algebraic approach to physical problems; in particular, we notice first that the C^*-algebra generated by the CAR or the CCR admits a cyclic representation of primary importance: the ordinary Fock-space representation! Every physicist knows indeed—and it must in any case be clear from our digression on Fock space in Chapter 1, Section 1— that every vector Ψ in Fock space can be approximated as closely as we want, in the strong topology, by vectors obtained from an arbitrary and fixed vector Ψ_0, e.g., the vacuum, by application of appropriate polynomials in the creation and annihilation operators. The occurrence of cyclic representations is in fact a much more general feature in the formalism developed so far; to see this consider again the proof of Theorem I.14: To each element R in $\mathfrak{R}$ we attribute a vector $\Phi(R) = \pi_\phi(R)\,\Phi(I)$, the representation space $\mathcal{H}_\phi$ being defined as the strong closure of these vectors; hence $\Phi(I)$ is cyclic for π_ϕ and we then have the following:

Corollary. *The GNS-construction associates to every state ϕ on $\mathfrak{R}$ a cyclic representation π_ϕ of $\mathfrak{R}$.*

We notice further that the only thing needed in the proof was that ϕ be extended to a mapping satisfying the following conditions:

(i) $$\phi: \mathfrak{R} \to \mathbb{C},$$

(ii) $$\langle \phi; \lambda R + \mu S \rangle = \lambda \langle \phi; R \rangle + \mu \langle \phi; S \rangle,$$

(iii) $$\langle \phi; R^* R \rangle \geqslant 0,$$

(iv) $$\langle \phi; I \rangle = 1.$$

† For a proof that this is possible see Dixmier [1964], Corollaire 1.8.3.
‡ Naimark [1964] 17.2 or Dixmier [1964] 2.2.7.

Conditions (i) to (iii) define what is called a *positive linear functional*† on the C^*-algebra $\mathfrak{R}$; a positive linear functional satisfying (iv), moreover, is said by extension to be a *state* on $\mathfrak{R}$.

In case ψ is only a positive linear functional on $\mathfrak{R}$, we define $\phi = \psi/\langle\psi; I\rangle$, which is a state on $\mathfrak{R}$, and proceed with the GNS construction by using ϕ instead of ψ; we can then rephrase our preceding corollary:

Corollary. *The GNS construction associates to each positive linear functional ψ on a C^*-algebra $\mathfrak{R}$ a cyclic representation π_ψ of $\mathfrak{R}$.*

Let $\pi: \mathfrak{R} \to \mathfrak{B}(\mathcal{H})$ be a representation of a C^*-algebra $\mathfrak{R}$; each vector Ψ in $\mathcal{H}$ defines, via the relation

$$\langle\psi; R\rangle = (\Psi, \pi(R)\Psi),$$

a positive linear functional on $\mathfrak{R}$, which is a state whenever Ψ is normalized to 1; every positive linear functional obtained in this way is said to be a *vector state associated* to the representation π.

Suppose now that we have an isomorphism U from $\mathcal{H}$ onto some Hilbert space $\mathcal{H}'$ (identical or not to $\mathcal{H}$); for all R in $\mathfrak{R}$ we form

$$\pi'(R) = U\pi(R)U^{-1},$$

which is clearly a representation of $\mathfrak{R}$ in $\mathcal{H}'$; we say in such a case that π and π' are *spatially* (*or unitarily*) *equivalent* and we denote this circumstance by $\pi \simeq \pi'$. Furthermore we note that Ψ in $\mathcal{H}$ and $U\Psi$ in $\mathcal{H}'$ generate, via $\pi(\mathfrak{R})$ and $\pi'(\mathfrak{R})$, respectively, the same functional on $\mathfrak{R}$; hence, if π is cyclic so is π'. This raises the question whether we get in this way all cyclic representations that generate the same functional; the answer is affirmative:

Theorem 1. *Let $\mathfrak{R}$ be a C^*-algebra, ψ, a positive linear functional on $\mathfrak{R}$ and π_ψ, the GNS representation associated with it. Then any cyclic representation $\pi': \mathfrak{R} \to \mathfrak{B}(\mathcal{H}')$ with cyclic vector Ψ' such that $\langle\psi; \mathrm{R}\rangle = (\Psi', \pi'(\mathrm{R})\Psi')$ for all R in $\mathfrak{R}$ is spatially equivalent to π_ψ.*

Proof. From $(\Psi, \pi_\psi(R)\Psi) = \langle\psi; R\rangle = (\Psi', \pi'(R)\Psi')$ we conclude first that $\mathrm{Ker}\pi_\psi = \mathrm{Ker}\pi'$ and that there exists an isometric linear mapping U from $\pi_\psi(\mathfrak{R})\Psi$ onto $\pi'(\mathfrak{R})\Psi'$ defined by $U\pi(R)\Psi \equiv \pi'(R)\Psi'$ for all R in $\mathfrak{R}$; since U is bounded (namely by 1) and $\pi_\psi(\mathfrak{R})\Psi$ (or $\pi'(\mathfrak{R})\Psi'$) is dense in $\mathcal{H}_\psi$ (or $\mathcal{H}'$), U can be extended to a unitary mapping from $\mathcal{H}$ to $\mathcal{H}'$; clearly $U\pi_\psi(R)U^{-1} = \pi'(R)$. ∎

† In the same way as we proved the lemma on page 60, we can check (Dixmier [1964], Proposition 2.1.4) that conditions (i) to (iii) imply

$$\|\phi\| \equiv \sup_{\|R\| \leqslant 1} |\langle\phi; R\rangle| = \langle\phi; I\rangle$$

and then ϕ is continuous.

In this sense the GNS representation π_ψ can be said to be canonically associated with ψ: up to unitary equivalence it is *the* cyclic representation with a cyclic vector generating ψ.

The proof of Theorem 1 can be expanded† to give an interesting criterion for the unitary equivalence of cyclic representations; to formulate this criterion in as concise a form as possible we introduce the following notation: we denote by $\mathfrak{B}_\pi$ the set of all vector states associated with a representation $\pi: \mathfrak{R} \to \mathfrak{B}(\mathcal{H})$ of a C^*-algebra $\mathfrak{R}$, i.e., the set of all functionals on $\mathfrak{R}$ defined by $\langle \psi; R \rangle = (\Psi, \pi(R)\Psi)$ with Ψ in $\mathcal{H}$ and $|\Psi| = 1$; we say that two representations are in the relation $\pi \lessapprox \rho$ whenever π is unitary equivalent to a subrepresentation‡ of ρ. We now have the following lemma:

Lemma. *Let ρ be an arbitrary representation of a C*-algebra $\mathfrak{R}$, π be a cyclic representation of $\mathfrak{R}$ with cyclic vector Ψ; we can assume without loss of generality that $|\Psi| = 1$. Let ψ be the state defined by $\langle \psi; \mathrm{R} \rangle = (\Psi, \pi(\mathrm{R})\Psi)$. Then the following three conditions are equivalent:*

(i) $$\psi \in \mathfrak{B}_\rho,$$

(ii) $$\pi \lessapprox \rho,$$

(iii) $$\mathfrak{B}_\pi \subseteq \mathfrak{B}_\rho.$$

Proof. (ii) $\Rightarrow$ (iii) $\Rightarrow$ (i) are trivial implications; $\psi \in \mathfrak{B}_\rho$ implies that there exists some Ψ' in $\mathcal{H}_\rho$ such that $\langle \psi; R \rangle = (\Psi', \rho(R)\Psi')$ for all R in $\mathfrak{R}$; we then conduct the proof of (i) $\Rightarrow$ (ii) along the same path as that followed in the proof of the preceding theorem, except for the replacement of $\mathcal{H}'$ by $\overline{\rho(\mathfrak{R})\Psi'}$ and in the conclusion of the proof π' by the restriction of ρ to the subspace $\overline{\rho(\mathfrak{R})\Psi'}$. ∎

This lemma then leads directly to the following theorem:

Theorem 2. *Let π and ρ be two cyclic representations of a C*-algebra $\mathfrak{R}$, with respective cyclic vectors Φ and Ψ; further let ϕ and ψ be the corresponding states on $\mathfrak{R}$. Then the following three conditions are equivalent:*

(i) $$\phi \in \mathfrak{B}_\rho; \qquad \psi \in \mathfrak{B}_\pi,$$

(ii) $$\pi \simeq \rho \qquad (\text{i.e., } \pi \lessapprox \rho \text{ and } \rho \lessapprox \pi),$$

(iii) $$\mathfrak{B}_\pi = \mathfrak{B}_\rho.$$

b. Irreducible Representations and Pure States

When commenting on the role played by cyclic representations in the early formulation of quantum field theory, we mentioned in passing that for the

† Kadison [1962], or Dixmier [1964], Complément 2.12.19.

‡ Let $\rho: \mathfrak{R} \to \mathfrak{B}(\mathcal{H})$ and suppose that there exists a nonzero subspace $\mathcal{H}_1$ of $\mathcal{H}$, stable under $\rho(\mathfrak{R})$. The restriction ρ_1 of ρ to $\mathcal{H}_1$ is said to be a *subrepresentation* of ρ.

Fock representation, which is cyclic, not only the vacuum but indeed every nonzero vector in Fock space is cyclic. This is due to a peculiarity of the Fock-space representation, namely that it is irreducible. We now want to substantiate this assertion and place it in its natural mathematical context.

Let $\pi: \mathfrak{R} \to \mathfrak{B}(\mathcal{H})$ be a representation of the C^*-algebra $\mathfrak{R}$; a subset $\mathfrak{M}$ of $\mathcal{H}$ is said to be *stable* with respect to $\pi(\mathfrak{R})$ if $\pi(R)\Psi$ is in $\mathfrak{M}$ for all R in $\mathfrak{R}$ and all Ψ in $\mathfrak{M}$; the representation is said to be *irreducible* if the only subspaces of $\mathcal{H}$ stable with respect to $\pi(\mathfrak{R})$ are $\{0\}$ and $\mathcal{H}$.

REMARK. Throughout this book we use the word *subspace* for a *closed* linear manifold $\mathfrak{M}$. We must warn the reader that this is in contradistinction with some of the literature in which subspaces are meant to be just linear manifolds and in which the closure property is explicitly specified when present. The definition of irreducibility given above is actually the definition of topological irreducibility in contrast to algebraic irreducibility, which is defined by replacing "subspaces" in the above definition with "linear manifolds"; algebraic irreducibility implies trivially topological irreducibility; the converse is not obvious but nevertheless is true for representations of C^*-algebras (see Kadison [1957b], Dixmier [1964] Corollaire 2.8.4). We therefore do not have to distinguish between algebraic and topological irreducibility in the context of this book.

Let $\pi: \mathfrak{R} \to \mathfrak{B}(\mathcal{H})$ be a representation of a C^*-algebra $\mathfrak{R}$; we denote by $\pi(\mathfrak{R})'$ the set of all bounded operators from $\mathcal{H}$ into itself which commute with all elements of $\pi(\mathfrak{R})$, i.e.,

$$\pi(\mathfrak{R})' = \{B \in \mathfrak{B}(\mathcal{H}) \mid \pi(R)B = B\,\pi(R) \quad \text{for all} \quad R \text{ in } \mathfrak{R}\},$$

and call this set the *commutant* of $\pi(\mathfrak{R})$; we define the *bicommutant* $\pi(\mathfrak{R})''$ of $\pi(\mathfrak{R})$ as the commutant of $\pi(\mathfrak{R})'$, and similarly, by recursion, $\pi(\mathfrak{R})^{(N)} = \{\pi(\mathfrak{R})^{(N-1)}\}'$. We notice that the *algebras* $\pi(\mathfrak{R})^{(N)}$ satisfy

(i) $\qquad B \in \pi(\mathfrak{R})^{(N)}$ implies $B^* \in \pi(\mathfrak{R})^{(N)}, \qquad (N \geqslant 0),$

(ii) $\qquad \pi(\mathfrak{R})^{(N)} = \pi(\mathfrak{R})^{(N+2)} = \{\pi(\mathfrak{R})^{(N)}\}'', \qquad (N \geqslant 1);$

these two properties (satisfied for all $N \geqslant 1$) are the defining properties of *von Neumann algebras;* we say that $\pi(\mathfrak{R})''$ is the *von Neumann algebra generated by the representation* $\pi(\mathfrak{R})$. We postpone our study of von Neumann algebras to Subsection e.

Lemma. *A representation* $\pi: \mathfrak{R} \to \mathfrak{B}(\mathcal{H})$ *of a* C**-algebra* $\mathfrak{R}$ *is irreducible if and only if* $\pi(\mathfrak{R})' = \{\lambda I\}$.

Proof. $\pi(\mathfrak{R})$ irreducible means exactly that the only projectors in $\pi(\mathfrak{R})'$ are 0 and I, which in turn is equivalent to $\pi(\mathfrak{R})' = \{\lambda I\}$, since a von Neumann algebra is generated by its projectors (see Theorem 9). ∎

REMARK. As a consequence of this lemma we see immediately that $\pi(\mathfrak{R})$ is irreducible if and only if every bounded self-adjoint operator on $\mathcal{H}$ is a self-adjoint element in $\pi(\mathfrak{R})''$, i.e., represents an observable "affiliated" to the representation $\pi(\mathfrak{R})$.

We now come back to the initial comment of this subsection and prove the following result:

Lemma. *A (nonzero) representation $\pi: \mathfrak{R} \to \mathfrak{B}(\mathcal{H})$ of a C*-algebra $\mathfrak{R}$ is irreducible if and only if every nonzero vector Ψ in $\mathcal{H}$ is cyclic.*

Proof. Suppose first that π is irreducible and let Ψ be any nonzero vector in $\mathcal{H}$; the linear manifold $\pi(\mathfrak{R})\Psi$ is stable with respect to $\pi(\mathfrak{R})$, hence, since $\pi(\mathfrak{R})$ is irreducible, $\pi(\mathfrak{R})\Psi$ is either 0 or dense in $\mathcal{H}$; the first alternative is excluded by the assumption that $\pi \neq 0$, hence $\pi(\mathfrak{R})\Psi$ is dense in $\mathcal{H}$; that is to say, Ψ is cyclic. For the converse we proceed *ad absurdo;* suppose that π is reducible; there then exists a proper subspace $\mathfrak{M}$ of $\mathcal{H}$, stable with respect to $\pi(\mathfrak{R})$. No vector Ψ in $\mathfrak{M}$ can possibly be cyclic for $\pi(\mathfrak{R})$; hence π must be irreducible. ▌

Corollary. *Let $\pi: \mathfrak{R} \to \mathfrak{B}(\mathcal{H})$ be an irreducible represenation of a C*-algebra $\mathfrak{R}$; the state ψ on $\mathfrak{R}$ defined by $\langle \psi; R \rangle = (\Psi, \pi(R)\Psi)$ for some fixed Ψ in $\mathcal{H}$ with $|\Psi| = 1$, defines Ψ up to a phase.*

Proof. Let Ψ and Φ be any two vectors in $\mathcal{H}$, normalized to 1; we form the linear manifolds $\pi(\mathfrak{R})\Psi$ and $\pi(\mathfrak{R})\Phi$ which are dense in $\mathcal{H}$, since in an irreducible representation every nonzero vector is cyclic; now suppose that Ψ and Φ generate the same state on $\mathfrak{R}$, i.e., that

$$(\Psi, \pi(R)\Psi) = (\Phi, \pi(R)\Phi) \quad \text{for all } R \text{ in } \mathfrak{R}.$$

We define the linear mapping U from $\pi(\mathfrak{R})\Psi$ to $\pi(\mathfrak{R})\Phi$ by

$$U\pi(R)\Psi = \pi(R)\Phi;$$

from the cyclicity of Φ and

$$(\Psi, \pi(R^*T)\Psi) = (\Phi, \pi(R^*T)\Phi)$$

we conclude first that if R and S are such that $\pi(R)\Psi = \pi(S)\Psi$, i.e., $\pi(R - S)\Psi = 0$, then $\pi(R - S)\Phi = 0$, hence $\pi(R)\Phi = \pi(S)\Phi$ so that U is well defined; second, we conclude from the same relation (with $R = T$) that $|U\pi(R)\Psi| = |\pi(R)\Psi|$, hence U is bounded by 1; furthermore, since $\pi(\mathfrak{R})\Psi$ is dense in $\mathcal{H}$, U can be extended to a unitary operator from $\mathcal{H}$ to itself. For all R and S in $\mathfrak{R}$ we now form

$$\begin{aligned}\pi(R)\, U\pi(S)\Psi &= \pi(R)\, \pi(S)\Phi = \pi(RS)\Phi \\ &= U\pi(RS)\Psi = U\pi(R)\, \pi(S)\Psi;\end{aligned}$$

i.e., $[\pi(R), U]\, \pi(S)\Psi = 0$; since $\pi(\mathfrak{R})\Psi$ is dense in $\mathcal{H}$, we conclude that U commutes with every $\pi(R)$, hence belongs to $\pi(\mathfrak{R})'$. Since $\pi(\mathfrak{R})$ is irreducible, $\pi(\mathfrak{R})' = \{\lambda I\}$, hence $U = \omega I$ with $|\omega| = 1$; we have, in particular, $\Phi = U\Psi = \omega\Psi$, which proves the assertion of the lemma. ∎

We remember that irreducible representations actually played a privileged role (not fully realized at the time) in the early formulation of quantum theories (see, for instance, the early postulates mentioned in the beginning of Chapter 1, Section 2); this is so true that the occurrence in some physical problems of representations which are not irreducible was recognized only at a very late date in the development of the theory, namely with the introduction of superselection rules† by Wick, Wightman, and Wigner [1952].

It is therefore interesting to determine, within the context of the algebraic approach, the circumstances that lead to the occurrence of irreducible representations. To do so we need the concept of pure state on a C^*-algebra; we introduce this concept in complete analogy with Segal algebras: we say that a state ϕ on a C^*-algebra $\mathfrak{R}$ is *pure* if it cannot be written as a convex linear combination of two other states on $\mathfrak{R}$. The physical meaning of a pure state on a C^*-algebra is given by the following lemma:

Lemma. *A state ϕ on a* C**-algebra $\mathfrak{R}$ is pure if and only if its restriction $\phi|_{\mathfrak{A}}$ to the Segal algebra $\mathfrak{A}$ of all self-adjoint elements of $\mathfrak{R}$ is pure.*

Sketch of the proof. We notice first that a state ϕ on $\mathfrak{R}$ is completely determined by its restriction to $\mathfrak{A}$, since ϕ is linear and any element R in $\mathfrak{R}$ can be written in a unique way as $R = R_+ + iR^-$ with $R_\pm$ in $\mathfrak{A}$; conversely, to any state ψ on $\mathfrak{A}$ we can associate uniquely a state ϕ on $\mathfrak{R}$ with $\langle\phi; R\rangle = \langle\psi; R_+\rangle + i\langle\psi; R_-\rangle$. The proof then proceeds *ad absurdo* in a straightforward manner. ∎

Again, in complete analogy with Segal algebra, we say that a state ϕ *dominates*‡ a state ψ if there exists a real number $\lambda > 1$ such that $(\lambda\phi - \psi)$ is a positive real linear functional on $\mathfrak{R}$; we then prove, exactly as in a Segal algebra (see p. 60), that a state ϕ on a C^*-algebra is pure if and only if it dominates no other state.

In connection with the axiomatic treatment of Chapter 1, Section 2, we might mention here that Størmer [1968b], extending a result first proved by Kadison and Singer [1959] for $\mathfrak{B}(\mathcal{H})$, obtained the following characterization of pure states; let ϕ be an arbitrary state on a C^*-algebra $\mathfrak{R}$ and denote by

† We refer again for this concept to Jauch's book [1968a].

‡ The definition of "domination" of a positive linear functional by another is not uniform throughout the literature; our choice was dictated by its meaning when restricted to states; we use the same definition as Naimark [1964]; Dixmier [1964] uses instead the following related notion: ϕ is said to *majorize* $\psi (\phi \geqslant \psi)$ if $(\phi - \psi)$ is positive.

$\mathfrak{A}_\phi$ the "definite set of ϕ" constituted by all self-adjoint elements A of $\mathfrak{R}$ such that $\langle\phi; A\rangle^2 = \langle\phi; A^2\rangle$; in cases in which $\mathfrak{R}$ has a unit and admits no one-dimensional representations ϕ is pure if and only if $\mathfrak{A}_\phi$ is maximal (under set inclusion) among the sets $\mathfrak{A}_\psi$, where ψ runs over all states on $\mathfrak{R}$; see also Kadison [1962].

We now want to prove that irreducible representations are precisely those generated, via the GNS construction, by pure states; to achieve this we need the following lemma:†

Lemma. *Let* $\pi:\mathfrak{R}\to\mathfrak{B}(\mathcal{H})$ *be the cyclic representation of the* C*-*algebra* $\mathfrak{R}$ *associated with the state* ϕ *on* $\mathfrak{R}$ *by the GNS construction; let* $\Phi(\mathrm{I})$ *be the corresponding cyclic vector. Then to every state* ψ *dominated by* ϕ *corresponds a* unique *element* B *in* $\mathfrak{B}(\mathcal{H})$ *such that*

(i) $\langle\psi; \mathrm{R}\rangle = (\mathrm{B}\Phi(\mathrm{I}), \pi(\mathrm{R})\,\mathrm{B}\Phi(\mathrm{I}))$ *for all* R *in* $\mathfrak{R}$,

(ii) $(\Psi, \mathrm{B}\Psi) \geqslant 0$ *for all* Ψ *in* $\mathcal{H}$,

(iii) $\mathrm{B}\in\pi(\mathfrak{R})'$;

(iv) $|\mathrm{B}\Phi(\mathrm{I})| = 1.$

Conversely, every B *in* $\mathfrak{B}(\mathcal{H})$ *satisfying the conditions* (ii) *to* (iv) *generates via the relation* (i) *a state* ψ *on* $\mathfrak{R}$ *dominated by* ϕ.

Proof. Let us prove first the "converse" part of the lemma; the mapping ψ from $\mathfrak{R}$ to $\mathbb{C}$ defined by (i) is linear. We form $\langle\psi; R^*R\rangle = (B\Phi(I), \pi(R^*R)\,B\Phi(I))$, and use the fact that π is a representation to get $\langle\psi; R^*R\rangle = |\pi(R)\,B\Phi(I)|^2 \geqslant 0$ which proves that ψ is positive; from (iv) $\langle\psi; I\rangle = 1$, hence ψ is a state on $\mathfrak{R}$. We still have to prove that ψ is dominated by ϕ; from the already established relation $\langle\psi; R^*R\rangle = |\pi(R)\,B\Phi(I)|^2$ and from the fact that B belongs to $\pi(\mathfrak{R})'$ and is bounded above by some λ which we can assume, without loss of generality, to be larger than 1, we get

$$\langle\psi; R^*R\rangle = |B\pi(R)\,\Phi(I)|^2 \leqslant \lambda^2\,|\pi(R)\,\Phi(I)|^2 = \lambda^2\langle\phi; R^*R\rangle,$$

which is to say that ψ is dominated by ϕ. We notice that we have not yet used condition (ii) or the fact that we are dealing with a cyclic representation; both are now used to prove the uniqueness of B. Suppose that there is in $\pi(\mathfrak{R})'$ another element, say C, which produces through (i) the same state ψ as B; we then have for all R_1 and R_2 in $\mathfrak{R}$

$$(B\Phi(I), \pi(R_1^*R_2)\,B\Phi(I)) = (C\Phi(I), \pi(R_1^*R_2)\,C\Phi(I)),$$

i.e.,

$$(\pi(R_1)\,\Phi(I), (B^*B - C^*C)\pi(R_2)\,\Phi(I)) = 0;$$

since $\Phi(I)$ is cyclic for π, this implies that $B^*B = C^*C$; since the positive square root of a positive operator is unique, we conclude that B is unique.

† Compare with Naimark [1964], Section 19.1, or Dixmier [1964], Proposition 2.5.1.

Let us now prove the direct part of the lemma. We define on $\mathfrak{R}_\phi \times \mathfrak{R}_\phi$ the following hermitian positive functional [,]:

$$[\Phi(R), \Phi(S)] \equiv \langle\psi; R^*S\rangle;$$

for this definition to make sense we first have to verify that the right-hand side depends on R and S only through $\Phi(R)$ and $\Phi(S)$. Suppose that S and S' in $\mathfrak{R}$ are ϕ-equivalent, i.e., $\langle\phi; R^*(S - S')\rangle = 0$ for all R in $\mathfrak{R}$; since ψ is dominated by ϕ, we have $\langle\psi; R^*(S - S')\rangle = 0$ which proves that the right-hand side depends on S only through $\Phi(S)$. The same proof can be repeated for R' ϕ-equivalent to R; now we again use the fact that ψ is dominated by ϕ to conclude that our functional [. . . , . . .] is bounded. We have indeed

$$|\langle\psi; R^*S\rangle|^2 \leqslant \langle\psi; R^*R\rangle\langle\psi; S^*S\rangle \leqslant \mu^2\langle\phi; R^*R\rangle\langle\phi; S^*S\rangle$$
$$= \mu^2\, |\Phi(R)|^2\, |\Phi(S)|^2;$$

[. . . , . . .] can therefore be extended by continuity to $\mathcal{H}_\phi \times \mathcal{H}_\phi$ for each Ψ in $\mathcal{H}_\phi$; the mapping $X \to [\Psi, X]$ is then a bounded linear functional on $\mathcal{H}_\phi$ and from Riesz's theorem we conclude that there exists some Ψ' in $\mathcal{H}_\phi$ such that $[\Psi, X] = (\Psi', X)$. We define the mapping Y from $\mathcal{H}_\phi$ to itself by $Y\Psi = \Psi'$ and then have $[\Psi, X] = (Y\Psi, X)$. Since [. . . , . . .] is bounded, hermitian, and positive, so is the operator Y. We now write

$$(Y\pi(R)\,\Phi(S), \Phi(T)) = (Y\Phi(RS), \Phi(T)) = \langle\psi; (RS)^*T\rangle = \langle\psi; S^*(R^*T)\rangle$$
$$= (Y\Phi(S), \Phi(R^*T)) = (Y\Phi(S), \pi(R)^*\,\Phi(T)) = (\pi(R)\; Y\Phi(S), \Phi(T));$$

from this equality and the fact that $\mathfrak{R}_\phi$ is dense in $\mathcal{H}_\phi$ we conclude that Y commutes with $\pi(R)$ for all R in $\mathfrak{R}$, i.e., that Y belongs to $\pi(\mathfrak{R})'$; its positive square root then satisfies the four conditions of the lemma. This concludes the proof of the lemma. ∎

With this lemma we can now prove the announced theorem which makes the connection between pure states and irreducible representations:

Theorem 3. *Let ϕ be a state on a* C*-*algebra $\mathfrak{R}$; the GNS representation of $\mathfrak{R}$ associated with ϕ is irreducible if and only if ϕ is pure.*

Proof. Suppose first that π_ϕ is irreducible and let ψ be any state dominated by ϕ; there then exists an element B of $\pi_\phi(\mathfrak{R})'$ which satisfies the four conditions of the preceding lemma; since, however, π_ϕ is irreducible, $\pi_\phi(\mathfrak{R})' = \{\lambda I\}$; hence $B = \lambda I$. From condition (ii) we get $\lambda \geqslant 0$ and from condition (iv), $\lambda = 1$; using (i), we get $\psi = \phi$. Hence, since ϕ dominates no other state, it is pure. Now suppose that ϕ is pure and let us prove that π_ϕ is irreducible. Suppose that B' is a positive element of $\pi_\phi(\mathfrak{R})'$ with $B'\Phi(I) = 0$; then $B'\Phi(R) = B'\pi_\phi(R)\,\Phi(I) = \pi_\phi(R)\,B'\Phi(I) = 0$. Since, however, B' is bounded and $\mathfrak{R}_\phi$, dense in $\mathcal{H}_\phi$, we can conclude that $B' = 0$. Hence we can assume without restriction that $B'\Phi(I) \neq 0$ for all nonzero positive elements of $\pi_\phi(\mathfrak{R})'$. For any such element B' we define $B = B'/|B\Phi(I)|$; B then satisfies the conditions (ii) to (iv) of the lemma and the state ψ associated with it by relation (i) is dominated by ϕ. Since, however, ϕ is pure $\psi = \phi$, from which we get for all R and S in $\mathfrak{R}$, $\langle \psi; R^*S \rangle = \langle \phi; R^*S \rangle$, i.e., $(\Phi(R), (B^2 - I)\,\Phi(S)) = 0$, or (since $\mathfrak{R}_\phi$ is dense in $\mathcal{H}_\phi$) $B^2 = I$, hence $B = I$. Consequently every positive element B' in $\pi(\mathfrak{R})'$ is of the form $B' = \lambda I$ with $\lambda = B'\Phi(I)$; since $\pi_\phi(\mathfrak{R})'$ is generated by its positive elements, $\pi_\phi(\mathfrak{R})' = \{\lambda I\}$, which implies that $\pi_\phi(\mathfrak{R})$ is irreducible. This concludes the proof of the theorem. ∎

Aside from the fundamental features mentioned up to this point, pure states on a C^*-algebra and the representations they induce via the GNS constructuon enjoy particular properties, two of which we want to quote here.

Corollary. *Every vector state associated with an irreducible representation $\pi: \mathfrak{R} \to \mathfrak{B}(\mathcal{H})$ of a C^*-algebra $\mathfrak{R}$ is pure.*

Proof. By definition ψ is a vector state associated with π if there exists in $\mathcal{H}$ a normalized vector Ψ such that $\langle \psi; R \rangle = (\Psi, \pi(R)\Psi)$ for all R in $\mathfrak{R}$; π irreducible implies (lemma, p. 84) that Ψ is cyclic. From Theorem 1 we know that π is spatially equivalent to the GNS representation associated with ψ. Since π is irreducible, so then is π_ψ, hence from the theorem just proved ψ is pure. ∎

REMARK. This corollary exhibits another peculiarity of the early postulates on quantum mechanics (say à la Dirac); the pure states were identified with the vector states of the realization considered, which at that stage of the development of the theory, was always assumed to be irreducible. Furthermore we have the following:

Corollary. *Let ϕ be a pure state on a C^*-algebra $\mathfrak{R}$; then the linear manifold $\mathfrak{R}_\phi$ appearing in the GNS construction is not only dense in $\mathcal{H}_\phi$ but equal to it.*

Proof. ϕ pure implies π_ϕ irreducible, hence "algebraically" irreducible; i.e., the only linear manifolds of $\mathcal{H}_\phi$ stable with respect to π_ϕ are $\{0\}$ and $\mathcal{H}_\phi$ itself. $\mathfrak{R}_\phi$, in particular, is a stable linear manifold in $\mathcal{H}_\phi$. Since it is dense in $\mathcal{H}_\phi$ the algebraic irreducibility of π_ϕ implies then that $\mathfrak{R}_\phi$ coincides with $\mathcal{H}_\phi$. ∎

When $\mathfrak{R}_\phi = \mathcal{H}_\phi$, we say that Φ is *strictly cyclic*; we should mention here that the converse of the above corollary is *not* true in general. However, Halperin [1967] generalized the lemma to the following assertion: $\mathfrak{R}_\phi = \mathcal{H}_\phi$ if and only if ϕ is a *finite* convex sum of pure states; i.e., $\phi = \sum_{i=1}^n \lambda_i \phi_i$ with $\lambda_i > 0$, $\sum_{i=1}^n \lambda_i = 1$, and ϕ_i pure. Furthermore he proved that such ϕ

determine the irreducible representations π_{ϕ_i} up to unitary equivalence. Specifically, if $\sum_{i=1}^{n} \lambda_i \phi_i = \phi = \sum_{j=1}^{m} \mu_j \psi_j$ (with all ϕ_i and ψ_j pure), then for each i $(1 \leqslant i \leqslant n)$ there exists an index $k(1 \leqslant k \leqslant m)$ such that $\pi_{\phi_i}(U^* A U) = \pi_{\psi_i}(A)$ for all A in $\mathfrak{R}$.

Glimm and Kadison [1960]† proved two remarkable results in this context and we want to quote them here in the following form:

Theorem 4. *Let ϕ and ψ be two* pure *states on a* C**-algebra* $\mathfrak{R}$, *and* π_ϕ, π_ψ *the representations of* $\mathfrak{R}$ *associated with them by the GNS construction. Then the following three propositions are in the logical relation*

$$(\mathrm{i}) \Rightarrow (\mathrm{ii}) \Leftrightarrow (\mathrm{iii}):$$

(i) $\|\phi - \psi\| < 2$ (*i.e. ϕ and ψ belong to the same "part"*),
(ii) π_ϕ *and* π_ψ *are spatially equivalent,*
(iii) *there exists a unitary element* U *in* $\mathfrak{R}$ *such that* $\langle \phi; \mathrm{R} \rangle = \langle \psi; \mathrm{U^* R\, U} \rangle$ *for all* R *in* $\mathfrak{R}$.

REMARKS

(a) From the results obtained up to this point the reader will find no difficulty in proving the implication (iii) ⇒ (ii); the other two implications call for a more sophisticated background. Incidentally, the proof of (iii) ⇒ (ii) is made slightly easier by the knowledge of the preceding corollary but the latter is not necessary to the completion of this proof.
(b) Proposition (i) is not a necessary condition for the validity of proposition (ii). The reader will convince himself that this is indeed the case by inspecting the C^*-algebra generated by the Pauli matrices (i.e., the algebra of all 2×2 matrices with complex entries; see below), where counterexamples are easy to come by.
(c) The implication (i) ⇒ (iii) plays an important role in our forthcoming discussion of automorphisms of C^*-algebras; this is actually our reason for mentioning this theorem here.

c. Examples

As an application of the concepts and techniques introduced in the last two subsections, we discuss first a trivial example drawn from the algebra of 2×2 matrices with complex entries; we then review the representation theory pertinent to the infinite, nonrelativistic free Bose gas.

Consider the algebra

$$\mathfrak{R} \equiv \{R = aI + b\sigma^x + c\sigma^y + d\sigma^z \mid a, b, c, d \in \mathbb{C}\},$$

† Corollaries 8 and 9.

where the operators I, σ^x, σ^y, and σ^z satisfy the usual composition laws:

$$I = I^2 = (\sigma^x)^2 = (\sigma^y)^2 = (\sigma^z)^2,$$

$$I\sigma^x = \sigma^x \qquad \text{(and similarly for } y \text{ and } z\text{)},$$

$$\sigma^x\sigma^y = i\sigma^z \qquad \text{(and similarly for the cyclic permutations of } x, y, z\text{)},$$

$$(\sigma^x)^* = \sigma^x \qquad \text{(and similarly for } y \text{ and } z\text{)}.$$

(The corresponding *real* algebra was discussed in Chapter 1, Section 2, Subsection g.) Suppose that this system were placed in a magnetic field B along the z-direction; we would then have for the Hamiltonian

$$H = -B\sigma^z$$

and for the canonical density matrix (at natural temperature $\beta = (kT)^{-1}$)

$$\rho = \frac{e^{\beta B\sigma^z}}{\mathrm{Tr}\, e^{\beta B\sigma^z}};$$

this density matrix generates the state

$$\langle\phi; R\rangle = a + dt$$

where

$$t = \tanh \beta B.$$

The zero temperature (i.e., $\beta = \infty$, $t = 1$) case is treated in complete analogy with the real case treated in I.2.g. For $0 \leqslant |t| < 1$, we easily get $\mathfrak{K}_\phi = 0$ so that the representation should be four-dimensional. We get indeed

$$\pi_\phi = \begin{pmatrix} \pi_0 & 0 \\ 0 & \pi_0 \end{pmatrix},$$

where

$$\pi_0(R) = \begin{pmatrix} a + d & b - ic \\ b + ic & a - d \end{pmatrix}.$$

The cyclic vector generating ϕ is

$$\Phi = \begin{pmatrix} u \\ 0 \\ 0 \\ v \end{pmatrix} \quad \text{with } u = \left(\frac{1+t}{2}\right)^{1/2} \quad \text{and} \quad v = \left(\frac{1-t}{2}\right)^{1/2}.$$

This representation provides a host of examples and counterexamples with which we can illustrate very simply most of the propositions given in Subsections a and b above; we leave this exercise to the reader's curiosity. In addition, we should like to mention the following nonaccidental properties of this representation:

(i) It is primary, i.e., $\pi(\mathfrak{R}) \cap \pi(\mathfrak{R})' = \{\lambda I\}$

(ii) The time evolution is generated by a Hamiltonian of the form

$$H_\phi = -B\begin{pmatrix} 0 & & & \\ & -2 & & \\ & & +2 & \\ & & & 0 \end{pmatrix},$$

such that $H_\phi\Phi = 0$ and the spectrum of H_ϕ is symmetric about the origin.

(iii) If ψ is a vector state in this representation, say, for instance, the state corresponding to

$$\Psi = \frac{1}{\sqrt{2}}\begin{pmatrix} 1 \\ 1 \\ 0 \\ 0 \end{pmatrix},$$

the ergodic everage

$$\langle\bar{\psi}; A\rangle = \lim_{T\to\infty} \frac{1}{T}\int_0^T \langle\psi_t; A\rangle\, dt = \sum_i \operatorname{Tr}(P_i \rho_\Psi P_i A)$$

(where P_i are the eigenprojectors of H_ϕ and ρ_Ψ is the density matrix corresponding to Ψ) is a vector state on this representation and is given here by

$$\bar{\Psi} = \begin{pmatrix} 1 \\ 0 \\ 0 \\ 1 \end{pmatrix}.$$

These properties are discussed in Section 2 of this chapter; they result quite generally from the fact that the canonical state ϕ satisfies the KMS condition (hence is, in particular, time-invariant); the third property has been pointed out to me in its proper generality by C. Radin.

We now want to discuss the proper Hilbert space description of the infinite, nonrelativistic free Bose gas.† Aside from the initial aim of further illustrating the techniques developed in Subsections a and b, this model is used repeatedly in the sequel as a prototype for many aspects of the general theory.

† The remaining part of this subsection is based on Araki and Woods [1963].

There are many reasons for considering the limit in which the volume of the system considered tends to infinity; the most obvious among them is that one of the main aims of statistical mechanics is to make statements about thermodynamical quantities, the latter being traditionally defined as bulk properties in this limit; there are more technical aspects to this general premise as well; for instance, we definitely know since the work of Lee and Yang [1952] that phase transitions manifestly appear only when this limit is duly taken (at least when a hard core is present).

We saw in Chapter 1, Section 1, that a Bose system is characterized by a field F and its canonical conjugate P which are both mappings from a "test function space"† $\mathcal{T}$ to some Hilbert space $\mathcal{H}$. In the Fock-space representation considered in Chapter 1 we saw that $P(g)$ and $F(f)$ are self-adjoint operators, admitting a common stable dense domain and satisfying the following canonical commutation rules (CCR) (see p. 15 with $\mathcal{T}$ now real):

$$[F(f), F(g)] = 0 = [P(f), P(g)],$$
$$[F(f), P(g)] = i(f, g)I.$$

To avoid domain questions, which are always somewhat cumbersome to carry on in an abstract treatment, we define the unitary operators

$$V(f) = e^{\{-iF(f)\}} \quad \text{and} \quad U(g) = e^{\{-iP(g)\}},$$

which inherit the following relations from the above commutation relations:

$$U(f_1)U(f_2) = U(f_1 + f_2),$$
$$V(g_1)V(g_2) = V(g_1 + g_2),$$
$$U(f)V(g) = V(g)U(f)e^{i(f,g)};$$

the third—and only nontrivial—of these relations can be obtained, for instance, from the Baker-Hausdorff formula.

This form of the CCR suggests the introduction of the operators

$$W(f, g) \equiv U(f)V(g)e^{-i(f,g)/2},$$

which satisfy the following composition law:

$$W(f_1, g_1)W(f_2, g_2) = W(f_1 + f_2, g_1 + g_2)e^{i\{(f_1,g_2)-(f_2,g_1)\}/2}.$$

We notice that

$$W(f, 0) = U(f) \quad \text{and} \quad W(0, g) = V(g),$$

so that the above relation summarizes the three preceding relations; we refer to this relation as the *Weyl's form of the CCR*. We notice further that as a consequence of this relation the complex linear manifold $\mathfrak{W}$ of all finite

† We shall henceforth assume that $\mathcal{T}$ is a dense linear manifold in the space of all *real* square-integrable functions on $\mathbb{R}^3$; for more details on the appropriate choice of $\mathcal{T}$ the reader is referred to Chapter 3, Section 1.

linear combinations

$$\sum_i c_i W(f_i, g_i)$$

is stable under the product of any two of its elements; hence it is an algebra that is moreover stable with respect to the involution corresponding to the hermitian adjunction. We have indeed

$$W(f, g)^* = W(-f, -g) \qquad [\text{and then } W(f, g)^* W(f, g) = I].$$

Finally the ordinary operator norm provides the norm on $\mathfrak{W}$ which then becomes a normed *-algebra, which we refer to as the *Weyl algebra*. On going through the proofs in this section and at the end of the last chapter once again the reader will convince himself that this algebra possesses all the structure needed for the application of the GNS technique. To use this technique† we first notice that a state ϕ on $\mathfrak{W}$ is entirely determined by the functional

$$\hat{\phi}(f, g) \equiv \langle \phi; W(f, g) \rangle,$$

which then satisfies the following characteristic properties:

$$\hat{\phi}(0, 0) = 1$$

$$\hat{\phi}(f, g)^* = \hat{\phi}(-f, -g)$$

$$\sum_{ij} c_i^* c_j \hat{\phi}(f_j - f_i, g_j - g_i) e^{i\{(f_j, g_i) - (f_i, g_j)\}/2} \geqslant 0,$$

which express, respectively, that the state ϕ is normalized to 1 and that it is hermitian and positive. As an illustration (which we use in the sequel) we mention that in the Fock-space representation of the Weyl algebra the functional $\hat{\phi}(f, g)$ obtained from the vacuum takes the form

$$\hat{\phi}_F(f, g) = (\Psi_0, W(f, g)\Psi_0)$$
$$= \exp\{-\tfrac{1}{4}(|f|^2 + |g|^2)\}.$$

For an arbitrary representation of the Weyl algebra $\mathfrak{W}$ (hence of the CCR) we define formally the "number operator" $N(\Omega)$ and the "density operator" $\rho(\Omega)$ for a bounded region Ω of the space $\mathbb{R}^3$ as

$$N(\Omega) = \tfrac{1}{2} \sum_i \{F(f_i)^2 + P(f_i)^2 - 1\} \quad \text{and} \quad \rho(\Omega) = \frac{N(\Omega)}{\Omega},$$

where $\{f_i\}$ is an orthonormal basis for the test-function space $\mathcal{T}_\Omega$ [which is to $\mathcal{T}$ what $\mathcal{L}^2(\Omega)$ is to $\mathcal{L}^2(\mathbb{R}^3)$]; we next suppose that the following "average density operator" exists in the representation considered:

$$\rho = \lim_{\Omega \to \infty} \rho(\Omega).$$

† For a mathematically complete account of the underlying general theory the reader is referred to the formalism developed in Section III.1.c and in particular to Theorem III.1.7.

We first want to argue that the Fock-space representation admits only $\rho = 0$ and is therefore inadequate for the description, in the thermodynamical limit, of even such a simple system as the free Bose gas at finite density. To this effect we form for any f and g of compact support† and any Ω which contains both supp (f) and supp (g)

$$e^{i\lambda\rho(\Omega)}\, W(f, g) = W\left(f \cos\frac{\lambda}{\Omega} + g \sin\frac{\lambda}{\Omega},\, g\cos\frac{\lambda}{\Omega} - f\sin\frac{\lambda}{\Omega}\right) e^{i\lambda\rho(\Omega)}$$

so that

$$\lim_{\Omega\to\infty} [e^{i\lambda\rho(\Omega)}, W(f, g)] = 0.$$

Since this equality holds for all f and g with compact support and any (real) c-number λ, we conclude that the "macroscopic" operator ρ (supposing that it exists at all) is "affiliated" with the commutant $\{W(f, g)\}'$ of our representation of the Weyl algebra. In particular, this operator reduces to a c-number whenever this representation is irreducible; we can therefore compute it by calculating its expectation value on any vector state of the irreducible representation considered; in particular, we see, on calculating $(\Psi_0, \rho\Psi_0)$ on the vacuum of the Fock representation, that $\rho = 0$ for this representation. This proves our previous statements on the limitations of Fock space for the description (in the thermodynamical limit) of even the simplest of all quantum statistical systems.

Furthermore we can use the same argument to show that the operator ρ exists and reduces to a c-number whenever we have a cyclic representation of the Weyl algebra with cyclic vector Ψ such that

$$\lim_{\Omega\to\infty} e^{iN(\Omega)/\Omega}\Psi = \omega\Psi \quad \text{with } |\omega| = 1.$$

We now want to construct the representation associated with the *ground state* of an infinite Bose gas of density $\rho \neq 0$. In line with the formalism developed so far the first task is to compute the corresponding functional $\hat{\phi}_{\rho,0}(f, g)$. To do so we notice that the ground state of this gas, when enclosed in a finite volume Ω (say a cubic box with periodic boundary conditions) consists of N particles ($N = \rho\Omega$), each with wave function $f_\Omega(x) = \Omega^{-1/2}$; the functional corresponding to this situation is

$$\hat{\phi}_{\Omega,\rho,0}(f, g) = (N!)^{-1}(a^*(f_\Omega)^N\Phi_0,\, W(f, g)\, a^*(f_\Omega)^N\Phi_0),$$

where Φ_0 is the Fock-space vacuum. This expression can be brought to the form

$$\hat{\phi}_{\Omega,\rho,0}(f, g) = \hat{\phi}_F(f, g) L_N\left\{\frac{[(f, f_\Omega)^2 + (g, g_\Omega)^2]}{2}\right\},$$

† The support of a function is the smallest closed subset outside of which it vanishes.

where $\hat{\phi}_F(f, g)$ is the Fock-vacuum functional introduced before and L_N is the Nth Laguerre polynomial. For each fixed f and g of compact support and each Ω containing both supp (f) and supp (g) we get

$$\begin{aligned}\lim_{\Omega\to\infty} \hat{\phi}_{\Omega,\rho,0}(f, g) &= \hat{\phi}_F(f, g) J_0\{[2\rho(\tilde{f}(0)^2 + \tilde{g}(0)^2)]^{1/2}\} \\ &= \hat{\phi}_F(f, g)\frac{1}{2\pi}\int d\theta \exp\{i(2\rho)^{1/2}[\tilde{f}(0)\cos\theta + \tilde{g}(0)\sin\theta]\} \\ &\equiv \hat{\phi}_{\rho,0}(f, g),\end{aligned}$$

where $\tilde{f}(0) = \int dx\, f(x)$ and similarly for g. The above expression is the functional we are looking for; we could now attempt to use it directly to carry over explicitly the GNS construction; instead, we exhibit a representation that generates this functional and use Theorem 1 to conclude that this representation is equivalent to the representation we would have obtained in the straightforward way.

We first form the Hilbert space

$$\mathcal{H}_{\rho,0} = \mathcal{H}_F \otimes \mathcal{L}^2(S^1),$$

where $\mathcal{H}_F$ is the Fock space and $\mathcal{L}^2(S^1)$ is the space of all functions on the unit circle that are square-integrable with respect to the usual Lebesque measure $(d\theta/2\pi)$. Next we define the operators $W_{\rho,0}(f, g)$ acting on this space by

$$W_{\rho,0}(f, g) = W_F(f, g) \otimes \exp\{i(2\rho)^{1/2}(\tilde{f}(0)A - \tilde{g}(0)B)\},$$

where $W_F(f, g)$ are the Weyl operators introduced previously on Fock space and A and B are the following operators acting on the space $\mathcal{L}^2(S^1)$:

$$\begin{aligned}(Af)(\theta) &= f(\theta)\cos\theta, \\ (Bf)(\theta) &= f(\theta)\sin\theta.\end{aligned}$$

Finally we form the vector

$$\Phi_{\rho,0} = \Phi_0 \otimes \Phi_0',$$

where Φ_0 is the Fock-space vacuum and $\Phi_0'(\theta) = 1$ for all θ. Since the operators A and B commute, it is trivial to verify that the operators $W_{\rho,0}(f, g)$ generate a representation of the Weyl algebra. From the cyclicity of the vacuum in Fock space and the cyclicity of Φ_0' in $\mathcal{L}^2(S^1)$ under the algebra generated by A and B we conclude that $\Phi_{\rho,0}$ is cyclic for this representation. We finally verify that we obtained the right representation by checking that

$$\hat{\phi}_{\rho,0}(f, g) = (\Phi_{\rho,0}, W_{\rho,0}(f, g)\Phi_{\rho,0}).$$

We now notice that

$$W_{\rho,0}(f, g) = \int d\theta \, W_\theta(f, g),$$

with

$$W_\theta(f, g) = W_F(f, g) \otimes \exp\{i(2\rho)^{1/2}(\tilde{f}(0)\cos\theta - \tilde{g}(0)\sin\theta)\},$$

provides a decomposition of the representation $W_{\rho,0}$ into a direct integral (see Dixmier [1964] 8.1) of irreducible representations; hence our representation is not irreducible; this means (Theorem 3) that the state $\hat{\phi}_{\rho,0}$ is not pure, a fact that we could have expected from the physical reason that the ground state of an infinite Bose gas of finite density contains an infinite number of particles in the zero-momentum state so that we can add or subtract any finite number of these particles from this state without changing its physical properties; mathematically, this fact manifests itself by the presence in the representation of the vectors

$$\Phi_n = \Phi_0 \otimes e^{-in\theta} \quad \text{with } n = 0, 1, 2, \ldots,$$

which form an orthonormal basis for the (infinite-dimensional) subspace of all the degenerate ground states. This fact has an exact equivalent in the BSC-theory of superconductivity.†

With these comments on the physical interpretation of the results obtained so far we close our discussion of the representation associated with the ground state of the infinite free Bose gas at finite density. For further information the reader is referred to the original work of Araki and Woods [1963]; a more recent account can be found in the review article by dell'Antonio [1968]. We only want to mention here that a calculation similar to that carried above but starting from the grand canonical density matrix for the finite free Bose gas at temperature $T \geqslant T_c$ (i.e., with no macroscopic occupation of the ground state) leads to the following functional:

$$\begin{aligned}\hat{\phi}_{\mu,\beta}(f, g) &= \lim_{\Omega\to\infty} \hat{\phi}_{\Omega,\mu,\beta}(f, g) \\ &= \hat{\phi}_F(f, g) \exp\left\{-\tfrac{1}{2}\int d\mathbf{k}\,\tilde{\rho}(k)[|\tilde{f}(\mathbf{k})|^2 + |\tilde{g}(\mathbf{k})|^2]\right\},\end{aligned}$$

where μ denotes the chemical potential, $\tilde{f}(\mathbf{k})$ (or $\tilde{g}(\mathbf{k})$) is the Fourier transform of $f(\mathbf{x})$ [or $g(\mathbf{x})$], and

$$\tilde{\rho}(k) = \{e^{-\beta[\mu-\omega(k)]} - 1\}^{-1} \quad \text{with } \omega(k) = \frac{k^2}{2m}.$$

† For a rapid survey the reader is referred back to Chapter 1, Section 1, Subsection f, where he will also find an extensive list of references on the physical nature and mathematical analysis of this model.

On introducing in this expression the explicit form of the Fock-vacuum functional, we get an expression of the form

$$\hat{\Phi}_{\mu,\beta}(f, g) = \exp\left\{-\tfrac{1}{4}\int d\mathbf{k}[\tilde{\alpha}(k)\,|\tilde{f}(\mathbf{k})|^2 + \tilde{\beta}(k)\,|\tilde{g}(\mathbf{k})|^2]\right\},$$

where $\alpha(k)$ and $\beta(k)$ are nonnegative functions of $k = |\mathbf{k}|$ and $\tilde{\alpha}(k)\,\tilde{\beta}(k) \geqslant 1$. Representations corresponding to this type of functional have recently been extensively studied by Klauder and Streit [1969] who also discuss† the possible dynamical situations compatible with this type of functional. We will, for the moment, retain here only the fact that these functionals lead again to reducible representations [unless $\alpha(k)\beta(k) = 1$ a.e.] and that whenever the set $\Gamma = \{k \mid \alpha(k)\,\beta(k) = 1\}$ is of measure zero the Hilbert space of these representations is of the form $\mathcal{H} = \mathcal{H}_F \otimes \mathcal{H}_F$; the latter results confirm those obtained by Araki and Woods [1963] for the free Bose gas with no macroscopic occupation of the ground state; the representation they constructed for this case is

$$W_\rho(f, g) = W([1 + \rho]^{1/2}f, [1 + \rho]^{1/2}g) \otimes W(\rho^{1/2}f, -\rho^{1/2}g),$$

where the products of the type ρf occurring in this expression mean $(\rho f)(k) = \tilde{\rho}(k)\tilde{f}(k)$. Finally, the case in which a macroscopic occupation of the ground state is present can be obtained as a combination of the two previous cases studied:

$$\mathcal{H} = \mathcal{H}_F \otimes \mathcal{H}_F \otimes \mathfrak{L}^2(S^1),$$

$$W_\rho'(f, g) = W_{\rho_1}(f, g) \otimes \exp\{i(2\rho_0)^{1/2}(\tilde{f}(0)A - \tilde{g}(0)B)\},$$

where ρ_0 and ρ_1 are defined by

$$\tilde{\rho}(k) = \tilde{\rho}_1(k) + \rho_0\delta(k) \quad \text{and} \quad \tilde{\rho}_1(k) \text{ continuous.}$$

d. Weak Topologies and Physical Equivalence of Representations

The aim of this subsection is to formulate in precise mathematical language—and to exploit—the fact that physical measurements determine in general the state of a physical system only within a limited degree of accuracy. The tools to achieve this aim are provided by the following coupled preliminary remarks:

1. Aside from the notion of spatial (or unitary) equivalence studied in Subsection a (see, in particular, the criteria provided by Theorem 2 and in

† The relevance of a study of these representations in general, aside from the application considered here and their appearance in the theory of the (generalized) free relativistic scalar field, is that functionals which satisfy the above conditions also occur in some particular *interacting* systems, such as the exactly soluble rotationally symmetric model constructed by Klauder [1965].

Subsection b, by Theorem 4), another notion of equivalence between representations has been discussed in the mathematical literature under the name of "weak equivalence"; in the physical literature this notion is referred to as that of "physical equivalence," since essentially it expresses the equivalence of representations with which we can associate the same set of states.

2. The above notion is closely related to the question of the physically most reasonable topology with which we should equip the dual $\mathfrak{R}^*$ (or $\mathfrak{A}^*$) of $\mathfrak{R}$ (or $\mathfrak{A}$); in Chapter 1, Section 2 (p. 61ff), we considered the weak*-topology on the dual $\mathfrak{A}^*$ of the Segal algebra $\mathfrak{A}$ of all observables and noticed that this topology, when restricted to the set $\mathfrak{S} \subseteq \mathfrak{A}^*$ of all states on $\mathfrak{A}$, was physically the natural one.

The first part of this subsection amplifies the latter remark and extends the notion of weak*-topology to the dual $\mathfrak{R}^*$ of a C^*-algebra $\mathfrak{R}$; strictly speaking this extension is not indispensable for the concept of physical equivalence, since the latter should, and actually will, be expressed in terms of $\mathfrak{A}$ and $\mathfrak{S}$ only; however we carry out this extension here for completeness as well as for the practical reason that we will have to use this topology on $\mathfrak{R}^*$ in later parts of this book. Most of the results mentioned in the forthcoming mathematical brush-up are standard in the study of Banach spaces and more generally of topological linear spaces; the reader curious about the proofs related to the general mathematical setting should consult Dunford and Schwartz [1957], Chapters II and V, Day [1962], Chapter I, or Bourbaki [1955], Chapter IV. Our review culminates in an approximation theorem (Theorem 6) and its corollary.

On the basis of this result, we proceed to the concept of physical equivalence, the definition of which immediately precedes Theorem 7; this theorem gives the general criteria for physical equivalence of two representations. It is followed by a detailed discussion of the meaning of the concept itself, which appears as an a posteriori justification of the C^*-algebraic approach: all faithful representations of any given C^*-algebra turn out to be physically equivalent. The relation between the concepts of physical equivalence and unitary equivalence is mentioned briefly in the same paragraph. Further criteria for physical equivalence are then given in special cases of physical interest, namely, cyclic representations (Theorem 8) and irreducible representations (corollaries to Theorems 7 and 8).

Let us now start with the extension of the weak*-topology from $\mathfrak{A}^*$ to $\mathfrak{R}^*$; in doing so we write in the main text the statements relative to $\mathfrak{R}$ and $\mathfrak{R}^*$, and as a reminder we keep within parentheses all statements relative to $\mathfrak{A}$ and $\mathfrak{A}^*$.

Let $\mathfrak{R}$ (or $\mathfrak{A}$) be a C^*-algebra (or a Segal algebra); the *dual* $\mathfrak{R}^*$ (or $\mathfrak{A}^*$) of $\mathfrak{R}$ (or $\mathfrak{A}$) is defined as the set of all continuous linear mappings ϕ from $\mathfrak{R}$

(or $\mathfrak{A}$) to $\mathbb{C}$ (or $\mathbb{R}$). Since $\mathfrak{R}$, $\mathfrak{A}$, $\mathbb{C}$, and $\mathbb{R}$ are normed linear spaces, the linear mappings ϕ are continuous if and only if they are bounded, i.e., if and only if

$$\|\phi\| \equiv \sup_{\substack{R\in\mathfrak{R}(\text{or }\mathfrak{A})\\ \|R\|\leqslant 1}} |\langle\phi; R\rangle| < \infty.$$

The definition $\langle\lambda\phi + \mu\psi; R\rangle \equiv \lambda\langle\phi; R\rangle + \mu\langle\psi; R\rangle$ clearly equips $\mathfrak{R}^*$ (or $\mathfrak{A}^*$) with the structure of a linear space over $\mathbb{C}$ (or $\mathbb{R}$); the expression $\|\cdots\|$ defined above is a norm on $\mathfrak{R}^*$ (or $\mathfrak{A}^*$), with respect to which it is complete. We similarly define the *bidual* $\mathfrak{R}^{**}$ (or $\mathfrak{A}^{**}$) or $\mathfrak{R}$ (or $\mathfrak{A}$) as the dual of $\mathfrak{R}^*$ (or $\mathfrak{A}^*$).

We next define a *conjugation* on $\mathfrak{R}^*$ by the relation

$$\langle\phi^*; R\rangle = \langle\phi; R^*\rangle^*$$

and say that an element ϕ of $\mathfrak{R}^*$ is *hermitian* whenever $\phi^* = \phi$; we denote by $\mathfrak{R}_h^*$ the set of all continuous, hermitian linear functionals from $\mathfrak{R}$ to $\mathbb{C}$ and we note the following:

Theorem 5. *Let $\mathfrak{R}$ be a C*-algebra and $\mathfrak{A}$ be the Segal algebra of all the self-adjoint elements of $\mathfrak{R}$; then $\mathfrak{R}_h^* = \mathfrak{A}^*$.*

Proof. We first notice that an element ϕ of $\mathfrak{R}^*$ is hermitian if and only if it sends $\mathfrak{A}$ into $\mathbb{R}$; for an arbitrary element ϕ of $\mathfrak{R}_h^*$ let us distinguish temporarily

$$\|\phi\|_{\mathfrak{R}} \equiv \sup_{\substack{R\in\mathfrak{R}\\ \|R\|\leqslant 1}} |\langle\phi; R\rangle|$$

$$\|\phi\|_{\mathfrak{A}} \equiv \sup_{\substack{R\in\mathfrak{A}\\ \|R\|\leqslant 1}} |\langle\phi; R\rangle|;$$

obviously $\|\phi\|_{\mathfrak{A}} \leqslant \|\phi\|_{\mathfrak{R}}$ so that the restriction of ϕ to $\mathfrak{A}$ is bounded, hence belongs to $\mathfrak{A}^*$; furthermore for all $\varepsilon > 0$ there exists R in $\mathfrak{R}$ such that $\|R\| \leqslant 1$ and $|\langle\phi; R\rangle| \geqslant \|\phi\| - \varepsilon$. We form $S = \{\langle\phi; R^*\rangle/|\langle\phi; R\rangle|\}R$; since ϕ is hermitian, $\langle\phi; S\rangle = |\langle\phi; R\rangle|$. Hence we can assume that there exists S in $\mathfrak{R}$ such that $\langle\phi; S\rangle \geqslant \|\phi\|_{\mathfrak{R}} - \varepsilon$ and $\|S\| \leqslant 1$; we now form $S_+ = \frac{1}{2}(S + S^*) \in \mathfrak{A}$; we then have

$$\langle\phi; S_+\rangle = \tfrac{1}{2}(\langle\phi; S\rangle + \langle\phi; S^*\rangle) = \langle\phi; S\rangle \geqslant \|\phi\|_{\mathfrak{R}} - \varepsilon$$

$$\|S_+\| = \tfrac{1}{2}\|S + S^*\| \leqslant \tfrac{1}{2}(\|S\| + \|S^*\|) = \|S\| \leqslant 1,$$

from which we conclude that $\|\phi\|_{\mathfrak{A}} \geqslant \|\phi\|_{\mathfrak{R}}$, hence $\|\phi\|_{\mathfrak{A}} = \|\phi\|_{\mathfrak{R}}$. This is to say that the norm of a *hermitian* element of $\mathfrak{R}^*$ is given by

$$\|\phi\| = \sup_{\substack{A\in\mathfrak{A}\\ \|A\|\leqslant 1}} |\langle\phi; A\rangle|.$$

Hence the restriction to $\mathfrak{A}$ of a hermitian element of $\mathfrak{R}^*$ is an *isometric* mapping from $\mathfrak{R}_h^*$ into $\mathfrak{A}^*$. Since an arbitrary element of $\mathfrak{R}^*$ is determined, through linearity, by its restriction to $\mathfrak{A}$, this mapping is injective; to show that it is also surjective we define for an arbitrary element ψ of $\mathfrak{A}^*$, $\langle\phi; R\rangle = \langle\psi; R_+\rangle + i\langle\psi; R_-\rangle$, which belongs to $\mathfrak{R}_h^*$; hence every element of $\mathfrak{A}^*$ can be regarded as the restriction to $\mathfrak{A}$ of an element of $\mathfrak{R}_h^*$. We have constructed a bijective isometric mapping from $\mathfrak{R}_h^*$ onto $\mathfrak{A}^*$; hence $\mathfrak{R}_h^*$ and $\mathfrak{A}^*$ are isomorphic Banach spaces, which is the assertion of the theorem. ∎

An element ϕ of $\mathfrak{R}_h^*$ (or $\mathfrak{A}^*$) is said to be *positive* whenever $\langle\phi; R^*R\rangle \geqslant 0$ for all R in $\mathfrak{R}$ (or $\mathfrak{A}$); we denote the set of all these elements by $\mathfrak{R}_+^*$ (or $\mathfrak{A}_+^*$); from the preceding theorem the identification of $\mathfrak{R}_+^*$ and $\mathfrak{A}_+^*$ is trivial. Furthermore we recall that, for all these elements, $\|\phi\| = \langle\phi; I\rangle$. A *state* ϕ on $\mathfrak{R}$ (or $\mathfrak{A}$) is a positive linear functional normalized to 1 (i.e., $\langle\phi; I\rangle = 1$). Hence the states on $\mathfrak{R}$ and $\mathfrak{A}$ are in one-to-one correspondence; we therefore denote by the same symbol $\mathfrak{S}$ the set of all these states.

We might also mention (Dixmier [1964] 2.6.4 and 12.3.4) that every hermitian continuous linear functional ϕ on a C^*-algebra can be written in a unique way as the difference of two positive linear functionals ϕ_1 and ϕ_2 in such a manner that $\|\phi\| = \|\phi_1\| + \|\phi_2\|$. Obviously, the objects of direct physical significance are the elements of the convex set $\mathfrak{S} = \mathfrak{A}_+^*/\mathbb{R}_+ = \mathfrak{R}_+^*/\mathbb{R}_+$; from them we can construct the elements of the positive cone $\mathfrak{A}_+^* = \mathfrak{R}_+^*$, then the real Banach space $\mathfrak{A}^* = \mathfrak{R}_h^*$, and finally the complex Banach space $\mathfrak{R}^*$; in physical applications, however, we should always remember that these structures draw their physical relevance from $\mathfrak{S}$ only.

The next question is whether the norm (or "metric") topology on $\mathfrak{R}^*$ is the right topology to be introduced on this set; as we have seen, the norm itself has a direct physical significance given by $\|\phi\| = \langle\phi; I\rangle$ for the elements of $\mathfrak{R}_+^*$. The question, however, is whether it leads to the correct mathematical expression of the physical notion of the neighborhood of a state. To say that a sequence $\{\phi_n\}$ of states converges in the norm topology toward a state ϕ means that for any $\varepsilon > 0$ there exists a positive integer $N(\varepsilon)$ such that $\|\phi_n - \phi\| < \varepsilon$ for all $n > N(\varepsilon)$; i.e., explicitly,

$$\sup_{\substack{A\in\mathfrak{A}\\ \|A\|\leqslant 1}} |\langle\phi_n - \phi; A\rangle| < \varepsilon \quad \text{for all } n > N(\varepsilon)$$

(see Theorem 5); the actual verification of this assertion *in the laboratory* has all the appearances of a physicist's nightmare, since, given ε, we should have to verify with that fixed ε that for all observables in $\mathfrak{A}$

$$|\langle\phi_n - \phi; A\rangle| < \varepsilon\, \|A\| \quad \text{for all } n > N(\varepsilon),$$

whereas computing the norm of A itself requires the knowledge of the expectation value of A on all states or at least on all dispersion-free states on A

(see Chapter 1, Section 2):

$$\|A\| = \sup_{\phi\in\mathfrak{S}} |\langle\phi;\, A\rangle| = \sup_{\phi\in\mathfrak{S}_A} |\langle\phi;\, A\rangle|.$$

It appears highly desirable then to have a concept of convergence which would be more accessible to experimental verification; it seems quite reasonable, from a physical point of view, to say that a sequence $\{\phi_n\}$ of states approximates a state ϕ if, given any observable A in $\mathfrak{A}$ and any $\varepsilon > 0$, we can find an integer $N(A, \varepsilon)$ such that

$$|\langle\phi_n - \phi;\, A\rangle| \leqslant \varepsilon \quad \text{for all } n \geqslant N(A, \varepsilon);$$

no further restriction is imposed by the extension of this condition from $\mathfrak{A}$ to $\mathfrak{R}$, that is to say, by requiring that $\{\phi_n\}$ converge to ϕ point-wise on $\mathfrak{R}$. Consequently it seems reasonable to equip $\mathfrak{R}^*$ (or $\mathfrak{A}^*$), considered as the extension of $\mathfrak{S}$, with the weakest possible topology such that the functions $\phi \to \langle\phi;\, R\rangle$ from $\mathfrak{R}^*$ (or $\mathfrak{A}^*$) to $\mathbb{C}$ (or $\mathbb{R}$) are continuous for all R in $\mathfrak{R}$ (or $\mathfrak{A}$). We notice now that the form $\langle \ldots;\ldots\rangle$ is bilinear in each of its variables and enjoys two remarkable properties:

1. For every nonzero ϕ in $\mathfrak{R}^*$ (or $\mathfrak{A}^*$) there exists R in $\mathfrak{R}$ (or $\mathfrak{A}$) such that $\langle\phi;\, R\rangle \neq 0$.
2. For every nonzero R in $\mathfrak{R}$ (or $\mathfrak{A}$) there exists ϕ in $\mathfrak{R}^*$ (or $\mathfrak{A}^*$) such that $\langle\phi;\, R\rangle \neq 0$.

These two facts are mathematically expressed by saying that the bilinear form $\langle \ldots;\ldots\rangle$ put $\mathfrak{R}^*$ (or $\mathfrak{A}^*$) and $\mathfrak{R}$ (or $\mathfrak{A}$) in a *duality relation*. This being the case, the topology we just defined† on $\mathfrak{R}^*$ (or $\mathfrak{A}^*$) is called the weak*-topology on $\mathfrak{R}^*$ (or $\mathfrak{A}^*$); a basis of neighborhoods for this topology is obtained by considering *all* sets of the form

$$N(\phi;\, \mathcal{S}, \varepsilon) = \{\psi \in \mathfrak{R}^* \text{ (or } \mathfrak{A}^*) \mid |\langle\phi - \psi;\, S\rangle| < \varepsilon\ \forall\, S \in \mathcal{S}\},$$

where ϕ is in $\mathfrak{R}^*$ (or $\mathfrak{A}^*$), $\mathcal{S}$ is a *finite* subset of $\mathfrak{R}$ (or $\mathfrak{A}$), and $\varepsilon > 0$. We notice in passing that the *weak-topology* on $\mathfrak{R}$ (or $\mathfrak{A}$) is defined analogously as the weakest topology such that the functions $R \to \langle\phi;\, R\rangle$ from $\mathfrak{R}$ (or $\mathfrak{A}$) to $\mathbb{C}$ (or $\mathbb{R}$) are continuous for all ϕ in $\mathfrak{R}^*$ (or $\mathfrak{A}^*$).

As a consequence of the similarity of the definition of the weak*-topology for $\mathfrak{R}^*$ and for $\mathfrak{A}^*$, all the results mentioned or proved in Chapter 1, Section 2 (p. 61 and ff), can be transposed verbatim to the case of C^*-algebras; for instance, we might mention that the discussion carried out there on the

† The nomenclature is not uniform throughout the mathematical literature; Bourbaki [1964] calls this topology the $\sigma(\mathfrak{R}^*, \mathfrak{R})$-topology and is followed in this, in particular, by Dixmier [1964]; Dunford and Schwartz [1957] call it the $\mathfrak{R}$-topology on $\mathfrak{R}^*$; Day [1962] however, uses the term employed here.

extension of partial states can be reproduced step by step in the case of C^*-algebras, with only the minor modification that the vectorial subspaces $\mathfrak{M}$ of $\mathfrak{R}$ on which we define partial states should now be self-adjoint (incidentally, another way to conduct this discussion would be to restrict the partial states under consideration to $\mathfrak{M} \cap \mathfrak{A}$ and then extend them to $\mathfrak{A}$ and to $\mathfrak{R}$). For the sake of completeness we transpose the general results of Chapter 1, Section 2, to the case of C^*-algebras as follows: $\mathfrak{R}^*$ (or $\mathfrak{A}^*$), equipped with its weak*-topology, is a locally convex linear topological space; the closed unit ball $\{\phi \in \mathfrak{R}^* \text{ (or } \mathfrak{A}^*) \mid \|\phi\| \leqslant 1\}$ in $\mathfrak{R}^*$ (or $\mathfrak{A}^*$) is compact in the weak*-topology (Alaoglu's theorem) and consequently a subset of $\mathfrak{R}^*$ (or $\mathfrak{A}^*$) is compact in the weak*-topology if and only if it is closed in the weak*-topology and bounded in the metric topology. An element ϕ of a subset $\mathfrak{K}$ of $\mathfrak{R}^*$ (or $\mathfrak{A}^*$) is said to be an *extremal point* of $\mathfrak{K}$ if it cannot be written as a proper convex combination $\lambda\phi_1 + (1 - \lambda)\phi_2$, with $0 < \lambda < 1$, of two elements $\phi_1 \neq \phi_2$ in $\mathfrak{K}$; furthermore, if $\mathfrak{K}$ is compact in the weak*-topology and convex, it is the w^*-closed convex hull of its extreme points (Krein-Milman theorem), i.e., $\mathfrak{K}$ can be obtained as the weak*-closure of all finite convex combinations of its extreme points. Another way to obtain the w^*-closed convex hull of a subset $\mathfrak{K}$ of $\mathfrak{R}^*$ (or $\mathfrak{A}^*$) containing 0 is via the bipolar theorem, which we now want to recall. The *polar set*† $\mathfrak{K}_p$ of a subset $\mathfrak{K}$ of $\mathfrak{R}^*$ (or $\mathfrak{A}^*$) is defined by

$$\mathfrak{K}_p = \{R \in \mathfrak{R} \text{ (or } \mathfrak{A}) \mid \mathbb{R}\text{e}\, (\langle \phi; R \rangle) \leqslant 1 \ \forall\ \phi \in \mathfrak{K}\};$$

the *bipolar set* $(\mathfrak{K}_p)^p$ of $\mathfrak{K}$ is then defined by

$$(\mathfrak{K}_p)^p = \{\phi \in \mathfrak{R}^* \text{ (or } \mathfrak{A}^*) \mid \mathbb{R}\text{e}\, \langle \phi; R \rangle \leqslant 1 \ \forall\ R \in \mathfrak{K}_p\};$$

the bipolar theorem then asserts that $(\mathfrak{K}_p)^p$ is the weak*-closed convex hull of $\{\mathfrak{K}, 0\}$, i.e., $(\mathfrak{K}_p)^p = {}^{w^*}\overline{\text{co}\{\mathfrak{K}, 0\}}$. We use this fact in the proof of Theorem 6 below.

In the proof of this theorem we shall need some topological notions that we now want to review briefly. We first recall that a *directed set* D is a partially ordered set (we denote the order relation by $\leqslant$) such that for each pair α, β of elements in D there exists an element γ in D that satisfies $\alpha \leqslant \gamma$ and $\beta \leqslant \gamma$. A *net* $\{x_\alpha \mid \alpha \in D\}$ on a set $\mathcal{S}$ is a mapping $x: D \to \mathcal{S}$, where D is a directed set. A net $\{x_\alpha\}$ on a topological space $\mathfrak{X}$ is said to converge to an element $x \in \mathfrak{X}$ if for every neighborhood U of x there is an $\alpha_U \in D$ such that $x_\alpha \in U$ for all $\alpha \geqslant \alpha_U$. The importance of nets is due to the following result:‡

Proposition. *Let $\mathfrak{X}$ be a topological space. A subset $\mathcal{S} \subseteq \mathfrak{X}$ is closed if and only if for every net $\{x_\alpha\}$ on $\mathcal{S}$ which converges to $x \in \mathfrak{X}$ we have $x \in \mathcal{S}$.*

† Here again the nomenclature is not standard throughout the literature; we are following Bourbaki [1964] Chapter IV, Section 1, N° 3, rather than Day [1962], for instance.

‡ For a proof, and a less sketchy review of these notions, see Choquet [1969], Section 4.

Clearly the notion of "net" is a generalization of the concept of "sequence"; this generalization is necessary whenever $\mathfrak{X}$ is not *first countable*, i.e., when it is not true that there exists a countable basis of neighborhoods of each point x in $\mathfrak{X}$. The reason why we have not yet had to introduce this generalization when we deal with a C^*-algebra $\mathfrak{R}$-equipped with the metric topology induced by the norm is precisely that metric spaces are first countable. We might note in this connection that a metric space $\mathfrak{X}$ is *second countable* (i.e., there exists a countable basis of neighborhoods for $\mathfrak{X}$) if and only if it is separable (i.e., there exists a countable subset $\{x_n\}$ of elements of $\mathfrak{X}$, which is dense in $\mathfrak{X}$). We now come back to the set $\mathfrak{S}$ of all states on a C^*-algebra $\mathfrak{R}$. The question then is whether $\mathfrak{S}$, equipped with its w^*-topology, is a metric space. The answer to this question is that this is indeed so if and only if $\mathfrak{R}$ (hence its self-adjoint part $\mathfrak{A}$) is separable. Furthermore, when this is the case $\mathfrak{S}$ is also separable, since it is compact.† Since, however, we also have to consider C^*-algebras that are *not* separable, we do *not* assume that $\mathfrak{S}$, with its w^*-topology, is a metric space and will work with nets instead of sequences.

Earlier we encountered the concept of vector states associated with a representation of a C^*-algebra; the importance of these states is emphasized by the following theorem:

Theorem 6. *Let $\mathfrak{H}$ be a complex Hilbert space and $\mathfrak{B}(\mathfrak{H})$, the set of all bounded operators acting on $\mathfrak{H}$, be equipped with the usual norm, involution, and algebraic operations. Let $\mathfrak{A}$ be a subset of self-adjoint elements of $\mathfrak{B}(\mathfrak{H})$ which inherits a structure of Segal algebra from $\mathfrak{B}(\mathfrak{H})$, with the symmetrized product defined by* $\mathrm{A} \circ \mathrm{B} = \frac{1}{2}(\mathrm{AB} + \mathrm{BA})$; *let $\mathfrak{S}$ be the set of all states on $\mathfrak{A}$ and $\mathfrak{V}$, the set of all vector states of $\mathfrak{A}$. Then* $\mathfrak{S} = \overline{{}^{w^*}\mathrm{co}(\mathfrak{V})}$.

Proof. We recall that with each vector Ψ in $\mathfrak{H}$, with $|\Psi| = 1$, we can associate a state $\langle\psi; A\rangle = (\Psi, A\Psi)$ on $\mathfrak{A}$; $\mathfrak{V}$ is the set of all states obtained in this way. The polar set of $\mathfrak{V}$ is

$$\mathfrak{V}_p = \{A \in \mathfrak{A} \mid (\Psi, A\Psi) \leqslant 1 \ \forall \, \Psi \in \mathfrak{H} \text{ with } |\Psi| = 1\}.$$

We recall that every self-adjoint operator A on $\mathfrak{H}$ can be written in a unique way as the difference $A_+ - A_-$ of two positive operators; clearly $\mathfrak{V}_p$ is identical to the set of all elements A in $\mathfrak{A}$ with $\|A_+\| \leqslant 1$. The bipolar set of $\mathfrak{V}$ is

$$(\mathfrak{V}_p)^p = \{\phi \in \mathfrak{A}^* \mid \langle\phi; A\rangle \leqslant 1 \ \forall \, A \in \mathfrak{A} \text{ with } \|A_+\| \leqslant 1\};$$

for each ϕ in $(\mathfrak{V}_p)^p$, each positive number λ, and each positive element A in $\mathfrak{A}$ we then have

$$1 \geqslant \langle\phi; I - \lambda A\rangle = \langle\phi; I\rangle - \lambda\langle\phi; A\rangle,$$

† For a proof of the last three statements above, see Dunford and Schwartz [1957] I.6.12, V.5.1, and I.6.15; see also Sakai [1965].

from which we conclude that $\langle\phi; A\rangle \geqslant 0$ (i.e., ϕ positive) and $\langle\phi; I\rangle \leqslant 1$ (i.e., $\|\phi\| \leqslant 1$). Hence $(\mathfrak{V}_p)^p$ is contained in the unit ball of $\mathfrak{A}_+^*$. We now prove that $(\mathfrak{V}_p)^p$ actually coincides with the unit ball in $\mathfrak{A}_+^*$ by noticing that for each positive ϕ with $\|\phi\| \leqslant 1$ and each A in $\mathfrak{A}$ with $\|A_+\| \leqslant 1$

$$\langle\phi; A\rangle = \langle\phi; A_+\rangle - \langle\phi; A_-\rangle \leqslant 1 - \langle\phi; A_-\rangle \leqslant 1.$$

On the other hand, we know from the bipolar theorem that $(\mathfrak{V}_p)^p$ is the weak* closed convex hull of $\{\mathfrak{V}, 0\}$; hence the unit ball in $\mathfrak{A}_+^*$ coincides with ${}^{w^*}\overline{\mathrm{co}(\mathfrak{V}, 0)}$. We finally recall that the states are precisely those elements in $\mathfrak{A}_+^*$ that are of norm 1, hence attribute the expectation value 1 to the element I in $\mathfrak{A}$; if, then, a net $\{\phi_\alpha\}$ of elements in the convex hull $\mathrm{co}(\mathfrak{V}, 0)$ of $\{\mathfrak{V}, 0\}$ is to approximate ϕ in $\mathfrak{S}$, we can assume without loss of generality that these ϕ_α are normalized to 1, hence belong to the convex hull $\mathrm{co}(\mathfrak{V})$ of $\mathfrak{V}$; this means that $\mathfrak{S}$ coincides with the w^* closure of $\mathrm{co}(\mathfrak{V})$. This concludes the proof of the theorem. ∎

Corollary. *With the assumptions of Theorem 6, let us denote by $\mathfrak{S}^{\mathrm{P}}$ the set of all pure states on $\mathfrak{A}$. Then $\mathfrak{S}^{\mathrm{P}} \subseteq {}^{w^*}\overline{\mathfrak{V}}$.*

Proof. $\mathfrak{A}^*$ is a locally convex Hausdorff space when equipped with the weak*-topology; $\mathfrak{S}$ is a compact convex subset of $\mathfrak{A}^*$; $\mathfrak{S}^P$ is the set of all extremal points of $\mathfrak{S}$. From the theorem we know that $\mathfrak{V}$ is such that ${}^{w^*}\overline{\mathrm{co}(\mathfrak{V})} = \mathfrak{S}$; these properties are exactly enough to prove the statement of the corollary (see Dixmier [1964], B.14, p. 355). ∎

REMARKS

(i) Theorem 6 can be rephrased by saying that for every state ϕ on $\mathfrak{A}$ there exists a net $\{\phi_\alpha\}$ of states ϕ_α obtained as finite convex combinations of vector states, with ϕ_α converging pointwise to ϕ on $\mathfrak{A}$. The corollary asserts that every pure state on $\mathfrak{A}$ can be approximated, pointwise on $\mathfrak{A}$, by a net of vector states on $\mathfrak{A}$; this remark shows in which sense the founders of quantum mechanics did not lose an essential part of the theory by considering only vector states and then density matrices: all states that we want to consider in a given representation can be obtained as weak*-limits of such states. As we have already pointed out, however, not every state on $\mathfrak{A}$ can be written as a *density matrix:*

$$\langle\phi; A\rangle = \mathrm{Tr}\, \rho A \quad \text{with} \quad 0 < \rho = \rho^* \in \mathfrak{B}(\mathcal{H}); \qquad \mathrm{Tr}\, \rho = 1,$$

as witnessed by the counterexample of the pure states which cannot be represented as vector states (see Chapter 1, Section 2, p. 64).

(ii) This theorem extends readily to states on any C^* subalgebra of $\mathfrak{B}(\mathcal{H})$. Kadison [1962] proved in addition that $\mathfrak{V}$ is norm-closed and that the norm closure of co$\mathfrak{B}$ is identical to the set of all states that can be written as density matrices; see also the second lemma on p. 118.

(iii) By looking at the proof again the reader will convince himself that it can be used to conclude that any continuous positive linear functional ϕ on a norm-closed self-adjoint algebra $\mathfrak{R}$ of bounded operators on $\mathcal{H}$ can be approximated in the w^*-topology by positive functionals of the form $\phi_n = \sum_{i=1}^{k_n} (\Psi_i^{(n)}, A\Psi_i^{(n)})$ with $\|\phi_n\| \leqslant \|\phi\|$; it should be remarked that it is in this form that the theorem was proved by Fell [1960] who also pointed out its relevance for the concept of weak equivalence of representations.

Let us now consider the set $\mathfrak{S}_\pi$ of all states on the C^*-algebra $\pi(\mathfrak{R})$, where $\mathfrak{R}$ is a given C^*-algebra and π is a given representation of $\mathfrak{R}$ by bounded operators on a Hilbert space $\mathcal{H}$. In a specific sense, which we now want to define, $\mathfrak{S}_\pi$ can be identified with a subset of the set $\mathfrak{S}$ of all states on $\mathfrak{R}$. For any arbitrary bounded linear functional ϕ on $\pi(\mathfrak{R})$ we define the linear functional $j_\pi(\phi)$ on $\mathfrak{R}$ by $\langle j_\pi(\phi); R\rangle \equiv \langle \phi; \pi(R)\rangle$; we notice immediately that $j_\pi(\phi)$ is bounded and positive whenever ϕ itself is positive, so that j_π is a positive map from $\pi(\mathfrak{R})^*$ into $\mathfrak{R}^*$; furthermore $j_\pi(\phi)$ vanishes on Ker π; conversely every (positive) bounded linear functional ϕ' on $\mathfrak{R}$ which vanishes on Ker π defines a (positive) bounded linear functional ϕ on $\pi(\mathfrak{R})$ via $\langle \phi; \pi(R)\rangle = \langle \phi'; R\rangle$ such that $\phi' = j_\pi(\phi)$; hence j_π is a positive bijective mapping from $\pi(\mathfrak{R})^*$ onto $(\text{Ker}\,\pi)^\perp \equiv \{\phi \in \mathfrak{R}^* \mid \langle \phi; R\rangle = 0\ \forall\ R \in \text{Ker}\,\pi\}$. It is easy to see that this mapping, restricted to $\pi(\mathfrak{R})_+^*$ is actually norm-preserving, so that its restriction to $\mathfrak{S}_\pi$ is a bijective mapping on $(\text{Ker}\,\pi)^\perp \cap \mathfrak{S}$. Furthermore the bijective mapping $j_\pi : \pi(\mathfrak{R})^* \to (\text{Ker}\,\pi)^\perp$ is bicontinuous for the weak*-topologies.† We have already noted that Ker π is a normed-closed two-sided ideal of $\mathfrak{R}$, so that in particular (i) $\pi(\mathfrak{R})$ is canonically isomorphic to $\mathfrak{R}/\text{Ker}\,\pi$ and (ii) Ker π is a weak-closed subspace of $\mathfrak{R}$, hence $((\text{Ker}\,\pi)^\perp)^\perp = \text{Ker}\,\pi$; then, taking Bourbaki's explicit notation for the weak*-topologies, we have‡ that the $\sigma((\text{Ker}\,\pi), \pi(\mathfrak{R}))$-topology on $(\text{Ker}\,\pi)$ is identical with the topology induced on Ker π by the $\sigma(\mathfrak{R}^*, \mathfrak{R})$-topology on $\mathfrak{R}^*$. As a result of this discussion, we see that when we write, for instance, $\mathfrak{S}_\pi = \overline{\text{co}(\mathfrak{B}_\pi)}^{w^*}$ we do not have to specify whether the w^*-topology (with respect to which the closure is made) is considered with respect to the representation π or abstractly, i.e., with respect to $\mathfrak{R}$.

Let us now consider, in line with the scheme set in Chapter 1, Section 2, the complete description $(\mathfrak{A}, \mathfrak{S}, \langle\, ;\, \rangle)$ of a physical system Σ, where the set of all observables can be identified with the set of all self-adjoint elements of a C^*-algebra $\mathfrak{R}$. Let $\pi : \mathfrak{R} \to \mathfrak{B}(\mathcal{H})$ be a representation of $\mathfrak{R}$; the triple

† Dixmier [1964], 2.11.6.
‡ Bourbaki [1964], IV.1.5.

$(\pi(\mathfrak{A}), \mathfrak{S}_\pi, j_\pi)$ then generates a partial description of Σ, which we refer to as *the partial description of Σ associated with the representation π*. It seems natural to say that two representations are physically equivalent whenever the descriptions associated with them are "the same." Several definitions of physical equivalence can be found in the literature, and this concept is often related to that of weak-equivalence.† The following lemma will help to show that (i) all these definitions actually coincide and (ii) they really do express in unambiguous terms what is meant by "physical equivalence."

Lemma. *For any two representations π and ρ of a C^*-algebra $\mathfrak{R}$ the following conditions are equivalent:*

(i) $j_\pi(\mathfrak{S}_\pi) \subseteq j_\rho(\mathfrak{S}_\rho)$.
(ii) *$Ker \rho \subseteq Ker\ \pi$.*
(iii) *Every state ϕ on $\mathfrak{R}$ that is a vector state for the representation π is the w*-limit of a net of finite convex combinations of states on $\mathfrak{R}$ which are vector states associated with the representation ρ.*
(iv) *Every density matrix associated with the representation π can be approximated, pointwise on $\mathfrak{R}$, as close as we want by a density matrix associated with the representation ρ.*

Proof. We prove successively that (i) is equivalent to (ii), (iii) and (iv). (ii) implies that $(\text{Ker}\ \pi)^\perp \subseteq (\text{Ker}\ \rho)^\perp$; hence $(\text{Ker}\ \pi)^\perp \cap \mathfrak{S} \subseteq (\text{Ker}\ \rho)^\perp \cap \mathfrak{S}$; that is to say $j_\pi(\mathfrak{S}_\pi) \subseteq j_\rho(\mathfrak{S}_\rho)$. Conversely, let R be an element of $\mathfrak{R}$ *not* contained in Ker π, i.e., such that $\pi(R) \neq 0$; since $\pi(\mathfrak{R})$ is a C^*-algebra, there exists an element ϕ in $\mathfrak{S}_\pi$ such that $\langle \phi; \pi(R)\rangle \neq 0$, i.e., $\langle j_\pi(\phi); R\rangle \neq 0$. Now (i) means that there exists ψ in $\mathfrak{S}_\rho$ such that $j_\rho(\psi) = j_\pi(\phi)$; for this ψ we have $0 \neq \langle j_\rho(\psi); R\rangle = \langle \psi; \rho(R)\rangle$. Hence $\rho(R) \neq 0$, i.e., $R \notin \text{Ker}\ \rho$; hence $\text{Ker}\ \rho \subseteq \text{Ker}\ \pi$. This achieves the proof that (i)$\Leftrightarrow$(ii). Written mathematically, rather than in words, (iii) is $j_\pi(\mathfrak{V}_\pi) \subseteq {}^{w^*}\overline{j_\rho(\text{co}\,\mathfrak{V}_\rho)}$; we know that for any representation ν of $\mathfrak{R}$, $\mathfrak{S}_\nu = {}^{w^*}\overline{\text{co}\,\mathfrak{V}_\nu}$ and that j_ν is bicontinuous for the weak*-topologies; therefore (iii) can be rewritten as $j_\pi(\mathfrak{V}_\pi) \subseteq j_\rho(\mathfrak{V}_\rho)$; this condition is then satisfied when (i) holds; conversely, on taking the w^*-closed convex hull of our last form of condition (iii) we receive (i). This concludes the proof that (i)$\Leftrightarrow$(iii). To prove the equivalence of (i) and (iv) we proceed along the same path by replacing $\mathfrak{V}_\nu$ with $\mathfrak{D}_\nu$ (for $\nu = \pi, \rho$; $\mathfrak{D}_\nu$ denotes the convex set of all density matrices associated with the representation ν) and noticing that $\mathfrak{V}_\nu \subseteq \mathfrak{D}_\nu \subseteq \mathfrak{S}_\nu$ and ${}^{w^*}\overline{\text{co}\,\mathfrak{V}_\nu} = \mathfrak{S}_\nu$ imply ${}^{w^*}\overline{\mathfrak{D}_\nu} = \mathfrak{S}_\nu$. This concludes the proof of the equivalence of the four conditions of the lemma. ∎

† See, for instance, Haag and Kastler [1964], Kastler [1964], and, for the mathematical origin of these discussions, Fell [1960].

When two representations satisfy any of the equivalent conditions of the lemma, we say that π is *weakly contained* in ρ; this concept is due to Fell [1960] who introduced it in a somewhat more general context, in which, instead of considering just one representation ρ, he considered a family of representations; however, we shall not need this extension in the sequel.

We now say in unambiguous terms that two representations π and ρ are *physically equivalent* whenever $j_\pi(\mathfrak{S}_\pi) = j_\rho(\mathfrak{S}_\rho)$, and we have the following theorem as an immediate consequence of the lemma:

Theorem 7. *Two representations π and ρ of a* C**-algebra are physically equivalent* if and only if *any of the four conditions of the preceding lemma holds true together with any of these four conditions in which the roles of π and ρ are exchanged.*

We should notice that this theorem gives its full meaning to the concept of physical equivalence: two representations π and ρ are physically equivalent if and only if $\mathfrak{S}_\pi$ and $\mathfrak{S}_\rho$ are identical as subsets of $\mathfrak{S}$ (i.e., via the canonical embeddings j_π and j_ρ) or *equivalently* if and only if Ker π = Ker ρ; we recall in this connection that, for any representation ν of $\mathfrak{R}$, $\nu(\mathfrak{R})$ is isomorphic to $\mathfrak{R}$/Ker ν; hence $\pi(\mathfrak{R})$ [resp. $\pi(\mathfrak{A})$] and $\rho(\mathfrak{R})$ [resp. $\rho(\mathfrak{A})$] are isomorphic. Consequently, the two partial descriptions of the physical system under consideration, associated respectively with π and with ρ, are really the same in the sense in which we defined (see Chapter 1, Section 2) the notion of the mathematical description of a physical system. We notice further, as a particular case of the theorem, that all faithful representations of a C^*-algebra are physically equivalent to one another and generate a description of the physical system under consideration that is identical to the original description $(\mathfrak{A}, \mathfrak{S}, \langle;\rangle)$; when $\mathfrak{R}$ is simple, we have already noticed that all its nonzero representations are faithful; therefore *all nonzero representations of a simple* C**-algebra are physically equivalent*. We finally remark, as a rephrasing of one pair of conditions of the theorem, that if we wish to attach some special importance to the density matrices associated with a given representation we can use the fact that, whenever two representations are physically equivalent, every weak*-neighborhood of any density matrix associated with one of these representations also contains a density matrix associated with the other representation. We recall in this connection that physical measurements determine at best w^*-neighborhoods in $\mathfrak{S}$ due to the inherent imprecision attached to any actual measurements.

We now want to discuss briefly the relation between physical and unitary equivalence. It should be clear that unitary equivalence implies physical equivalence; indeed we have immediately that two representations π and ρ which are unitarily equivalent have the same kernel, hence are physically equivalent (Theorem 7). An alternative way to see this implication is to use

Theorem 2 to conclude from the unitary equivalence of π and ρ that $j_\pi(\mathfrak{B}_\pi) = j_\rho(\mathfrak{B}_\rho)$; then, taking the weak*-closed hull of this relation, we get $j_\pi(\mathfrak{S}_\pi) = j_\rho(\mathfrak{S}_\rho)$, which is to say that π and ρ are physically equivalent. This second way of proceeding suggests that physical equivalence is a *weaker* requirement than unitary equivalence, since the requirement $\{j_\mu(\mathfrak{S}_\mu) \supseteq j_\rho(\mathfrak{B}_\rho)$ and $j_\rho(\mathfrak{S}_\rho) \supseteq j_\mu(\mathfrak{B}_\mu)\}$ seems less stringent than the requirement $\{j_\rho(\mathfrak{B}_\rho) = j_\mu(\mathfrak{B}_\mu)\}$; actually, there are examples of representations that are physically equivalent without being unitarily equivalent. One of the most relevant is the case of the canonical anticommutation relations: the C^*-algebra generated by the CAR is simple,† so that all its representations are physically equivalent. We saw, however, in Chapter 1, Section 1, that in the same physical problem (namely, the BCS model) several unitarily inequivalent representations of this algebra might occur; we shall actually see later on that "many" unitarily inequivalent irreducible representations of this algebra do exist. This counterexample of representations, which are physically equivalent, though not unitarily equivalent, suggests two further remarks. First, we should keep in mind that the physics attached to the various representations of the CAR occurring in the BCS model is different, although these representations are "physically equivalent"; this should be taken as an indication of the limitations of the concept of physical equivalence. The latter means just what is contained in Theorem 7, and we should be seriously wary of semantic extrapolations. Second, since we know that even irreducible representations can be physically equivalent without being unitarily equivalent, it is of some interest to discuss whether the notion of physical equivalence takes some special form in the case of irreducible representations; this will be done in part by the next corollary to Theorem 7 and completed by the corollary to Theorem 8.

Corollary. *Two irreducible representations π and ρ are physically equivalent if and only if* $\overline{^{w^*}j_\pi(\mathfrak{B}_\pi)} = \overline{^{w^*}j_\rho(\mathfrak{B}_\rho)}$.

Proof. Suppose first that $\overline{^{w^*}j_\pi(\mathfrak{B}_\pi)} = \overline{^{w^*}j_\rho(\mathfrak{B}_\rho)}$; then $j_\pi(\mathfrak{B}_\pi) \subseteq j_\rho(\mathfrak{S}_\rho)$ and $j_\rho(\mathfrak{B}_\rho) \subseteq j_\pi(\mathfrak{S}_\pi)$; by Theorem 7 this implies that π and ρ are physically equivalent. Now suppose that π and ρ are physically equivalent, so that by Theorem 7 in particular Ker π = Ker ρ. Let ϕ be a vector state associated with the representation π (i.e., $\phi \in \mathfrak{B}_\pi$); since π is irreducible $j_\pi(\phi)$ is pure (see the first corollary to Theorem 3); since Ker $\rho \subseteq$ Ker π, $j_\pi(\phi)$ can be considered

† See below: Theorem III.2.2 (Guichardet [1966], Proposition 3.1; also Glimm [1960] Theorem 5.1, and Shale and Stinespring [1964]); Doplicher and Powers [1968] showed, furthermore, that the same holds true for the algebra of all quasilocal observables for a free Fermi-Dirac field, i.e., for the C^*-algebra generated by the even polynomials in the field operators with test-functions of compact support; see below, Theorem IV.2.3 (Størmer [1970]).

as a state on $\rho(\mathfrak{R})$, i.e., there exists a state $\psi \in \mathfrak{S}_\rho$ such that $j_\rho(\psi) = j_\pi(\phi)$: we verify *ad absurdo* that ψ is pure. We then have $j_\pi(\phi) \in j_\rho(\mathfrak{S}_\rho^P)$; from the corollary to Theorem 6 $\mathfrak{S}_\rho^P \subseteq {}^{w^*}\overline{\mathfrak{B}}_\rho$: furthermore, since j_ρ is continuous, we get $j_\pi(\phi) \in {}^{w^*}\overline{j_\rho(\mathfrak{B}_\rho)}$, i.e., since ϕ was arbitrary in $\mathfrak{B}_\pi : j_\pi(\mathfrak{B}_\pi) \subseteq {}^{w^*}\overline{j_\rho(\mathfrak{B}_\rho)}$. Exchanging the roles of π and ρ, we get the statement of the corollary. ▮

To conclude our discussion of the concept of physical equivalence, we now want to see its implications for the representations we are particularly interested in, namely, those obtained via the GNS construction from states. To do so we need the following lemma:

Lemma. *Let $\mathfrak{R}$ be a C*-algebra, ϕ, a state on $\mathfrak{R}$, and π_ϕ, the GNS representation associated with ϕ. Then Ker π_ϕ is the largest norm-closed two-sided ideal contained in Ker ϕ.*

Proof. We have already seen that Ker π_ϕ is a two-sided, norm-closed ideal of $\mathfrak{R}$; clearly Ker $\pi_\phi \subseteq$ Ker ϕ. Now suppose that $\mathfrak{K}$ is a norm-closed, two-sided ideal contained in Ker ϕ; for all K in $\mathfrak{K}$, R^*KS is also contained in $\mathfrak{K}$, hence in Ker ϕ, so that $\langle \phi; R^*KS\rangle = 0$; by the GNS construction this is $(\Phi(R), \pi_\phi(K)\,\Phi(S)) = 0$. Since $\{\Phi(R) \mid R \in \mathfrak{R}\}$ is dense in $\mathcal{H}_\phi$, this implies $\pi_\phi(K) = 0$, i.e., $\mathfrak{K} \subseteq$ Ker π_ϕ, which proves the lemma. ▮

With the information gathered up to this point, this lemma allows a simple proof of the following lemma:

Lemma. *Let $\mathfrak{R}$ be a C*-algebra, ϕ, a state on $\mathfrak{R}$, ρ, an arbitrary representation of $\mathfrak{R}$, and π, the representation of $\mathfrak{R}$ associated with ϕ by the GNS construction. Then the following conditions are equivalent:*

(i) $$\phi \in {}^{w^*}\overline{j_\rho(co\mathfrak{B}_\rho)}$$

(ii) $$Ker\rho \subseteq Ker\phi$$

(iii) $$Ker\rho \subseteq Ker\pi$$

(iv) $$j_\pi(\mathfrak{B}_\pi) \subseteq {}^{w^*}\overline{j_\rho(co\mathfrak{B}_\rho)}$$

(v) $$j_\pi(\mathfrak{D}_\pi) \subseteq {}^{w^*}\overline{j_\rho(\mathfrak{D}_\rho)}$$

(vi) $$j_\pi(\mathfrak{S}_\pi) \subseteq j_\rho(\mathfrak{S}_\rho)$$

Proof. (i) implies $\phi \in j_\rho(\mathfrak{S}_\rho)$, hence (ii); by the preceding lemma (ii) $\Rightarrow$ (iii); by the lemma preceding Theorem 7, (iii) implies (iv), (v), and (vi), each of these conditions, in turn, trivially implying (i). ▮

As an immediate consequence of this lemma we can extend Theorem 7 to the following form:

Theorem 8. *The GNS representations π and ρ, respectively associated with two states ϕ and ψ of a C*-algebra $\mathfrak{R}$, are physically equivalent* if and only if *any of the six conditions of the preceding lemma holds true* together *with any of these six conditions in which the roles of* (ϕ, π) *and* (ψ, ρ) *are exchanged.*

Corollary. *Two irreducible representations* π *and* ρ *of a* C^**-algebra* $\mathfrak{R}$ *are physically equivalent if and only if there exists at least one* $\phi \in \mathfrak{V}_\pi$ *and one* $\psi \in \mathfrak{V}_\rho$ *such that* $j_\pi(\phi) \in \overline{j_\rho(\mathfrak{V}_\rho)}^{w^*}$ *and* $j_\rho(\psi) \in \overline{j_\pi(\mathfrak{V}_\pi)}^{w^*}$.

Proof. Suppose first that there exists $\phi \in \mathfrak{V}_\pi$ such that $j_\pi(\phi) \in \overline{j_\rho(\mathfrak{V}_\rho)}^{w^*}$; π irreducible implies that the vector Φ corresponding to ϕ is cyclic (see lemma, p. 84) and π is unitarily equivalent to π_ϕ (Theorem 1). From the lemma preceding Theorem 8 we then get that π_ϕ is weakly contained in ρ; since π is equivalent to π_ϕ, π is also weakly contained in ρ. Exchanging the roles of π and ρ, we see that physical equivalence follows from the condition of the corollary. That this condition is necessary for physical equivalence follows directly from the corollary to Theorem 7. ∎

To complete our survey of the concept of physical equivalence we should emphasize that the considerations developed in this subsection have been strictly restricted to those aspects of the descriptions of physical systems that involved the global representations of the algebra of observables and their associated states; no other structure, such as the relation of these descriptions to the space-time properties of the system considered, has been invoked. In the latter connection we might point out that the concept of physical equivalence historically entered the realm of the C^*-algebraic approach as a somewhat marginal outgrowth of an investigation by Borchers, Haag, and Schroer [1963] on the existence of a vacuum in local quantum field theories: a comparison of their Theorem IV and the corollary to Theorem 8 above would be instructive. We feel, however, that applications of this kind are best postponed at this point as some of the tools (e.g., action of symmetry groups) and postulates (e.g., on the local structure of the algebra considered) are yet to be introduced.

e. von Neumann Algebras and Quasi-Equivalence of Representations

In Subsection b we encountered the concept of the von Neumann algebra $\pi(\mathfrak{R})''$ generated by a representation $\pi: \mathfrak{R} \to \mathfrak{B}(\mathcal{H})$ of a C^*-algebra $\mathfrak{R}$. This construct provides a useful tool in the study of the representations of C^*-algebras; the aim of this subsection is to review those properties of von Neumann algebras that appear to be most relevant for this purpose.

Given a complex Hilbert space $\mathcal{H}$ and the C^*-algebra $\mathfrak{B}(\mathcal{H})$ of all bounded operators on $\mathcal{H}$, equipped with the usual algebraic operations, the operator-norm and the involution provided by the adjunction, we define a *von Neumann* *algebra* $\mathfrak{N}$ as a *-subalgebra of $\mathfrak{B}(\mathcal{H})$ such that $\mathfrak{N} = \mathfrak{N}''$; explicitly we then have

(i) $A, B \in \mathfrak{N}$: $\lambda, \mu \in \mathbb{C} \Rightarrow \lambda A + \mu B \in \mathfrak{N}$, and $AB \in \mathfrak{N}$

(ii) $A \in \mathfrak{N} \Rightarrow A^* \in \mathfrak{N}$

(iii) $B \smile \smile A \ \forall A \in \mathfrak{N} \Rightarrow B \in \mathfrak{N}$,

where $A \smile B$ means that the commutator $[A, B] = AB - BA$ vanishes, and $B \smile \smile A$ means $B \smile C$ for all C such that $C \smile A$.

Some examples might help: trivially $\{\lambda I\}$ and $\mathfrak{B}(\mathcal{H})$ are von Neumann algebras; we notice that they are C^*-algebras as well; this raises the question whether every von Neumann algebra is a concrete C^*-algebra, i.e., a C^*-subalgebra of $\mathfrak{B}(\mathcal{H})$ for the Hilbert space $\mathcal{H}$ on which it is defined; we shall see that the answer is yes, which makes von Neumann algebras special types of C^*-algebras so that all theorems proved for C^*-algebras apply as well to von Neumann algebras. The converse, however, is not true; consider, for instance, the C^*-subalgebra $\mathfrak{A}$ of $\mathfrak{B}(\mathcal{H})$ which consists of all compact operators. Its bicommutant $\mathfrak{A}''$ is $\mathfrak{B}(\mathcal{H})$ so that $\mathfrak{A}$ is properly contained in $\mathfrak{A}''$ whenever $\mathcal{H}$ is infinite dimensional; in this connection we might take notice† of the fact that the bidual $\mathfrak{A}^{**}$ of $\mathfrak{A}$ is isomorphic, as a Banach space, to $\mathfrak{B}(\mathcal{H})$. This is no accident and is actually an illustration of a very general theorem which, as we shall see in the sequel, allows us to associate canonically a von Neumann algebra with every abstract C^*-algebra.

An abstract characterization of von Neumann algebras can be given as follows‡: a *W^*-algebra* is a C^*-algebra which is the dual of some Banach space; we can then show that every W^*-algebra has a faithful representation as a von Neumann algebra defined on some Hilbert space. This way to approach von Neumann algebras, however elegant it might be from the mathematical point of view, seems to have little directly interpretable physical meaning and we shall not pursue this issue any further.

For the general study of von Neumann algebras the reader is referred to Dixmier [1957] and [1964, Appendix A], Sakai [1962], or Schwartz [1967].

We now want to start our survey by recalling without proof the following lemma and its corollary:

Lemma. *A bounded normal operator* T *on a complex Hilbert space* $\mathcal{H}$ *uniquely determines a regular countably additive self-adjoint spectral measure* P *on the Borel sets of the complex plane, which vanishes on the resolvant set* ρ(T) *of* T *and has the property that*

$$f(T) = \int_{\sigma(T)} f(\lambda)\, dP_\lambda \ \forall f \in \mathcal{C}(\sigma(T)).$$

† See Schatten [1950]; von Neumann and Schatten [1946]; or Dixmier [1950].
‡ See, for instance, Sakai [1962].

This result, known as the *spectral theorem* for normal operators, is proved† in almost any treatise that has some connection with functional analysis; we mention here only that the standard proof nowadays makes use of the GNS representation theorem for commutative C^*-algebras (see Theorem 1.9). The proper understanding of this result calls for some definitions that we want to recall briefly.

A bounded operator T on a Hilbert space is said to be *normal* if $TT^* = T^*T$; in particular, self-adjoint or unitary operators are normal, and we consider the particularization of the above lemma to self-adjoint operators in the next corollary. A spectral measure generalizes to projection-valued measures the situation reviewed on p. 57 for real-valued measures; specifically, let $\mathcal{S}$ be a σ-algebra (i.e., a σ-ring with identity) and let $\mathcal{P}(\mathcal{H})$ be the set of all projectors on $\mathcal{H}$; a *self-adjoint spectral measure* is a homomorphism $P: \mathcal{S} \to \mathcal{P}(\mathcal{H})$ such that $P(1) = I$ (and evidently $P(\phi) = 0$); we say that P is *countably additive* if for every sequence $\{S_i \mid i = 1, 2, \ldots\}$of disjoint elements of $\mathcal{S}$

$$P\left(\bigcup_{i=1}^{\infty} S_i\right) = \sum_{i=1}^{\infty} P(S_i)$$

(the convergence of the sum being understood equivalently in the strong or weak operator topologies); the regularity of a self-adjoint spectral measure defined on the Borel sets $\mathcal{S}(\mathfrak{E})$ of a locally compact Hausdorff space $\mathcal{X}$ is defined again equivalently with respect to the strong or weak operator topologies as the natural generalization from the case of real-valued measures; we say, indeed, that P is *regular* if for every Ψ in $\mathcal{H}$, every Borel set S and every $\varepsilon > 0$ there exists a closed set $F \subseteq S$ and an open set $G \supseteq S$ such that $|P(S')\Psi| < \varepsilon$ for every Borel set S' contained in $G - F$. We then recall that the *resolvant set* $\rho(T)$ of T is defined as the open set constituted by the complex numbers such that $(\lambda I - T)^{-1}$ exists as a bounded operator *on* $\mathcal{H}$. The *spectrum* $\sigma(T)$ of T is defined as the complement of $\rho(T)$, $\mathcal{C}(\sigma(T))$ denotes the C^*-algebra of all (bounded) continuous complex-valued functions on $\sigma(T)$, and we further denote by $\mathcal{C}^*(T)$ the smallest C^*-subalgebra of $\mathfrak{B}(\mathcal{H})$ generated by I and T; $\mathcal{C}^*(T)$ is commutative, since T is normal. We can then use the C^*-equivalent of Theorem 1.9 and define $f(T)$ as the element of $\mathcal{C}^*(T)$ corresponding to $f \in \mathcal{C}(\sigma(T))$ via the canonical isometric *-isomorphism from $\mathcal{C}^*(T)$ to $\mathcal{C}(\sigma(T))$. Finally we define for every f in $\mathcal{C}(\sigma(T))$ the integral of the right-hand side of the formula appearing in the lemma in complete analogy with the definition of the corresponding integral for real-valued measures, with only the supplementary requirement that we understand the limiting procedure involved in this extension to be carried out in the uniform operator topology. This completes the list of definitions pertinent to the

† See, for instance, Dunford and Schwartz [1957] X.2.4, or Naimark [1964] IV.17.4.

understanding of the lemma. For bounded self-adjoint operators this spectral theorem implies the following well-known† result:

Corollary. *Let* $\mathcal{H}$ *be a complex Hilbert space,* $\mathrm{mI} \leqslant \mathrm{A} \leqslant \mathrm{MI}$ *(with* $\mathrm{m}, \mathrm{M} \in \mathbb{R}$*) be a self-adjoint operator acting on* $\mathcal{H}$*. Then there exists a unique family* $\{\mathrm{P}_\lambda \mid \lambda \in \mathbb{R}\}$ *of projectors* (i.e., $\mathrm{P}_\lambda^* = \mathrm{P}_\lambda = \mathrm{P}_\lambda^2$) *that satisfies the following conditions:*

(i) $\mathrm{P}_\lambda \mathrm{P}_\mu = \mathrm{P}_\lambda$ *whenever* $\lambda \leqslant \mu$;
(ii) $\mathrm{P}_\lambda = 0$ *for* $\lambda < \mathrm{m}$ *and* $\mathrm{P}_\lambda = \mathrm{I}$ *for* $\lambda \geqslant \mathrm{M}$;
(iii) $\mathrm{P}_\lambda \Psi$ *is a semicontinuous (from the right) function of* λ *for all* $\Psi \in \mathcal{H}$;
(iv) $\mathrm{P}_\lambda \smile\smile \mathrm{A}$;
(v) *if* $\mathrm{f}(\lambda)$ *is continuous on* $[\mathrm{m}, \mathrm{M}]$ $\mathrm{f}(\mathrm{A}) = \int_{\mathrm{m}}^{\mathrm{M}} \mathrm{f}(\lambda)\, d\mathrm{P}_\lambda$.

In particular $\mathrm{A} = \int_{\mathrm{m}}^{\mathrm{M}} \lambda\, d\mathrm{P}_\lambda$*; for each value of* λ, P_λ *is the strong limit of a sequence of polynomials in* A.

We are now equipped to start our study of von Neumann algebras with the following result:

Theorem 9. *Let* $\mathfrak{N}$ *be a von Neumann algebra; let us denote by* $\mathcal{I}$ *the set of all projectors in* $\mathfrak{N}$ *and by* $\mathcal{U}$ *the set of all unitary operators in* $\mathfrak{N}$*. Then* $\mathfrak{N} = \mathcal{I}'' = \mathcal{U}''$*.*

Proof. $\mathcal{I} \subseteq \mathfrak{N}$ implies $\mathcal{I}'' \subseteq \mathfrak{N}'' = \mathfrak{N}$; to prove $\mathfrak{N} \subseteq \mathcal{I}''$ assume first that $A = A^*$ in $\mathfrak{N}$. Let $\mathfrak{M}$ be the von Neumann algebra generated by A: $\mathfrak{M} = \{A\}'' \subseteq \mathfrak{N}'' = \mathfrak{N}$. Now let $\{P_\lambda\}$ be the spectral family associated with A by the corollary to the preceding lemma; $P_\lambda \smile\smile A$, i.e., $P_\lambda \in \mathfrak{M}$; hence $P_\lambda \in \mathcal{I}$. For any X in $\mathcal{I}'$ we have, in particular, $X \smile P_\lambda$; hence by the preceding lemma $X \smile A$, i.e., $A \in \mathcal{I}''$. Since any B in $\mathfrak{N}$ can be written as a finite linear combination of self-adjoint elements in $\mathfrak{N}$ ($\mathfrak{N}$ is a *-algebra!), we conclude that $\mathfrak{N} \subseteq \mathcal{I}''$; together with the first assertion of the proof this implies that $\mathfrak{N} = \mathcal{I}''$. To prove that $\mathfrak{N} = \mathcal{U}''$ we prove that every element A in $\mathfrak{N}$ can be written as a finite linear combination of unitary operators in $\mathfrak{N}$. Since $\mathfrak{N}$ is a *-algebra, we can assume without loss of generality that $A = A^*$ and $\|A\| < 1$. We then have $I - A^2 \geqslant 0$; $(I - A^2)^{1/2}$ still belongs to $\mathfrak{N}$ (this can be seen from the spectral theorem). We then form $U = A + i(I - A^2)^{1/2}$ and notice that U is a unitary operator in $\mathfrak{N}$; furthermore $\frac{1}{2}(U + U^*) = A$. From this we conclude that $\mathfrak{N} \subseteq \mathcal{U}''$, and since $\mathcal{U}'' \subseteq \mathfrak{N}$ we have $\mathfrak{N} = \mathcal{U}''$. ∎

In some sense this theorem indicates that a von Neumann algebra is generated by its projectors as well as by its unitary elements. Theorem 10 clarifies this statement. For any subset $\mathcal{M}$ in $\mathfrak{B}(\mathcal{H})$ let us now denote by $\mathfrak{F}(\mathcal{M})$ the *-algebra generated from $\mathcal{M}$ by forming all finite linear combinations and all finite products of elements in $\mathcal{M}$ and in $\mathcal{M}^* \equiv \{M^* \mid M \in \mathcal{M}\}$.

† See, for instance, Riesz and Sz.-Nagy [1955] or Akhiezer and Glazman [1961].

From the proof of Theorem 9 we see that $\mathfrak{F}(\mathfrak{U}) = \mathfrak{R}$, so that in a *-algebraic sense $\mathfrak{R}$ is actually generated by its unitary elements. In general, the situation is not so simple with $\mathfrak{F}(\mathfrak{T})$.† There is, however, a general result which we want to mention (see remark following Theorem 10). For this we need a few topological properties of $\mathfrak{B}(\mathcal{H})$ which we now want to review briefly.

At least five useful topologies can be defined on $\mathfrak{B}(\mathcal{H})$; all are distinct when $\mathcal{H}$ is infinite-dimensional and coincide otherwise; each turns out to be important in some applications.

The *uniform topology* is the topology induced on $\mathfrak{B}(\mathcal{H})$ by the operator norm; it is also referred to as the "operator-norm" topology or the "metric topology" of $\mathfrak{B}(\mathcal{H})$; it is the topology we considered up to now when we spoke of $\mathfrak{B}(\mathcal{H})$ as being a C^*-algebra.

The *strong* (*operator*) *topology* is defined as the weakest topology on $\mathfrak{B}(\mathcal{H})$ for which the mappings $B \to B\Psi$ from $\mathfrak{B}(\mathcal{H})$ to $\mathcal{H}$ (the latter being equipped with its strong topology) are continuous; a basis of neighborhoods for this topology is obtained by considering *all* sets of the form

$$N(A; \mathcal{S}, \varepsilon) = \{B \in \mathfrak{B}(\mathcal{H}) \mid |(B - A)\Psi_i| < \varepsilon \ \forall \Psi_i \in \mathcal{S}\},$$

where $A \in \mathfrak{B}(\mathcal{H})$, $\mathcal{S}$ is a *finite* set of vectors in $\mathcal{H}$, and $\varepsilon > 0$.

The *weak* (*operator*) *topology* is defined as the weakest topology on $\mathfrak{B}(\mathcal{H})$ for which the mappings $B \to B\Psi$ from $\mathfrak{B}(\mathcal{H})$ to $\mathcal{H}$ (the latter being equipped with its weak topology) are continuous; equivalently it is the weakest topology for which the mappings $B \to (\Psi, B\Phi)$ from $\mathfrak{B}(\mathcal{H})$ to $\mathbb{C}$ are continuous; a basis of neighborhoods for this topology is obtained by considering *all* sets of the form

$$N(A; \mathcal{S}, \varepsilon) = \{B \in \mathfrak{B}(\mathcal{H}) \mid |(\Psi_i, (B - A)\Phi_i)| < \varepsilon \ \forall (\Psi_i, \Phi_i) \in \mathcal{S}\},$$

where $A \in \mathfrak{B}(\mathcal{H})$, $\mathcal{S}$ is a *finite* set of couples of vectors in $\mathcal{H}$, and $\varepsilon > 0$.

To define the last two topologies we introduce the set $\mathfrak{H}$ of all sequences $\bar{\Psi} \equiv \{\Psi_i\}$ of vectors in $\mathcal{H}$ with $|\bar{\Psi}|^2 \equiv \sum_i |\Psi_i|^2 < \infty$; for each B in $\mathfrak{B}(\mathcal{H})$ we then define the mapping $\bar{B}$ from $\mathfrak{H}$ to $\mathfrak{H}$ by $\bar{B}\bar{\Psi} \equiv \bar{B}\{\Psi_i\} \equiv \{B\Psi_i\}$.

The *ultrastrong topology* is defined‡ as the weakest topology for which the mappings $B \to \bar{B}\bar{\Psi}$ from $\mathfrak{B}(\mathcal{H})$ to $\mathfrak{H}$ (the latter being equipped with the topology derived from the norm defined above) are continuous; a basis of neighborhoods for this topology is obtained by considering *all* sets of the form

$$N(A; \bar{\Psi}, \varepsilon) = \{B \in \mathfrak{B}(\mathcal{H}) \mid |(\bar{B} - \bar{A})\bar{\Psi}| < \varepsilon\},$$

where $A \in \mathfrak{B}(\mathcal{H})$, $\bar{\Psi} \in \mathfrak{H}$, and $\varepsilon > 0$.

† See Fillmore and Topping [1967] and references quoted therein.
‡ The ultrastrong topology on $\mathfrak{B}(\mathcal{H})$ is sometimes referred to as the strongest topology on $\mathfrak{B}(\mathcal{H})$; we shall avoid this somewhat misleading nomenclature.

The *ultraweak topology* is defined as the weakest topology for which the mappings $B \to (\overline{\Psi}, \bar{B}\overline{\Phi}) \equiv \sum_i (\Psi_i, B\Phi_i)$ from $\mathfrak{B}(\mathcal{H})$ to $\mathbb{C}$ are continuous; equivalently, it is the weakest topology for which the mappings $B \to \bar{B}\overline{\Psi}$ from $\mathfrak{B}(\mathcal{H})$ to $\mathfrak{H}$ (the latter being equipped with its weak topology) are continuous; a basis of neighborhoods for this topology is obtained by considering *all* sets of the form

$$N(A; \mathcal{S}, \varepsilon) = \{B \in \mathfrak{B}(\mathcal{H}) \mid |((\bar{B} - \bar{A})\overline{\Psi}, \overline{\Phi})| < \varepsilon \ \forall\ (\overline{\Psi}, \overline{\Phi}) \in \mathcal{S}\},$$

where $A \in \mathfrak{B}(\mathcal{H})$, $\mathcal{S}$ is a *finite* set of couples of elements in $\mathfrak{H}$, and $\varepsilon > 0$.

The relations between these topologies on $\mathfrak{B}(\mathcal{H})$ are summarized by the following diagram:

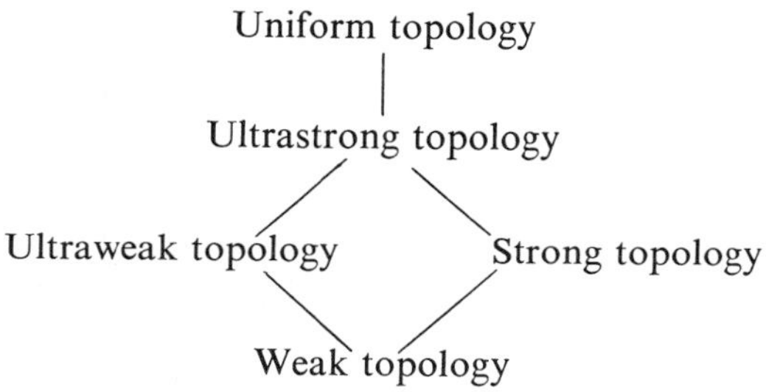

An ascending line between two topologies τ_1 and τ_2 means that τ_1 is weaker than τ_2. We shall use the facts that on the *bounded* subsets of $\mathfrak{B}(\mathcal{H})$ (a) the strong and ultrastrong topologies coincide and (b) the weak and ultraweak topologies coincide. We mention that the mappings $A \to AB$ (with B fixed!) and $B \to AB$ (with A fixed!) are continuous for all five topologies; however (except for the uniform topology), the mappings $(A, B) \to AB$ are *not* continuous.† Also we should note that the mapping $A \to A^*$ is continuous for the uniform, the ultraweak, and the weak operator topologies but *not* for the strong nor the ultrastrong topologies. This remark concludes our survey of these topologies; the following theorem shows their relevance in the study of von Neumann algebras.

Lemma. *Let $\mathfrak{M}$ be a *-subalgebra of $\mathfrak{B}(\mathcal{H})$ satisfying the condition $\mathcal{H} = \overline{\{R\Psi \mid R \in \mathfrak{M}, \Psi \in \mathcal{H}\}}$. Then every element in the bicommutant $\mathfrak{M}''$ of $\mathfrak{M}$ is the ultrastrong limit of a net on $\mathfrak{M}$.*

Proof. Dixmier [1957], Lemma 1.3.6, or Naimark [1964], VII.34.2. ∎

† Incidentally, this is a typical example in which we should distinguish between nets and sequences (see p. 102 ff); in this respect the few remarks made at the beginning of Section 84 in Riesz and Nagy [1955] can be misleading if this distinction is not realized.

Theorem 10 (*von Neumann density theorem*†). *Let $\mathfrak{M}$ be a *-subalgebra of $\mathfrak{B}(\mathcal{H})$ which satisfies the condition $\mathcal{H} = \overline{\{R\Psi \mid R \in \mathfrak{M}, \Psi \in \mathcal{H}\}}$. Then $\mathfrak{M}''$ is the closure in $\mathfrak{B}(\mathcal{H})$ of $\mathfrak{M}$ with respect to the weak operator topology, the ultraweak topology, the strong operator topology, and the ultrastrong topology.*

Proof. The partial ordering relation between the four topologies considered, as described by the diagram preceding the lemma, implies

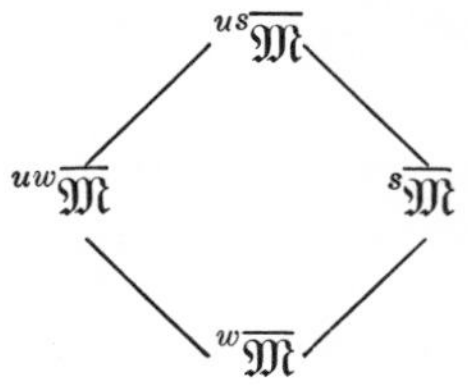

An ascending line between ${}^{\tau_1}\overline{\mathfrak{M}}$ and ${}^{\tau_2}\overline{\mathfrak{M}}$ means that ${}^{\tau_2}\overline{\mathfrak{M}} \subseteq {}^{\tau_1}\overline{\mathfrak{M}}$. $\mathfrak{M}''$ contains $\mathfrak{M}$, and being the commutant of $\mathfrak{M}'$ it is weakly closed; hence $\mathfrak{M}'' \supseteq {}^{w}\overline{\mathfrak{M}}$. From the lemma we have on the other hand ${}^{us}\overline{\mathfrak{M}} \supseteq \mathfrak{M}''$. Combining all these inclusions, we get $\mathfrak{M}'' = {}^{us}\overline{\mathfrak{M}} = {}^{s}\overline{\mathfrak{M}} = {}^{uw}\overline{\mathfrak{M}} = {}^{w}\overline{\mathfrak{M}}$, which is the statement of the theorem. ∎

REMARKS. Every von Neumann algebra $\mathfrak{N}$ contains the identity operator, hence satisfies the hypothesis of the theorem; since, by definition $\mathfrak{N} = \mathfrak{N}''$, we conclude from the theorem that $\mathfrak{N}$ is closed with respect to the weak operator topology, hence with respect to any of the five topologies introduced above on $\mathfrak{B}(\mathcal{H})$; in particular, any von Neumann algebra $\mathfrak{N}$ is closed with respect to the norm topology and is a C^*-algebra. From the theorem we conclude also that every *-subalgebra of $\mathfrak{B}(\mathcal{H})$ containing the identity operator and closed with respect to the weak operator topology (the strong operator topology, the ultraweak topology, or the ultrastrong topology) is a von Neumann algebra, thus providing alternative definitions of von Neumann algebras. We further notice that if $\mathfrak{N}$ is a von Neumann algebra, $\mathfrak{P}$, the set of all its projectors, and $\mathcal{F}(\mathfrak{P})$, the set generated from $\mathfrak{P}$ by the *-algebraic operations, then $\mathcal{F}(\mathfrak{P})$ is a *-algebra satisfying the hypothesis in Theorem 10 so that ${}^{\tau}\overline{\mathcal{F}(\mathfrak{P})} = \mathcal{F}(\mathfrak{P})''$ (where τ denotes any of the four topologies mentioned in the theorem); on the other hand, we know from Theorem 9 that $\mathfrak{P}'' = \mathfrak{N}$, hence $\mathcal{F}(\mathfrak{P})'' = \mathfrak{N}$. Combining these consequences of Theorems 9 and 10, we get ${}^{\tau}\overline{\mathcal{F}(\mathfrak{P})} = \mathfrak{N}$; that is to say, every element of $\mathfrak{N}$ is the τ-limit of polynomials in the projectors of $\mathfrak{N}$.

† We shall also use in the sequel another "density theorem" (Kaplanski [1951]; see also Dixmier [1957] I.3.5): Let $\mathfrak{M} \subseteq \mathfrak{N}$ be *-subalgebras of $\mathfrak{B}(\mathfrak{H})$ with $\mathfrak{M}$ strongly dense in $\mathfrak{N}$; then the unit ball $\mathfrak{M}_1$ of $\mathfrak{M}$ is strongly dense in the unit ball $\mathfrak{N}_1$ of $\mathfrak{N}$.

Theorems 9 and 10 and their aforementioned consequences (in particular, that every von Neumann algebra equipped with the weak operator topology is generated by its projectors) are the cornerstones of many applications of von Neumann algebras. In connection with our interest in an axiomatic formulation of quantum theory we might mention that von Neumann [1936], starting from an abstract Jordan algebra with distributive symmetrized product $A \circ B$, mimicked the weak operator topology with a topology τ that satisfied the following axioms:

(i) *The τ-topology is compatible with the structure of $\mathfrak{A}$ as a real vector space; in particular* $A + B$ *is a τ-continuous function of both of its variables, and* $-A$, $\frac{1}{2}A$ *are τ-continuous functions of* A;
(ii) $A \circ B$ *is a τ-continuous function of* A *(for fixed* B*)*;
(iii) *if* A_n, B_n *are positive elements of* $\mathfrak{A}$, *with* $\tau\text{-}\lim_{n\to\infty}(A_n + B_n) = 0$, *then* $\tau - \lim_{n\to\infty} A_n = 0$;
(iv) *The set* $\mathfrak{D}$ *of positive elements of* $\mathfrak{A}$ *is τ-closed.*
(v) *The unit ball* $\mathfrak{U}$ *of* $\mathfrak{A}$ *is τ-separable and τ-compact.*

These axioms replace the topological axioms imposed in the formulation of the formalism presented in Chapter 1, Section 2; they also appear as a generalization of the concept of von Neumann algebras in at least two aspects: *first* they deal with an abstract structure, with no reference to a realization as an algebra of operators acting on a Hilbert space, and *second* the equivalent of the ordinary product AB does not appear, and only the (physically meaningful) symmetrized product $A \circ B$ is needed. With this structure alone, which seems to capture some of the essential features of von Neumann algebras, von Neumann [*loc. cit.*] was able to derive most of the results currently attached to the theory of von Neumann algebras in the sense described in the beginning of this subsection: functional analysis (and in particular the spectral theory) can be carried out in complete analogy to the usual theory in Hilbert space without appealing to elements lying outside $\mathfrak{A}$ itself; the concept of commutativity, including its relation with the spectral decomposition of observables, can be discussed completely; finally the contact with the "proposition-calculus" approach to quantum theory can be carried much further than has been done in Subsection I.2.e, all physically relevant propositions being identifiable with idempotents of $\mathfrak{A}$. All these things are not directly possible within the pure C^*-algebraic approach, since a C^*-algebra $\mathfrak{R}$ in general does not contain "enough" projectors to generate $\mathfrak{R}$ back from them. This situation suggests the question whether a von Neumann algebra $\mathfrak{N}$ can be associated in general, in some canonical fashion, with any given C^*-algebra $\mathfrak{R}$. As first pointed out by Sherman [1950], this is indeed possible†; to state this fact in the general form given by Dixmier

† See also Takeda [1954] and Grothendiek [1957].

[1964] we need some preliminary definitions and results which we now want to list, referring for the proofs (which are somewhat involved and do not seem to belong here) to Dixmier [1957, Chapter 1, Section 3].

Let $\mathfrak{N}$ be a von Neumann algebra; we saw that $\mathfrak{N}$ is a C^*-subalgebra of $\mathfrak{B}(\mathcal{H})$; let $\mathfrak{N}^*$ be the dual of $\mathfrak{N}$, i.e., the set of all linear functionals $\phi : \mathfrak{N} \to \mathbb{C}$ which are continuous with respect to the norm topology on $\mathfrak{N}$ (see Subsection d above).

Lemma. *For ϕ in $\mathfrak{N}^*$ the following conditions are equivalent:*

(i) *ϕ is continuous on $\mathfrak{N}$, the latter being equipped with the weak operator topology;*

(ii) *ϕ is continuous on $\mathfrak{N}$, the latter being equipped with the strong operator topology;*

(iii) *There exist $\{\lambda_i \geqslant 0 \mid i = 1, 2, \ldots, n < \infty\}$ and $\{\Phi_i, \Phi'_i\}$ in $\mathcal{H}$ such that $(\Phi_i, \Phi_j) = \delta_{ij} = (\Phi'_i, \Phi'_j)$; $\|\phi\| = \sum_{i=1}^n \lambda_i$; $\langle \phi; A \rangle = \sum_{i=1}^n \lambda_i (\Phi_i, A\Phi'_i)$ for all A in $\mathfrak{N}$.*

We denote by $\mathfrak{N}_\sim$ the linear manifold of $\mathfrak{N}^*$ formed by all elements of $\mathfrak{N}^*$ satisfying any of the three conditions of the lemma. If ϕ is a positive linear functional on $\mathfrak{N}$ (hence belongs to $\mathfrak{N}^*$, since $\mathfrak{N}$ is a C^*-algebra), we can write in the third condition of the lemma: $\Phi_i = \Phi'_i$ for $i = 1, 2, \ldots, n < \infty$; in particular, a state ϕ on $\mathfrak{N}$ is weakly (or strongly) continuous on $\mathfrak{N}$ if and only if it belongs to the convex hull $\mathrm{co}(\mathfrak{V})$ of the set $\mathfrak{V}$ of all vector states on $\mathfrak{N}$.

Lemma. *For ϕ in $\mathfrak{N}^*$ the following conditions are equivalent:*

(i) *ϕ is continuous on $\mathfrak{N}$, the latter being equipped with the ultraweak topology;*

(ii) *ϕ is continuous on $\mathfrak{N}$, the latter being equipped with the ultrastrong topology;*

(iii) *There exist $\{\Phi_i, \Phi'_i \mid i = 1, 2, \ldots\}$ in $\mathcal{H}$ with $\sum_i |\Phi_i|^2 < \infty$; $\sum_i |\Phi'_i|^2 < \infty$ such that $\langle \phi; A \rangle = \sum_i (\Phi_i, A\Phi'_i)$.*

We denote by $\mathfrak{N}_*$ the linear manifold of $\mathfrak{N}^*$ formed by all elements of $\mathfrak{N}^*$ satisfying any of the three conditions of the lemma. If ϕ is a positive linear functional on $\mathfrak{N}$, it belongs to $\mathfrak{N}_*$ if and only if there exists a positive operator ρ in $\mathfrak{B}(\mathcal{H})$ with Tr $\rho < \infty$ such that

$$\langle \phi; A \rangle = \mathrm{Tr}\, \rho A;$$

furthermore we can prove that this condition is equivalent to

$$\langle \phi; \sum_i P_i \rangle = \sum_i \langle \phi; P_i \rangle$$

for every family $\{P_i\}$ of mutually orthogonal projectors in $\mathfrak{N}$; we refer to this condition as "complete additivity." We will also refer to the positive elements of $\mathfrak{N}_*$ as *normal* positive functionals on $\mathfrak{N}$ (see also p. 127).

Lemma. *In the uniform topology of* $\mathfrak{N}^*$ *(inherited from the uniform topology of* $\mathfrak{N}$*), the closure of* $\mathfrak{N}_\sim$ *is* $\mathfrak{N}_*$*. The bilinear form* $\langle ; \rangle$ *from* $\mathfrak{N}^* \times \mathfrak{N}$ *to* $\mathbb{C}$ *restricted to* $\mathfrak{N}_* \times \mathfrak{N}$ *is such that* $\mathfrak{N}$*, equipped with its norm topology is the dual of* $\mathfrak{N}_*$ *(in the usual sense of Banach spaces).*

We see then that every von Neumann algebra $\mathfrak{N}$ is a C^*-algebra, which, as a Banach space, is the dual of a Banach space (namely, the Banach space $\mathfrak{N}_*$ of all ultraweakly continuous functionals on $\mathfrak{N}$), thus proving that $\mathfrak{N}$ is a W^*-algebra; $\mathfrak{N}_*$ is called the *predual* of $\mathfrak{N}$.

From the physical point of view we should notice that as a consequence of this lemma the uniformly closed convex hull $\overline{\overline{co(\mathfrak{B})}}$ of all vector states on $\mathfrak{N}$ is identical to the set of all states on $\mathfrak{N}$ which are ultraweakly continuous, hence representable by density matrices.

An interesting illustration of these lemmata is provided by† the von Neumann algebra $\mathfrak{B}(\mathcal{H})$. We denote by $\mathfrak{F}(\mathcal{H})$ the set of all operators in $\mathfrak{B}(\mathcal{H})$ which are of finite rank; they form a *-subalgebra of $\mathfrak{B}(\mathcal{H})$. The uniform closure of $\mathfrak{F}(\mathcal{H})$ consists of the C^*-algebra $\mathfrak{A}(\mathcal{H})$ of all compact operators in $\mathfrak{B}(\mathcal{H})$ and is a two-sided, closed*-ideal of $\mathfrak{B}(\mathcal{H})$. We further introduce the set $\mathfrak{L}(\mathcal{H})$ of all Hilbert-Schmidt operators on $\mathcal{H}: A \in \mathfrak{L}(\mathcal{H})$ if and only if $\mathrm{Tr}\, A^*A < \infty$; we easily verify that $\mathfrak{F}(\mathcal{H}) \subset \mathfrak{L}(\mathcal{H}) \subset \mathfrak{A}(\mathcal{H})$ and that $\mathfrak{L}(\mathcal{H})$ is a Banach *-algebra with respect to the norm $\|A\|_{\mathfrak{L}} = (\mathrm{Tr}\, A^*A)^{1/2}$. This norm furthermore derives from a scalar product $(A, B) \equiv \mathrm{Tr}\, A^*B$ and $\mathfrak{L}(\mathcal{H})$ is a *Hilbert-algebra;*‡ furthermore $\mathfrak{L}(\mathcal{H})$ is the closure of $\mathfrak{F}(\mathcal{H})$ with respect to the Hilbert-Schmidt norm $\|\cdots\|_{\mathfrak{L}}$. We finally define the trace-class $\mathfrak{T}(\mathcal{H})$ as the set of all elements of $\mathfrak{B}(\mathcal{H})$ that can be obtained as products of two elements in $\mathfrak{L}(\mathcal{H})$; since $\mathfrak{L}(\mathcal{H})$ is a two-sided *-ideal of $\mathfrak{B}(\mathcal{H})$, we have $\mathfrak{F}(\mathcal{H}) \subset \mathfrak{T}(\mathcal{H}) \subset \mathfrak{L}(\mathcal{H})$ and $\mathfrak{T}(\mathcal{H})$ is a two-sided *-ideal of $\mathfrak{B}(\mathcal{H})$; furthermore we verify that $\mathfrak{T}(\mathcal{H})$ is a Banach *-algebra under the norm $\|T\|_{\mathfrak{T}} = \mathrm{Tr}\,\{(T^*T)^{1/2}\}$ and is the closure of $\mathfrak{F}(\mathcal{H})$ with respect to the topology induced by this norm. The relations between these algebras can be summarized by the following table:

$$\mathfrak{T}(\mathcal{H}) = {}^{\mathfrak{T}}\overline{\mathfrak{F}(\mathcal{H})}; \; \mathfrak{L}(\mathcal{H}) = {}^{\mathfrak{L}}\overline{\mathfrak{F}(\mathcal{H})}; \; \mathfrak{A}(\mathcal{H}) = {}^{\mathfrak{B}}\overline{\mathfrak{F}(\mathcal{H})},$$

$$\mathfrak{F}(\mathcal{H}) = \mathfrak{B}(\mathcal{H})_\sim; \; \mathfrak{T}(\mathcal{H}) = \mathfrak{B}(\mathcal{H})_* = \mathfrak{A}(\mathcal{H})^*,$$

$$\mathfrak{B}(\mathcal{H})^* = \mathfrak{B}(\mathcal{H})_* \oplus \mathfrak{A}(\mathcal{H})^\perp; \; \mathfrak{T}(\mathcal{H})^* = \mathfrak{B}(\mathcal{H}),$$

† These results are due to Schatten [1950] and Dixmier [1950]; see also Takeda [1954a, b, c]; they are summarized in Rickart [1960] Appendix, Section 1 and Dixmier [1957] Exercise I.3.6.

‡ A *Hilbert-algebra* $\mathfrak{R}$ is a *-algebra equipped with a scalar product that satisfies the following conditions: (a) $\{\mathfrak{R}, (,)\}$ is a prehilbert space, (b) $(R, S) = (S^*, R^*)$, (c) $(RS, T) = (S, R^*T)$, (d) at fixed R, $S \to RS$ is continuous, these conditions being understood for all R, S, and T in $\mathfrak{R}$.

where $\mathfrak{A}(\mathcal{H})^{\perp}$ denotes the set of all elements of $\mathfrak{B}(\mathcal{H})^*$ which vanish on $\mathfrak{A}(\mathcal{H})$; the first two Banach space isomorphisms of the second line are provided by the mapping $\langle \rho; \ldots \rangle = \mathrm{Tr}\,(\rho \ldots)$ and so is the last isomorphism of the third line. From this table we conclude in particular that the double dual $\mathfrak{A}(\mathcal{H})^{**}$ of the C^*-algebra $\mathfrak{A}(\mathcal{H})$ is isomorphic to the von Neumann algebra $\mathfrak{B}(\mathcal{H}) = \mathfrak{A}(\mathcal{H})''$, a fact we have already mentioned and the significance of which we will presently see.

Let us now return to the general case and consider an arbitrary C^*-algebra $\mathfrak{R}$ and its set $\mathfrak{S}$ of states. We recall that to every state ϕ in $\mathfrak{S}$ corresponds, via the GNS construction, a Hilbert space $\mathcal{H}_\phi$, a represenation $\pi_\phi : \mathfrak{R} \to \mathfrak{B}(\mathcal{H}_\phi)$, and a cyclic vector Φ in $\mathcal{H}_\phi$ such that $\langle \phi; R \rangle = (\Phi, \pi_\phi(R)\Phi)$, and that every cyclic representation of $\mathfrak{R}$ can be obtained in this way (provided that we identify any two representations that are unitarily equivalent). We now form the Hilbert space

$$\mathcal{H}_u = \bigoplus_{\phi \in \mathfrak{S}} \mathcal{H}_\phi$$

and the representation $\pi_u : \mathfrak{R} \to \mathfrak{B}(\mathcal{H})_u$ defined by the requirement that for each R in $\mathfrak{R}$ $\pi_u(R)$ is the element of $\mathfrak{B}(\mathcal{H}_u)$ which induces $\pi_\phi(R)$ on $\mathcal{H}_\phi$; this representation is then the direct Hilbert sum $\pi_u = \bigoplus_{\phi \in \mathfrak{S}} \pi_\phi$ and is referred to as the *universal representation* of $\mathfrak{R}$. Its "universal" character appears in the immediate sequel. We first notice that every cyclic representation of $\mathfrak{R}$ is unitarily equivalent to a direct summand of π_u. Furthermore, π_u is faithful; let indeed R in $\mathfrak{R}$ be different from 0. There exists then an element ϕ in $\mathfrak{S}$ such that $\langle \phi; R \rangle \neq 0$; let Φ be the vector in $\mathcal{H}_u$ such that $\langle \phi; R \rangle = (\Phi, \pi_\phi(R)\Phi) = (\Phi, \pi_u(R)\Phi)$; we then have $\pi_u(R)\Phi \neq 0$, hence $\pi_u(R) \neq 0$. Incidentally this remark shows that every abstract C^*-algebra $\mathfrak{R}$ has a faithful representation as a C^*-algebra of operators acting on a Hilbert space. We now form the von Neumann algebra $\pi_u(\mathfrak{R})''$, which for simplicity we denote as $\mathfrak{R}''$; we refer to this algebra as the (universal) *enveloping von Neumann algebra* of $\mathfrak{R}$. The intimate relation between a C^*-algebra and its enveloping von Neumann algebra is best exhibited by first considering the states of $\mathfrak{R}$ in relation to those of $\mathfrak{R}''$. We already know that to every state ϕ in $\mathfrak{S}$ corresponds a unit vector Φ in $\mathcal{H}_u$ such that $\langle \phi; R \rangle = (\Phi, \pi_u(R)\Phi)$; we define the extension $\tilde{\phi}$ of ϕ to $\mathfrak{R}''$ by $\langle \tilde{\phi}; S \rangle = (\Phi, S\Phi)$ for all S in $\mathfrak{R}''$. Since $\tilde{\phi}$ is a vector state on the von Neumann algebra $\mathfrak{R}''$, it is continuous on $\mathfrak{R}''$, the latter being equipped with either of the following four topologies: weak-operator, strong operator, ultraweak, or ultrastrong. By continuity $\tilde{\phi}$ is the *unique* weakly (or strongly, ultraweakly or ultrastrongly) continuous state on $\mathfrak{R}''$ which, when restricted to $\mathfrak{R}$, coincides with ϕ. Conversely, if ψ is a normal (hence ultrastrongly, or equivalently ultraweakly, continuous) state on $\mathfrak{R}''$, its restriction to $\pi_u(\mathfrak{R})$ is a state ϕ on $\pi_u(\mathfrak{R})$; since π_u is faithful, ϕ acually is a state on $\mathfrak{R}$ itself. We then conclude from the above discussion that $\psi = \tilde{\phi}$, and in particular that ψ is also weakly continuous on $\mathfrak{R}''$. We then have constructed an isomorphism $\phi \to \tilde{\phi}$ from the set $\mathfrak{S}$ of all states on $\mathfrak{R}$ to the set $\mathfrak{S}_*$ of all normal states on $\mathfrak{R}''$, the latter also being weakly continuous on $\mathfrak{R}''$. This remark then reduces the study of all the states of a C^*-algebra $\mathfrak{R}$ to the study of all the normal states on its enveloping von Neumann algebra. The mapping $\phi \to \tilde{\phi}$ extends easily to a norm preserving isomorphism from the dual $\mathfrak{R}^*$ of $\mathfrak{R}$ to the predual $(\mathfrak{R}'')_*$ of $\mathfrak{R}''$. We recall that by *predual* of $\mathfrak{R}''$ we mean that $\mathfrak{R}''$, equipped with its norm topology and considered as a Banach space, is the dual of the (closed!) subspace $(\mathfrak{R}'')_*$ of $(\mathfrak{R}'')^*$, the dual of $\mathfrak{R}''$. From the isomorphism between $\mathfrak{R}^*$ and $(\mathfrak{R}'')_*$ we conclude that $\mathfrak{R}''$ is isomorphic as a Banach space to the bidual $\mathfrak{R}^{**}$ of $\mathfrak{R}$; the isometric isomorphism $\Lambda : \mathfrak{R}'' \to \mathfrak{R}^{**}$ is indeed provided by forming for each S in $\mathfrak{R}''$ the continuous linear functional $\Lambda(S) : \mathfrak{R}^* \to \mathbb{C}$ defined by $\langle \phi; \Lambda(S) \rangle = \langle \tilde{\phi}; S \rangle$. Furthermore, since every ultrastrongly continuous linear functional on $\mathfrak{R}''$ has been seen to be also weakly continuous, the above defining equality shows that the mapping Λ from $\mathfrak{R}''$ (equipped with its weak operator topology) to $\mathfrak{R}^{**}$ [equipped with its weak *-topology $\sigma(\mathfrak{R}^{**}, \mathfrak{R}^*)$] is bicontinuous; finally it again follows from this equality that $\Lambda \circ \pi_u$ is the canonical injection of $\mathfrak{R}$ into $\mathfrak{R}^{**}$. One more aspect of the "universality" of $\mathfrak{R}''$ is that a representation of $\mathfrak{R}''$ can be canonically associated with any representation of $\mathfrak{R}$ itself. To see this it is sufficient to consider the case of a cyclic representation $\pi : \mathfrak{R} \to \mathfrak{B}(\mathcal{H})$. Since π is cyclic, it is (unitarily equivalent to) a direct summand π_ϕ of π_u; let P_ϕ be the projector from $\mathcal{H}_u$ to the space $\mathcal{H}_\phi$ of the representation π_ϕ. Since $\mathcal{H}_\phi$ is stable with respect to π_u, P_ϕ belongs to $\pi_u(\mathfrak{R})'$; hence π is (unitarily equivalent to) the representation $\pi_u(\mathfrak{R})P_\phi$ of $\mathfrak{R}$. Let us denote by τ any of the four topologies: ultrastrong, ultraweak, weak-operator, strong-operator. We have seen that the product AB is τ-continuous in A for fixed B; hence the mapping $\pi_u(R) \to \pi_u(R)P_\phi$ can be extended to a τ-continuous representation of the τ-closure $\mathfrak{R}''$ of $\pi_u(\mathfrak{R})$ onto $\mathfrak{R}''P_\phi = \pi_u(\mathfrak{R})''P_\phi = (\pi_u(\mathfrak{R})P_\phi)''$, since P_ϕ belongs to $\pi_u(\mathfrak{R})'$. The latter is (unitarily equivalent to) $\pi(\mathfrak{R})''$; hence $\pi(\mathfrak{R})''$ is a τ-continuous representation of $\mathfrak{R}''$. By continuity it is the only τ-continuous representation of $\mathfrak{R}''$ in $\mathfrak{B}(\mathcal{H})$ which, when reduced to $\mathfrak{R}$, coincides with $\pi(\mathfrak{R})$. We can summarize the information gained so far on the universal enveloping von Neumann algebra of a C^*-algebra:

Theorem 11. *Let $\mathfrak{R}$ be a C*-algebra; let $\pi_u : \mathfrak{R} \to \mathfrak{B}(\mathcal{H}_u)$ be its universal representation; $\mathfrak{R}'' \equiv \pi_u(\mathfrak{R})''$, its universal enveloping von Neumann algebra; $\mathfrak{R}^*$ (resp. $\mathfrak{R}^{**}$) the dual (resp. bidual) of $\mathfrak{R}$; $\mathfrak{S}$, the set of all states on $\mathfrak{R}$; $\mathfrak{S}_*$, the set of all normal states on $\mathfrak{R}''$; and $\mathfrak{R}''_*$, the predual of $\mathfrak{R}''$. Then*

(i) *π_u is faithful.*
(ii) *To every state ϕ in $\mathfrak{S}$ corresponds a unit vector Φ in $\mathcal{H}_u$ and a unique*

weakly continuous extension $\tilde{\phi}$ *of* ϕ *from* $\mathfrak{R}$ *to* $\mathfrak{R}''$, *given by* $\langle\tilde{\phi}; \mathrm{S}\rangle = (\Phi, \mathrm{S}\Phi)$. *The mapping* $\phi \to \tilde{\phi}$ *sends* $\mathfrak{S}$ *onto* $\mathfrak{S}_*$ *and extends to an isometric isomorphism from* $\mathfrak{R}^*$ *to* $\mathfrak{R}''_*$.

(iii) *The mapping* $\Lambda: \mathfrak{R}'' \to \mathfrak{R}^{**}$, *defined by* $\langle\phi; \Lambda(\mathrm{S})\rangle = \langle\tilde{\phi}; \mathrm{S}\rangle$ *with* ϕ *in* $\mathfrak{R}^*$ *and* S *in* $\mathfrak{R}''$, *enjoys the following properties:*

(a) *It is an isometric isomorphism between* $\mathfrak{R}''$ *and* $\mathfrak{R}^{**}$ *(both being equipped with their natural Banach space structure).*
(b) *It is bicontinuous when* $\mathfrak{R}''$ *is equipped with its weak-operator topology and* $\mathfrak{R}^{**}$ *is equipped with the weak*-topology* $\sigma(\mathfrak{R}^{**}, \mathfrak{R}^*)$.
(c) $\Lambda \circ \pi_{\mathrm{u}}: \mathfrak{R} \to \mathfrak{R}^{**}$ *is the canonical injection of* $\mathfrak{R}$ *into its bidual* $\mathfrak{R}^{**}$.

(iv) *For any representation* $\pi: \mathfrak{R} \to \mathfrak{B}(\mathcal{H})$ *there exists a unique ultraweakly (or ultrastrongly, weakly or strongly) continuous extension* $\tilde{\pi}$ *of* π *from* $\mathfrak{R}$ *to* $\mathfrak{R}''$, *and* $\tilde{\pi}(\mathfrak{R}'')$ *coincides with* $\pi(\mathfrak{R})''$.

Proof. See the discussion preceding the statement of the theorem, or Dixmier [1964], Corollary 12.1.3 and Proposition 12.1.5. ∎

REMARK. From the physical point of view it is interesting to note that part (iv) of the theorem persists† if we replace the requirement that π be an homomorphism for the C^*-algebraic structure of $\mathfrak{R}$ with the requirement that it be a Jordan homomorphism, namely, that π be a linear, *-preserving mapping from $\mathfrak{R}$ to $\mathfrak{B}(\mathcal{H})$ such that $\pi(R^2) = \pi(R)^2$ for all R in $\mathfrak{R}$ (or equivalently $\pi(R \circ S) = \frac{1}{2}\{\pi(R)\,\pi(S) + \pi(S)\,\pi(R)\}$ for all R, S in $\mathfrak{R}$); we note further that it is actually sufficient to require the latter condition only on the Jordan algebra $\mathfrak{A}$ of all self-adjoint elements of $\mathfrak{R}$.

The concept of the universal enveloping von Neumann algebra $\mathfrak{R}''$ of a C^*-algebra $\mathfrak{R}$ is closely related to the notion of quasi-equivalence of representations of $\mathfrak{R}$, which we discuss presently. Let us consider two representations $\pi_i: \mathfrak{R} \to \mathfrak{B}(\mathcal{H}_i)$ $(i = 1, 2)$ of $\mathfrak{R}$; let $\tilde{\pi}_i$ be the unique ultraweakly continuous extension of π_i to $\mathfrak{R}''$. We have just seen that $\tilde{\pi}_i(\mathfrak{R}'') = \pi_i(\mathfrak{R})''$. We say that the two representations π_1 and π_2 of $\mathfrak{R}$ are *quasi-equivalent* whenever there exists an *-isomorphism α from $\pi_1(\mathfrak{R})''$ to $\pi_2(\mathfrak{R})''$ such that $\pi_2(R) = \alpha\pi_1(R)$ for all R in $\mathfrak{R}$. This last condition *implies* that Ker π_1 = Ker π_2, hence that π_1 and π_2 are *physically equivalent*; quasi-equivalence adds to the requirement of physical equivalence (namely, Ker π_1 = Ker π_2), the condition that the resulting *-algebraic isomorphism between $\pi_1(\mathfrak{R})$ and $\pi_2(\mathfrak{R})$ extends to their respective (ultra-) weak closures $\tilde{\pi}_1(\mathfrak{R}'')$ and $\tilde{\pi}_2(\mathfrak{R}'')$. The isomorphism α furthermore satisfies the condition $\tilde{\pi}_2(S) = \alpha\tilde{\pi}_1(S)$ for all S in $\mathfrak{R}''$; indeed, since $\mathfrak{R}''$ is the ultraweak closure of $\mathfrak{R}$ [identified in the sequel with $\pi_u(\mathfrak{R})$ for simplicity's sake] there exists for each S in $\mathfrak{R}''$ a net $\{R_\gamma\}$ of elements of $\mathfrak{R}$

† Kadison [1965].

converging ultraweakly to S; since $\tilde{\pi}_i$ are ultraweakly continuous, we have $\tilde{\pi}_i(S) = uw \lim \pi_i(R_\gamma)$. Furthermore, since† any isomorphism α between two von Neumann algebras is bicontinuous for the ultraweak (and ultrastrong) topologies, we have $\tilde{\pi}_2(S) = uw \lim \pi_2(R_\gamma) = uw \lim \alpha\pi_1(R_\gamma) = \alpha\, uw \lim \pi_1(R_\gamma) = \alpha\tilde{\pi}_1(S)$. Consequently the condition $\tilde{\pi}_2(S) = \alpha\tilde{\pi}_1(S)$ for all S in $\mathfrak{R}''$ is both necessary and sufficient for the quasi-equivalence of the representations π_1 and π_2 of $\mathfrak{R}$; the interest in this condition is that it is equivalent to the requirement $\text{Ker}\,\tilde{\pi}_1 = \text{Ker}\,\tilde{\pi}_2$. Now considering $\mathfrak{R}''$ as a C^*-algebra (see p. 119) we can apply the criterion in Theorem 7 to the representations $\tilde{\pi}_1$ and $\tilde{\pi}_2$ of $\mathfrak{R}''$ and conclude that they are physically equivalent if and only if the representations π_1 and π_2 of $\mathfrak{R}$ are quasi-equivalent. From the discussion in Subsection d we know several equivalent conditions for $\tilde{\pi}_1$ and $\tilde{\pi}_2$ to be physically equivalent; these conditions are based entirely on the C^*-algebraic structure of $\mathfrak{R}''$. We now want to take advantage of the fact that $\mathfrak{R}''$ is also a von Neumann algebra. We first notice that $\tilde{\pi}_2(S) = \alpha\tilde{\pi}_1(S)$ for all S in $\mathfrak{R}''$ has the following consequence: let $\tilde{\phi}_2$ be a normal state on $\tilde{\pi}_2(\mathfrak{R}'')$ and define the linear functional $(\alpha^*\tilde{\phi}_2)$ on $\tilde{\pi}_1(\mathfrak{R}'')$ by $\langle\alpha^*\tilde{\phi}_2; \tilde{\pi}_1(S)\rangle = \langle\tilde{\phi}_2; \alpha\tilde{\pi}_1(S)\rangle$; since α is ultraweakly continuous, $(\alpha^*\tilde{\phi}_2)$ inherits the ultraweak continuity of $\tilde{\phi}_2$; since α is an isomorphism and $\tilde{\phi}_2$ is positive, so is $(\alpha^*\tilde{\phi}_2)$; $(\alpha^*\tilde{\phi}_2)$ is clearly normalized to 1, hence is a normal state on $\tilde{\pi}_1(\mathfrak{R}'')$. We conclude then that the isomorphism α induces a bijective mapping α^* from the set $(\mathfrak{S}_2)_*$ of all normal states on $\tilde{\pi}_2(\mathfrak{R}'')$ to the set $(\mathfrak{S}_1)_*$ of all normal states on $\tilde{\pi}_1(\mathfrak{R}'')$; we notice incidentally that α^* can be extended to an isometric isomorphism between the Banach spaces $\tilde{\pi}_1(\mathfrak{R}'')_*$ and $\tilde{\pi}_2(\mathfrak{R}'')_*$, respective preduals of the von Neumann algebras $\tilde{\pi}_1(\mathfrak{R}'')$ and $\tilde{\pi}_2(\mathfrak{R}'')$. We recall (see second lemma on p. 118) that every $\tilde{\phi}_i$ in $(\mathfrak{S}_i)_*$ can be written in the form $\langle\tilde{\phi}_i; \tilde{\pi}_i(S)\rangle = \text{Tr}\,\{\rho_i\tilde{\pi}_i(S)\}$, where ρ_i is a density matrix on the representation space $\mathcal{H}_i$; i.e., $\rho_i = \rho_i^* > 0$, $\text{Tr}\,\rho_i = 1$. Adjoining this to the last but one remark, we conclude that to any density matrix ρ_j ($j = 2, 1$) on $\mathcal{H}_j$ corresponds at least one density matrix ρ_i ($i = 1, 2$) on $\mathcal{H}_i$ such that $\text{Tr}\,\rho_1\tilde{\pi}_1(S) = \text{Tr}\,\rho_2\tilde{\pi}_2(S)$ for all S in $\mathfrak{R}''$ and in particular for all S in $\mathfrak{R}$. We therefore see that the quasi-equivalence of the representations π_1 and π_2 of $\mathfrak{R}$ implies that the set of all density-matrix states on $\pi_1(\mathfrak{R})$ *coincides* with the set of all density-matrix states on $\pi_2(\mathfrak{R})$. Let us now consider the converse situation in which we suppose that to any density-matrix ρ_i on $\mathcal{H}_i$ ($i = 1, 2$) corresponds a density matrix ρ_j on $\mathcal{H}_j$ ($j = 2, 1$) such that $\text{Tr}\,\rho_1\pi_1(R) = \text{Tr}\,\rho_2\pi_2(R)$ for all R in $\mathfrak{R}$; by continuity we conclude the existence of a bijection $\alpha^*: (\mathfrak{S}_2)_* \to (\mathfrak{S}_1)_*$ such that $\langle\alpha^*\tilde{\phi}_2; \tilde{\pi}_1(S)\rangle = \langle\tilde{\phi}_2; \tilde{\pi}_2(S)\rangle$ for all S in $\mathfrak{R}''$. Suppose now that K belongs to $\text{Ker}\,\tilde{\pi}_1$. We have $0 = \langle\tilde{\phi}_2; \tilde{\pi}_2(K)\rangle = \text{Tr}\,\rho_2\tilde{\pi}_2(K)$ for all density-matrices ρ_2 on $\mathcal{H}_2$; this implies

† See Dixmier [1957], I.4.3., Corollary 2 to Theorem 2.

that $\tilde{\pi}_2(K) = 0$; i.e., K belongs to Ker $\tilde{\pi}_2$. By exchanging the roles of $\tilde{\pi}_1$ and $\tilde{\pi}_2$, we get Ker $\tilde{\pi}_1$ = Ker $\tilde{\pi}_2$ so that $\tilde{\pi}_1$ and $\tilde{\pi}_2$ are physically equivalent and π_1 and π_2 are quasiequivalent.

In the above paragraph we used the "physical" term of "density-matrix states" on $\pi_i(\mathfrak{R})$; it is clear that the set of all these states coincides with the set of all ultraweakly (and ultrastrongly) continuous states on $\pi_i(\mathfrak{R})$. As such, the elements of this set can be extended in a unique way to ultraweakly (and ultrastrongly) continuous, hence normal, states on $\tilde{\pi}_i(\mathfrak{R}'')$; all normal states on $\tilde{\pi}_i(\mathfrak{R}'')$ are obtainable in this way. We shall henceforth denote by the same symbol $(\mathfrak{S}_i)_*$ the set of all density matrix states on $\pi_i(\mathfrak{R})$ and the set of all normal states on $\tilde{\pi}_i(\mathfrak{R}'')$; we recall that as a consequence of the lemma on p. 119 $(\mathfrak{S}_i)_* = \overline{\overline{\text{co}\ \mathfrak{V}_{\tilde{\pi}_i}}}$.

We can summarize the results obtained up to this point on the notion of quasi-equivalence with the following theorem:

Theorem 12. *Let* $\mathfrak{R}$ *be a* C*-*algebra*; $\mathfrak{R}''$, *its universal enveloping von Neumann algebra*; π_{i} (i = 1, 2), *two representations of* $\mathfrak{R}$; $\tilde{\pi}_{\mathrm{i}}(\mathfrak{R}'') = \pi_{\mathrm{i}}(\mathfrak{R})''$; $(\mathfrak{S}_{\mathrm{i}})_*$ (*or* $\mathfrak{V}_{\mathrm{i}}$), *the set of all density-matrix* (*or vector*) *states associated with the representation* π_{i}; *and* j_{i}, *the natural injection of the set* $\mathfrak{S}_{\mathrm{i}}$ *of all states on* $\pi_{\mathrm{i}}(\mathfrak{R})$ *into the set* $\mathfrak{S}$ *of all states on* $\mathfrak{R}$. *Then the following four conditions are equivalent*.

(i) π_1 *and* π_2 *are quasi-equivalent*.

(ii) $\tilde{\pi}_1$ *and* $\tilde{\pi}_2$ *are physically equivalent*.

(iii) *There exists a bijective mapping* α^* *from* $(\mathfrak{S}_2)_*$ *to* $(\mathfrak{S}_1)_*$ *such that* $\langle\alpha^*\phi_2; \pi_1(\mathrm{R})\rangle = \langle\phi_2; \pi_2(\mathrm{R})\rangle$ *for all* ϕ_2 *in* $(\mathfrak{S}_2)_*$ *and all* R *in* $\mathfrak{R}$.

(iv) $\mathrm{j}_1(\overline{\overline{\text{co}\ \mathfrak{V}_1}}) = \mathrm{j}_2(\overline{\overline{\text{co}\ \mathfrak{V}_2}})$.

Proof. See the discussion preceding the statement of the theorem. ∎

The equivalence between conditions (i) and (iii) in the case of faithful representations was first proved by Takeda [1954]; the key to the general case can be obtained from Takeda's result by noticing, as we did above, that both conditions imply separately that Ker π_1 = Ker π_2, thus providing an alternative proof for that part of the theorem.

Theorem 12 seems to be as close as one can possibly get to an interpretation of the concept of quasi-equivalence in physical terms; the mathematical significance of this concept is discussed in detail in Dixmier [1964] Section 5.3, to which the reader is referred for further information along this line. We only want to mention that the concept of quasi-equivalence for C^*-algebras, as we defined it above, can be made to coincide with the concept of quasi-equivalence as it appears in the theory of group representations;† indeed, if

† For this theory the reader is referred to the lecture notes by Mackey [1955].

$\pi: \mathfrak{R} \to \mathfrak{B}(\mathcal{H})$ is a representation of a C^*-algebra $\mathfrak{R}$, a *subrepresentation* of π is defined as the representation of $\mathfrak{R}$ obtained by restricting $\pi(\mathfrak{R})$ to a *nonzero* subspace of $\mathcal{H}$, stable with respect to $\pi(\mathfrak{R})$. We denote by $\mathcal{S}_\pi$ the set of all subrepresentations of π and note that $\mathcal{S}_\pi$ reduces to a single element, namely π, whenever π is irreducible. Two representations π_1 and π_2 are said to be *disjoint* (which we denote by $\pi_1 \between \pi_2$) whenever ρ_1 in $\mathcal{S}_1$ and ρ_2 in $\mathcal{S}_2$ implies that ρ_1 and ρ_2 are *not* unitarily equivalent; we note also that, if π_1 and π_2 are irreducible, to say that they are disjoint amounts to saying that they are unitarily inequivalent. We denote by $\mathcal{Q}(\pi_1, \pi_2)$ for any two representations π_1 and π_2 of $\mathfrak{R}$ the set of all subrepresentations of π_1 that are disjoint from π_2; Dixmier [1964] shows† that two representations π_1 and π_2 of a C^*-algebra $\mathfrak{R}$ are quasi-equivalent if and only if $\mathcal{Q}(\pi_1, \pi_2)$ and $\mathcal{Q}(\pi_2, \pi_1)$ are empty, thus rejoining the theory of representations of groups. As a consequence of this property of quasi-equivalent representations of a C^*-algebra we notice that two quasi-equivalent *irreducible* representations of a C^*-algebra are unitarily equivalent. The converse is evidently true, since it holds as well for general representations; suppose, indeed, that two representations π_1 and π_2 are unitarily equivalent. This unitary equivalence extends by continuity to a unitary equivalence between $\pi_1(\mathfrak{R})''$ and $\pi_2(\mathfrak{R})''$, to define the isomorphism $\alpha: \pi_1(\mathfrak{R})'' \to \pi_2(\mathfrak{R})''$ such that $\pi_2(R) = \alpha\pi_1(R)$ for all R in $\mathfrak{R}$; this establishes the quasi-equivalence of π_1 and π_2.

We conclude our discussion of quasi-equivalence by remarking that we have, as of now, encountered three notions of equivalence between representations of a C^*-algebra, namely, unitary (i.e., spatial) equivalence, quasi-equivalence, and physical equivalence; each one implies the next in the order in which we have listed them; unitary and quasi-equivalence coincide for irreducible representations and quasi-equivalence and physical equivalence coincide for von Neumann algebras. Throughout this section we have given physical criteria for each of these notions in terms of the states associated with the representations considered.

f. Traces and Types

C^*-algebras, their representations as algebras of operators acting on a Hilbert space, and, in particular, von Neumann algebras come in a wide variety of forms. A remarkable classification of von Neumann algebras in mutually exclusive "types" was worked out at an early stage of the theory by Murray and von Neumann [1936] for the special case of "factors"; their classification is complete in the sense that a factor necessarily belongs to one type. This classification is based on the properties of the range of a

† See Proposition 5.3.1, which he incidentally establishes for any *-algebra $\mathfrak{R}$.

"dimension-function" which generalizes the usual notion of trace when the latter is restricted to the projectors of the von Neumann algebra under consideration.† An extension of the classification of factors to a classification of general von Neumann algebras has been attempted on the basis of a generalized notion of trace, which we consider in the sequel. The "types" of general von Neumann algebras, obtained in this way, coincide with the Murray and von Neumann types in the case of factors; they are again mutually exclusive. The classification, however—in contrast to the case of factors—is not exhaustive, i.e., a general von Neumann algebra is not necessarily of one type. The interest of this classification is that it is nevertheless always possible to "decompose" in a canonical way a von Neumann algebra into von Neumann algebras of definite types. The classification of von Neumann algebras can next be extended in a natural way to a classification of representations of a general C^*-algebra.

The organization of this subsection is the following: we first state briefly the essential properties of the trace operation on $\mathfrak{B}(\mathcal{H})$; we then define the abstract notion of trace on a C^*-algebra and use it for the classification of von Neumann algebras. Specialization to the case of factors is discussed next and we close with the classification of representations of C^*-algebras, with emphasis on its meaning in terms of the states that generate the representation considered. Several examples are listed at the end of this subsection and discussed in more detail later on; the aim of these examples is to substantiate the assertion that the mathematical classification of representations has an explicit physical counterpart.

The format of this book does not allow space for a complete mathematical treatment of this classification and we state mainly the basic definitions and the principal results. For a more complete exposition the reader is referred to Naimark [1964] (Sections 35 to 38) and Dixmier [1957] (Chapter III, Section 2) for the case of factors, to Dixmier [1957] (Chapter I, Sections 6 and 8) for general von Neumann algebras, and to Dixmier [1964] (Sections 5 and 6) for the definition and properties of generalized traces and the types of representation of C^*-algebras; other sections in Dixmier [1957 and 1964] contain many results that pertain to the aspects of *-algebras discussed in this subsection.

Now let $\mathcal{H}$ be a complex Hilbert space; let $\mathfrak{A}$ be the set of all self-adjoint elements in $\mathfrak{B}(\mathcal{H})$, let $\mathfrak{A}^+$ be the set of all positive elements in $\mathfrak{B}(\mathcal{H})$, and let $\{\Psi_i\}$ be an orthonormal basis in $\mathcal{H}$. We define a *trace* ψ on $\mathfrak{A}^+$ by

$$\langle \psi; A\rangle = \sum_i (\Psi_i, A\Psi_i).$$

† The fact that this classification uses only the projectors suggests a similar classification of abstract orthomodular lattices (Holland [1968]).

The following properties of ψ are well-known:†

(o) $\psi:\mathfrak{A}^+ \to [0, \infty]$.
(i) $\langle\psi; A + B\rangle = \langle\psi; A\rangle + \langle\psi; B\rangle \ \forall A, B \in \mathfrak{A}^+$.
(ii) $\langle\psi; \lambda A\rangle = \lambda\langle\psi; A\rangle \quad \lambda \geqslant 0, \forall A \in \mathfrak{A}^+$.
(iii) $\langle\psi; RR^*\rangle = \langle\psi; R^*R\rangle \ \forall R \in \mathfrak{B}(\mathcal{H})$.
(iii′) $\langle\psi; UAU^*\rangle = \langle\psi; A\rangle \ \forall A \in \mathfrak{A}^+$ and $\forall$ unitary U in $\mathfrak{B}(\mathcal{H})$.
(iv) $\langle\psi; A\rangle = 0$ and $A \in \mathfrak{A}^+$ implies $A = 0$.
(v) $\langle\psi; A\rangle = \sup_{\substack{B\in\mathfrak{A}^+ \\ B\leqslant A;\langle\psi;B\rangle<\infty}} \langle\psi; B\rangle \ \forall A \in \mathfrak{A}^+$
(v′) For every A in $\mathfrak{A}^+$, with $A \neq 0$, there exists B in $\mathfrak{A}^+$, with $A \geqslant B \neq 0$, such that $\langle\psi; B\rangle < \infty$.

REMARK. Proposition iii′ shows in particular why we define ψ on $\mathfrak{A}^+$ and not on $\mathfrak{B}(\mathcal{H})$ itself; suppose indeed that A is a self-adjoint operator with a (nondegenerate) discrete spectrum consisting of the terms of a conditionally convergent series; then, on rearranging the order of the corresponding eigenvectors $\{\Psi_i\}$ (which can be achieved by a unitary transformation of $\mathcal{H}$) we can make the trace converge to any arbitrary real number so that the trace would not satisfy (iii′) and actually could not be defined at all without reference to a particular orthonormal basis in $\mathcal{H}$ when $\mathcal{H}$ is infinite-dimensional.

A further property which we now want to express carefully is enjoyed by ψ. Under the relation $A \leqslant B$ [defined here by $(\Psi, A\Psi) \leqslant (\Psi, B\Psi)$ for all Ψ in $\mathcal{H}$] $\mathfrak{A}$ becomes a partially ordered set; let $\mathcal{F}$ be a subset‡ of $\mathfrak{A}$, *directed* by this relation. We say that $\mathcal{F}$ is *bounded above* if there exists A in $\mathfrak{A}$ such that $A \geqslant F$ for all F in $\mathcal{F}$; we denote by ${}^w\overline{\mathcal{F}}$ and ${}^s\overline{\mathcal{F}}$, respectively, the weak and strong closure of $\mathcal{F}$; we know that if $\mathcal{F}$ is a directed set, bounded above in $\mathfrak{A}$, then ${}^s\overline{\mathcal{F}}$ contains an upper bound of $\mathcal{F}$; furthermore, if $\mathcal{F}$ is a directed set in $\mathfrak{A}$ and if there exists an X in ${}^w\overline{\mathcal{F}}$ such that $X \geqslant F$ for all F in $\mathcal{F}$, then§ X is the least upper bound of $\mathcal{F}$. As a consequence we see that if $\mathfrak{N}$ is a von Neumann algebra and $\mathfrak{N}^+$ is the set of all its positive elements every directed set $\mathcal{F}$ in $\mathfrak{N}^+$, bounded above in $\mathfrak{B}(\mathcal{H})$, has an upper bound in $\mathfrak{N}$. Now let $\mathfrak{N}$ and $\mathfrak{M}$ be two von Neumann algebras (e.g., $\mathfrak{B}(\mathcal{H})$ and $\mathbb{C}$); a positive mapping ψ from $\mathfrak{N}$ to $\mathfrak{M}$ is said to be *normal* whenever

(vi) for every directed set $\mathcal{F} \subseteq \mathfrak{N}^+$, bounded above in $\mathfrak{N}^+$: $\sup_{F\in\mathcal{F}} \langle\psi; F\rangle = \langle\psi; \sup_{F\in\mathcal{F}} F\rangle$;

this definition justifies the use made previously of the term "normal" for an ultraweakly or ultrastrongly continuous positive linear functional on a von

† See, for instance, Dixmier [1957], I.6.6: Theorem 5.
‡ In the terminology introduced on p. 102 $\mathcal{F}$ is a directed set; Dixmier [1957] and [1964] speaks of an "ensemble filtrant croissant."
§ See Vigier [1946] or Dixmier [1957], Appendix II.

Neumann algebra;† we can now express a further property of the trace defined above by saying that it is *normal*. The last property of the trace we now want to recall is that

(vii) $\langle\psi; P\rangle = \dim P\mathcal{H}$ for all projectors P in $\mathfrak{B}(\mathcal{H})$.

To generalize the notion of *trace* to an arbitrary C^*-algebra $\mathfrak{R}$, whose set of positive elements we denote by $\mathfrak{R}^+ \equiv \mathfrak{A}^+$, we retain as axioms the properties (o) to (iii) above (in the latter R now runs over $\mathfrak{R}$). We say that a trace is *faithful* if it satisfies property (iv), that it is *semifinite* if it satisfies condition (v), and, in particular, that it is *finite* if $\langle\psi; A\rangle < \infty$ for all A in $\mathfrak{A}^+$.

If, in addition $\mathfrak{N}$ is a von Neumann algebra, we can equivalently‡ take for axioms of the trace the properties (o), (i), (ii), and (iii′), where U now runs on the unitary elements of $\mathfrak{N}$. A trace on a von Neumann algebra is said to be *normal* if it satisfies property (vi) with $\mathfrak{M}^+ = \mathbb{R}^+$; for normal traces on a von Neumann algebra, conditions (v) and (v′) are equivalent.§

A von Neumann algebra $\mathfrak{N}$ is said to be *finite* if for every A in $\mathfrak{N}^+$ with $A \neq 0$ there exists a finite normal trace ψ on $\mathfrak{N}^+$ such that $\langle\psi; A\rangle \neq 0$. For instance, the von Neumann algebra $\mathfrak{B}(\mathcal{H})$ is finite when $\mathcal{H}$ is finite-dimensional. Since every vector state on an abelian von Neumann algebra $\mathfrak{N}$ satisfies condition (iii) trivially, it is a trace on $\mathfrak{N}$; it is also a finite trace; furthermore, it is ultraweakly continuous, hence normal. We conclude therefore that every abelian von Neumann algebra is finite. A von Neumann algebra that is not finite is said to be *infinite*; in particular, a von Neumann algebra is said to be *properly infinite* if $\psi = 0$ is the only finite normal trace on $\mathfrak{N}^+$. Using property (iii′), we conclude that the von Neumann algebra $\mathfrak{B}(\mathcal{H})$ is properly infinite when $\mathcal{H}$ is infinite-dimensional.

Let $\mathfrak{N}$ be a von Neumann algebra and P, a projector in $\mathfrak{N}'$. We denote by $\mathfrak{N}_P$ the von Neumann algebra obtained by restricting $\mathfrak{N}$ to the subspace $P\mathcal{H}$, stable with respect to $\mathfrak{N}$.

Lemma. *Let $\mathfrak{N}$ be a von Neumann algebra; there exists a projector* F *in the center $\mathfrak{N} \cap \mathfrak{N}'$ of $\mathfrak{N}$ such that*

(i) $\mathfrak{N}_{\mathrm{F}}$ *is finite*,
(ii) $\mathfrak{N}_{\mathrm{I-F}}$ *is properly infinite*,
(iii) *for any projector* E *in* $\mathfrak{N} \cap \mathfrak{N}'$, $\mathfrak{N}_{\mathrm{E}}$ *finite* (*or properly infinite*) *implies* $\mathrm{E} \leqslant \mathrm{F}$ (*or* $\mathrm{E} \leqslant \mathrm{I} - \mathrm{F}$).

Proof. Dixmier [1957] I.6.7, Proposition 8. ∎

† See Dixmier [1957] I.4.2, Theorem 1.
‡ For a proof see Dixmier [1957], I.6.1, Corollary 1 to Proposition 1.
§ See Dixmier [1964], Appendix A.28.

A von Neumann algebra $\mathfrak{N}$ is said to be *semifinite* if for every A in $\mathfrak{N}^+$ with $A \neq 0$ there exists a semifinite normal trace ψ on $\mathfrak{N}^+$ such that $\langle \psi; A \rangle \neq 0$; it is then† possible to find a semifinite normal trace on $\mathfrak{N}^+$ that is faithful; every finite von Neumann algebra is obviously semifinite, but the converse is certainly not true, since the von Neumann algebra $\mathfrak{B}(\mathcal{H})$ is semifinite, whatever the dimension of $\mathcal{H}$ might be. The commutant of a semifinite von Neumann algebra is a semifinite von Neumann algebra.‡ A von Neumann algebra is said to be *purely infinite* if $\psi = 0$ is the only semifinite normal trace on $\mathfrak{N}^+$; every purely infinite von Neumann algebra is obviously properly infinite, but the converse is not true, as shown by the example of $\mathfrak{B}(\mathcal{H})$ with $\mathcal{H}$ infinite-dimensional. Many instances of purely infinite von Neumann algebras are encountered in physical applications.

Lemma. *Let $\mathfrak{N}$ be a von Neumann algebra; there exists a projector $\mathrm{E}_{\mathrm{III}}$ in $\mathfrak{N} \cap \mathfrak{N}'$ such that*

(i) $\mathfrak{N}_{\mathrm{I}-\mathrm{E}_{\mathrm{III}}}$ *is semifinite,*
(ii) $\mathfrak{N}_{\mathrm{E}_{\mathrm{III}}}$ *is purely infinite,*
(iii) *for any projector* E *in* $\mathfrak{N} \cap \mathfrak{N}'$, $\mathfrak{N}_{\mathrm{E}}$ *semifinite (or purely infinite) implies* $\mathrm{E} \leqslant \mathrm{I} - \mathrm{E}_{\mathrm{III}}$ (*or* $\mathrm{E} \leqslant \mathrm{E}_{\mathrm{III}}$).

Furthermore $FE_{\mathrm{III}} = 0$.

Proof. Dixmier [1957], I.6.7, Proposition 8. ▌

A von Neumann algebra $\mathfrak{N}$ is said to be of *type I* (or *discrete*) if it is isomorphic to a von Neumann algebra $\mathfrak{M}$ with abelian commutant; since $\mathfrak{M}'$ is abelian, it is finite, hence semifinite; so is its commutant $\mathfrak{M}'' = \mathfrak{M}$; hence we have that an algebra of type I is semifinite. A von Neumann algebra is said to be *continuous* if there exists no projector P in its center such that $P \neq 0$ and $\mathfrak{N}_P$ is discrete.

Lemma. *Let $\mathfrak{N}$ be a von Neumann algebra; there exists a projector E_{I} in the center $\mathfrak{N} \cap \mathfrak{N}'$ of $\mathfrak{N}$ such that*

(i) $\mathfrak{N}_{\mathrm{E}_{\mathrm{I}}}$ *is discrete,*
(ii) $\mathfrak{N}_{\mathrm{I}-\mathrm{E}_{\mathrm{I}}}$ *is continuous,*
(iii) *for any projector* E *in* $\mathfrak{N} \cap \mathfrak{N}'$, $\mathfrak{N}_{\mathrm{E}}$ *discrete (or continuous) implies that* $\mathrm{E} \leqslant \mathrm{E}_{\mathrm{I}}$ (or $\mathrm{E} \leqslant \mathrm{I} - \mathrm{E}_{\mathrm{I}}$).

Furthermore $\mathrm{E}_{\mathrm{I}}\mathrm{E}_{\mathrm{III}} = 0$.

Proof. Dixmier [1957], I.8.1, Corollary 1 to Proposition 1 and Corollary 1 to Proposition 2. ▌

From $E_{\mathrm{I}}E_{\mathrm{III}} = 0$ we conclude in particular that if $E_{\mathrm{III}} = I$ then $E_{\mathrm{I}} = 0$, i.e., $I - E_{\mathrm{I}} = I$; that is to say, every purely infinite von Neumann algebra is

† See Dixmier [1957], I.6.7, Proposition 9(i).
‡ See Dixmier [1957], I.6.8, Corollary 1.

continuous. Had we exchanged in the above reasoning the roles of E_{I} and E_{III} we would evidently have found the result already pointed out, namely that every discrete von Neumann algebra is semifinite.

Since for an arbitrary von Neumann algebra $\mathfrak{N}$, $\mathfrak{N} \cap \mathfrak{N}'$ is abelian, we can form the following Boolean lattice of projectors in $\mathfrak{N} \cap \mathfrak{N}'$:

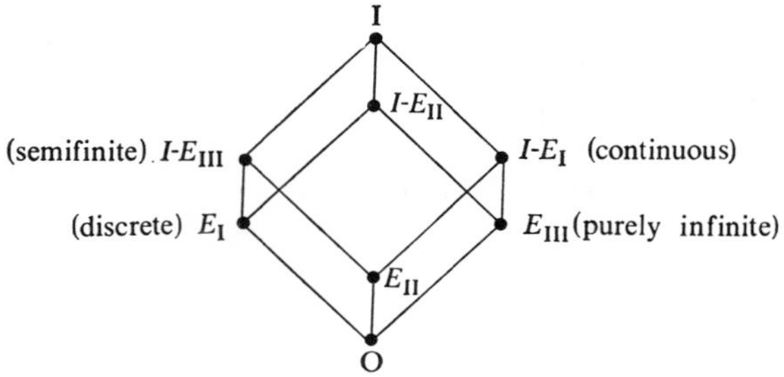

This lattice expresses the following facts:

1. The projector E_{II} [defined as the intersection $I - (E_{\mathrm{I}} + E_{\mathrm{III}})$ of the projectors $I - E_{\mathrm{I}}$ and $I - E_{\mathrm{III}}$] is the largest projector in $\mathfrak{N} \cap \mathfrak{N}'$ such that $\mathfrak{N}_E$ is semifinite and continuous;
2. The projector $E_{\mathrm{I}} + E_{\mathrm{II}}$ is the largest projector in $\mathfrak{N} \cap \mathfrak{N}'$ such that $\mathfrak{N}_E$ is semifinite.
3. The projector $E_{\mathrm{III}} + E_{\mathrm{II}}$ is the largest projector in $\mathfrak{N} \cap \mathfrak{N}'$ such that $\mathfrak{N}_E$ is continuous;
4. The projectors E_{I}, E_{II}, and E_{III} are mutually orthogonal and add to the identity.

These facts suggest the following definitions: a von Neumann algebra is said to be *type II* if it is semifinite and continuous, and *type III* if it is purely infinite (hence continuous). We see from the above lattice that a von Neumann algebra can be at most one of the three types, I, II, and III, and that an arbitrary von Neumann algebra can be canonically decomposed into three von Neumann algebras, $\mathfrak{N}_{E_{\mathrm{I}}}$, $\mathfrak{N}_{E_{\mathrm{II}}}$, and $\mathfrak{N}_{E_{\mathrm{III}}}$, which are, respectively, types I, II, and III.

We can pursue this decomposition one step further by considering the following Boolean lattice of projectors in $\mathfrak{N} \cap \mathfrak{N}'$:

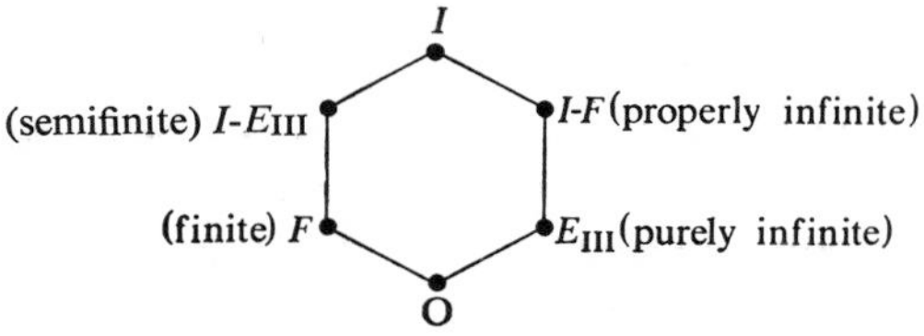

We find first that the algebras of type I can be further decomposed into two mutually disjoint classes: those that are finite (denoted I_n $(n < \infty)$) and those that are properly infinite (denoted I_∞). Similarly, the von Neumann algebras of type II can be further separated into two mutually disjoint classes: those that are finite and those that are properly infinite; the former are said to be *type* II_1, the latter, *type* II_∞; the projector E_{II} is then the sum of two orthogonal projectors $E_{II}(1)$ and $E_{II}(\infty)$ such that $E_{II}(1)$ [resp. $E_{II}(\infty)$] is the largest projector E in $\mathfrak{N} \cap \mathfrak{N}'$ such that $\mathfrak{N}_E$ is continuous and finite [resp. continuous and properly infinite]. This remark concludes the classification of general von Neumann algebras, and we summarize the situation as described up to this point, as follows:

Theroem 13. *Let* $\mathfrak{N}$ *be an arbitrary von Neumann algebra; then there exists a partition of the identity by five projectors* $E_I(n)$, $E_I(\infty)$, $E_{II}(1)$, $E_{II}(\infty)$, E_{III} *in* $\mathfrak{N} \cap \mathfrak{N}'$ *such that the restriction of* $\mathfrak{N}$ *to the corresponding subspaces is a von Neumann algebra that is, respectively, type* I_n *(discrete, finite),* I_∞ *(discrete, properly infinite),* II_1 *(continuous, finite),* II_∞ *(continuous, properly infinite), and* III *(purely infinite).*

The following results are useful in applications; let $\mathfrak{N}$ be a von Neumann algebra; if E is a projector in $\mathfrak{N}'$, we define $\mathfrak{N}_E$ as above; if E is a projector in $\mathfrak{N}$, we define the elements of the von Neumann algebra $\mathfrak{N}_E$ as those operators obtained as the restriction of B to $E\mathcal{H}$, where B runs over the set of all elements of $\mathfrak{N}$ such that $BE = EB = B$. In both cases $\mathfrak{N}_E$ inherits from $\mathfrak{N}$ the following properties: (a) it is semifinite; (b) it is purely infinite (i.e., type III); (c) it is discrete (i.e., type I); (d) it is continuous; (e) it is type II; (f) it is finite; and (g) it is type II_1. For the proofs see Dixmier [1957]; specifically, for (a) and (f) see I.6.8, Proposition 11; for (b) see I.6.8, Corollary 4; for (c) and (d) see I.8.3, Proposition 4; (e) follows from (a) and (d); and (g) follows from (e) and (f). The same rules are true for $\mathfrak{N}'$, except for the last two which are not true in general. For the proofs see Dixmier [1957]; specifically, for (a) see I.6.8, Corollary 1; for (b) see I.6.8, Corollary 3; for (c) see I.8.2, Theorem 1; for (d) see I.8.2, Corollary 1; (e) follows from (a) and (d). Note, however, that (f) hence (g) are also true for $\mathfrak{N}'$ when there exists a finite set $\{\Psi_1, \ldots, \Psi_n\}$ in $\mathcal{H}$ cyclic with respect to $\mathfrak{N}$ (Dixmier [1957] II.2, Exercise 2).

A von Neumann algebra $\mathfrak{N}$ is said to be a *factor* whenever its center is trivial (i.e., $\mathfrak{N} \cap \mathfrak{N}' = \{\lambda I\}$). The only projectors in the center of a factor are then 0 and I; consequently an arbitrary factor is *necessarily* of one type and the above classification is therefore exhaustive in the particular case of factors.

The case of factors presents another interesting peculiarity: the seventh and last property of the ordinary trace on $\mathfrak{B}(\mathcal{H})$ (i.e., its use to "measure" the dimension of the subspace associated to a projector) can be generalized to give some further insight into the nomenclature used in the classification of von Neumann algebras; this theory is now presented, as far as possible for general von Neumann algebras, to emphasize some peculiarities proper to the case of factors.

Let us first recall that a partial isometry on a Hilbert space is a linear operator $U:\mathcal{H} \to \mathcal{H}$ which is isometric on a subspace $\mathcal{M}$ of $\mathcal{H}$ and zero on its orthogonal complement $\mathcal{M}^{\perp}$, i.e., $|U\Psi| = |\Psi|$ for all Ψ in $\mathcal{M}$ and $U\Psi = 0$ for all Ψ in $\mathcal{M}^{\perp}$; the following five conditions are then equivalent:

(i) $U =$ a partial isometry,

(ii) $U^*U = P$ is a projector,

(iii) $UU^* = Q$ is a projector,

(iv) $UU^*U = U$,

(v) $U^*UU^* = U^*$;

P is then the projector on $\mathcal{M}$ and Q is the projector on $U\mathcal{M}$; $UPU^* = Q$ and $\dim P = \dim Q$.

We now say that two projectors P and Q acting on a Hilbert space are *equivalent* with respect to a von Neumann algebra $\mathfrak{N} \subseteq \mathfrak{B}(\mathcal{H})$ if there is a partial isometry U in $\mathfrak{N}$ such that $U^*U = P$ and $UU^* = Q$; we denote this situation by $P \sim Q \pmod{\mathfrak{N}}$ or simply by $P \sim Q$ when no confusion is likely to occur. Clearly $P \sim Q \pmod{\mathfrak{N}}$ implies that P and Q belong to $\mathfrak{N}$.

We say that a projector P in a von Neumann algebra is *finite* if $Q \in \mathfrak{N}$, $Q\mathcal{H} \subseteq P\mathcal{H}$, and $Q \sim P \pmod{\mathfrak{N}}$ imply together that $P = Q$. It is easy to see† that if $\mathfrak{N}_P$ is finite (as a von Neumann algebra), then P is finite in the above sense; the converse is also true, but much harder to prove.‡ We say that a projector is *infinite* when it is not finite.

For two projectors P and Q in a von Neumann algebra $\mathfrak{N}$ we say that P is *not larger* than Q (which we denote by $P \leqslant Q$) if there exists a projector R in $\mathfrak{N}$ with $P \sim R$ and $R \subseteq Q$ (i.e., $R\mathcal{H} \subseteq Q\mathcal{H}$); we say that P is *smaller* than Q (which we denote by $P < Q$) if $P \leqslant Q$ without P being equivalent to Q. Clearly $P \leqslant Q$ and $Q \leqslant P$ imply $P \sim Q$, and $P \leqslant Q$ with $Q \leqslant R$ implies $P \leqslant R$, so that the relation $\leqslant$ partially orders the set $\mathcal{P}$ of all projectors of a von Neumann algebra. Furthermore we have the following theorem.

Theorem 14 (Comparison Theorem). *Let $\mathfrak{N}$ be a von Neumann algebra, $\mathfrak{Z}$ its center, and* P *and* Q, *two projectors of $\mathfrak{N}$; then there exists a projector* R *in $\mathfrak{Z}$ such that* $PR \leqslant QR$ *and* $Q(I - R) \leqslant P(I - R)$.

Proof. Dixmier [1957], III.1.2, Theorem 1. ▌

The principal interest for us in this theorem lies in the particular case in which $\mathfrak{N}$ is a *factor*; then the projector R, the existence of which is asserted

† Dixmier [1957], III.2.3, Proposition 4.

‡ See, for instance, Dixmier [1957], III.8.6, Corollary 1.

by the theorem, can be only 0 or I. If $R = I$, then $P \preccurlyeq Q$, and if $R = 0$ then $Q \preccurlyeq P$; consequently the relation $\preccurlyeq$ totally orders the set $\mathfrak{P}$ of all projectors of a factor. This property, which does not hold in general for an arbitrary von Neumann algebra, is the key to the power of the definition of a relative dimension for a factor, a notion that we are now to discuss.

Let $\mathfrak{N}$ be a factor, $\mathfrak{P}$, the set of all projectors in $\mathfrak{N}$, and ψ, a faithful normal trace on $\mathfrak{N}^+$; if $\mathfrak{N}$ is semifinite, ψ is supposed to be semifinite; the restriction of ψ to $\mathfrak{P}$ is said to be a *relative dimension d* on $\mathfrak{P}$. In this connection let us consider the case in which $\mathfrak{N}$ is a von Neumann algebra and ψ is a normal trace on $\mathfrak{N}^+$; then there exist two projectors $E \geqslant F$, both in $\mathfrak{N} \cap \mathfrak{N}'$ such that ψ induces a faithful semifinite trace on $\mathfrak{N}^+_{E-F}$, an identically zero trace on $\mathfrak{N}^+_F$, and a trace ψ on $\mathfrak{N}^+_{I-E}$ such that $\langle\psi; S\rangle = +\infty$ for all nonvanishing, elements in $\mathfrak{N}^+_{I-E}$; for a proof see Dixmier [1957], I.6.1, Corollary 2. In particular, if $\mathfrak{N}$ is a factor, the couple (F, E) can take only the values $(0, 0)$, $(0, I)$, or (I, I); the third case would imply $\mathfrak{N}_F = \mathfrak{N}$, hence $\psi = 0$, and is then excluded if $\psi \neq 0$ is imposed. Suppose, in addition, that $\mathfrak{N}$ is purely infinite; the second case is excluded as well because it means that $\mathfrak{N}_{E-F} = \mathfrak{N}$ and would then imply that ψ is a normal, nonvanishing, semifinite trace on $\mathfrak{N}^+$ which is impossible by definition of a purely infinite von Neumann algebra. We therefore conclude that a faithful normal trace on a purely infinite factor takes the value $+\infty$ on every nonvanishing element of $\mathfrak{N}^+$. In addition to this, on any factor $\mathfrak{N}$ two normal, faithful semifinite traces are proportional to one another.† Since a factor is either semifinite or purely infinite, we conclude from these two remarks that *the relative dimension is determined up to a positive, finite multiplicative constant.* Since ψ is a faithful trace, we have from the defining properties (o), (ii), and (iv) of ψ

(a) $$d(P) \geqslant 0 \;\forall\, P \in \mathfrak{P}; \qquad d(P) = 0 \Rightarrow P = 0.$$

If P and Q are orthogonal, i.e., $PQ = 0$, then $P + Q$ also belong to $\mathfrak{P}$, and from the defining property (i) of the trace we have

(b) $$PQ = 0 \Rightarrow d(P + Q) = d(P) + d(Q).$$

Suppose now that $P \sim Q$ and let us denote by U the partial isometry that provides the equivalence; we then have $U^*U = P$ and $UU^* = Q$, which, when written in the defining property (iii) of the trace ψ, give

(c) $$P \sim Q \Rightarrow d(P) = d(Q).$$

Furthermore it is easy to see‡ that $d(P)$ is finite (or infinite) if P is finite (or infinite); consequently

(d) $$d(P) = +\infty \quad \text{if and only if } P \text{ is infinite.}$$

† Dixmier [1957], I.6.4, Corollary to Theorem 3.

‡ See, for instance, Dixmier [1957], III.2.7, Proposition 13(i) and (ii).

We finally notice that $P \preccurlyeq Q$ implies that $d(P) \leqslant d(Q)$; then using property (c), we have

(e) $$P < Q \text{ with } P \text{ finite, implies } d(P) < d(Q).$$

Dixmier also mentions that if P and Q are finite with $d(P) = d(Q)$ [or $d(P) < d(Q)$], then $P \sim Q$ (or $P \preccurlyeq Q$). Properties (a) to (e) are often† taken as the defining properties of the relative dimension.

The connection with the classification of factors previously established is given by the following theorem:

Theorem 15. *Let* $\mathfrak{N}$ *be a factor;* $\mathfrak{T}$ *be the set of all its projectors and* d, *a relative dimension on* $\mathfrak{T}$. *The following cases then occur:*

(i) *If* $\mathfrak{N}$ *is type* I, *we can, by a suitable choice of a multiplicative constant, make* $\mathrm{d}(\mathfrak{T})$ *coincide either with* $\{0, 1, 2, \ldots, \mathrm{n}\}$ *or with* $\{0, 1, 2, \ldots, \infty\}$ *(in the former case* $\mathfrak{N}$ *is said to be type* $\mathrm{I_n}$, *in the latter, type* I_∞).

(ii) *If* $\mathfrak{N}$ *is type* II_1, *we can, by a suitable choice of a multiplicative constant, make* $\mathrm{d}(\mathfrak{T})$ *coincide with the interval* $[0, 1]$.

(iii) *If* $\mathfrak{N}$ *is type* II_∞, $\mathrm{d}(\mathfrak{T}) = [0, +\infty]$.

(iv) *If* $\mathfrak{N}$ *is type* III, $\mathrm{d}(\mathfrak{T})$ *takes only the two values* 0 *and* $+\infty$.

The proof that the range $d(\mathfrak{T})$ of the relative dimension d can be only the five types (I_n, I_∞, II_1, II_∞, and III) was first given by Murray and von Neumann [1936] (Theorem VIII), who also showed that this characterization of factors is invariant under algebraic isomorphisms. A standard textbook exposition of these results (following essentially the lines of the original investigation) can be found in Naimark [1964]; the connection with the general classification of von Neumann algebras is given in Dixmier [1957] (see, in particular, III.2.7, Proposition 14) with examples of each type of factor.

The origin of the terms used in the classification of von Neumann algebras is now made more intuitive: Murray and von Neumann characterized factors by the properties of the range of the relative dimension function that these factors admit: discrete ($\mathrm{I}_n, \mathrm{I}_\infty$) versus continuous ($\mathrm{II}_1, \mathrm{II}_\infty$, III); finite ($\mathrm{I}_n, \mathrm{II}_1$) versus infinite (i.e., for factors, properly infinite) ($\mathrm{I}_\infty, \mathrm{II}_\infty$, III); semifinite ($\mathrm{I}_n, \mathrm{I}_\infty, \mathrm{II}_1, \mathrm{II}_\infty$) versus purely infinite (III).

A projector P in a von Neumann algebra $\mathfrak{N}$ is said to be *minimal* (or a *point* or *atom*) if $P \neq 0$, and if Q in $\mathfrak{N}$ with $Q\mathcal{H} \subseteq P\mathcal{H}$ implies that Q is either 0 or P; it is clear from the above classification of factors and from the general properties of the relative dimension that a factor admits points if and only if it is discrete. This incidentally is also true‡ for general von Neumann algebras. This circumstance shows a serious limitation of those

† See, for instance, Naimark [1964], VII.36.4, Theorem 1.
‡ See Dixmier [1957], I.8.3, Corollary 3.

proposition-calculus approaches to axiomatic quantum theory that want to assume the existence of points; the physicist might be interested in the fact that von Neumann had already suggested, perhaps for the wrong reason, however, that types other than I might occur in physics: he especially advocated type II_1 on the ground that a relative dimension, normalized to 1, exists on these factors (as well as on factors of type I_n), hence allows the definition of a finite, uniform a priori probability on the set $\mathfrak{T}$ of all projectors of the factors of type II_1. As we shall see later on, other factors do actually occur as well in particular physical models.

On considering a representation $\pi: \mathfrak{R} \to \mathfrak{B}(\mathcal{H})$ of a C^*-algebra $\mathfrak{R}$, we define the *type of representation* as follows: we say that π is type I (II or III) if the von Neumann algebra $\pi(\mathfrak{R})''$ [or equivalently $\pi(\mathfrak{R})'$] is type I (II or III). From this definition and that of quasi-equivalence we easily see that the type of representation is invariant under quasi-equivalence; from the definition of discrete von Neumann algebras we see that π is type I if and only if it is quasi-equivalent to a representation ρ of $\mathfrak{R}$ such that $\rho(\mathfrak{R})'$ is abelian. Furthermore, π is *factorial* [or *primary*, i.e., $\pi(\mathfrak{R})''$ is a factor] *of type I* if and only if it is quasi-equivalent to an irreducible representation of $\mathfrak{R}$. All the notions introduced in this subsection, such as the subdivision of type II into types II_1 and II_∞ and semifiniteness, extend in a similar fashion from von Neumann algebras to representations of C^*-algebras. We mention finally that a C^*-algebra is said to be type I if all its representations are type I.

We now want to turn our attention to those representations of a C^*-algebra that are associated with states and see in particular what can be said in general about the characterization of factor types of cyclic representations. We first give the preliminary definitions and the proofs relative to Theorem 16 on an alternative characterization of primary representations.

We recall that for an arbitrary representation $\pi: \mathfrak{R} \to \mathfrak{B}(\mathcal{H})$ we denoted by $\mathfrak{V}_\pi$ the set of all states ψ on $\mathfrak{R}$ that can be written as $\langle \psi; R \rangle = (\Psi, \pi(R)\Psi)$, $\Psi \in \mathcal{H}$; by extension, if ϕ is a state on $\mathfrak{R}$, we denote by $\mathfrak{V}_\phi$ the set of all vector states with respect to the GNS representation π_ϕ. Incidentally, we might mention at this point that an abstract characterization of $\mathfrak{V}_\phi$ has been given by Kadison [1962]†: $\mathfrak{V}_\phi$ is the closure in $\mathfrak{R}^*$ (equipped with its norm-topology) of the set of all states ψ of the form $\langle \psi; R \rangle = \langle \phi; S^*RS \rangle$, where S runs over all elements of $\mathfrak{R}$ such that $\langle \phi; S^*S \rangle = 1$.

Theorem 2 and its lemma suggest the following definitions: we say that two states ψ and ϕ on $\mathfrak{R}$ are in the relation $\psi \leqslant \phi$ whenever $\mathfrak{V}_\psi \subseteq \mathfrak{V}_\phi$; i.e., equivalently $\pi_\psi \leqslant \pi_\phi$ or $\psi \in \mathfrak{V}_\phi$. We write $\psi \sim \phi$ whenever $\psi \leqslant \phi$ and $\phi \leqslant \psi$. Clearly the relation $\leqslant$ partially orders $\mathfrak{S}$ with respect to the equivalence relation $\sim$.

† See, in particular, his Theorem C; see, alternatively, Dixmier [1964], II.2.12.19.

Now consider the set $\mathfrak{B}_\phi$ equipped with the partial ordering relation $\leqslant$ which it inherits from $\mathfrak{S}$ and suppose that π_ϕ is a primary representation. Then let $\psi_i (i = 1, 2)$ be two elements in $\mathfrak{B}_\phi$ and let Ψ_i $(i = 1, 2)$ be any two vectors in $\mathcal{H}_\phi$ such that $\langle \psi_i; R \rangle = (\Psi_i, \pi_\phi(R)\Psi_i)$ for all R in $\mathfrak{R}$. Furthermore let P_i be the projector on $\overline{\pi_\phi(\mathfrak{R})\Psi_i}$. We recall that $P_i \in \pi_\phi(\mathfrak{R})'$; since π_ϕ is primary, $\pi_\phi(\mathfrak{R})'$ is a factor. We now tie this up with the notion of relative equivalence defined on p. 132 on the set $\mathfrak{P}$ of all projectors of a von Neumann algebra and recall that the relation $\leqslant$ defined on $\mathfrak{P}$ (p. 132) totally orders $\mathfrak{P}$ whenever the von Neumann algebra considered is a factor. We then have either $P_1 \leqslant P_2$ or $P_2 \leqslant P_1$; take, for instance, the former case. This relation implies the existence of a projector Q and a partial isometry U, both in $\pi_\phi(\mathfrak{R})'$ such that $Q\mathcal{H}_\phi \subseteq P_2\mathcal{H}_\phi$ and $UP_1U^* = Q$; this partial isometry then implements a unitary equivalence between the representations $\pi_\phi(\mathfrak{R})P_1$ and $\pi_\phi(\mathfrak{R})Q$, the latter obviously being a subrepresentation of $\pi_\phi(\mathfrak{R})P_2$. Since we already know that $\pi_\phi(\mathfrak{R})P_i$ is unitarily equivalent to π_{ψ_i}, we conclude that π_{ψ_1} is unitarily equivalent to a subrepresentation of π_{ψ_2} and $\psi_1 \leqslant \psi_2$. The relation $P_2 \leqslant P_1$ implies similarly that $\psi_2 \leqslant \psi_1$. Hence the relation $\leqslant$ totally orders $\mathfrak{B}_\phi$ when π_ϕ is primary. The converse is also true; suppose, indeed, that $\leqslant$ totally orders $\mathfrak{B}_\phi$. Then let P be a projector in $\pi_\phi(\mathfrak{R})' \cap \pi_\phi(\mathfrak{R})''$, with $P \neq 0, I$, and let Φ be the cyclic vector in $\mathcal{H}_\phi$ corresponding to ϕ; $P \neq 0, I$ clearly implies $P\Phi \neq 0$ and $(I - P)\Phi \neq 0$. Let ψ_1 and ψ_2 be the two states in $\mathcal{H}_\phi$ respectively generated by $\Psi_1 = P\Phi/|P\Phi|$ and $\Psi_2 = (I - P)\Phi/|(I - P)\Phi|$; since P belongs to $\pi_\phi(\mathfrak{R})'$, it reduces $\pi_\phi(\mathfrak{R})$ and Ψ_1 is cyclic for the representation $\pi_\phi(\mathfrak{R})P$ which is then unitarily equivalent to π_{ψ_1}. The same reasoning is used for π_{ψ_2}; by assumption we have either $\psi_1 \leqslant \psi_2$ or $\psi_2 \leqslant \psi_1$. Consider the former case first; π_{ψ_1} is then unitarily equivalent to a subrepresentation of π_{ψ_2} so that $\pi_\phi(\mathfrak{R})P$ is unitarily equivalent to a subrepresentation of $\pi_\phi(\mathfrak{R})(I - P)$. Let U be the unitary operator implementing this equivalence:

$$U\pi_\phi(R)P = \pi_\phi(R)(I - P)U \quad \text{for all } R \text{ in } \mathfrak{R},$$

and, in particular, (for $R = I$), $V \equiv UP = (I - P)U$; by introducing this V in the preceding relation we get $V\pi_\phi(R) = \pi_\phi(R)V$ for all R in $\mathfrak{R}$; i.e., $V \in \pi_\phi(\mathfrak{R})'$. On the other hand, $V^*V = P$ and $VV^* = (I - P)$, so that V is a partial isometry in $\pi_\phi(\mathfrak{R})'$; furthermore, we have $V = VV^*V = (I - P)V = (I - P)VV^*V = (I - P)VP$. We now use the fact that by hypothesis P also belongs to $\pi_\phi(\mathfrak{R})''$, hence commutes with V, so that the last relation above implies $V = 0$ and therefore $P = 0$; this contradicts the assumption $P \neq 0$, and we are left with $\psi_2 \leqslant \psi_1$. This, however, would imply, in the same way as above, that $P = I$, again in contradiction with the assumption $P \neq I$. Consequently, $\pi_\phi(\mathfrak{R})' \cap \pi_\phi(\mathfrak{R})''$ contains only two projectors: 0 and I; since a von Neumann algebra is generated by its projectors, this implies that $\pi_\phi(\mathfrak{R})' \cap \pi_\phi(\mathfrak{R})'' = \{\lambda I\}$, hence $\pi_\phi(\mathfrak{R})''$ is a factor! Therefore we have proved the following theorem:†

Theorem 16. *Let $\mathfrak{R}$ be a C*-algebra; ϕ, a state on $\mathfrak{R}$; π_ϕ, the GNS representation of $\mathfrak{R}$ obtained from ϕ; and $\mathfrak{V}_\phi$, the set of all states on $\mathfrak{R}$ which can be represented as vector states on $\pi_\phi(\mathfrak{R})$; then π_ϕ is primary if and only if the relation $\leqslant$ totally orders $\mathfrak{V}_\phi$ with respect to the equivalence relation $\sim$.*

From the remark made earlier that a factor is discrete if and only if it admits minimal projections, we now prove the following theorem:

Theorem 17. *With the notation of Theorem* 16, *suppose that π_ϕ is primary; furthermore, let $\mathfrak{S}^P$ denote the set of all pure states on $\mathfrak{R}$. Then π_ϕ is type* I *if and only if $\mathfrak{V}_\phi \cap \mathfrak{S}^P$ is nonempty.*

Proof. Suppose first that π_ϕ is primary and type I, i.e., that $\pi_\phi(\mathfrak{R})''$, hence $\pi_\phi(\mathfrak{R})'$ is discrete. There then exists at least one minimal projector P in $\pi_\phi(\mathfrak{R})'$; form the subrepresentation $\pi_\phi(\mathfrak{R})P$. Since $P \neq 0$, we already know that $P\Phi \neq 0$, where Φ is the cyclic vector corresponding to ϕ and $P\Phi$ is cyclic for the subrepresentation considered. If the latter were not irreducible, there would be a projector Q in $\pi_\phi(\mathfrak{R})'$ with $0 \neq Q \subset P$ and this would contradict the minimality of P; hence $\pi_\phi(\mathfrak{R})P$ is irreducible and the state ψ in $\mathfrak{V}_\phi$ defined by $\langle\psi; R\rangle = (\Phi, P\pi_\phi(R)\Phi)/|P\Phi|^2$ is pure. Suppose now that there is a vector Ψ in $\mathcal{H}_\phi$ such that $\langle\psi; R\rangle = (\Psi; \pi_\phi(R)\Psi)$ is pure on $\mathfrak{R}$; let P_Ψ be the projector on $\overline{\pi_\phi(\mathfrak{R})\Psi}$. We have seen that P_Ψ belongs to $\pi_\phi(\mathfrak{R})'$. If there existed a projector Q in $\pi_\phi(\mathfrak{R})'$ such that $0 \neq Q \subset P_\Psi$, $\pi_\phi(\mathfrak{R})P_\Psi$ would not be irreducible in contradiction with the assumption that ψ is pure. The proof of the theorem is completed. ∎

As a consequence of this theorem we see that whenever π_ϕ is factorial, and type II or III, no vector state with respect to π_ϕ is pure and no pure state on $\mathfrak{R}$ can be written as a vector state on π_ϕ. This evidently goes far beyond the usual rules of traditional quantum mechanics (e.g., see the first corollary to Theorem 3); compare this result also with the corollary to Theorem 6.

A state ϕ on $\mathfrak{R}$ is said to be *minimal* if ψ in $\mathfrak{S}$ and $\psi \leqslant \phi$ imply $\psi \sim \phi$. Suppose, then, that ϕ is minimal and dim $\mathcal{H}_\phi > 1$; for any P in $\pi_\phi(\mathfrak{R})'$ with $P \neq 0$ we form the vector $P\Phi$ (where Φ is the cyclic vector corresponding to ϕ) and the corresponding state ψ; $\psi \in \mathfrak{V}_\phi$, hence $\psi \leqslant \phi$. Since ϕ is minimal $\psi \sim \phi$; there exists a unitary U such that $U\pi_\phi(R)PU^* = \pi_\phi(R)$ for all R in $\mathfrak{R}$. We then form $V = UP$ and get $V \in \pi_\phi(\mathfrak{R})'$, $V^*V = P$, and $VV^* = I$, i.e., $P \sim I$ [mod $\pi_\phi(\mathfrak{R})$]. We conclude then that if ϕ is minimal (and dim $\mathcal{H}_\phi > 1$), every nonzero projector in $\pi_\phi(\mathfrak{R})'$ is equivalent to the identity; this implies

† For Theorems 16 to 19 see Kadison [1962], who, however, works with nonnormalized "states" (i.e., positive, linear functionals on $\mathfrak{R}$).

first of all that $\pi_\phi(\mathfrak{R})'$ is primary, since all projectors in $\pi_\phi(\mathfrak{R})'$ are comparable. Two cases can now occur: either $\pi_\phi(\mathfrak{R})' = \{\lambda I\}$, in which case $\pi_\phi(\mathfrak{R})$ is irreducible and ϕ is pure; or $\pi_\phi(\mathfrak{R})' \neq \{\lambda I\}$, hence, since all its projectors are equivalent to I, $\pi_\phi(\mathfrak{R})'$ is a factor of type III; so also is $\pi_\phi(\mathfrak{R})$. We should like to be able to discriminate between these two cases by using the properties of ϕ itself; to do so let us further consider the first case. We say that a state is *primary* if the GNS representation π_ϕ it generates is primary; a primary state ϕ is said to be *maximal* if every primary state $\psi \in \mathfrak{S}$ such that $\phi \leqslant \psi$ is necessarily equivalent to ϕ. We can see rather easily† that a state ϕ with dim $\mathcal{H}_\phi > 1$ is pure if and only if it is primary and not maximal; consequently the requirement that ϕ be both miminal and maximal (together with the trivial requirement that dim $\mathcal{H}_\phi > 1$) leaves us only with type III primary representations. Conversely, if π_ϕ is primary and type III, ϕ has to be minimal and maximal; we then have a characterization of states generating type III primary representations. By elimination we can characterize as well the cases in which π_ϕ is primary and type II: first, ϕ cannot be minimal, for then it would be either type I or III; suppose, further, that there existed $\psi \leqslant \phi$, with ψ minimal; if this were the case, π_ψ would be type I or III and unitarily equivalent to a subrepresentation of π_ϕ, which is impossible since π_ϕ is assumed to be type II. Conversely, if ϕ is such that $\psi \leqslant \phi$ implies ψ not minimal, we exclude first the possibility that ϕ itself is minimal and then that π_ϕ is primary and type III; second, we exclude the possibility that there exists a pure (hence minimal) state ψ such that $\psi \leqslant \phi$, and that π_ϕ is primary and type I, so that the above condition indeed is a characterization of type II representation, provided that π_ϕ is assumed to be primary. We can summarize the results obtained so far by the following theorem:

Theorem 18. *With the notation in Theorem* 16, *suppose that dim* $\mathcal{H}_\phi > 1$*; then the following two conditions are equivalent:*

(i) π_ϕ *is primary and type* III,
(ii) ϕ *is minimal and maximal.*

Furthermore, if ϕ *is primary, the following two conditions are equivalent:*

(iii) π_ϕ *is type* II,
(iv) $\psi \leqslant \phi$ *implies that* ψ *is* not *minimal.*

This characterization of primary representations and their types was first obtained by Kadison [1962], who also gave the following characterization of all primary representations belonging to any of the three types, I_∞, II_∞, and III:

Theorem 19. *A primary state* ϕ *on a* C*-*algebra* $\mathfrak{R}$ *is infinite if and only if* $\mathfrak{B}_\phi$ *contains an infinite family of states* ϕ_n *such that* $\|\phi_n - \phi_m\| = 2$ *(unless* $n = m$) *and* $\phi_n \sim \phi_m$ $\forall$ n, m.

† See, for instance, Kadison [1962].

Combes [1967] gave the following characterization of primary states:

Theorem 20. *Let $\mathfrak{R}$ be a C*-algebra; $\mathfrak{R}^{**}$ be the bidual of $\mathfrak{R}$; $\mathfrak{Z}$, the center of $\mathfrak{R}^{**}$; ϕ, a state on $\mathfrak{R}$, and $\tilde{\phi}$, the unique normal extension of ϕ to $\mathfrak{R}^{**}$. Then the following three conditions are equivalent:*

(i) *ϕ is primary,*
(ii) *$\tilde{\phi}$ restricted to $\mathfrak{Z}$ is pure,*
(iii) $\langle\tilde{\phi}; \mathrm{RZ}\rangle = \langle\phi; \mathrm{R}\rangle\langle\phi; \mathrm{Z}\rangle$ *for all* R *in* $\mathfrak{R}$ *and* Z *in* $\mathfrak{Z}$.

Furthermore two primary states ϕ_1 and ϕ_2 are quasi-equivalent if and only if their extensions $\tilde{\phi}_1$ and $\tilde{\phi}_2$ coincide on $\mathfrak{Z}$.

Combes also mentioned as a consequence of this theorem that in the norm-topology of $\mathfrak{R}^*$ the following subsets are closed: $\mathfrak{S}^F$ (the set of all primary states), $\mathfrak{S}^F_{\mathrm{I}}$, $\mathfrak{S}^F_{\mathrm{II}}$, and $\mathfrak{S}^F_{\mathrm{III}}$ (the sets of all primary states, respectively of types I, II, and III); in particular, ϕ belongs to $\mathfrak{S}^F_{\mathrm{I}}$ if it is the norm-limit of a sequence of pure states on $\mathfrak{R}$.

Although the mathematical literature on primary states and their types is enormous, our interest in these states is motivated by their occurrence and meaning in physical applications.

Consider first the example of the Bose-gas treated in Subsection c above. A simple inspection of the results shows that (a) for the ground state ($T = 0$), π_ϕ'' is a direct integral of irreducible representations; (b) for the grand-canonical equilibrium ($T \neq 0$) and no macroscopic occupation of the ground state (which happens when $T > T_c$), π_ϕ'' is a factor; (c) for the grand-canonical equilibrium ($T \neq 0$) with macroscopic occupation of the ground state ($T < T_c$), π_ϕ'' is a direct integral of factors. The types† of these representations are I_∞ in case (a) and III in cases (b) and (c).

Verboven and Verbeure‡ studied in detail the similar case of a system of countably many harmonic oscillators with linear interactions of van Hove's type (see Chapter 1, Section 1), mimicking either an EM field or an Einstein crystal. Here evidently no phase transition occurs and all thermodynamical representations are primary; they are type I_∞ for $T = 0$ and either type I_∞ (EM field) or III (Einstein crystal) for $T \neq 0$.

Araki and Woods' [1963] analysis of the Bose-gas has been transposed to Fermi gas by Araki and Wyss [1964] with essentially the same results as far as factor types are concerned. For an interesting account of these results see the review article by dell'Antonio [1968] which also extends this analysis to cases of interest in the relativistic quantum theory of (free) fields; we shall come back to these simple models in Chapter 4.

† See Araki [1964c].
‡ See Verboven [1966a] and [1966b]; Verbeure and Verboven [1966], [1967a], and [1967b]; See also Segal [1962].

To complete this list of results of the simplest possible models of interest to statistical mechanics we can cite Jelinek's [1968] analysis of the BCS-model. In this case the following representations are obtained: for $T = 0$, a direct integral of irreducible representations; for $T_c > T > 0$, a direct integral of type III primary representations; for $\infty > T > T_c$, a type III primary representation; and finally for $T = \infty$, a type II_1 primary representation.

Even from such a rapid survey of these results at least two remarkable features emerge which we now want to mention, postponing their discussion, however, until Subsections e and f of the next section (see also Subsection IV.2.b).

First, it appears that the occurrence of primary representations is linked to the existence of a unique thermodynamical phase at the temperature considered. Second, type III representations seem to be the general rule, except for the particular cases $T = 0$ (type I_∞) and $T = \infty$ (type II_1).

Both facts can be understood on the basis of the observation made by Haag, Hugenholtz, and Winnink [1967] that it is to be expected that the thermodynamical limit of (grand) canonical equilibrium states leads to a state which still satisfies a properly transcribed version of the Kubo [1957]-Martin-Schwinger [1959] analytic condition first formulated for thermal Green functions. Aside from the Bose-gas, which served as a model for HHW's investigation, the thermodynamical states of several systems† have been explicitly proved to satisfy this condition. Taking it as a criterion for recognizing equilibrium states pertaining to a given temperature, we might conjecture further that pure thermodynamical phases are exactly those of these states that cannot be decomposed further into states still satisfying this condition; this interpretation of the "extremal KMS states" has also been checked for a system showing phase transitions and seems indeed to be correct.‡ It has been used as well to describe the breaking of euclidian symmetry occurring in crystallization.§ We later prove (independently of the above interpretation) that extremal KMS states generate primary representations via the GNS construction, thus explaining (according to the above interpretation) the first of the two features mentioned, namely that primary representations do occur when only one thermodynamical phase can exist at the temperature considered. A judicious use of the KMS condition has also enabled Hugenholtz [1967] to explain on general grounds the second of the two features mentioned, namely the natural occurrence of type III factors in statistical mechanics. We shall see in the sequel that, for different reasons,

† See, for instance, the one-dimensional, infinite, quantum spin lattice with finite-range interaction studied by Araki [1969].

‡ Emch and Knops [1970].

§ Emch, Knops and Verboven [1970].

type III factors occur repeatedly in general local theories and even more generally when the system considered possesses a symmetry group with certain asymptotic properties.

g. Σ^*-Algebras and Connections with Other Approaches

This subsection is based on the papers of Davies [1968] and Plymen [1968a, b]; since the material to be surveyed is not indispensable to the subsequent developments of the general theory, as presented in this book, we merely summarize the results of physical significance and give only references for the proofs.

The results mentioned in the last two subsections suggest that there would be a definite mathematical advantage in being able to regard, on phenomenological grounds, the observables as self-adjoint elements of a W^*-algebra; indeed, the requirement that a C^*-algebra $\mathfrak{N}$ be a W^*-algebra, hence faithfully representable as a von Neumann algebra, would make it possible to have entirely within $\mathfrak{N}$ such tools as the spectral projection theorem, which is at the basis of our use of traditional functional analysis. If this requirement were justified by physical arguments, it would be much easier to establish the connection between the algebraic and other axiomatic approaches to physical theories such as the proposition-calculus approach.

It seems, however, difficult to find a physical argument for requiring that the C^*-algebra $\mathfrak{R}$, generated by the observables, be itself a W^*-algebra, i.e., that $\mathfrak{R}$, as a Banach space, be the dual of some Banach space. Actually some well known physical examples suggest that this particular circumstance is not realized in general; for instance, the C^*-algebra of the canonical commutation relations is *not* the W^*-algebra of all bounded observables in Fock space, although its Fock-space realization is faithful and irreducible. In this respect the much simpler case of classical mechanics is worth considering in some detail, for it illustrates clearly the problem at hand and provides a good guide to its solution in the general quantum case.

Let $\mathfrak{R}$ be a commutative (but not necessarily separable) C^*-algebra. A trivial extension of Theorem I.2.9 shows that $\mathfrak{R}$ is faithfully realized (i.e., isomorphic, both algebraically and metrically) by the C^*-algebra $\mathfrak{C}(\Gamma)$ of all continuous functions $f: \Gamma \to \mathbb{C}$, where Γ is a compact Hausdorff space. Incidentally, the compactness of Γ is linked to our implicit assumption that $\mathfrak{R}$ has a unit; however, should we want to insist on $\mathfrak{R}$ not having a unit, we would get that Γ is only a *locally* compact Hausdorff space and that $\mathfrak{C}(\Gamma)$ is to be understood as the C^*-algebra generated by the complex-valued continuous functions on Γ which vanish outside compact subsets. This generalization would only make somewhat heavier the notation in the subsequent argument, without adding to our understanding of what is coming up; we therefore assume tacitly that $\mathfrak{R}$ has a unit and that Γ is compact. In general $\mathfrak{C}(\Gamma)$ is *not* a W^*-algebra. Actually, to insist that $\mathfrak{C}(\Gamma)$ be a W^*-algebra would impose undue restriction on Γ; for instance, we know† that every W^*-algebra is an *AW^*-algebra*, that is, a C^*-algebra which is also a *Baer ring* i.e., a ring such that any two (hence all three) of the following conditions are satisfied:

(i) $\mathfrak{R}$ possesses a unit;
(ii) for any $\mathcal{S} \subseteq \mathfrak{R}$, $\mathfrak{R}_r(\mathcal{S}) \equiv \{R \in \mathfrak{R} \mid SR = 0 \; \forall \; S \in \mathcal{S}\}$ is of the form $E\mathfrak{R}$ where E is an idempotent of $\mathfrak{R}$;
(iii) for any $\mathcal{S} \subseteq \mathfrak{R}$, $\mathfrak{R}_l(\mathcal{S}) \equiv \{R \in \mathfrak{R} \mid RS = 0 \; \forall \; S \in \mathcal{S}\}$ is of the form $\mathfrak{R}E$, where E is an idempotent of $\mathfrak{R}$.

An alternative definition of an AW^*-algebra is that $\mathfrak{R}$ is a C^*-algebra that satisfies the following conditions: (a) in the set of projections, any collection of orthogonal projections has a least upper bound and (b) any *maximal commutative* (i.e., $\mathcal{S} = \mathcal{S}' \cap \mathfrak{R}$) self-adjoint subalgebra $\mathcal{S}$ of $\mathfrak{R}$ is generated by its projections. Now, it is known that $\mathfrak{C}(\Gamma)$ is an AW^*-algebra if and only if Γ is a *Stonean space*, i.e., if and only if the closure of every open set in Γ is both open and closed. Incidentally, this property of Γ is equivalent‡ to the following property: every uniformly bounded, increasing directed set of real-valued continuous functions on Γ has a continuous function as a least upper bound. This is evidently a severe restriction on Γ that would actually exclude usual classical mechanics from this approach.

Consequently it seems hopelessly restrictive to assume from the start that $\mathfrak{R}$ be a W^*-algebra.

The next question then is, given a C^*-algebra $\mathfrak{R}$, is it possible to associate canonically some W^*-algebra with $\mathfrak{R}$? We already have (Theorem 11) a positive answer to this question; indeed, with every C^*-algebra $\mathfrak{R}$ we can associate its (universal) enveloping von Neumann algebra $\mathfrak{R}''$, which is isometrically isomorphic to the double dual $\mathfrak{R}^{**}$ of $\mathfrak{R}$. The trouble with $\mathfrak{R}^{**}$ is that it is so terribly large; for instance, in the commutative case just discussed we saw (Theorem I.2.9) that $\mathfrak{S}$ can be realized by the set of all normalized regular Borel measures on Γ. From this we can construct $\mathfrak{R}^*$ and try to figure out, in simple terms, what its dual $\mathfrak{R}^{**}$ looks like; the latter turns out to be such a distressingly large space that we would rather like to avoid having to deal with it. The same is true, and probably even more disastrously so, for the noncommutative case.

† For a textbook exposition of this and related structures, see, for instance, Kaplanski [1968].

‡ Stone [1949], Sakai [1956].

The question then is whether it is really a W^*-algebra containing $\mathfrak{R}$ that we are looking for or whether we would not rather settle for any C^*-algebra $\mathfrak{X}$ that satisfies more modest conditions. In loose terms these conditions, for instance, could read as follows

(i) $\mathfrak{X}$ should be canonically defined from $\mathfrak{R}$ (to require no additional physical assumptions on $\mathfrak{R}$);
(ii) $\mathfrak{X}$ should contain $\mathfrak{R}$;
(iii) it should be possible to have a projection spectral theorem within $\mathfrak{X}$;
(iv) $\mathfrak{X}$ should not be too wild, so that we could possibly work with it in a useful way.

These conditions are admittedly quite vague but can be made more precise; specifically, the following requirement is attributed† to Mackey: "Find a canonical C^*-algebra $\mathfrak{X}$ containing $\mathfrak{R}$ such that when $\mathfrak{R}$ is commutative and separable $\mathfrak{X}$ determines, and is determined by, the Borel structure underlying the topology on the spectrum $\hat{\mathfrak{R}}$ of $\mathfrak{R}$. The point is that in classical mechanics the set $\mathfrak{A}$ of observables determines the Borel structure underlying the topology of phase space." Stated in this way, the problem is evidently linked with that of finding a natural realization of Mackey's own axioms;‡ we shall come to this point later. For the time being we want to remark that, stated in another way, what we are looking for is actually a noncommutative generalization of probability theory. In Mackey's axiomatization we consider essentially one observable at a time, so that the subalgebra of interest is generated by one observable; in this context the separability assumptions are quite harmless and it becomes natural to speak of the *Borel* structure. There is, however, no compelling reason for assuming in general that the compact Hausdorff space Γ, associated with a commutative $\mathfrak{R}$, is separable. This is naturally not to say that Γ is never separable; actually, we know§ that if the compact Hausdorff space Γ is metrizable it is separable and furthermore that Γ is metric if and only if $\mathfrak{R} \sim \mathfrak{C}(\Gamma)$ is separable. The latter circumstance does actually occur in some physical applications (see Chapter 4); we just do not at this point want to rule out the possible cases in which Γ is *not* separable. Furthermore this point of view has the advantage of making the structure of the present endeavor appear more clearly. We therefore do not assume that Γ is separable; aside from this minor modification, we now want to show that Mackey's hint can be followed literally. To do so we need some topological notions that we now want to recall.

We first remind the reader that we defined (p. 57) the *Borel sets* of a locally compact Hausdorff space Γ as the elements of the σ-ring $\mathcal{S}(\mathcal{E})$ generated by all

† Plymen [1968a].
‡ Mackey [1963].
§ See, for instance, Dunford and Schwartz [1957] I.6.12, I.6.10, and IV.13.16.

compact subsets of Γ. A subset S of a topological space is said to be a G_δ if there exists a sequence $\{U_n\}$ of open sets such that $S = \bigcap_{n=1}^{\infty} U_n$. We now define the *Baire sets* of a locally compact Hausdorff space Γ as the elements of the σ-ring $\mathcal{S}(\mathcal{E}_\delta)$ generated by all *compact* G_δ subsets of Γ. Of interest here is the fact that the class of all Baire sets is the minimal σ-ring which contains enough sets to describe the topology of Γ; specifically,† if $\mathcal{U}$ is any class of open sets that generates the topology of Γ, the σ-ring $\mathcal{S}(\mathcal{U})$ generated by $\mathcal{U}$ contains every Baire set, and every open set in Γ is the union of open Baire sets. Clearly $\mathcal{S}(\mathcal{E}_\delta) \subseteq \mathcal{S}(\mathcal{E})$; furthermore we recall‡ that if Γ is separable every compact subset of Γ is a G_δ, so that in this case Borel sets and Baire sets coincide.

In view of this simple result it seems that if we are to recover the topological structure of Γ from an algebra of functions on Γ (which shall then be our "generalized observables") the natural candidate is the C^*-algebra $\Sigma(\Gamma)$ of all complex-valued, bounded, and Baire-measurable functions on Γ; since Γ is compact, hence Baire-measurable, to say that f is Baire-measurable means simply that $f^{-1}(M)$ is a Baire set whenever M is a Borel set in $\mathbb{C}$. We now remark§ that every continuous function on Γ is Baire measurable, so that $\mathfrak{C}(\Gamma)$ is a sub-C^*-algebra of $\Sigma(\Gamma)$; furthermore, if $\mathcal{S}$ is a σ-ring of subsets of Γ, the following conditions are equivalent:

(i) $f \in \mathfrak{C}(\Gamma)$ is measurable with respect to $\mathcal{S}$;
(ii) $\mathcal{S}$ contains every Baire set;
(iii) $\mathcal{S}$ contains every compact G_δ.

This exhibits the meaning of $\mathcal{S}(\mathcal{E}_\delta)$ as the smallest σ-ring with respect to which the continuous functions are measurable. Hence $\Sigma(\Gamma)$ indeed appears to be the natural candidate for the purpose we are pursuing.

The next question is to ask for the properties of $\Sigma(\Gamma)$ which are transposable to the noncommutative case. To this effect we first notice‖ that if $\{f_n \mid n = 1, 2, \ldots\}$ is a sequence of functions in $\mathfrak{C}(\Gamma)$ and f belongs to $\mathfrak{C}(\Gamma)$ the following conditions are equivalent:

(i) $\lim_{n\to\infty} \langle\phi; f_n\rangle = \langle\phi; f\rangle \ \forall\, \phi \in \mathfrak{C}(\Gamma)^*$;
(ii) there exists $K < \infty$ such that $\|f_n\| \leqslant K$ for all n and $\lim_{n\to\infty} f_n(\gamma) = f(\gamma) \ \forall\, \gamma \in \Gamma$.

We now pick up the second condition and notice¶ that the class of all Baire functions is the smallest class of functions on Γ that contains $\mathfrak{C}(\Gamma)$ and is

† See, for instance, Berberian [1965], Lemma 56.1 and Corollary to Theorem 56.1.
‡ For a proof see Halmos [1950], Theorem 50.E.
§ See, for instance, Berberian [1965], Theorem 56.2 and its first corollary.
‖ See, for instance, Dunford and Schwartz [1957], Corollary IV.6.4.
¶ See, for instance, Berberian [1965], second corollary to Theorem 56.2.

closed under *sequential pointwise* limits. Hence, if we intend to keep condition (ii), which indeed has a simple operational meaning [since with each $\gamma \in \Gamma$ is associated a pure state on $\mathfrak{C}(\Gamma)$ and conversely; see the two lemmata on p. 61)], then $\Sigma(\Gamma)$ again appears to be the simplest choice. Finally we notice that the regular Borel measure μ associated with the state ϕ on $\mathfrak{C}(\Gamma)$ by Theorem I.2.9 (or more specifically by the Riesz representation theorem) can evidently be restricted to $\mathcal{S}(\mathcal{E}_\delta)$ so that $\int f(\gamma)\, d\mu(\gamma)$ makes sense for every f in $\Sigma(\Gamma)$; that is to say that ϕ extends to a state $\tilde{\phi}$ on $\Sigma(\Gamma)$ defined by

$$\langle \tilde{\phi}\,; f\rangle = \int f(\gamma)\, d\mu(\gamma)$$

and satisfying the property that if $\{f_n\}$ is a bounded sequence of elements in $\Sigma(\Gamma)$ converging pointwise to $f \in \Sigma(\Gamma)$, then

$$\lim_{n\to\infty} \langle \tilde{\phi}; f_n\rangle = \langle \tilde{\phi}; f\rangle.$$

This again is our condition (i), now extended to $\Sigma(\Gamma)$. Conversely, if $\{f_n\}$ is a sequence of elements in $\Sigma(\Gamma)$ such that the above condition is satisfied for all ϕ in $\mathfrak{S}(\mathfrak{C}(\Gamma))$, it results from the uniform boundedness theorem that there exists $K < \infty$ such that $\|f_n\| < K$ for all n and f_n converges pointwise to an element f in $\Sigma(\Gamma)$.

Following Davies [1968] (see also Plymen [1968a]), we abstract from the preceding remarks the following structure of $\Sigma(\Gamma)$:

I. $\Sigma(\Gamma)$ is a C^*-algebra.

II. The set $\mathcal{F}$ of all ordered pairs $\{f_n, f\}$, where f_n is a sequence of elements of $\Sigma(\Gamma)$ and f is an element of $\Sigma(\Gamma)$, contains a privileged subset $\mathcal{G}$ (here the set of all uniformly bounded, pointwise convergent sequences and their limits) called the set of *convergent sequences*, which has the following properties:

(i) $\{f_n, f\} \in \mathcal{G} \Rightarrow \exists\, K < \infty$ such that $\|f_n\| \leqslant K\ \forall\, n$,

(ii) $\{f_n, f\} \in \mathcal{G} \Rightarrow \{f_n^*, f^*\} \in \mathcal{G}$,
$\{f_n, f\}, \{f'_n, f'\} \in \mathcal{G}$ and $\lambda \in \mathbb{C} \Rightarrow$
$\{f_n + f'_n, f + f'\} \in \mathcal{G}$ and $\{\lambda f_n, \lambda f\} \in \mathcal{G}$,

(iii) $\{f_n, f\} \in \mathcal{G}$ and $f' \in \Sigma(\Gamma) \Rightarrow$
$\{f_n f', f f'\} \in \mathcal{G}$.

III. The set of all states on $\Sigma(\Gamma)$ contains a subset $\mathfrak{S}_\sigma$ such that

(i) $\{f_n, f\} \in \mathcal{G} \Rightarrow \langle \phi; f_n\rangle \to \langle \phi; f\rangle\ \forall\, \phi \in \mathfrak{S}_\sigma$,

(ii) $\{f_n\} \in \Sigma(\Gamma)$ and $\langle \phi; f_n\rangle$ converge $\forall\, \phi \in \mathfrak{S}_\sigma$
$\Rightarrow \exists f \in \Sigma(\Gamma)$ such that $\{f_n, f\} \in \mathcal{G}$,

(iii) $f \neq 0, f \in \Sigma(\Gamma)\ \exists\, \phi \in \mathfrak{S}_\sigma$ such that $\langle \phi; f\rangle \neq 0$.

A C^*-algebra that possesses all these properties is called a Σ^*-*algebra*.

The next point in the program is evidently to show that $\Sigma(\Gamma)$ is not a pathology and that it will make sense in particular to speak of a Σ^*-algebra of operators acting on a Hilbert space.

The reader will remember that in Subsection II.1.d we noticed that a distinction must be made in general between nets and sequences and that in general again a subset $\mathcal{S}$ of a topological space $\mathfrak{X}$ is closed if and only if for every *net* $\{x_\alpha\}$ on $\mathcal{S}$ which converges to $x \in \mathfrak{X}$ we have $x \in \mathcal{S}$. This is true in particular for the weak-operator topology on $\mathfrak{B}(\mathcal{H})$. We are therefore speaking of a new topology, the *sequentially weak-operator* topology, when we say that a subset $\mathcal{S} \subseteq \mathfrak{B}(\mathcal{H})$ is *σ-closed* if and only if for every *sequence* $\{R_n\}$ in $\mathcal{S}$ which converges to R in $\mathfrak{B}(\mathcal{H})$ with respect to the weak-operator topology we have $R \in \mathcal{S}$. (Incidentally, we recall† that every *sequence* $\{R_n\}$ that converges in the weak-operator topology is necessarily uniformly bounded). For an arbitrary set $\mathcal{S}$ in $\mathfrak{B}(\mathcal{H})$ we denote by $\sigma(\mathcal{S})$ the σ-closure of $\mathcal{S}$ [i.e., the smallest σ-closed subset of $\mathfrak{B}(\mathcal{H})$ containing $\mathcal{S}$]. We define a *concrete Σ^*-algebra* $\mathfrak{R}$ as a C^*-algebra of operators acting on a Hilbert space $\mathcal{H}$ such that $\sigma(\mathfrak{R}) = \mathfrak{R}$; the subset $\mathcal{G}$ of $\mathfrak{R}$ associated with this Σ^*-algebra will be the collection of all *sequences* in $\mathfrak{R}$ that are convergent for the weak-operator topology.

Theorem 21. *Let $\mathfrak{R}$ be a* C**-algebra of operators acting on some Hilbert space $\mathcal{H}$.*

(i) *$\sigma(\mathfrak{R})$ is a concrete* Σ^*-algebra*;*
(ii) *if $\mathfrak{R}$ is separable, $\sigma(\mathfrak{R})$ has a unit;*
(iii) *If $\mathcal{H}$ is separable, $\sigma(\mathfrak{R})$ coincides with the von Neumann algebra $\mathfrak{R}''$ generated by $\mathfrak{R}$;*
(iv) *if the center of $\mathfrak{R}''$ is countably decomposable* (*i.e., each family of mutually orthogonal projectors in $\mathfrak{R}'' \cap \mathfrak{R}'$ is countable*) *and $\mathfrak{R}$ is norm-separable, then $\sigma(\mathfrak{R}) = \mathfrak{R}''$.*

Proof. For (i) and (ii) see Davies [1968]; for (iii) and (iv) see Kadison [1968]. ∎

We now say that $\pi:\Sigma \to \mathfrak{B}(\mathcal{H})$ is a *σ-representation* of the Σ^*-algebra Σ if π is a representation for its C^*-algebraic structure and $\pi:\mathcal{G}(\Sigma) \to \mathcal{G}(\mathfrak{B}(\mathcal{H}))$, i.e., $\{\pi(R_n)\}$ converges to $\pi(R)$ in the weak-operator topology whenever $\{R_n, R\}$ is a σ-convergent sequence in Σ.

As for C^*- and W^*-algebras, we can show (Davies [1968]) that no distinction need be made between abstract and concrete Σ^*-algebras; specifically, every Σ^*-algebra admits a faithful σ-representation.

We can now associate canonically with any abstract C^*-algebra $\mathfrak{R}$ a Σ^*-algebra called the *σ-envelope* of $\mathfrak{R}$ and denoted $\mathfrak{R}^\sigma$ in a manner similar

† See, for instance, Riesz and Sz.-Nagy [1955], No. 84.

to the way we defined the enveloping von Neumann algebra $\mathfrak{R}''$ of $\mathfrak{R}$; specifically, we define $\mathfrak{R}^\sigma$ as the σ-closure of the universal representation of $\mathfrak{R}$ (see p. 120). In general $\mathfrak{R}^\sigma$ will be properly contained in the enveloping von Neumann algebra of $\mathfrak{R}$ and is easier to handle; for instance, if Γ is a compact Hausdorff space, $\mathfrak{C}(\Gamma)^\sigma = \Sigma(\Gamma)$. As another example (in which, however, the σ-envelope and the enveloping von Neumann algebra coincide), let $\mathfrak{R}$ be the C^*-algebra of all compact operators on a separable Hilbert space; then $\mathfrak{R}^\sigma = \mathfrak{B}(\mathcal{H})$.

In spite of being smaller in general than the von Neumann enveloping algebra, the σ-envelope retains many of the interesting features of the latter (see Theorem 11). In particular, if $\pi: \mathfrak{R} \to \mathfrak{B}(\mathcal{H})$ is a representation of the C^*-algebra $\mathfrak{R}$, π admits a unique extension π^σ to a σ-representation of $\mathfrak{R}^\sigma$ in $\mathfrak{B}(\mathcal{H})$ and every σ-representation of $\mathfrak{R}^\sigma$ arises in this way. Finally, if ϕ is a state on $\mathfrak{R}$, ϕ admits a unique extension $\tilde{\phi}$ to a σ-state on $\mathfrak{R}^\sigma$ and every σ-state on $\mathfrak{R}^\sigma$ arises in this way.

We recall that our initial motivation in trying to extend $\mathfrak{R}$ was to find a C^*-algebra (of "generalized observables") such that the latter would admit a spectral theorem. Davies [1968] proved in this connection that the spectral measure associated with a self-adjoint element of a concrete Σ^*-algebra still belongs to Σ; this result is a direct consequence of the fact that if Σ is a concrete Σ^*-algebra and Ω is a compact Hausdorff space (e.g., the spectrum of $A = A^* \in \Sigma$) then any C^*-representation $\pi: \mathfrak{C}(\Omega) \to \Sigma$ admits a unique extension to a σ-representation of $\Sigma(\Omega)$ into Σ.

The last question we want to examine in this subsection is the following: given the C^*-algebra $\mathfrak{R}$ generated by the observables on a physical system, we might be tempted to interpret the self-adjoint elements of the σ-envelope $\mathfrak{R}^\sigma$ of $\mathfrak{R}$ as "generalized observables"; supposing then that the "physics" is actually contained in the Jordan algebra generated by these generalized observables, can we possibly relate this approach to other axiomatic approaches to physical theories? Plymen [1968a, b] provided a positive answer to this question for two alternative approaches; we summarize his results in the following two theorems.

Theorem 22. *Let Σ be a Σ^*-algebra; $\mathfrak{S}_\sigma$, the set of all σ-states on Σ; and $\mathfrak{O}$, the set of all self-adjoint elements of Σ. Let $\mathcal{S}$ denote the σ-ring of all Borel sets on $\mathbb{R}$ and $\mathcal{T}$, the set of all projectors in Σ.*

(i) *For each $\phi \in \mathfrak{S}_\sigma$ and each $\mathrm{A} \in \mathfrak{O}$ there exists a unique probability measure $\mathrm{M} \to \mathrm{p}(\phi, \mathrm{A}, \mathrm{M})$ on the spectrum of A such that $\langle \phi; \mathrm{A}^n \rangle = \int \lambda^n \, d\mathrm{p}(\phi, \mathrm{A}, \lambda)$ for all $n = 1, 2, \ldots$. In particular $\mathrm{p}(\phi, \mathrm{A}, \varnothing) = 0$, $\mathrm{p}(\phi, \mathrm{A}, \mathbb{R}) = 1$, and $\mathrm{p}(\phi, \mathrm{A}, \bigcup_{i=1}^\infty \mathrm{M}_i) = \sum_{i=1}^\infty \mathrm{p}(\phi, \mathrm{A}, \mathrm{M}_i)$ for any sequence of mutually disjoint $\mathrm{M}_i \in \mathcal{S}$;*

(ii) *If* $p(\phi, A, M) = p(\phi, A', M)$ *for all* $\phi \in \mathfrak{S}_\sigma$ *and all* $M \in \mathcal{S}$, *then* $A = A'$; *similarly, if* $p(\phi, A, M) = p(\phi', A, M)$ *for all* $A \in \mathfrak{O}$ *and all* $M \in \mathcal{S}$, *then* $\phi = \phi'$.

(iii) *For any* $A \in \mathfrak{O}$ *and any real bounded Borel function f on* $\mathbb{R}$ *there exists* $B \in \mathfrak{O}$ *such that* $p(\phi, B, M) = p(\phi, A, f^{-1}(M))$ *for all* $\phi \in \mathfrak{S}_\sigma$ *and all* $M \in \mathcal{S}$.

(iv) *For any sequence* $\{\phi_i\} \subseteq \mathfrak{S}_\sigma$ *and any sequence* $\{\lambda_i\} \subset [0, 1]$ *with* $\sum_{i=1}^{\infty} \lambda_i = 1$, *there exists* $\phi \in \mathfrak{S}_\sigma$ *such that* $p(\phi, A, M) = \sum_{i=1}^{\infty} \lambda_i p(\phi_i, A, M)$ *for all* $A \in \mathfrak{O}$ *and all* $M \in \mathcal{S}$.

(v) *For any sequence* $\{P_n\}$ *of mutually orthogonal elements in* $\mathcal{T}$ *there exists* $P \in \mathcal{T}$ *such that* $\langle \phi; P \rangle = \sum_{n=1}^{\infty} \langle \phi; P_n \rangle$ *for all* $\phi \in \mathfrak{S}_\sigma$.

(vi) *For any compact real spectral measure* E *in* Σ *there exists a self-adjoint element* $A \in \Sigma$ *such that* $\chi_M(A) = E(M)$ *for all* $M \in \mathcal{S}$.

(vii) *For any* $P \neq 0$ *in* $\mathcal{T}$, *there exists a* ϕ *in* $\mathfrak{S}_\sigma$ *such that* $\langle \phi; P \rangle = 1$.

The reader will recognize in the properties (i) to (vii) the axioms put forward by Mackey [1963], except for two modifications; the first one, manifested in (iii), is due to the fact that Plymen restricts his attention to *bounded* observables; the second is much more serious: Mackey's Axiom 7 has been dropped altogether, as is physically required by an axiomatization that wants to cover systems with superselection rules. Consequently Plymen has shown by this theorem that if $\mathfrak{R}$ is the C^*-algebra of observables on a physical system then the Segal algebra $\mathfrak{O}$ of all self-adjoint elements of the σ-envelope $\mathfrak{R}^\sigma$ of $\mathfrak{R}$, together with the set of all σ-states on $\mathfrak{R}^\sigma$ (hence, by extension, all states on $\mathfrak{R}$), satisfies the essentials of Mackey's axioms. It is then a simple corollary of this theorem that the same situation occurs for the Segal algebra of all self-adjoint elements of the enveloping von Neumann algebra $\mathfrak{R}''$ of $\mathfrak{R}$ together with the set of all normal states on $\mathfrak{R}''$. The point is, however, that $\mathfrak{R}^\sigma$ is in general much smaller than $\mathfrak{R}''$ so that the self-adjoint elements of $\mathfrak{R}^\sigma$ have a better chance to qualify as bona-fida "generalized observables."

Plymen [1968b] also proved the following result:

Theorem 23. *The set* $\mathcal{T}$ *of all projectors in a* Σ^**-algebra* Σ *(with unit), equipped with the usual partial ordering relation and complementation, is a* σ*-complete, orthocomplemented, weakly modular lattice which is atomic if* Σ *is a separable* C^**-algebra of type I.*

We should notice that it is not claimed that $\mathcal{T}$ is a complete lattice but only that it is a σ-complete lattice, i.e., that every *sequence* $\{P_n\}$ in $\mathcal{T}$ has a greatest lower bound in $\mathcal{T}$. The assertion that $\mathcal{T}$ is weakly modular means that if P and Q in $\mathcal{T}$ are in the relation $P \leqslant Q$ the sublattice of $\mathcal{T}$ generated by P and Q is a Boolean algebra (i.e., is isomorphic to the algebra of all subsets of a set). By "atomic" we mean not only that for every Q in $\mathcal{T}$ there exists an *atom* P (i.e., an element $P \in \mathcal{T}$ such that $R \in \mathcal{T}$ and $R \leqslant P$ implies either $R = 0$ or

$R = P$) such that $P \leqslant Q$, but also that if P is an atom in $\mathfrak{T}$ and Q is an arbitrary element in $\mathfrak{T}$ then $R \in \mathfrak{T}$ and $Q \leqslant R \leqslant Q \cup P$ implies either $R = Q$ or $R = Q \cup P$.

This theorem establishes that if $\mathfrak{R}$ is the C^*-algebra of the observables on a physical system then the set of all projectors in the σ-envelope $\mathfrak{R}^\sigma$ of $\mathfrak{R}$ satisfies the essential postulates of the proposition-calculus approach to physical theories (e.g., see Piron [1964]).

We have briefly reviewed the connections that exist between the C^*-algebraic approach (developed from Segal's ideas), the noncommutative probability approach (illustrated by Mackey's axioms), and the proposition-calculus approach (developed from Birkhoff and von Neumann's idea and crystallized in Piron's axioms). These are only some of the possible approaches to physical theories, and many variations are still possible, since none of these axiomatizations is completely free from unphysical postulates (usually involving some idealized measuring procedures that cannot be completely carried out in the laboratory). For the reader curious about alternate routes we mention here a mere sampling from the recent literature on the subject: Gunson [1967], Ludwig [1968], Dähn [1968], Pool [1968a, b], Pedersen [1969c], Roberts and Roepstorff [1969], Edwards [1970], Davies and Lewis [1970], and the references quoted therein.

SECTION 2 SYMMETRIES AND SYMMETRY GROUPS

OUTLINE. In many aspects of modern physics symmetry considerations play a central role in establishing general properties of the physical systems under investigation. We have in mind particularly the important consequences of space-time symmetries, ergodic theory, the study of invariant states, and spontaneous symmetry breaking. We present in this section the formalism pertinent to the treatment of these problems in the framework of the C^*-algebraic approach.

In this formalism Jordan automorphisms appear to lead to the proper mathematical description of the physical concept of symmetry. We illustrate this point with a generalization of Wigner's theorem and with a discussion of the duality between the Heisenberg and Schrödinger pictures.

We examine the notions of space- and time-averages of observables and states in the general context of "amenable" groups of symmetries.

With the purpose of studying systems that admit a topological group of symmetries we construct a covariant representation theory. In doing so we pay special attention to the representations associated with invariant states. On the one hand, this theory is a lead toward quantum generalizations of classical ergodic theory; on the other hand, it provides a basis from which we shall subsequently discuss certain aspects of the statistical mechanics of phase transitions, of nonequilibrium statistical mechanics, and of the quantum theory of fields. Further tools for the study of equilibrium statistical mechanics are described in the subsection on the algebraic formulation of the Kubo-Martin-Schwinger condition as well as in the last subsection of this section.

a. Definition of a Symmetry

Once any description of a physical system Σ has been chosen, we know intuitively what a symmetry should be; in loose terms a symmetry is a transformation of Σ that preserves the structure of the description considered; this formulation is admittedly vague and our first task is to make it more precise in the framework adopted here, namely, that of the algebraic formalism. To achieve this in meaningful contact with what we already know from more traditional theories let us first mention two preliminary and familiar illustrations of the general concept of symmetry.

In the early days of quantum theory the description considered consisted of a complex Hilbert space $\mathcal{H}$, the set $\mathfrak{A}$ of all (bounded) self-adjoint operators on $\mathcal{H}$, which was identified with the set of all (bounded) observables on Σ, and the set $\mathfrak{B}$ of all states ϕ on $\mathfrak{A}$ of the form $\langle \phi; A\rangle = (\Phi, A\Phi)$, where Φ runs over all vectors of $\mathcal{H}$ of norm 1; it is clear that in this description the elements ϕ of $\mathfrak{B}$ are in one-to-one correspondence with the sets $\{\omega\Phi \mid \omega \in \mathbb{C}; |\omega| = 1\}$, the latter being referred to as *unit rays* in $\mathcal{H}$. We denote by the same symbol ϕ the elements in $\mathfrak{B}$ and the unit rays in $\mathcal{H}$; furthermore these are also in one-to-one correspondence with the one-dimensional projectors in $\mathfrak{B}(\mathcal{H})$. The projector P_ϕ corresponding to the state ϕ in $\mathfrak{B}$ is then interpreted as the observable corresponding to the statement: "the system Σ is in state ϕ"; given two states ϕ and ψ in $\mathfrak{B}$, the expectation value $\langle \phi; P_\psi\rangle = |(\Phi, \Psi)|^2$ is called the *transition probability* between the states ϕ and ψ.

It was soon realized† that in this setting a symmetry could be defined as an injective mapping ν from $\mathfrak{B}$ onto itself, which preserves the transition probabilities between states. As a result of this definition, we get (*Wigner's theorem*) that every symmetry can be unitarily or antiunitarily implemented, i.e., implemented by a mapping $U: \mathcal{H} \to \mathcal{H}$ such that for each Ψ and Φ in $\mathcal{H}$ and each λ in $\mathbb{C}$

$$\text{(i)} \qquad \{U\Phi \mid \Phi \in \phi\} = \nu[\phi],$$
$$\text{(ii)} \qquad U(\Phi + \Psi) = U\Phi + U\Psi,$$

† Wigner [1931]; see also the systematic account given by Bargmann [1964].

and either (unitary case)

(iii) $$U\lambda\Phi = \lambda U\Phi,$$
(iv) $$(U\Phi, U\Psi) = (\Phi, \Psi),$$

or (antiunitary case)

(iii′) $$U\lambda\Phi = \lambda^* U\Phi$$
(iv′) $$(U\Phi, U\Psi) = (\Phi, \Psi)^*.$$

The second illustration we want to mention here is taken from the proposition-calculus approach to physics†; in contrast to our previous illustration it emphasizes the transformation properties of a certain class of observables rather than the transformation properties of a certain class of states. In this formalism a symmetry is defined as a bijective mapping α of the set $\mathfrak{T}$ of all propositions on Σ onto itself and such that (a) $\alpha[P] \leqslant \alpha[Q]$ whenever $P \leqslant Q$ and (b) $\alpha[P'] = \alpha[P]'$; i.e., $\alpha[I - P] = I - \alpha[P]$ for all P and Q in $\mathfrak{T}$. These two properties are sufficient to ensure that all other logical relations between propositions are preserved by a symmetry. If $\mathfrak{T}$ is realized as the set of all projectors in some $\mathfrak{B}(\mathcal{H})$ with $\mathcal{H}$ complex, the conclusion of Wigner's theorem can be reached again, and $\alpha[P] = UPU^{-1}$.

Generalizations of Wigner's theorem to the case in whch $\mathcal{H}$ is constructed on the real numbers or on the real quaternions have been obtained in both illustrations mentioned, and the properties of symmetries have been discussed in cases in which the system considered presents discrete superselection rules; for more details on these aspects of the theory of symmetries the reader is referred to the papers of Bargmann and Emch and Piron already mentioned.

Let us now consider the case of an algebraic description of a physical system. If the C^*-algebra $\mathfrak{R}$ underlying the description were to be taken as the fundamental object of the theory, it would seem natural to define a symmetry as a mapping α of $\mathfrak{R}$ onto itself, which preserves the whole structure of $\mathfrak{R}$, namely, a C^*-automorphism, i.e.,

(i) $$\alpha[\lambda R + \mu S] = \lambda\alpha[R] + \mu\alpha[S],$$
(ii) $$\alpha[RS] = \alpha[R]\alpha[S],$$
(iii) $$\alpha[R^*] = \alpha[R]^*$$

for all R and S in $\mathfrak{R}$ and all complex λ and μ. Because of the special properties of a C^*-algebra (in particular $\|R^*R\| = \|R\|^2$), it is superfluous‡ to assume that $\|\alpha[R]\| = \|R\|$, since this property follows from the other three.

We should remember, however, from the arguments developed in Chapter 1, Section 2, that $\mathfrak{R}$ itself is *not* of immediate physical relevance and that the

† See Emch and Piron [1964].
‡ See, for instance, Dixmier [1964], 1.3.7.

fundamental object of the theory is the Jordan algebra $\mathfrak{A}$ of all self-adjoint elements of $\mathfrak{R}$. We must then formulate the defining properties of a symmetry with respect to $\mathfrak{A}$ alone. We accordingly assume that α is a bijective mapping of $\mathfrak{A}$ onto itself such that for all A and B in $\mathfrak{A}$ and all real λ and μ

(i) $$\alpha[\lambda A + \mu B] = \lambda\alpha[A] + \mu\alpha[B],$$

(ii) $$\alpha[A^2] = \alpha[A]^2;$$

the formula $A \circ B = \frac{1}{2}\{(A + B)^2 - A^2 - B^2\}$, together with (i), implies that (ii) is equivalent to

(iii) $$\alpha[A \circ B] = \alpha[A] \circ \alpha[B].$$

Without any physical restriction, we can now extend α from $\mathfrak{A}$ to $\mathfrak{R}$: we first decompose R in $\mathfrak{R}$ as $R = R_+ + iR_-$ with $R_+ = \frac{1}{2}(R + R^*)$ and $R_- = (1/2i)(R - R^*)$; we then define†

$$\alpha[R] = \alpha[R_+] + i\alpha[R_-]$$

and obtain in this way a bijective mapping from $\mathfrak{R}$ onto itself such that for all R and S in $\mathfrak{R}$ and all complex λ and μ

(i) $$\alpha[\lambda R + \mu S] = \lambda\alpha[R] + \mu\alpha[S],$$

(ii) $$\alpha[R^*] = \alpha[R]^*,$$

(iii) $$\alpha[RS + SR] = \alpha[R]\alpha[S] + \alpha[S]\alpha[R].$$

A bijective mapping that satisfies these three conditions is called a *Jordan*-automorphism;* we notice in particular that a Jordan*-automorphism maps onto itself the Jordan algebra $\mathfrak{A}$ of all self-adjoint elements of $\mathfrak{R}$ and that this mapping then satisfies all the properties previously required from a symmetry. Furthermore a Jordan*-automorphism is determined in general by its restriction to $\mathfrak{A}$.

Clearly any C^*-automorphism is in particular a Jordan*-automorphism; the converse, however, is not true and Jordan*-automorphisms actually provide a genuine and physically relevant generalization of C^*-automorphisms. To see this let us return to our first illustration of the concept of symmetry: let $\mathfrak{R} = \mathfrak{B}(\mathcal{H})$; the operator U obtained from Wigner's theorem implements a symmetry in the sense just described via the relation

$$\alpha[A] = UAU^{-1} \quad \text{for all } A \text{ in } \mathfrak{A};$$

† As much as it deals with unobservable quantities, this convention is a priori perfectly legitimate and nothing of physical relevance could possibly be attached to it; in particular, this convention does not prejudice in favor of either unitary or antiunitary symmetries (see below). We might, however, want to choose in some cases to define $\alpha[R] = \alpha[R_+] - i\alpha[R_-]$; the point at issue here is linked to the formal difference between Wigner's [1931] original definition of time-reversal and Schwinger's [1951]; from the physical point of view this convention seems to be neither better nor worse than ours. We therefore chose ours for its mathematical convenience. The author is indebted to Mr. J. Wolfe for attracting his attention to this point of formal discrepancy in the physical literature.

we can now extend α to a Jordan*-automorphism of $\mathfrak{B}(\mathcal{H})$ by the procedure indicated above and get

$$\alpha[R] = \alpha[R_+] + i\alpha[R_-] = UR_+U^{-1} + iUR_-U^{-1} = U(R_+ \pm R_-)U^{-1},$$

depending on whether U is unitary or antiunitary; in the former $\alpha[R] = URU^{-1}$ and α is a C^*-automorphism; in the latter, however, $\alpha[R] = UR^*U^{-1}$ and then $\alpha[RS] = \alpha[S]\alpha[R]$, i.e., α reverses the order of the terms in the product. It is still a Jordan*-automorphism of $\mathfrak{R}$ but it is not a C^*-automorphism: a Jordan*-automorphism that satisfies the above property is said to be a C^*-antiautomorphism. Since the usual time-reversal operator is antiunitary, it implements a C^*-antiautomorphism of $\mathfrak{B}(\mathcal{H})$, providing a concrete proof that Jordan*-automorphisms are a physically relevant and mathematically genuine generalization of C^*-automorphisms.

The conclusion of Wigner's theorem suggests the question whether it is possible to determine the cases in which a symmetry is either a C^*-automorphism or a C^*-antiautomorphism; for an isolated symmetry the best general answer to this question seems to be the following result†:

Theorem 1. *A linear, adjoint-preserving mapping* α *of a* C*-*algebra* $\mathfrak{R}$ *onto a* C*-*algebra* $\mathfrak{B}$ *of operators acting on some Hilbert space* $\mathcal{H}$ *is a Jordan*-homomorphism if and only if there exists a projector* P *in* $\mathfrak{B}'' \cap \mathfrak{B}'$ *such that* $\alpha[\mathrm{RS}]\mathrm{P} = \alpha[\mathrm{R}]\alpha[\mathrm{S}]\mathrm{P}$ *and* $\alpha[\mathrm{RS}](\mathrm{I} - \mathrm{P}) = \alpha[\mathrm{S}]\alpha[\mathrm{R}](\mathrm{I} - \mathrm{P})$ *for all* R *and* S *in* $\mathfrak{R}$.

Now let $\pi: \mathfrak{R} \to \mathfrak{B}(\mathcal{H})$ be an arbitrary representation of $\mathfrak{R}$ and α, any Jordan*-automorphism of $\mathfrak{R}$; for all R in $\mathfrak{R}$ we form $\pi(\alpha[R])$ which induces a Jordan*-automorphism α_π of $\pi(\mathfrak{R})$ if and only if α maps the kernel of π into itself. As an application of the theorem, we see that if π is primary, i.e., $\pi(\mathfrak{R})'' \cap \pi(\mathfrak{R})' = \{\lambda I\}$, $P = 0$ or I; hence α_π is either a C^*-automorphism or a C^*-antiautomorphism.

We now want to establish the connection between our definition of symmetries as Jordan*-automorphisms of $\mathfrak{R}$ and a possible definition as transformations of the set $\mathfrak{S}$ of all states on $\mathfrak{R}$.

Let $\mathfrak{R}$ be a C^*-algebra and α, a Jordan*-automorphism of $\mathfrak{R}$; for any state ϕ on $\mathfrak{R}$ we define the linear functional $\alpha^*[\phi]$ from $\mathfrak{R}$ to $\mathbb{C}$ by $\langle\alpha^*[\phi]; R\rangle = \langle\phi; \alpha[R]\rangle$. We recall that for every R in $\mathfrak{R}$ there exists A in $\mathfrak{A}$ such that $R^*R = A^2$. Since α is a Jordan automorphism, $\alpha[R^*R] = \alpha[A^2] = \alpha[A]^2$ is positive and $\langle\alpha^*[\phi]; R^*R\rangle = \langle\phi; \alpha[A]^2\rangle \geqslant 0$; i.e., $\alpha^*[\phi]$ is positive. Since $\alpha[I] = I$ and $\langle\alpha^*[\phi]; I\rangle = 1$, then $\alpha^*[\phi]$ is again a state. Since α is bijective, every state in $\mathfrak{S}$ can be obtained as the image through α^* of some state; α^* appears then as a bijective mapping of $\mathfrak{S}$ onto itself.

† See Kadison [1965], Theorem 2.6; see also Kadison [1951], Theorem 10, and Størmer [1965].

From the linearity of α we conclude that α^* is an *affine* mapping, i.e., $\alpha^*[\lambda\phi_1 + (1 - \lambda)\phi_2] = \lambda\alpha^*[\phi_1] + (1 - \lambda)\alpha^*[\phi_2]$ for all ϕ_1 and ϕ_2 in $\mathfrak{S}$ and all λ in $[0, 1]$; this implies in particular that α^* is a bijection of the set of all pure states onto itself. Finally, we read from the definition of the w^*-topology of $\mathfrak{R}^*$ (see Chapter 2, Section 1, Subsection e) that if a state ϕ in $\mathfrak{S}$ is approximated in this topology by a net $\{\phi_\alpha\}$ of states in $\mathfrak{S}$, $\{\alpha^*[\phi_\alpha]\}$ will approximate $\alpha^*[\phi]$ in this same topology; that is to say, α^* is w^*-continuous.

Physicists, accustomed as they are to passing freely from the "Heisenberg picture" to the "Schrödinger picture," and vice versa, will naturally ask at this point whether the above-mentioned properties of the transformation α^* on the states would be strong enough to provide an alternative definition of symmetries. This is indeed the case, as we shall presently see.

Let ν be an affine, w^*-continuous, bijective mapping from $\mathfrak{S}$ onto itself. With future applications in mind, and also to emphasize the role played by the w^*-continuity of ν, we shall introduce this property only at the end of the forthcoming argument. We first extend ν by linearity to a bijective linear mapping from $\mathfrak{R}^*$ onto itself. We denote the extended mapping by the same symbol ν. For each element S of the double dual $\mathfrak{R}^{**}$ of $\mathfrak{R}$ we define the bounded linear functional $\nu^*[S]$ from $\mathfrak{R}^*$ to $\mathbb{C}$ by $\langle\phi; \nu^*[S]\rangle = \langle\nu\phi; S\rangle$; hence $\nu^*[S]$ can be identified as an element of $\mathfrak{R}^{**}$. As such, its norm (see Theorem II.1.11) is given by $\sup_{\phi\in\mathfrak{S}} |\langle\phi; \nu^*[S]\rangle|$; then recalling that ν is bijective, we conclude that $\|\nu^*[S]\| = \|S\|$. Hence ν^* appears as a linear isometric mapping of $\mathfrak{R}^{**}$ onto itself. Consider $\langle\phi; \nu^*[S^*]\rangle = \langle\nu[\phi]; S^*\rangle = \langle\nu[\phi]; S\rangle^* = \langle\phi; \nu^*[S]\rangle^* = \langle\phi; \nu^*[S]^*\rangle$ valid for all ϕ in $\mathfrak{S}$, from which we conclude† that $\nu^*[S^*] = \nu^*[S]^*$ for all S in $\mathfrak{R}^{**}$. Hence ν^* maps the set of all hermitian elements of $\mathfrak{R}^{**}$ onto itself. Let us now identify $\mathfrak{R}^{**}$ with the universal enveloping von Neumann algebra $\mathfrak{R}''$ of $\mathfrak{R}$. Clearly $\nu^*[I] = 1$; we now want to show that ν^* is positive. For any positive A in $\mathfrak{R}''$ we can define $A' = A/\|A\|$, hence assume without loss of generality in what follows that $\|A\| = 1$; for such A we have $\|\nu^*(A - I)\| = \|A - I\| \leqslant 1$ since ν^* is norm preserving, and since $\nu^*[I] = I$ we conclude that $\|\nu^*[A] - I\| \leqslant 1$ and $\nu^*[A] \geqslant 0$. Since ν^* is linear, it preserves order. From Kadison's generalized Schwartz inequality‡ we then have $\nu^*[A]^2 \leqslant \nu^*[A^2]$ for all self-adjoint elements A in $\mathfrak{R}''$. Since ν^* is bijective, there exists for each self-adjoint element A in $\mathfrak{R}''$ a self-adjoint element B in $\mathfrak{R}''$ such that $\nu^*[B] = \nu^*[A]^2 \leqslant \nu^*[A^2]$. Since the inverse μ^* of ν^* exists and satisfies the same properties, we conclude from this inequality that $B \leqslant A^2$. On the other hand,

† Incidentally, the fact that ν^* preserves adjoints follows directly from the fact that $\mathfrak{R}''$ is a C^*-algebra and that ν^* is a linear map carrying the identity into itself and preserving the norm (of the normal elements of $\mathfrak{R}''$) (see Kadison [1951], Lemma 8).

‡ See Kadison [1952], Theorem 1.

$B = \mu^*[\nu^*[A]^2] \geqslant (\mu^*[\nu^*[A]])^2 = A^2$. Combining these two inequalities, we get $B = A^2$, hence $\nu^*[A]^2 = \nu^*[B] = \nu^*[A^2]$. Since ν^* is linear, it is then a Jordan automorphism of the set of all self-adjoint elements of $\mathfrak{R}''$; by linearity we conclude then that ν^* is a Jordan*-automorphism of $\mathfrak{R}''$. Up to this point we did not use the fact that ν was supposed to be w^*-continuous. If ν is now w^*-continuous on $\mathfrak{S}$, so then is its linear extension to $\mathfrak{R}^*$; for any R in $\mathfrak{R}$ (but not necessarily in $\mathfrak{R}^{**}$) and any net $\{\phi_\alpha\}$ of elements of $\mathfrak{R}^*$ converging to ϕ in the w^*-topology we have the following: $\langle\phi_\alpha; \nu^*[R]\rangle = \langle\nu[\phi_\alpha]; R\rangle$ converges to $\langle\nu[\phi]; R\rangle$, since ν is w^*-continuous. Consequently $\langle\phi_\alpha; \nu^*[R]\rangle$ converges to $\langle\phi; \nu^*[R]\rangle$ and therefore $\nu^*[R]$ is a w^*-continuous linear functional from $\mathfrak{R}^*$ to $\mathbb{C}$. This implies† that $\nu^*[R]$ belongs to $\mathfrak{R}$. Hence the requirement that ν be w^*-continuous implies that the restriction α of ν^* to $\mathfrak{R}$ maps $\mathfrak{R}$ onto itself; furthermore α inherits from ν^* its structure as a Jordan*-automorphism. This completes the proof of the following theorem:

Theorem 2. *Let $\mathfrak{R}$ be a C*-algebra and $\mathfrak{S}$ be the convex set of all states on $\mathfrak{R}$. Then the relation $\langle\phi; \alpha[R]\rangle = \langle\nu[\phi]; \mathrm{R}\rangle$ for all ϕ in $\mathfrak{S}$ and all* R *in $\mathfrak{R}$ associates a Jordan*-automorphism α of $\mathfrak{R}$ to any affine, w^*-continuous, bijective mapping ν of $\mathfrak{S}$ onto itself, and conversely.*

REMARKS. This theorem establishes the equivalence for an isolated symmetry of the "Schrödinger picture" and the "Heisenberg picture." Alternatively, it provides a definition of symmetries in terms of $\mathfrak{S}$ only. The proof as given above follows in essence that of Kadison [1965], who actually proved the theorem under a weaker assumption, namely, that $\mathfrak{S}$ be replaced by convex subsets $\mathfrak{T}$ and $\nu[\mathfrak{T}]$ of $\mathfrak{S}$, "full" with respect to $\mathfrak{R}$ (i.e., $R \geqslant 0$ if $\langle\phi; R\rangle \geqslant 0\ \forall\ \phi \in \mathfrak{T}$ or $\nu[\mathfrak{T}]$).

We might also note, as a byproduct of the proof of the theorem, that if α is a Jordan*-automorphism of $\mathfrak{R}$ it extends‡ to a Jordan*-automorphism of the enveloping von Neumann algebra of $\mathfrak{R}$. In spite of this and of part (iv) in Theorem II.1.11, we should point out that a Jordan*-automorphism of a concrete C^*-algebra $\mathfrak{R}$ does not necessarily extend to the weak-closure $\mathfrak{R}''$ of $\mathfrak{R}$. The reason for this peculiarity is that the extension $\tilde{\alpha}_u$ of α to the universal enveloping von Neumann algebra $\pi_u(\mathfrak{R})'' = \mathfrak{R}^{**}$ does not necessarily map onto itself the kernel of the representation $\pi: \mathfrak{R}^{**} \to \mathfrak{R}''$. When dealing with some specific Jordan*-automorphisms of a concrete C^*-algebra $\mathfrak{R}$, we can in some cases prove indirectly that α actually extends to a Jordan-*-automorphism $\tilde{\alpha}$ of the bicommutant $\mathfrak{R}''$ of $\mathfrak{R}$. When this happens to be the

† See, for instance, Dunford and Schwartz [1957], V.3.9.

‡ This can actually be seen directly: first we transfer trivially α from $\mathfrak{R}$ to its universal representation π_u, which is faithful; the resulting Jordan*-automorphism α_u is bicontinuous for the weak-operator topology on $\pi_u(\mathfrak{R})$ and can then be extended by continuity to the weak-operator closure of $\pi_u(\mathfrak{R})$, i.e., to the universal enveloping von Neumann algebra of $\mathfrak{R}$.

case, we see from the proof of Theorem 1 that $\tilde{\alpha}$ is a C^*-automorphism (or C^*-antiautomorphism) of $\mathfrak{R}''$ whenever α is a C^*-automorphism (or a C^*-antiautomorphism) of $\mathfrak{R}$. We shall see in the next subsection (Theorem 4) that many physically relevant symmetries turn out to be C^*-automorphisms and not just Jordan*-automorphisms. Consequently the extension of α to $\tilde{\alpha}$, when it is possible, is an appreciable technical advantage, since the study of C^*-automorphisms of von Neumann algebras is well developed. As an indication of the status of the latter theory we now list some standard results, but give only bibliographical references for the proofs.

Result 1. *Every Jordan*-automorphism α of a von Neumann algebra $\mathfrak{N}$ is bicontinuous for the ultraweak and the ultrastrong topologies; furthermore, the restriction of α to any bounded subset of $\mathfrak{N}$ is bicontinuous for the weak and strong topologies.*

Proof. This result is obtained as a trivial modification of Dixmier [1957], I.4.3, Theorem 2 and Corollary 1. ∎

A C^*-isomorphism α from a von Neumann algebra $\mathfrak{N}_1$ ($\subseteq \mathfrak{B}(\mathcal{H}_1)$) onto a von Neumann algebra $\mathfrak{N}_2$ ($\subseteq \mathfrak{B}(\mathcal{H}_2)$) is said to be *spatial* if there exists an isometric isomorphism U from $\mathcal{H}_1$ onto $\mathcal{H}_2$ such that $\alpha[R] = URU^{-1}$ for all R in $\mathfrak{N}_1$. There are several particular cases in which spatial isomorphisms exhaust all C^*-isomorphisms and we now want to review some of them.

A vector $\Psi \in \mathcal{H}$ is said to be *cyclic* (resp. *separating*) for a von Neumann algebra $\mathfrak{N} \subseteq \mathfrak{B}(\mathcal{H})$ if $\overline{\mathfrak{N}\Psi} = \mathcal{H}$ (resp. $R\Psi = 0$, $R \in \mathfrak{N}$ imply $R = 0$).

Result 2. *Let* $\mathfrak{N}_i \subseteq \mathfrak{B}(\mathcal{H}_i)$ (i = 1, 2) *be two von Neumann algebras; suppose that there exist* Ψ_i *in* $\mathcal{H}_i$ *such that* Ψ_i *is both cyclic and separating for* $\mathfrak{N}_i$. *Then every C^*-isomorphism α from* $\mathfrak{N}_1$ *onto* $\mathfrak{N}_2$ *is spatial.*

Proof. See Dixmier [1957], III.1.4, Theorem 3. ∎

Result 3. *Let* $\mathfrak{N}_i \subseteq \mathfrak{B}(\mathcal{H}_i)$ (i = 1, 2) *be two von Neumann algebras; suppose that there exists an involution* J_i *of* $\mathcal{H}_i$ *such that* $J_i\mathfrak{N}_iJ_i = \mathfrak{N}_i'$ *and* $J_iZJ_i = Z^*$ *for every Z in the center* $\mathfrak{Z}_i$ *of* $\mathfrak{N}_i$. *Then every* C^**-isomorphism α from* $\mathfrak{N}_1$ *onto* $\mathfrak{N}_2$ *is spatial.*

Proof. See Dixmier [1957], III.1.5, Theorem 6. ∎

REMARK. Since a *standard* von Neumann algebra can be defined (Dixmier [1957], III.1.5, Corollary to Theorem 6) as a semifinite von Neumann algebra satisfying the conditions imposed in the preceding result, the latter admits as a consequence that every C^*-isomorphism between standard von Neumann algebras is spatial.

Result 4. *Every* C^**-automorphism α of a* finite (*resp.* discrete) *von Neumann algebra* $\mathfrak{N}$, *such that* $\alpha[Z] = Z$ *for all Z in the center* $\mathfrak{Z}$ *of* $\mathfrak{N}$, *is spatial* (*resp. implemented by a unitary element* U *in* $\mathfrak{N}$).

Proof. See Dixmier [1957], III.6.4, Corollary (or III.3.2, Corollary to Proposition 4). ▮

REMARK. As a consequence of this result we have that every C^*-automorphism of a factor of type II_1 (resp. type I) is spatial (resp. is implemented by a unitary element $U \in \mathfrak{N}$).

The condition that the automorphism leaves the center elementwise invariant can be relaxed for certain discrete von Neumann algebras.

Result 5. *Let* $\mathfrak{N}_i$ $(i = 1, 2)$ *be two von Neumann algebras with abelian commutants. Then every* C^**-isomorphism from* $\mathfrak{N}_1$ *onto* $\mathfrak{N}_2$ *is spatial.*

Proof. See Dixmier [1957], III.3.2, Corollary to Proposition 3. ▮

REMARK. This result appears as a particular case of a more general theorem which we shall quote after some preliminary definitions.

A von Neumann algebra $\mathfrak{N}$ is said to be *homogeneous* if there exists in $\mathfrak{N}$ a family $\{P_\kappa \mid \kappa \in K\}$ of projectors such that (a) $P_\kappa P_\lambda = 0$ if $\kappa \neq \lambda$, (b) $P_\kappa \sim P_\lambda$, (c) $\sum_{\kappa \in K} P_\kappa = I$, and (d) for each P_κ, $\mathfrak{N}_{P_\kappa}$ is abelian. Card K turns out to be an algebraic invariant characteristic of $\mathfrak{N}$ and is called the *multiplicity* of $\mathfrak{N}$ (or of $\mathfrak{N}'$); its intuitive meaning is given by the fact that a von Neumann algebra is homogeneous of multiplicity Card K if and only if it is spatially isomorphic to a von Neumann algebra of the form $\mathfrak{A} \otimes \mathfrak{B}(\mathcal{H})$ with $\mathfrak{A}$ abelian and $\dim \mathcal{H} = \text{Card K}$ (for these notions see Dixmier [1957], III.2.1).

Result 6. *Let* $\mathfrak{N}_i$ $(i = 1, 2)$ *be two von Neumann algebras with homogeneous commutants; suppose further that* $\mathfrak{N}_1$ *and* $\mathfrak{N}_2$ *have the same multiplicity. Then every* C^**-automorphism from* $\mathfrak{N}_1$ *onto* $\mathfrak{N}_2$ *is spatial.*

Proof. See Dixmier [1957], III.2.3, Proposition 3. ▮

REMARK. It is clear that this implies Result 5, since every abelian von Neumann algebra is homogeneous and of multiplicity 1.

Result 7. *Let* $\mathfrak{N}$ *be a von Neumann algebra with properly infinite commutant. Every* C^**-automorphism* α *of* $\mathfrak{N}$ *such that* $\alpha[Z] = Z$ *for all Z in the center* $\mathfrak{Z}$ *of* $\mathfrak{N}$ *is spatial.*

Proof. See Dixmier [1957], III.8.6, Corollary 8. ▮

REMARK. This result has several interesting consequences which we now discuss briefly. First we recall that if $\mathfrak{N}$ is a type III von Neumann algebra then so is its commutant (see p. 131), which is then purely, hence properly, infinite. Consequently Result 7 implies that every C^*-automorphism of a type III von Neumann algebra is spatial, provided that it leaves the center elementwise invariant. The latter condition can be dropped if the type III von Neumann algebra considered acts in a separable Hilbert space.† In any case the condition on the action of the automorphism on the center of the algebra considered is redundant in the case of a factor, and we therefore conclude that every C^*-automorphism of a factor of type III is spatial. For type II_∞ factors the situation is not so simple. Suppose that $\mathfrak{N}$ is a factor of type II_∞; we require in addition that there exists a finite set M in $\mathcal{H}$, separating for $\mathfrak{N}$. Then‡ M is a finite cyclic set for $\mathfrak{N}'$. Since $\mathfrak{N}$ is a factor, so is $\mathfrak{N}'$, which is then either finite or properly infinite. If $\mathfrak{N}'$ were finite, so, however, would $\mathfrak{N}$ be, since there exists a finite cyclic set for $\mathfrak{N}'$ (see p. 131) and this would contradict the hypothesis. Hence our condition is sufficient to ensure that $\mathfrak{N}'$ is properly infinite, and we can then conclude that every C^*-automorphism of a type II_∞ factor that admits a finite separating set is spatial. The assumption that $\mathfrak{N}$ is a factor is, moreover, not essential, as can be seen from the lemma on p. 128: we indeed know from this lemma that there exists a projector E in $\mathfrak{N} \cap \mathfrak{N}'$ such that $\mathfrak{N}'_E$ is finite and $\mathfrak{N}'_{I-E}$ is properly infinite. We see furthermore that the set EM is cyclic in $E\mathcal{H}$ with respect to $\mathfrak{N}'$, hence to $\mathfrak{N}'_E$. Consequently $(\mathfrak{N}'_E)' = \mathfrak{N}''_E = \mathfrak{N}_E$ is finite; since E is in the center of $\mathfrak{N}$, we conclude again from the same lemma that $E \leqslant F$, where $(I - F)$ is the largest projector in the center of $\mathfrak{N}$ such that $\mathfrak{N}_{(I-F)}$ is properly infinite. Since $\mathfrak{N}$ itself is properly infinite by assumption, $F = I$ and then $E = 0$; that is to say, $\mathfrak{N}'$ is properly infinite. We therefore conclude that the requirement in Result 7, namely, that $\mathfrak{N}'$ be properly infinite, can be replaced by the requirements that $\mathfrak{N}$ itself be properly infinite, provided that it admits a finite separating set. This concludes our discussion of Result 7.

A von Neumann algebra $\mathfrak{N}$ is said to be *uniform* if there exists in $\mathfrak{N}$ a family $\{P_\kappa \mid \kappa \in K\}$ of projectors such that (a) $P_\kappa P_\lambda = 0$ for $\kappa \neq \lambda$, (b) $P_\kappa \sim P_\lambda$, (c) $\sum_{\kappa\in K} P_\kappa = I$, and (d) P_κ finite; then Card K (when infinite) is characteristic of $\mathfrak{N}$ and is called the *order* of $\mathfrak{N}$ (see Dixmier [1957], III.3, Exercise 6).

Result 8. *Let* $\mathfrak{N}_i$ (i = 1, 2) *be two von Neumann algebras with properly infinite, uniform commutants of the same order. Then every* C^**-automorphism from* $\mathfrak{N}_1$ *onto* $\mathfrak{N}_2$ *is spatial.*

The study of C^*-antiautomorphisms (and more generally of C^*-anti-isomorphisms) can for the most part be reduced to that of C^*-isomorphisms by the following remark. Let $\mathfrak{N}_i \subseteq \mathfrak{B}(\mathcal{H}_i)$ be two von Neumann algebras; suppose that there exists a C^*-anti-isomorphism α from $\mathfrak{N}_1$ onto $\mathfrak{N}_2$. Let K be an antiunitary operator from $\mathcal{H}_1$ onto itself. We define the von Neumann algebra $\mathfrak{N}_1^K \subseteq \mathfrak{B}(\mathcal{H}_1)$ as $\mathfrak{N}_1^K \equiv \{KR^*K^{-1} \mid R \in \mathfrak{N}_1\}$. Let us denote by κ the

† See Dixmier [1957], III.8.6, Corollary 7; or [1964], Appendix A, Result No. 51.
‡ See Dixmier [1957], I.1.4, Proposition 5 and its corollary.

resulting spatial anti-isomorphism from $\mathfrak{N}_1$ onto $\mathfrak{N}_1^K$. We can then define the mapping $\alpha^\kappa = \alpha \circ \kappa^{-1}$ from $\mathfrak{N}_1^K$ onto $\mathfrak{N}_2$. Clearly α^κ is a C^*-isomorphism so that any C^*-anti-isomorphism can be written without loss of generality as the product of a spatial C^*-anti-isomorphism and a C^*-isomorphism.

Little is known in general about spatial implementation of an isolated C^*-automorphism of an arbitrary concrete C^*-algebra aside from the more or less immediate consequences of the results mentioned earlier in this chapter. There is, however, a particular case of singular physical relevance, which we now want to discuss.

Theorem 3. *Let α be a* C**-automorphism of a* C**-algebra $\mathfrak{R}$; for any state ϕ on $\mathfrak{R}$, invariant under α^*, α can be unitarily implemented in the GNS representation $\pi_\phi: \mathfrak{R} \to \mathfrak{B}(\mathcal{H}_\phi)$ associated with ϕ in such a way that the cyclic vector $\Phi(I)$ is invariant under the corresponding unitary operator.*

Proof. We first recall that $K \in \operatorname{Ker} \pi_\phi$ if and only if $\langle \phi; R^*KS \rangle = 0$ for all R and S in $\mathfrak{R}$. We then have $\langle \phi; R^*\alpha^{-1}[K]S \rangle = \langle \alpha^*[\phi]; R^*\alpha^{-1}[K]S \rangle = \langle \phi; \alpha[R^*]K\alpha[S] \rangle = 0$ so that α maps onto itself the kernel of π_ϕ and therefore α_ϕ is properly defined by $\alpha_\phi[\pi_\phi(R)] = \pi_\phi(\alpha[R])$ for all R in $\mathfrak{R}$. A similar reasoning shows that $\Phi(R) \to \Phi(\alpha[R])$ uniquely defines a mapping from the set $\pi_\phi(\mathfrak{R})\,\Phi(I)$, dense in $\mathcal{H}_\phi$, into itself; this mapping is clearly isometric, hence can be extended to an unitary operator U_ϕ on $\mathcal{H}_\phi$. From its definition we immediately get $U_\phi\Phi(I) = \Phi(I)$ and $U_\phi\pi_\phi(R)U_\phi^{-1}\Phi(S) = U_\phi\pi_\phi(R)\Phi(\alpha^{-1}[S]) = U_\phi\Phi(R\alpha^{-1}[S]) = \Phi(\alpha[R]S) = \pi_\phi(\alpha[R])\Phi(S)$ for all R and S in $\mathfrak{R}$. Then $\alpha_\phi[\pi_\phi(R)] = U_\phi\pi_\phi(R)U_\phi^{-1}$. ∎

REMARK. Theorem 3 assumes that α is a C^*-automorphism; let us replace this condition with the requirement that α is a C^*-antiautomorphism. The constructive argument of the proof can be reproduced for this case with only one modification: we now define the *antiunitary* operator U_ϕ by $U_\phi\Phi(R) = \Phi(\alpha[R^*])$; as a result, we again get $\alpha_\phi[\pi_\phi(R_+ + iR_-)] = \alpha_\phi[\pi_\phi(R_+)] + i\alpha_\phi[\pi_\phi(R_-)]$, which reflects, in the representation π_ϕ, the convention we originally chose for extending the Jordan automorphism α from $\mathfrak{A}$ to $\mathfrak{R}$. Hence, in spite of the essential replacement of "unitary" by "antiunitary," the conclusion of the theorem remains the same in the case in which α is a C^*-antiautomorphism.

b. Symmetry Groups

Given a description $(\mathfrak{R}, \mathfrak{S}, \langle;\rangle)$ of a physical system Σ, we denote by $\mathcal{A}$ the set of all symmetries of Σ. We recall (see Theorem 2) that an element of $\mathcal{A}$ is a Jordan*-automorphism α of $\mathfrak{R}$, or, equivalently, an affine, weak*-continuous bijective mapping ν of $\mathfrak{S}$ onto itself, and that ν extends naturally to a weak*-continuous linear mapping of $\mathfrak{R}^*$ onto itself.

We now want to equip $\mathcal{A}$ with a physically reasonable topology; starting again with the view that a physically meaningful statement must involve only expectation values of observables, calculated on states, we infer that a suitable topology on $\mathcal{A}$ is provided by the basis of neighborhoods obtained by considering all sets of the form

$$N(\alpha; \mathcal{F}, \varepsilon) = \{\beta \in \mathcal{A} \mid |\langle \phi_n; \alpha[A_n] - \beta[A_n]\rangle| < \varepsilon\} \ \forall \ (A_n, \phi_n) \in \mathcal{F}$$

for any $\varepsilon > 0$ and any *finite* family $\mathcal{F}$ of couples (A_n, ϕ_n) in $\mathfrak{R} \times \mathfrak{S}$ (or, equivalently, in $\mathfrak{R} \times \mathfrak{R}^*$); this topology on $\mathcal{A}$ is then the topology $\mathcal{A}$ inherits from the weak*-topology on $\mathfrak{R}^*$. It is only a technical matter to verify that this topology, together with the usual composition law of Jordan*-automorphisms, equips $\mathcal{A}$ with the structure of a topological group. Given a description $(\mathfrak{R}, \mathfrak{S}, \langle;\rangle)$ of a physical system Σ, we define the *action* of a topological group G on Σ as a homomorphism (in the sense of topological groups) α from G into $\mathcal{A}$; in particular, we require that $\langle \phi; \alpha_g[A]\rangle$ be a continuous, real-valued function of g in G for all A in $\mathfrak{A}$ and all ϕ in $\mathfrak{S}$. When this homomorphism is given, we say (though it is, strictly speaking, an abuse of language) that G is a *symmetry group* for the system considered.

We now want to show that we can rule out the possibility that α_g (g in G) is anything but a C^*-automorphism when G is a *connected* symmetry group. First we prove a continuity statement:

Lemma. *Let* G *be a connected symmetry group of* $(\mathfrak{R}, \mathfrak{S}, \langle;\rangle)$ *and* $\pi : \mathfrak{R} \to \mathfrak{B}(\mathcal{H})$, *an arbitrary representation of* $\mathfrak{R}$. *Then* $\pi_g[A] \equiv \pi \circ \alpha_g[A]$ *is strongly continuous in* g *for all* A *in* $\mathfrak{R}$.

Proof. Every vector Ψ in $\mathcal{H}$, normalized to 1, generates a state ψ on $\mathfrak{R}$ so that $(\Psi, \pi_g[A]\Psi) = \langle \psi; \alpha_g[A]\rangle$ is a continuous function of g for all A in $\mathfrak{R}$. The condition that Ψ be normalized can clearly be dispensed with. By polarization we get that $(\Psi, \pi_g[A]\Phi)$ is continuous in g for all Ψ, Φ in $\mathcal{H}$ and all A in $\mathfrak{R}$, i.e., $\pi_g[A]$ is continuous in g for the weak-operator topology on $\mathfrak{B}(\mathcal{H})$. For each Ψ in $\mathcal{H}$ and each self-adjoint A in $\mathfrak{R}$ we then form

$$\|(\pi_g[A] - \pi(A))\Psi\|^2 = (\pi_g[A]\Psi, \pi_g[A]\Psi) - (\pi_g[A]\Psi, \pi(A)\Psi) - (\pi(A)\Psi, \pi_g[A]\Psi) + (\pi(A)\Psi, \pi(A)\Psi);$$

since π is a representation and α_g is a Jordan*-automorphism, we have that $\pi_g[A]^*\pi_g[A] = \pi_g[A^2]$ for all self-adjoint elements A in $\mathfrak{R}$. Because of the weak continuity of $\pi_g[A]$ for all A, we then have that the right-hand side of the above equality goes to zero as g goes to the identity in G. This proves that $\pi_g[A]\Psi$ converges strongly to $\pi(A)\Psi$ as g tends to the identity in G for all Ψ in $\mathcal{H}$ and all A in $\mathfrak{A}$. The restriction $A \in \mathfrak{A}$ is removed by noticing that π_g is linear and by remembering that every element in $\mathfrak{R}$ can be written as a finite linear combination of elements in $\mathfrak{A}$. Finally, using the group property, we can translate to any element g in G the continuity obtained so far at $g = e$. This concludes the proof of the lemma. ∎

REMARK. This lemma extends trivially to each connected subset of G separately in cases in which G is not connected.

Under the assumptions of the lemma let us now consider the particular case in which π is primary, i.e., $\pi(\mathfrak{R})''$ is a factor; by Theorem 1 we know that π_g is either a C^*-homomorphism of $\mathfrak{R}$ onto $\pi(\mathfrak{R})$ or a C^*-antihomomorphism; we want to prove that if G is connected the latter case would violate the continuity just established. We can assume without loss of generality that $\pi(\mathfrak{R})$ is not abelian (indeed, if $\pi(\mathfrak{R})$ were abelian, no distinction would be necessary between C^*-homomorphisms and C^*-antihomomorphisms). There then exist A and B in $\mathfrak{R}$ such that $|(\Psi, \{\pi(A)\,\pi(B) - \pi(B)\,\pi(A)\}\Phi)| = \delta > 0$ for some Ψ and Φ in $\mathcal{H}$. For each $\varepsilon > 0$ we define the neighborhood $N(e; \varepsilon)$ of the identity in G as the intersection of the three neighborhoods:

$$N_1(e; \varepsilon) = \{g \in G \mid |(\Psi, \pi_g[AB]\Phi) - (\Psi, \pi(AB)\Phi)| < \varepsilon\},$$
$$N_2(e; \varepsilon) = \{g \in G \mid |(\pi_g[A] - \pi(A))\Phi| < \varepsilon\},$$
$$N_3(e; \varepsilon) = \{g \in G \mid |(\pi_g[B^*] - \pi(B^*))\Psi| < \varepsilon\}$$

it is then easy to see that for each g in $N(e; \varepsilon)$ we can introduce enough counterterms to make

$$|(\Psi, \{\pi(A)\,\pi(B) - \pi(B)\,\pi(A)\}\Phi)| < \varepsilon(1 + |\pi(A)\Phi| + |\pi(B^*)\Psi| + \varepsilon)$$

if π_g were a C^*-anti-isomorphism. Since ε can be made as small as we want, the right-hand side of the above inequality could be made smaller than δ in contradiction with the choice of A, B, Ψ, and Φ; π_g has therefore to be a C^*-isomorphism and we have then proved the following lemma, the extension from $N(e; \varepsilon)$ to G being trivial:

Lemma. *If in addition to the assumptions of the preceding lemma we suppose that π is primary, then π_g is a C^*-isomorphism for all g in G.*

We are now ready for the proof of the announced theorem.

Theorem 4. *If G is a connected symmetry group for $(\mathfrak{R}, \mathfrak{S}, \langle;\rangle)$, then $\{\alpha_g \mid g \in G\}$ is a weakly continuous family of C^*-automorphisms of $\mathfrak{R}$.*

Proof. For each pure state ϕ in $\mathfrak{S}$, π_ϕ is irreducible, hence a fortiori primary; from the preceding lemma we therefore have, for any vector Φ in $\mathcal{H}_\phi$, $(\Phi, \pi_g[RS]\Phi) = (\Phi, \pi_g[R]\pi_g[S]\Phi)$ and, in particular, $\langle\phi; \alpha_g[RS] - \alpha_g[R]\alpha_g[S]\rangle = 0$. We now recall that $\mathfrak{S}$ is the closed convex hull (in the w^*-topology) of $\mathfrak{S}_P$, the set of all pure states on $\mathfrak{R}$, so that $\langle\phi; R\rangle = 0$ for all ϕ in $\mathfrak{S}_P$ implies that $R = 0$ (a fact referred to by saying that $\mathfrak{S}_P$ is a separating family of states). We can therefore conclude that $\alpha_g[RS] = \alpha_g[R]\alpha_g[S]$ for all R and S in $\mathfrak{R}$; weak continuity follows from the definition of a symmetry group. This concludes the proof of the theorem. ∎

From this theorem we see that if π is an arbitrary representation of $\mathfrak{R}$ we have $\pi_g[RS] = \pi \circ \alpha_g[RS] = \pi(\alpha_g[R]\,\alpha_g[S]) = \pi \circ \alpha_g[R]\pi \circ \alpha_g[S] = \pi_g[R]\pi_g[S]$, provided that G is connected. In particular, if α_g maps Ker π into itself for all g in G, then $\alpha_g^\pi : \pi(\mathfrak{R}) \to \pi(\mathfrak{R})$ is well defined by $\alpha_g^\pi\pi[A] = \pi_g[A]$ and we have that $\{\alpha_g^\pi \mid g \in G\}$ is a weakly continuous family of C^*-automorphism of the C^*-algebra $\pi(\mathfrak{R})$.

While going through the proofs given above the reader might have noticed that we used our assumptions in a rather weak way; hence he might have surmised that the theorem could hold true under somewhat more general circumstances; this is indeed the case, since the proofs, as presented, are actually a slightly simplified version of Kadison's original arguments. Kadison [1965] was mainly interested in establishing under the weakest possible conditions the existence of a Hamiltonian that would generate the time-evolution in some specific representations. In order to present Kadison's main assumptions in a convincing light, let us consider first the case in which we actually have a Hamiltonian theory. Suppose, indeed, that in a given *faithful* representation π of $\mathfrak{R}$ we have $\pi(\alpha_t[R]) = U_t\pi(R)U_t^*$ for all R in $\mathfrak{R}$ and all times t with $U_t = \exp(+iHt)$. First we prove the following result:

Proposition. *Let* $\pi : \mathfrak{R} \to \mathfrak{B}(\mathcal{H})$ *be an arbitrary representation of a* C**-algebra* $\mathfrak{R}$*;* Φ *and* Ψ*, any two normalized vectors in* $\mathcal{H}$*; and* ϕ *and* ψ*, the corresponding states on* $\mathfrak{R}$*; then* $\|\phi - \psi\| \leqslant 2\,\|\Phi - \Psi|$.

Proof.
$$\begin{aligned}|\langle\phi - \psi; R\rangle| &\leqslant |(\Phi - \Psi, \pi(R)\Phi)| + |(\Psi, \pi(R)(\Phi - \Psi))| \\ &\leqslant |\Phi - \Psi| \cdot \|\pi(R)\| \cdot |\Phi| + |\Psi| \cdot \|\pi(R)\| \cdot |\Phi - \Psi| \\ &\leqslant 2\,\|R\| \cdot |\Phi - \Psi|; \text{ hence}\end{aligned}$$
$$\|\phi - \psi\| \equiv \sup_{\substack{R\in\mathfrak{R}\\ \|R\|\leqslant 1}} |\langle\phi - \psi; R\rangle| \leqslant 2\,|\Phi - \Psi|. \quad ∎$$

This result, joined to the fact that U_t is a strongly continuous one-parameter group of unitary operators, i.e., $|(U_t - I)\Phi| \to 0$ as $t \to 0$, implies that $\|(\alpha_t^* - I)\phi\| \to 0$ as $t \to 0$ for all states in $\mathfrak{B}_\pi$. This condition implies the weaker property that for all ϕ in $\mathfrak{B}_\pi$ and all R in $\mathfrak{R} : |\langle\phi; (\alpha_t - I)R\rangle|$ goes to zero as t approaches zero. We can therefore conclude that the mapping $\nu_t \equiv \alpha_t^*$ enjoys the following properties:

(i) for each t, ν_t is an affine w^*-continuous, bijective mapping of co $\mathfrak{B}_\pi$ onto itself;

(ii) $\langle\nu_t\phi; R\rangle$ is continuous in t, for all R in $\mathfrak{R}$ and all ϕ in co $\mathfrak{B}_\pi$;

(iii) $\overline{{}^{w^*}\text{co}\,\mathfrak{V}_\pi} = \mathfrak{S}$;
(iv) $\mathfrak{V}_\pi$ (hence a fortori co $\mathfrak{V}_\pi$) is a separating family of states for $\mathfrak{R}$.

(We notice that the last two properties, which are equivalent, come from the assumption that π is faithful.) Any triple $(\mathfrak{R}, \mathfrak{S}_0, t \to \nu_t)$ satisfying the above assumptions, with $\mathfrak{S}_0$ in place of co $\mathfrak{V}_\pi$, is called by Kadison a *dynamical system*; the main difference in our treatment is that we assumed throughout that $\mathfrak{S}_0$ was $\mathfrak{S}$ itself.† It is for his general dynamical systems, with the additional assumption that $\mathfrak{S}_0$ contains the vector states of some separating family of factor representations of $\mathfrak{R}$, that Kadison [1965] (his Theorem 3.4) was able to prove the result of our Theorem 4. ∎

Suppose now that, in a given representation $\pi: \mathfrak{R} \to \mathfrak{B}(\mathcal{H})$, $\alpha_g: \text{Ker}\,\pi \to \text{Ker}\,\pi$ for all g in G so that α_g^π is defined. Under the general assumptions made so far we might wish to establish the following properties:

(a) for each g in G there exists a unitary operator U_g acting on $\mathcal{H}$, such that $\alpha_g^\pi[\pi(R)] = U_g \pi(R) U_g^*$ for all R in $\mathfrak{R}$;
(b) U_g is a strongly continuous representation of G;
(c) if G is a Lie group, the generators H_i of its representation U_G are "affiliated" to $\pi(\mathfrak{R})$, i.e., the spectral projectors of each of the self-adjoint operators H_i are contained in $\pi(\mathfrak{R})''$.

These questions are difficult to answer in general; to get a measure of this difficulty we recall that in the discussion following the proof of Theorem 4 we saw that (a) implies that ν_g maps onto itself the set $\mathfrak{V}_\pi$ of all vector states for the representation considered and that (b) implies the strong continuity condition $\|(\nu_g - I)\phi\| \to 0$ as g approaches the identity in G. Both conditions [which, incidentally, imply that α_g extends from $\pi(\mathfrak{R})$ to $\pi(\mathfrak{R})''$] seem to be stringent in general, and to date only a few particular cases have been found that satisfy them; see, for instance, Kadison [1965], dell'Antonio [1966], or Montvay [1965]. One case of particular relevance to physics is given by the following theorem:

Theorem 5. *Let* G *be a connected symmetry group for* $(\mathfrak{R}, \mathfrak{S}, \langle;\rangle)$; ϕ, *a state on* $\mathfrak{R}$ *invariant under the action of* G; *and* π_ϕ, *the GNS representation associated to* ϕ. *There then exists a strongly continuous unitary representation* $\{U_\phi(g) \mid g \in G\}$ *of* G *in* $\mathcal{H}_\phi$ *such that* $\pi_\phi(\alpha_g[R]) = U_\phi(g)\pi_\phi(R)U_\phi(g)^*$ *for all* g *in* G *and all* R *in* $\mathfrak{R}$.

Proof. Since G is connected, we know from Theorem 4 that, for each g in G, α_g is a C^*-automorphism of $\mathfrak{R}$. By Theorem 3 we then know that for all g

† Incidentally, Kadison's generalization is necessary in the context of the Σ^*-algebraic formulation (see Subsection II.1.g. and Plymen [1968a]). We shall have to make use of this generalization in Chapter 4.

in G there exists a unitary operator $U_\phi(g)$ acting on $\mathcal{H}_\phi$ and such that, for all R in $\mathfrak{R}$: $\pi_\phi(\alpha_g[R]) = U_\phi(g)\,\pi(R)U_\phi(g)^*$ and $U_\phi(g)\,\Phi(R) = \Phi(\alpha_g[R])$. Therefore, we have immediately $U_\phi(g_1)\; U_\phi(g_2)\; \Phi(R) = U_\phi(g_1)\; \Phi(\alpha_{g_2}[R]) = \Phi(\alpha_{g_1}\alpha_{g_2}[R]) = \Phi(\alpha_{g_1g_2}[R]) = U_\phi(g_1g_2)\,\Phi(R)$. Since we have by construction $\{\Phi(R) \mid R \in \mathfrak{R}\}$ dense in $\mathcal{H}_\phi$ we can conclude that $U_\phi(g_1g_2) = U_\phi(g_1)\,U_\phi(g_2)$ for all g_1 and all g_2 in G. We then have only to prove the fact that $U_\phi(g)$ is strongly continuous in g. First we notice that for all R in $\mathfrak{R}$:

$$\begin{aligned}|(U_\phi(g) - I)\,\Phi(R)| &= |(U_\phi(g) - I)\,\pi(R)\,\Phi(I)| \\ &= |[U_\phi(g)\,\pi(R)\,U_\phi^*(g) - \pi(R)]\,\Phi(I)| \\ &= |(\pi(\alpha_g[R]) - \pi(R))\,\Phi(I)| \\ &= |(\pi_g[R] - \pi(R))\,\Phi(I)|,\end{aligned}$$

which, from the lemma on p. 160, goes to zero as g approaches the identity in G. We therefore have that $U_\phi(g)$ is strongly continuous on $\{\Phi(R) \mid R \in \mathfrak{R}\}$. We now use the fact that this linear manifold is dense in $\mathcal{H}_\phi$: for any $\varepsilon > 0$ and any Ψ in $\mathcal{H}_\phi$ there exists R in $\mathfrak{R}$ such that $|\Psi - \Phi(R)| < \varepsilon/3$ and for that R there exists, by our preceding result, a neighborhood $N(e)$ of the identity in G such that $|(U_\phi(g) - I)\,\Phi(R)| < \varepsilon/3$. We then have

$$\begin{aligned}|(U_\phi(g) - I)\Psi| &\leqslant |(U_\phi(g) - I)(\Psi - \Phi(R))| + |(U_\phi(g) - I)\,\Phi(R)| \\ &\leqslant 2\,|\Psi - \Phi(R)| + |(U_\phi(g) - I)\,\Phi(R)| < \varepsilon\end{aligned}$$

for all g in $N(e)$. Hence $U_\phi(g)$ is indeed strongly continuous on $\mathcal{H}_\phi$ itself. This concludes the proof of the theorem. ▮

A representation satisfying the conclusion of this theorem is called a *covariant representation* (π, U) of $(\mathfrak{R}, G)$.

c. Amenable Groups

In several physical problems we are interested in quantities of the form $\overline{\langle\phi;\, \alpha_g[A]\rangle}^G$, where ϕ is a state, A is an observable, G is a symmetry group and $\overline{}^G$ denotes some "invariant averaging process" with respect to G. Traditional examples of such constructions are embodied in most of the formulations of statistical mechanics. One instance is the usual "ergodic average" $\lim_{T\to\infty} (1/2T) \int_{-T}^{T} dt\, f(t)$; another is the definition of some of the macroscopic observables as space averages.

In this subsection we first give a definition of what will be understood by the term "invariant mean" over a group. We are then concerned with the existence question and prove, as an illustration of the concepts introduced, that the euclidian group in three dimensions is amenable; finally, we discuss some of those properties of invariant means that might be of relevance for physical applications.

From the definition of G as a symmetry group on a physical system we know that for every state ϕ in $\mathfrak{S}$ and every observable A in $\mathfrak{A}$ (or every element R in $\mathfrak{R}$) $\langle\phi;\, \alpha_g[A]\rangle$ (or $\langle\phi;\, \alpha_g[R]\rangle$) is a real- (or complex-) valued, continuous bounded function of g. It seems appropriate therefore to consider for our definition of "mean" the C^*-algebra $\mathfrak{C}(G)$ of all complex-valued, continuous bounded functions on G equipped with the usual operations: $f^*(g) = f(g)^*$, $(\Sigma_n \lambda_n f_n)(g) = \Sigma_n \lambda_n f_n(g)$, $(f_1 f_2)(g) = f_1(g) f_2(g)$, and $\|f\| = \sup_{g\in G} |f(g)|$. We notice that $\mathfrak{C}(G)$ possesses an identity, namely, $u(g) = 1$, for all g in G. We now define two ways in which G acts on $\mathfrak{C}(G)$: to each element h in G we associate the continuous *-automorphisms $h[\ \ ]$ and $[\ \ ]h$ of $\mathfrak{C}(G)$ defined respectively by

$$(h[f])(g) = f(hg),$$

$$([f]h)(g) = f(gh).$$

We now define a *mean* η on G as a state on $\mathfrak{C}(G)$, i.e.,

(i) η is a linear mapping from $\mathfrak{C}(G)$ to $\mathbb{C}$,
(ii) $\eta(f) \geqslant 0$ if $f \geqslant 0$,
(iii) $\eta(u) = 1$.

We say that a mean η on G is *left-invariant* (or *right-invariant*) if $\eta(g[f]) = \eta(f)$ (or $\eta([f]g) = \eta(f)$) for all f in $\mathfrak{C}(G)$ and all g in G. A mean is simply said to be *invariant* if it is *both* left- and right-invariant.

The existence of at least one left- or right-invariant mean characterizes a class of topological groups that is called *amenable*. We restrict ourselves in this subsection to the study of those amenable groups that are also locally compact; this mathematically nontrivial restriction, however, is natural in the context of this book, since most of the groups considered in physics are either discrete or Lie groups and as such are locally compact; this restriction has the further advantage that the class of locally compact amenable groups has been extensively studied in the mathematical literature (for reviews see Day [1957], Hewitt and Ross [1963], Pier [1965], Effros and Hahn [1968], or Greenleaf [1969]).

We recall first that a fundamental property of locally compact groups is the existence of a left-invariant *measure*, unique up to a multiplicative constant (there is evidently also a right-invariant measure, unique up to a multiplicative constant). This measure, called the *Haar measure*, is finite [i.e. $\mu(G) < \infty$] if and only if G is compact; in the latter case the integral with respect to the Haar measure μ,

$$\frac{1}{\mu(G)} \int_G f(g)\, d\mu(g),$$

provides a left-invariant mean over G; we therefore see that *every compact group is amenable*, so that locally compact amenable groups appear as a natural generalization of compact groups.

We now want to establish the following criterion for amenabilty, due to Dixmier [1950]:

Criterion 1. G *is amenable if and only if for any finite sequence* $\{f_n\}$ *of real-valued functions in* $\mathfrak{C}(G)$ *and any finite sequence* $\{g_n\}$ *in* G *we have that* $\sum_{n=1}^{N} (g_n[f_n] - f_n)(g) \geqslant a$ *for all* $g \in G$ *implies* $a \leqslant 0$.

Proof. Let us denote by $\mathfrak{C}_{\mathbb{R}}(G)$ the real vector space formed by all real-valued functions in $\mathfrak{C}(G)$; let $\mathfrak{D}$ be the vectorial subspace of $\mathfrak{C}_{\mathbb{R}}(G)$ of all functions of the form $h \equiv \sum_{n=1}^{N} (g_n[f_n] - f_n)$, with N finite, f_n in $\mathfrak{C}_{\mathbb{R}}(G)$ and $\{g_n\}$ in G. We notice immediately that if G is amenable then for any left-invariant mean η on G we have $\eta(h) = 0$ for all h in $\mathfrak{D}$; furthermore, since η is a *positive* functional on $\mathfrak{C}(G)$, the lower bound of any h in $\mathfrak{D}$ cannot exceed 0, which proves the *necessity* of the condition. Conversely, let $\mathfrak{E}$ denote the convex set $\{f \in \mathfrak{C}_{\mathbb{R}}(G) \mid \inf_{g\in G} f(g) > 0\}$; then the condition of the criterion is that $\mathfrak{D} \cap \mathfrak{E}$ be empty. By Hahn-Banach's theorem there exists a positive linear functional η on $\mathfrak{C}_{\mathbb{R}}(G)$, which vanishes on $\mathfrak{D}$; η then satisfies, in particular, the condition $\eta(g[f]) = \eta(f)$ for all f in $\mathfrak{C}_{\mathbb{R}}(G)$, i.e., η is a left-invariant positive functional on $\mathfrak{C}_{\mathbb{R}}(G)$. We can normalize it so that $\eta(I) = 1$; the extension of η to a left-invariant mean reduces to the trivial extension $\eta(f_1 + if_2) \equiv \eta(f_1) + i\eta(f_2)$ of the left-invariant positive linear functional η on $\mathfrak{C}_{\mathbb{R}}(G)$ to the complexification of the latter, namely, $\mathfrak{C}(G)$. This remark concludes the proof of sufficiency. ∎

We now want to use this criterion to prove† that every abelian group is amenable. Let G be an abelian (topological) group; we first need to introduce some notation. Let $A = \{g_i \mid i = 1, 2, \ldots, n\}$ be a finite but otherwise arbitrary sequence of elements in G; for every positive integer p we consider the set $P = \{\lambda\}$ of all integers λ such that $1 \leqslant \lambda \leqslant p$ and the following subset A_p of G:

$$A_p = \{g_1^{\lambda_1} g_2^{\lambda_2} \cdots g_n^{\lambda_n} \mid g_i \in A, \lambda_i \in P\}.$$

† Dixmier [1950] gave a rather detailed sketch of this proof; we present here a slight elaboration of Dixmier's arguments; see also Greenleaf [1969].

Since G is abelian the order in which the elements g_i of A enter the products in A_p is irrelevant and the number $N(A_p)$ of elements in A_p is bounded above by p^n. Suppose now that for some $\varepsilon > 0$ we had $N(A_{p+1}) \geqslant (1 + \varepsilon)N(A_p)$ for all p's; we would then have $N(A_{p+1}) \geqslant (1 + \varepsilon)^p$, which increases faster in p than $(p + 1)^n$, the previously recognized upper bound for $N(A_{p+1})$. We conclude therefore that for any $\varepsilon > 0$ there exists at least an integer p such that $N(A_{p+1}) \leqslant (1 + \varepsilon) N(A_p)$. Furthermore for every g in A the subset gA_p of A_{p+1} has $N(A_p)$ elements. We denote by $(gA_p - A_p)$ the subset of all elements in gA_p which are not in A_p and by $(A_p - gA_p)$ the subset of all elements in A_p which are not in gA_p; clearly these two sets have the same number of elements: $N(gA_p - A_p) = N(A_p) - N(A_p \cap gA_p) \leqslant N(A_{p+1}) - N(A_p)$. For any finite set A of elements of G and any finite set $\{f_i \mid i = 1, 2, \ldots, n\}$ of elements in $\mathfrak{C}_{\mathbb{R}}(G)$ we form $\sum_{i=1}^{N} (g_i[f_i] - f_i)$ and denote by a the lower bound of this function. Suppose $a > 0$; we then have

$$\begin{aligned} aN(A_p) &\leqslant \sum_{g\in A_p} \sum_{i=1}^{n} (g_i[f_i] - f_i)(g) \\ &= \sum_{i=1}^{n} \left(\sum_{g\in g_iA_p} - \sum_{g\in A_p} \right) f_i(g) \\ &\leqslant \sum_{i=1}^{n} \left(\sum_{g\in g_iA_p - A_p} + \sum_{g\in A_p - g_iA_p} \right) |f_i(g)| \\ &\leqslant 2n\{N(A_{p+1}) - N(A_p)\}M, \end{aligned}$$

where M denotes the upper bound of the $|f_i(g)|$. We know that for each $\varepsilon > 0$ there exists a p such that $N(A_{p+1}) < (1 + \varepsilon)N(A_p)$, so that for this p the above expression is majorized by $2n\, N(A_p)\varepsilon M$ and then $a \leqslant 2nM\varepsilon$; which contradicts $a > 0$, since ε is arbitrary. This concludes the proof of our assertion that *every abelian group is amenable*, since it satisfies the criterion previously established.

The following result is useful in some applications:

Proposition. *Let* G *be a locally compact group and let* μ *be its Haar measure; then* G *is amenable if for every finite sequence* $\{g_i \mid i = 1, 2, \ldots, n\}$ *of elements of* G *and every* $\varepsilon > 0$ *there exists a compact subset* A *in* G *such that* $0 < \mu(A) < \infty$ *and* $\mu(g_iA - A) < \varepsilon\mu(A)$ *for all* $i = 1, 2, \ldots, n$.

Proof. The idea of the proof is to show that a group which satisfies the above condition also satisfies the criterion already established. Consider the function $h(g)$ defined by $h = \sum_{i=1}^{n} (g_i[f_i] - f_i)$ with g_i in A and f_i in $\mathfrak{C}_{\mathbb{R}}(G)$.

Let a be its lower bound; on integrating h over A, we get

$$\begin{aligned}
a\mu(A) &\leqslant \sum_{i=1}^{n} \int_A (g_i[f_i] - f_i)(g)\, d\mu(g) \\
&= \sum_{i=1}^{n} \left\{ \int_A f(g_i g)\, d\mu(g) - \int_A f_i(g)\, d\mu(g) \right\} \\
&= \sum_{i=1}^{n} \left\{ \int_{g_i A} - \int_A \right\} f_i(g)\, d\mu(g) \\
&= \sum_{i=1}^{n} \left\{ \int_{g_i A - A} - \int_{A - g_i A} \right\} f_i(g)\, d\mu(g) \\
&\leqslant \sum_{i=1}^{n} \left\{ \int_{g_i A - A} + \int_{A - g_i A} \right\} |f_i(g)|\, d\mu(g) \\
&\leqslant \sum_{i=1}^{n} 2\mu(g_i A - A) M \\
&< 2nM\varepsilon\mu(A)
\end{aligned}$$

and, since $\varepsilon > 0$ is arbitrary, we conclude that $a \leqslant 0$, so that the criterion is satisfied and the group is amenable. ▌

This proposition now allows us to prove that the "euclidian group in three dimensions" is amenable; this group, defined in physical terms as the group $\mathbb{E}^3$ of all rigid transformations of the three-dimensional real space $\mathbb{R}^3$, has the following mathematical structure: it is the semidirect product of the group T^3 of all translations in $\mathbb{R}^3$ by the group 0^3 of all rotations in $\mathbb{R}^3$. Explicitly, this means that a homomorphism $R \to R[a]$ from 0^3 onto the group of automorphisms of T^3 is given and that every element g in $\mathbb{E}^3$ can be written as a pair (a, R) with a in T^3 and R in 0^3, the composition law being given by

$$(a_1, R_1)(a_2, R_2) = (a_1 + R_1[a_2], R_1 R_2),$$

where $R[a]$ is the translation obtained from a by rotating it by R. The topology of $\mathbb{E}^3$ is that of the cartesian product of the topological spaces underlying the groups T^3 and 0^3, the latter being equipped with their natural topologies as transformation groups of $\mathbb{R}^3$. Now let (a_i, R_i) be a finite sequence of elements in $\mathbb{E}^3$ and $\varepsilon > 0$ be an arbitrary positive number; it is geometrically evident that it is always possible to find a sphere S in $\mathbb{R}^3$ of finite radius but large enough to satisfy the condition $\mu(a_i S - S) < \varepsilon\mu(S)$ for $i = 1, 2, \ldots, n$, where μ is the usual Lebesgue measure on $\mathbb{R}^3$. The subset $A = (S, 0^3)$ of $\mathbb{E}^3$ satisfies the conditions of the proposition just proved and so *the euclidian group* $\mathbb{E}^3$ *is amenable.*

We proved the above result with elementary tools; a deeper study would have made it appear as a particular manifestation of the general theory of locally compact amenable groups, of which we now want to mention some results without proofs.

Proposition. *A locally compact group* G, *with Haar measure* μ, *is amenable if and only if for every compact set* $K \subseteq G$ *and every* $\varepsilon, \delta > 0$ *there exists a compact set* $E \subseteq G$ *and a measurable subset* $A \subseteq K$, *such that* $\mu(E) > 0$, $\mu(A) < \delta$, *and*† $\mu(E \,\Delta\, g E) < \varepsilon\mu(E)$ *for all* g *in the complement of* A *with respect to* K.

We note in particular that this property is trivially satisfied for G compact (write $E = G$ and $A = \varnothing$).

Many alternative definitions of amenability have been proposed for locally compact groups and we quote here just a few of those that are related to representation theory. Suppose, indeed, that $g \to U_g$ is a weakly continuous representation of G by unitary operators acting in some Hilbert space $\mathcal{H}$. For an arbitrary vector Ψ in $\mathcal{H}$ let us construct the function $f(g) = (\Psi, U_g\Psi)$ which is obviously continuous in g. Furthermore we notice that for any finite family $\{g_n\}$ of elements in G, and any finite sequence $\{z_n\}$ of elements in $\mathbb{C}$

$$\sum_{m,n} f(g_m^{-1} g_n) z_m z_n^* \geqslant 0;$$

this property is referred to by saying that $f(g)$ is a *continuous function of positive type*. Conversely, if $f(g)$ is any continuous function of positive type, there exists a weakly continuous unitary representation of G on some Hilbert space $\mathcal{H}$ and a vector Ψ in $\mathcal{H}$, cyclic for $\{U_g \mid g \in G\}$, such that $f(g) = (\Psi, U_g\Psi)$; furthermore, f determines U up to unitary equivalence. This result‡ reminds us of the GNS construction for C^*-algebras; this is indeed no accident, as we shall presently see.

Consider the set $\mathfrak{L}^1(G)$ of all μ-measurable functions $f: G \to \mathbb{C}$ such that

$$\|f\|_1 = \int_G |f(g)| \, d\mu(g) < \infty.$$

We first equip this set with the structure of a linear vector space on the complex field by defining on it the usual composition laws of complex-valued functions. The involution, however, is defined by $f^*(g) = f(g^{-1})^*\Delta(g^{-1})$, where $\Delta(g)$ is the real-valued, positive function defined by $d\mu(gh) = \Delta(h)\, d\mu(g)$ and referred to as the *modular function* on G with respect to the

† $A \,\Delta B \equiv \{x \in A \cup B \mid x \notin A \cap B\}$.
‡ See, for instance, Dixmier [1964], 13.4.5.

left-invariant Haar measure μ. We finally introduce the convolution product:

$$(f_1 * f_2)(g) = \int_G f_1(h) f_2(h^{-1}g)\, d\mu(h).$$

We now come back to the case in which we are given an arbitrary weakly continuous representation $g \to U_g$ of G by unitary operators acting on a Hilbert space $\mathcal{H}$. For each f in $\mathfrak{L}^1(G)$ we now define the operator

$$\pi_U(f) = \int_G U_g f(g)\, d\mu(g);$$

clearly $|\pi_U(f)\Psi| \leqslant \|f\|_1 \cdot |\Psi|$ for all f in $\mathfrak{L}^1(G)$ and all Ψ in $\mathcal{H}$, so that $\pi_U(f)$ belongs to $\mathfrak{B}(\mathcal{H})$; furthermore $\pi_U(f^*)$ is equal to the adjoint of $\pi_U(f)$. Finally $\pi_U(f)$ is linear in f and $\pi_U(f_1 * f_2) = \pi_U(f_1)\, \pi_U(f_2)$. We have therefore obtained a representation of the involutive normed algebra $\mathfrak{L}^1(G)$ by bounded linear operators acting on $\mathcal{H}$.

An example might be in order at this point. Consider the Hilbert space $\mathfrak{L}^2(G)$ of all μ-measurable, square-integrable functions on G with respect to the Haar measure; for each Φ in $\mathfrak{L}^2(G)$ and each h in G we define the element $U_h\Phi$ of $\mathfrak{L}^2(G)$ by

$$(U_h\Phi)(g) = \Phi(h^{-1}g).$$

We easily check that the mapping $h \to U_h$ provides a weakly continuous representation of G by unitary operators acting on $\mathfrak{L}^2(G)$. This representation is called the *left-regular* representation of G. From this representation we construct, by the process indicated above, a representation of $\mathfrak{L}^1(G)$ and we notice that it takes the particular form

$$\pi_U(f)\Phi = f * \Phi,$$

so that for this representation $\pi_U(f) = 0$ implies in particular $f * \Phi = 0$ for all Φ in $\mathfrak{K}(G)$ (= the set of all continuous functions on G, with compact support), which in turn implies that $f = 0$. Hence the representation associated with the left-regular representation G is injective.

Let us now come back to the general case and denote by $\mathfrak{U}_G$ the set of *all* weakly continuous representations of G by unitary operators. We can now define a *new* norm on $\mathfrak{L}^1(G)$ by $\|f\| = \sup_{U \in \mathfrak{U}_G} \|\pi_U(f)\|$. With the help of the preceding example we verify immediately that $\|f\| = 0$ only happens for $f = 0$; furthermore, since $\|\pi_U(f)\| \leqslant \|f\|_1$ for all U in $\mathfrak{U}_G$, we have $\|f\| \leqslant \|f\|_1$. The completion of $\mathfrak{L}^1(G)$ with respect to this norm is called the *C^*-algebra of the group G* and is denoted by $\mathfrak{C}^*(G)$. We can prove† that there

† For an indication of how this is achieved see Dixmier [1964], 13.9.3.

exists a one-to-one correspondence between the weakly continuous unitary representations of G and the nondegenerate† representations of $\mathcal{C}^*(G)$.

Equipped with the above notation and results, we can now state—in addition to criterion 1—the following criteria for a locally compact group G to be amenable:

Criterion 2 (or 2′). *Every continuous function of positive type (or the identity function* $u: G \to 1$*) on* G *is the uniform limit on every compact subset of* G *of functions of the form* $k * \tilde{k}$ *with* k *in* $\mathfrak{K}(G)$ *and* $\tilde{k}(g) = k(g^{-1})^*$.

Criterion 3. *For every weakly continuous unitary representation of* G *the corresponding representation of* $\mathcal{C}^*(G)$ *is weakly contained (in the sense of* Fell, *see Subsection* II.1d.) *in the representation of this algebra associated with the left-regular representation of* G.

Criterion 3′. *The representation of* $\mathcal{C}^*(G)$ *associated with the left-regular representation of* G *is faithful.*

Criterion 4. *For every* f *in* $L^1(G)$

$$\|f\| = \sup_{U \in \mathfrak{U}} \|\pi_U(f)\| \quad and \quad \|f\|_\rho = \sup_{\substack{g \in L^2(G) \\ \|g\|_2 \leq 1}} \|f * g\|_2 \ coincide.$$

Criterion 5. *For every continuous function* p *of positive type on* G *the mapping* $f \to \int_G f(g)\, p(g)\, d\mu(g)$ *defines a positive linear functional on* $\mathcal{L}^1(G)$, *continuous with respect to the norm* $\|\cdots\|_\rho$.

Criterion 6 (or 6′). *For every compact* K *in* G *and every* $\varepsilon > 0$ *there exists an* f *in* $\mathcal{L}^1(G)$ [or $\mathcal{L}^2(G)$] *with* $f > 0$ *and* $\|f\|_1 = 1$ *(or* $\|f\|_2 = 1$*) such that* $\|g[f] - f\|_1 < \varepsilon$ *(or* $\|g[f] - f\|_2 < \varepsilon$*) for all* g *in* K.

The relation, on the one hand, between the representations of G and the continuous functions of positive type on G and, on the other hand, the representations of $\mathcal{L}^1(G)$ naturally invite the suspicion that the dual space‡ of $\mathcal{L}^1(G)$ might play a role in the definition of means, hence of amenable groups; this is indeed the case, and we can replace (in the definitions given in the beginning of this section) $\mathfrak{C}(G)$ with $\mathcal{L}^\infty(G)$ (and consequently "positive" with "essentially positive", etc.) without changing the content of the theory. This

† We recall that $\pi: \mathfrak{R} \to \mathfrak{B}(\mathcal{H})$ is said to be *nondegenerate* whenever $\{\pi(R)\Psi \mid R \in \mathfrak{R}, \Psi \in \mathcal{H}\}$ is dense in $\mathcal{H}$.

‡ We recall that the dual space of $\mathcal{L}^1(G)$ can be identified with $\mathcal{L}^\infty(G)$, the space of all μ-essentially bounded functions from G to $\mathbb{C}$; for this, and the basic properties of $\mathcal{L}^\infty(G)$, the reader is referred, for instance, to Dunford and Schwartz [1957], Chapter IV.

remark makes some proofs easier; as an example, we mention the following criteria for amenability of locally compact groups:

Criterion 7 [or 7′ or 7″]. *There exists a mean* η *on* $\mathfrak{L}^{\infty}(G)$ *such that* $\eta(\Phi_1 * f) = \eta(f)$ [*or* $\eta(f * \Phi_2) = \eta(f)$ *or* $\eta(\Phi_1 * f * \Phi_2) = \eta(f)$] *for all* f in $\mathfrak{L}^{\infty}(G)$ *and all* Φ_1, Φ_2 *in* $\mathfrak{L}^1(G)$ *with* $\Phi_i > 0$ *and* $\|\Phi_i\|_1 = 1$.

To conclude this brief review we mention that the class of all locally compact amenable groups enjoys the following functorial properties: if G is a locally compact amenable group, every closed subgroup H of G is amenable; furthermore, if H is invariant in G, then G/H is again amenable. Conversely, if G is a locally compact group and possesses an invariant closed subgroup H such that both H and G/H are amenable, then so is G.

We proved earlier that every compact group and every abelian (locally compact) group is amenable; from the above properties we now understand the general reason why the euclidean group is amenable and conclude in addition that every one of its closed subgroups is also amenable.

We should point out, however, that there are many groups in physics that are not amenable: no noncompact semisimple Lie group is amenable; in particular, the homogeneous and the inhomogeneous Lorentz groups are not amenable.

In concluding this subsection we should perhaps insist again on the fact that we defined amenable groups as those groups for which *at least one* invariant mean exists on $\mathfrak{C}(G)$. Changing the topology on G or even changing $\mathfrak{C}(G)$ to another space of functions on G might change the nature of the *existence* problem completely and more generally the more delicate problem of *uniqueness* which we did not even touch here. We should mention in this connection that in most cases of physical interest uniqueness of a mean on $\mathfrak{C}(G)$ is not at all guaranteed; it is, however, remarkable, as we shall see in the next subsection, that this lack of uniqueness does not affect several of the physical consequences of the fact that we are dealing with amenable groups.

d. Invariant and Extremal Invariant States and Asymptotic Abelianness

In this subsection we are concerned with the structure of the set $\mathfrak{S}_G$ of all invariant states on a physical system and we discuss in some detail various cases in which the physical system under consideration possesses some particular properties known under the general name of asymptotic abelianness; we also study the meaning and relevance of some clustering properties familiar from field theory and statistical mechanics and show their connection with an extremal property of the corresponding states.

Lemma. *Let* G, *amenable, be a symmetry group for* $(\mathfrak{R}, \mathfrak{S}, \langle ; \rangle)$; *then the set* $\mathfrak{S}_G = \{\phi \in \mathfrak{S} \mid \nu_g \phi = \phi \ \forall\, g \in G\}$ *is not empty.*

Proof. Consider for each ϕ in $\mathfrak{S}$ and each R in $\mathfrak{R}$ the expression $\langle\phi;\, \alpha_g[R]\rangle$. Considered as a function of g, the latter is continuous and bounded (namely, by $\|R\|$); we can therefore form for any invariant mean η on G the expression $\eta\langle\phi;\, \alpha_g[R]\rangle$ which defines for each ϕ in $\mathfrak{S}$ a mapping, denoted $\eta\phi$, from $\mathfrak{R}$ to $\mathbb{C}$, given by $\langle\eta\phi;\, R\rangle \equiv \eta\langle\phi;\, \alpha_g[R]\rangle$. Since α_g, ϕ, and η are positive and linear, so is $\eta\phi$; furthermore $\langle\eta\phi;\, I\rangle = 1$, so that $\|\eta\phi\| = 1$ and $\eta\phi$ is a *state* on $\mathfrak{R}$. Moreover, since η is an invariant mean, we have

$$\langle\eta\phi;\, \alpha_h[R]\rangle = \eta\langle\phi;\, \alpha_{gh}[R]\rangle = \langle\eta\phi;\, R\rangle;$$

that is to say, $\eta\phi$ is invariant under the action of G. We have therefore proved the lemma by explicitly exhibiting an element in $\mathfrak{S}_G$. ∎

REMARKS. It should first be noticed that the last step in the proof depends essentially on the fact that $\mathfrak{R}$ is assumed to possess a unit I. If this were not the case, we could prove only that $\|\eta\phi\| < 1$, but we would not be able to rule out the possibility that $\|\eta\phi\|$ is strictly smaller than 1 nor even that $\eta\phi$ is zero.

It is actually easy to construct a mathematical counterexample of this lemma if $\mathfrak{R}$ is not assumed to have a unit. Let, $\mathfrak{R}$ be the C^*-algebra of all compact operators acting on a (separable) Hilbert space $\mathcal{H}$; furthermore, let H, be a (bounded) self-adjoint operator acting on $\mathcal{H}$. We now make the requirement that the spectrum of H be strictly continuous, i.e., $H\Psi = a\Psi$ with $\Psi \in \mathcal{H}$ implies $\Psi = 0$. We now define from H:

$$\alpha_t[R] = U_t R U_{-t} \quad \text{with } U_t = e^{+iHt} \quad \text{for all } t \text{ in } \mathbb{R} \text{ and all } R \text{ in } \mathfrak{R}.$$

Since U_t is bounded and $\mathfrak{R}$ is a two-sided ideal of $\mathfrak{B}(\mathcal{H})$, $\alpha_t[R]$ still belongs to $\mathfrak{R}$ for every R in $\mathfrak{R}$ and every t in $\mathbb{R}$. Clearly $\{\alpha_t\}$ satisfies all the conditions required from it to be a symmetry group of $\mathfrak{R}$. Now suppose that the dynamical system just defined admits an invariant state ϕ. We recall (see p. 119) that with *every* state ϕ on $\mathfrak{R}$ corresponds a density matrix on $\mathcal{H}$, i.e., a positive, self-adjoint operator ρ with $\operatorname{Tr}\rho = 1$. To assume that ϕ is invariant amounts to assuming that

$$\langle\phi;\, \alpha_t[R]\rangle = \operatorname{Tr} \rho U_t R U_{-t} = \operatorname{Tr} U_{-t}\rho U_t R = \operatorname{Tr} \rho R = \langle\phi;\, R\rangle$$

for all t in $\mathbb{R}$ and all R in $\mathfrak{R}$; we notice now that $\mathfrak{R}$ contains all finite rank operators and that the latter already separate $\mathfrak{B}(\mathcal{H})$. We can therefore conclude from the above equality that U_t commute with ρ. At this point we recall that ρ is a positive, trace-class operator, hence a positive element in $\mathfrak{R}$. This implies† that the spectrum of ρ is composed only of a discrete collection $\{\lambda_i\}$ of positive eigenvalues, each of the λ_i (with the possible exception of $\lambda_0 = 0$)

† See, for instance, Riesz and Sz.-Nagy [1955], No. 93.

having a finite multiplicity. Since U_t commutes with ρ, each of the eigen-projectors P_i of ρ belongs to the commutant of $\{U_t\}$. Consequently each of the P_i ($\neq P_0$) reduces $\{U_t\}$, hence its generator H, to a *finite*-dimensional subspace $P_i\mathcal{H}$ of $\mathcal{H}$. There exists therefore an orthonormal basis $\{\Psi_{i,n}\}$ in $P_i\mathcal{H}$ such that $U_t\Psi_{i,n} = e^{+ia_nt}\Psi_{i,n}$, i.e., $H\Psi_{i,n} = a_n\Psi_{i,n}$; we now recall our initial assumption on the spectrum of H and conclude that $\Psi_{i,n} = 0$; hence $P_i = 0$ for all $P_i \neq P_0$, so that $\rho = 0$. The dynamical system just defined does not admit any invariant state. We can generalize these considerations in essentially two directions (which can evidently be combined): *first*, we can assume that H has some discrete eigenvalues but that its spectrum still contains some continuous part; *second*, we can extend $\mathfrak{R}$ to $\mathfrak{B}(\mathcal{H})$ but restrict our attention to the set $\mathfrak{S}_0$ of all normal states on $\mathfrak{B}(\mathcal{H})$, to obtain a dynamical system in the sense of Kadison (see p. 163). The same conclusion still holds and this simple counter-example can be exploited† to exhibit some of the limitations of usual *Liouville space techniques* in nonequilibrium statistical mechanics.

We now come back to our remarks on the lemma proper, assuming again that $\mathfrak{R}$ has a unit and that $\mathfrak{S}$ is the set of all states on $\mathfrak{R}$, so that $\mathfrak{S}_G$ is not empty.

Our second remark is that the mapping $\eta:\mathfrak{S} \to \mathfrak{S}_G$ constructed in the proof of the lemma is clearly surjective and affine; the state $\eta\phi$ is said to be the *mean* of ϕ.

Our third remark is that in analogy with the construction performed in the proof of the lemma we can define a mapping η from $\mathfrak{R}$ to its double dual $\mathfrak{R}^{**}$ by $\eta\langle\phi;\, \alpha_g[R]\rangle = \langle\phi;\, \eta R\rangle$ and we verify that $\nu_g^*[\eta R] = \eta R$, where ν_g^* denotes the natural extension of α_g to $\mathfrak{R}^{**}$ for each g in G; we come back to these generalized observables ηR later in this subsection.

For the time-being we shall be satisfied to know that $\mathfrak{S}_G$ is not empty in at least some cases of physical interest (time evolution, space translations, etc.). We can then proceed with our study of the set $\mathfrak{S}_G$ of all invariant states with respect to a symmetry group G by noticing that it is a w^*-compact, convex subset of $\mathfrak{R}^*$. We can then transfer to $\mathfrak{S}_G$ the discussion of $\mathfrak{S}$ carried out in Chapter 1, Section 2; we see in particular that $\mathfrak{S}_G$ is the w^*-closed convex hull of its extreme points—the *extremal G-invariant states*‡—and prove that a G-invariant state is extremal in $\mathfrak{S}_G$ if and only if it dominates no other G-invariant state.

Lemma. *Let* G *be a symmetry group for* $(\mathfrak{R}, \mathfrak{S}, \langle;\rangle)$ *and let* ϕ *be in* $\mathfrak{S}_G$; *let* $\{\pi_\phi, \mathrm{U}_\phi\}$ *be the covariant representation associated with* ϕ *by Theorem* 5.

† See Emch [1966a,b] and, for the Liouville space techniques, the references quoted therein.
‡ These states are also called *ergodic states* by Ruelle [1969a] who follows Segal [1951] in this usage.

To every ψ *in* $\mathfrak{S}_G$, *dominated by* ϕ, *corresponds a unique* B *in* $\mathfrak{B}(\mathcal{H}_\phi)$ *with*

(i) $\langle\psi; \mathrm{R}\rangle = (\mathrm{B}\Phi, \pi_\phi(\mathrm{R})\mathrm{B}\Phi)$

(ii) $(\Psi, \mathrm{B}\Psi) \geqslant 0$ *for all* Ψ *in* $\mathcal{H}_\phi$

(iii) $\mathrm{B} \in \mathfrak{N}'_\phi \equiv \pi_\phi(\mathfrak{R})' \cap \mathrm{U}_\phi(\mathrm{G})'$

(iv) $|\mathrm{B}\Phi| = 1$;

Conversely every B *satisfying* (ii) *to* (iv) *generates via the relation* (i), *now taken as a definition of* ψ, *a state* ψ *in* $\mathfrak{S}_G$ *dominated by* ϕ.

Proof. We already know from the lemma on p. 86 that $\psi \leqslant \lambda\phi$ implies the existence of a unique B that satisfies the conditions (i), (ii), (iv), and B in $\pi_\phi(\mathfrak{R})'$; for the direct part of the lemma it is then sufficient to prove that this B belongs to $U_\phi(G)'$. To see this we note that ψ in $\mathfrak{S}_G$ implies

$$(\pi_\phi(R)\Phi,\, U_\phi(g)\, B^2 U_\phi(g)^{-1}\, \pi_\phi(S)\Phi) = (\pi_\phi(R)\Phi,\, B^2\, \pi_\phi(S)\Phi)$$

for all g in G and all R, S in $\mathfrak{R}$. Since Φ is cyclic for $\pi_\phi(\mathfrak{R})$, we have $(U_\phi(g)\, BU_\phi(g)^{-1})^2 = B^2$; furthermore, since B is the unique positive square-root of B^2 and $U_\phi(g)\, BU_\phi(g)^{-1}$ is positive, we have $U_\phi(g)\, BU_\phi(g)^{-1} = B$ for all g in G and therefore B belongs to $U_\phi(G)'$. This concludes the proof of the direct part of the lemma. The converse part is trivial. ∎

If G, amenable, is a symmetry group for $(\mathfrak{R}, \mathfrak{S}, \langle;\rangle)$, we say that an arbitrary state ϕ in $\mathfrak{S}$ is *η-clustering* with respect to the invariant mean η on G if for all A, B in $\mathfrak{A}$ we have

$$\eta\langle\phi; \alpha_g[A]B\rangle = \eta\langle\phi; \alpha_g[A]\rangle\langle\phi; B\rangle;$$

we see then that a G-invariant state ϕ is η-clustering if and only if

$$\eta\langle\phi; \alpha_g[A]B\rangle = \langle\phi; A\rangle\langle\phi; B\rangle.$$

Extremal G-invariant states and G-invariant η-clustering states play quite an important role in physical applications; these two notions are actually closely related, as we shall presently see.

To do so we return to the construction used in the proof of the first lemma of this subsection; we want to interpret $\eta\phi$ as the average with respect to G of the state ϕ. Similarly, we interpret ηA as the average with respect to G of the observable A. If, for instance G is the group representing the time evolution of the physical system under consideration, $\eta\phi$ would be a "time-averaged" state. If G is the translation group T^3, ηA would be interpreted as the macroscopic observable corresponding to the space average of the observable A.

Now let us introduce a heuristic argument that we shall be able to straighten out in the sequel (see, in particular, Chapter 4). If $\mathfrak{R}$ is the C^*-algebra obtained from actual laboratory experiments, the essentially local character of the latter would suggest that the self-adjoint elements of $\mathfrak{R}$ are quasi-local entities. Now, the observables relative to two causally disjoint regions of space commute; this fact is picked up when we average the commutation relation $[A, B]$ over all possible translations of B. Consequently, we would expect that the quasi-local observables in $\mathfrak{R}$ commute with the nonlocal observables of $\eta\mathfrak{R}$. We formalize this condition as $\eta: \mathfrak{R} \to \mathfrak{R}'' \cap \mathfrak{R}'$. It should now be realized that this condition, which seems natural when applied to space translations, is much more questionable when applied to time translations. It would nevertheless be useful to formalize the concept introduced here without reference to the interpretation of G as the space translations, leaving open the interpretation of G and the question whether the action of a particular group G on a particular physical system satisfies the condition considered; this procedure gives us the advantage of deriving the consequences of the condition itself, independently of its interpretation. We therefore introduce the following definition (which, incidentally, avoids mentioning explicitly the universal representation $\pi_u(\mathfrak{R})$ and the von Neumann algebras $\mathfrak{R}''$ and $\mathfrak{R}'$ acting on $\mathcal{H}_u$):

Definition. *The system consisting of a description* $(\mathfrak{R}, \mathfrak{S}, \langle ; \rangle)$, *an amenable symmetry group* G, *and an invariant mean* η *on* G *is said to be* η-asymptotically abelian *whenever*

$$\eta\langle\phi; \alpha_g[A]B - B\alpha_g[A]\rangle = 0$$

for all ϕ *in* $\mathfrak{S}$ *and all* A, B *in* $\mathfrak{A}$.

This notation was introduced in this form by Doplicher, Kadison, Kastler, and Robinson [1967]; several related concepts have appeared in the literature and are discussed in the sequel. For the original papers see, for instance, Ruelle [1966], Doplicher, Kastler, and Robinson [1966], Kastler and Robinson [1966], Lanford and Ruelle [1967], Størmer [1967], Doplicher, Kastler, and Størmer [1969], and Narnhofer [1970]. Many applications of these concepts to problems in statistical mechanics have been made; see, for instance, Haag, Hugenholtz, and Winnink [1967], Hugenholtz [1967], Robinson [1968], Kastler, Haag, and Michel [1968], and Emch, Knops, and Verboven [1968a, b, 1970].

We notice that for any locally compact, noncompact amenable group the condition for η-asymptotic abelianness is satisfied whenever one of the following two stronger conditions is satisfied:

$$\langle\phi; \alpha_g[A]B - B\alpha_g[A]\rangle \to 0 \quad \text{as} \quad g \to \infty, \forall\, \phi \in \mathfrak{S},\ A, B \in \mathfrak{R},$$

$$\|\alpha_g[A]B - B\alpha_g[A]\| \to 0 \quad \text{as} \quad g \to \infty\ \forall\, A, B \in \mathfrak{R},$$

which are referred to, respectively, as *weak-* and *norm-asymptotic abelianness.* Størmer [1967] proposed and exploited the following much weakened form of norm-asymptotic abelianness: for each self-adjoint element A in $\mathfrak{R}$ there exists a countable sequence $\{g_n(A)\}$ of elements of G such that

$$\lim_{n\to\infty} \| [\alpha_{g_n(A)}[A], B] \| = 0 \quad \text{for all } B \text{ in } \mathfrak{R}.$$

In connection with the above three conditions, we should notice that their formulation does not require the amenability of G, so that they could be used, if satisfied, for nonamenable groups such as the Lorentz group; they all express in their particular way that the observables A and B "tend to commute" when A is transported "far away" from B.

The specific consequences of strengthening the condition of η-asymptotic abelianness to that of norm-asymptotic abelianness are derived in Chapter 4, Section 1, once the local structure of the theory has been precisely defined. For the purposes of the present subsection η-asymptotic abelianness proves to be rich enough; actually, it is even convenient to introduce the following weakened forms of this condition to elucidate its role in various applications.

Definition. *The system consisting of a description* $(\mathfrak{N}, \mathfrak{S}, \langle ; \rangle)$, *an amenable symmetry group* G, *and an invariant mean* η *on* G *is said to be* G-abelian on the state ϕ *in* $\mathfrak{S}_G$ *whenever*

$$\eta\langle \psi; \alpha_g[A]B - B\alpha_g[A] \rangle = 0$$

for all A, B *in* $\mathfrak{A}$ *and all* ψ *in* $\mathfrak{B}_\phi^\circ$, *where* $\mathfrak{B}_\phi^\circ$ *denotes the set of all states of the form* $\langle \psi; R \rangle = (\Psi, \pi_\phi(R)\Psi)$, *with* Ψ *in* $E_\phi \mathcal{H}_\phi$; *it is further said to be* η-abelian on the state ϕ *in* $\mathfrak{S}_G$ *whenever*

$$\eta\langle \phi; R^*(\alpha_g[A]B - B\alpha_g[A])R \rangle = 0$$

for all A, B *in* $\mathfrak{A}$, *and all* R *in* $\mathfrak{R}$. *Finally, we simply say that the system is* G-abelian (*or* η-abelian) *when it is* G-abelian (*or* η-abelian) *on all* ϕ *in* $\mathfrak{S}_G$.

Incidentally, to say that a system is G-abelian simply means that

$$\eta\langle \phi; \alpha_g[A]B - B\alpha_g[A] \rangle = 0 \quad \text{for all } \phi \text{ in } \mathfrak{S}_G.$$

The notion of G-abelianness was introduced by Lanford and Ruelle [1967] in a somewhat different form. Theorem 6 establishes the needed connection. We first prove the following:

Lemma. *Let* $g \to U_g$ *be a weakly continuous representation of the amenable group* G *by unitary operators acting in some Hilbert space* $\mathcal{H}$; *for any invariant mean* η *on* G, *and any* Φ, Ψ *in* $\mathcal{H}$; *we have*

$$\eta(\Phi, U_g\Psi) = (\Phi, E_0\Psi),$$

where E_0 *is the projector on the subspace of* $\mathcal{H}$ *defined by* $U_g\Psi = \Psi$ *for all* g *in* G.

Proof. By assumption $(\Phi, U_g\Psi)$ is continuous in g and bounded (namely, by $\|\Phi\| \cdot \|\Psi\|$), so that $[\Phi, \Psi] \equiv \eta(\Phi, U_g\Psi)$ is defined. Since U_g and η are linear, $[\Phi, \Psi]$ defines a bilinear functional on $\mathcal{H}$, which is obviously bounded; from Riesz's theorem there exists then a linear bounded operator E acting on $\mathcal{H}$ such that $[\Phi, \Psi] = (\Phi, E\Psi)$. We have $(\Phi, E_0E\Psi) = \eta(E_0\Phi, U_g\Psi) = \eta(\Phi, E_0\Psi) = (\Phi, E_0\Psi)$, i.e., $E_0E = E_0$; conversely, $(\Phi, U_hE\Psi) = \eta(U_h^*\Phi, U_g\Psi) = \eta(\Phi, U_{hg}\Psi) = (\Phi, E\Psi)$, i.e., $U_hE\Psi = E\Psi$, hence $E_0E = E$. Combining these two results, we get $E_0 = E$, which proves the lemma. ∎

It should be noticed that as a consequence of this lemma the average $\eta(\Phi, U_g\Psi)$ does *not* depend on the mean η on G chosen to compute it.

Theorem 6. *Let* G *be an amenable symmetry group for* $(\mathfrak{R}, \mathfrak{S}, \langle;\rangle)$. *For any* ϕ *in* $\mathfrak{S}_G$ *consider the covariant representation* $(\pi_\phi(\mathfrak{R}), U_\phi(G))$ *described in Theorem* 5; *let* E_ϕ *be the projector on the subspace of* $\mathcal{H}_\phi$ *formed by all vectors invariant under* $U_\phi(G)$. *Furthermore let* $\mathfrak{N}_\phi$ *be the von Neumann algebra generated by* $(\pi_\phi(\mathfrak{R}), U_\phi(G))$. *A necessary and sufficient condition for* G-*abelianness on* ϕ *is that* $E_\phi\pi_\phi(\mathfrak{R})E_\phi$ *(or equivalently* $E_\phi\mathfrak{N}_\phi E_\phi$*) is abelian.*

Proof. We first notice that $E_\phi\pi_\phi(\mathfrak{R})E_\phi$ abelian is equivalent to $E_\phi\pi_\phi(\mathfrak{R})''E_\phi$ abelian; furthermore, every finite product of elements of $\pi_\phi(\mathfrak{R})$ and $U_\phi(G)$ can be reduced to the form $\pi_\phi(R)\, U_\phi(g)$ so that $\mathfrak{N}_\phi E_\phi$ is the closure (in the strong-operator topology) of the set $\{\sum_i c_i\pi_\phi(R_i)\, U_\phi(g_i)\}E_\phi = \{\sum_i c_i\pi_\phi(R_i)\}E_\phi$. Hence, we get, upon taking the strong-operator closure of both sides, $\mathfrak{N}_\phi E_\phi = \pi_\phi(\mathfrak{R})''E_\phi$. We then have that the two conditions $E_\phi\pi_\phi(\mathfrak{R})E_\phi$ abelian and $E_\phi\mathfrak{N}_\phi E_\phi$ abelian are equivalent. We now prove the connection with G-abelianness. With ϕ in $\mathfrak{S}_G$ and Ψ in $\mathcal{H}_\phi$ such that $|\Psi| = 1$ and $U_\phi(g)\Psi = \Psi$ or all g in G, we have by virtue of the preceding lemma

$$\eta\langle\psi; [\alpha_g[R], S]\rangle = (\Psi, [E_\phi\pi_\phi(R)E_\phi, E_\phi\pi_\phi(S)E_\phi]\Psi),$$

which proves the sufficiency. Conversely, if the condition of G-abelianness is satisfied, we have for all Ψ in $\mathcal{H}_\phi$

$$(\Psi, [E_\phi\pi_\phi(R)\, E_\phi, E_\phi\pi_\phi(S)E_\phi]\Psi) = 0,$$

since either $E_\phi\Psi = 0$ or $E_\phi\Psi$, when properly normalized, generates a state ψ in $\mathfrak{B}^\circ_\phi$. We then get by polarization

$$(\Psi, [E_\phi\pi_\phi(R)E_\phi, E_\phi\pi_\phi(S)E_\phi]\Psi') = 0, \qquad \Psi, \Psi' \text{ in } \mathcal{H}_\phi,$$

which is to say that $E_\phi\pi_\phi(\mathfrak{R})E_\phi$ is abelian; this concludes the proof of the theorem. ∎

We now notice that the condition of the theorem, namely, that $E_\phi\pi_\phi(\mathfrak{R})E_\phi$ be abelian, is independent of the mean chosen to formulate the original condition of G-abelianness, so that the latter is actually independent of any particular choice of η; i.e., if the system $(\mathfrak{R}, \mathfrak{S}, \alpha)$ is G-abelian on ϕ with respect to one given mean η on G, it is G-abelian on ϕ with respect to all means η on G; furthermore, the statement of the new condition does not even require that G be amenable. It is in this form that Lanford and Ruelle [1967] introduced the concept of G-abelianness. We further notice that our original condition for G-abelianness implies that $\inf_{g\in G} |\langle\phi;\ [\alpha_g[A], B]\rangle| = 0$, which again neither involves η nor requires that G be amenable; the latter condition has actually been shown (Lanford and Ruelle [1967] or Doplicher, Kadison, Kastler, and Robinson [1967]) to be sufficient to ensure that $E_\phi\pi_\phi(\mathfrak{R})E_\phi$ is abelian. The last remark has the following application: suppose that G, nonamenable (e.g., the Lorentz group) is a symmetry group for $(\mathfrak{R}, \mathfrak{S}, \langle;\rangle)$ and that this system is H-abelian for some amenable subgroup H of G (e.g., the translation group); we then have $\eta\langle\phi;\ [\alpha_h[R], S]\rangle = 0$ and $\inf_{h\in H} |\langle\phi;\ [\alpha_h[R], S]\rangle| = 0$ for all R, S in $\mathfrak{R}$ and all ϕ in $\mathfrak{S}_H$. Consequently $\inf_{g\in G} |\langle\phi;\ [\alpha_g[R], S]\rangle| = 0$ for all ϕ in $\mathfrak{S}_G$ ($\subseteq \mathfrak{S}_H$) and our system is also G-abelian.

We further remark that the condition that $E_\phi\mathfrak{N}_\phi E_\phi$ be abelian, encountered in the theorem, has the following consequence that will play an important role in the sequel:

Corollary. *In the notation of Theorem* 6, *for any* ϕ *in* $\mathfrak{S}_G$, $E_\phi\mathfrak{N}_\phi E_\phi$ *abelian implies* $\mathfrak{N}_\phi'$ *abelian; in particular,* $\mathfrak{N}_\phi'$ *is abelian when the system considered is* G-*abelian* (*on* ϕ *in* $\mathfrak{S}_G$).

Proof. This corollary follows from the fact that the commutant $\mathfrak{N}_\phi$ of the von Neumann algebra $\mathfrak{N}_\phi'$ contains an abelian projector† (E_ϕ) which has in its range a vector (Φ) cyclic with respect to $\mathfrak{N}_\phi$. Let us now see how this actually works out. We first consider the von Neumann algebra $\mathfrak{N}_{E_\phi} = E_\phi\mathfrak{N}_\phi E_\phi$ as a von Neumann algebra acting within $E_\phi\mathcal{H}_\phi$; since the vector Φ, associated with the state ϕ by the GNS construction is cyclic in $\mathcal{H}_\phi$ with respect to $\pi_\phi(\mathfrak{R})$, it is a fortiori cyclic with respect to $\mathfrak{N}_\phi$ and $\overline{\mathfrak{N}_{E_\phi}\Phi} = E_\phi\mathcal{H}_\phi$. Consequently we can use the fact‡ that an abelian von Neumann algebra, namely $\mathfrak{N}_{E_\phi}$, with cyclic vector, namely Φ, is equal to its commutant $\mathfrak{N}_{E_\phi} = \mathfrak{N}_{E_\phi}'$, so that $\mathfrak{N}_{E_\phi}'$ is abelian. We further recall§ that $\mathfrak{N}_{E_\phi}' = (\mathfrak{N}_\phi')_{E_\phi}$, so that the von Neumann algebra

$$(\mathfrak{N}_\phi')_{E_\phi} = \{E_\phi T E_\phi \mid T \in \mathfrak{N}_\phi'\}$$

is abelian. We finally consider the mapping $T \to E_\phi T E_\phi$ from $\mathfrak{N}_\phi'$ onto $(\mathfrak{N}_\phi')_{E_\phi}$; since $T \in \mathfrak{N}_\phi'$ and $E_\phi \in \mathfrak{N}_\phi$, this mapping is a *-algebraic homomorphism. Furthermore, if $E_\phi T E_\phi = 0$ for some T in $\mathfrak{N}_\phi'$ we have for all N in $\mathfrak{N}_\phi$

$$TN\Phi = NT\Phi = NTE_\phi\Phi = NE_\phi TE_\phi\Phi = 0.$$

† A projector E in a von Neumann algebra $\mathfrak{N}$ is said to be an *abelian projector* if $E\mathfrak{N}E$ is abelian.

‡ See Dixmier [1957], I.6.3, Corollary 2.

§ See Dixmier [1957], I.2.1, Proposition 1.

Since Φ is cyclic under $\mathfrak{N}_\phi$, this means that $T = 0$; hence we conclude that the mapping $T \to E_\phi T E_\phi$ is injective. Since we already know that it is a surjective homomorphism, we conclude now that $\mathfrak{N}'_\phi$ is isomorphic to $(\mathfrak{N}'_\phi)_{E_\phi}$. The purpose of the first part of the proof was precisely to show that the latter is abelian; consequently $\mathfrak{N}'_\phi$ is abelian. The second part of the corollary follows trivially from the theorem. ▌

We now want to turn our attention to the requirement of η-abelianness that, we recall, implies G-abelianness in particular and is itself satisfied as a sequel of the stronger requirement of η-asymptotic abelianness from which we started these considerations. We first establish the following result:

Lemma. *Let* G, *amenable, be a symmetry group for* $(\mathfrak{R}, \mathfrak{S}, \langle ; \rangle)$ *and* η *be an invariant mean on* G. *For every covariant representation* $(\pi(\mathfrak{R}), \mathrm{U}(\mathrm{G}))$ *there exists a mapping* η_π *from* $\mathfrak{R}$ *to* $\pi(\mathfrak{R})'' \cap \mathrm{U}(\mathrm{G})'$ *such that*

(i) $\eta(\Phi, \mathrm{U}(g)\, \pi(\mathrm{R})\, \mathrm{U}(g)^{-1}\Psi) = (\Phi, \eta_\pi(\mathrm{R})\Psi)\ \forall\ \Phi, \Psi \in \mathcal{H}_\pi$,
(ii) $\eta_\pi(\mathrm{R})\mathrm{E}_0 = \mathrm{E}_0\eta_\pi(\mathrm{R}) = \mathrm{E}_0\pi(\mathrm{R})\mathrm{E}_0\ \forall\ \mathrm{R} \in \mathfrak{R}$, *where* E_0 *is the projector on* $\{\Phi \in \mathcal{H}_\pi \mid \mathrm{U}(g)\Phi = \Phi\ \forall\ g \in \mathrm{G}\}$.

Proof. For every Ψ in the representation space $\mathcal{H}_\pi$ with $|\Psi|^2 = 1$ we form $\langle \psi; R \rangle = (\Psi, \pi(R)\Psi)$, which is a state on $\mathfrak{R}$ so that $\langle \psi; \alpha_g[R] \rangle$ is continuous in g; hence $(\Phi, U(g)\, \pi(R)\, U(g)^{-1}\Psi)$ is also continuous in g (by polarization) for all Φ, Ψ in $\mathcal{H}_\pi$. This expression is bounded; hence $\eta(\Phi, U(g)\, \pi(R)\, U(g)^{-1}\Psi)$ makes sense and defines, via the Riesz theorem, a bounded operator $\eta_\pi(R)$ acting on $\mathcal{H}_\pi$, and satisfying (i) in the lemma. For any X in $\pi(\mathfrak{R})'$ we have

$$\begin{aligned}(\Phi, X\eta_\pi(R)\Psi) &= \eta(X^*\Phi, \pi(\alpha_g[R])\Psi) = \eta(\Phi, \pi(\alpha_g[R])X\Psi) \\ &= (\Phi, \eta_\pi(R)X\Psi)\end{aligned}$$

for all Φ, Ψ in $\mathcal{H}_\pi$ and all R in $\mathfrak{R}$; we conclude, then, that $\eta_\pi(R)$ belongs to $\pi(\mathfrak{R})''$. Using, in addition, the fact that the mean η is invariant, we prove in the same way that $\eta_\pi(R)$ belongs to $U(G)'$. Hence η_π actually maps $\mathfrak{R}$ into $\pi(\mathfrak{R})'' \cap U(G)'$ as was to be proved. Finally, $\eta_\pi(R) \in U(G)'$ implies $\eta_\pi(R)E_0 = E_0\eta_\pi(R)$ and for every Φ, Ψ in $\mathcal{H}_\pi$

$$\begin{aligned}(\Phi, \eta_\pi(R)E_0\Psi) &= \eta(E_0\Phi, U(g)\, \pi(R)\, U(g)^{-1}E_0\Psi) \\ &= (\Phi, E_0\, \pi(R)E_0\Psi),\end{aligned}$$

which proves (ii) in the lemma. ▌

We have already noted that some physical results are unaffected by the possible nonuniqueness of η. Here is another;† let ϕ be a G-invariant state, Φ, the corresponding cyclic vector, and S any element in $\pi_\phi(\mathfrak{R})'$ such that $|S\Phi| = 1$; we then form the state $\langle \psi; R \rangle = (\Phi, S^*\, \pi(R)S\Phi)$ and notice as

† The author is indebted to Dr. C. Radin for this remark.

a consequence of our lemma, that $\eta\psi$, as a state on $\mathfrak{R}$, is now independent of the mean η used to compute it. This result is of particular interest when Φ is not only cyclic with respect to $\pi_\phi(\mathfrak{R})$ but also with respect to $\pi_\phi(\mathfrak{R})'$ (see Subsection e). If the latter condition is satisfied, we conclude from the lemma that $\eta_\phi(R)$ does not depend either on the mean η used to compute it, so that (see Theorem 7) η-abelianness on ϕ is, in this case, independent of the mean used to formulate it.

Theorem 7. *Let* G, *amenable, be a symmetry group for* $(\mathfrak{R}, \mathfrak{S}, \langle;\rangle)$; *for any* ϕ *in* $\mathfrak{S}_G$ *consider the covariant representation* $(\pi_\phi(\mathfrak{R}), U_\phi(G))$ *described in Theorem* 5. *Let* η_ϕ *be the mapping associated with it by the preceding lemma, and let* $\mathfrak{N}_\phi$ *be the von Neumann algebra generated by* $(\pi_\phi(\mathfrak{R}), U_\phi(G))$. *A necessary and sufficient condition for the system to be* η*-abelian on* ϕ *is* $\eta_\phi(\mathfrak{R}) \subseteq \pi_\phi(\mathfrak{R})'' \cap \pi_\phi(\mathfrak{R})' \cap U_\phi(G)'$.

Proof. We already know from the lemma that $\eta_\phi(\mathfrak{R})$ is contained in $\pi_\phi(\mathfrak{R})'' \cap U_\phi(G)'$, so that the essence of the condition is to impose in addition that $\eta_\phi(\mathfrak{R})$ be contained in $\pi_\phi(\mathfrak{R})'$. For any R_1, R_2, R, and S in $\mathfrak{R}$ we have

$$\eta\langle\phi; R_1^*[\alpha_g[R], S]R_2\rangle = (\pi_\phi(R_1)\Phi, [\eta_\phi(R), \pi_\phi(S)]\, \pi_\phi(R_2)\Phi),$$

from which we immediately see that $[\eta_\phi(R), \pi_\phi(S)] = 0$ implies

$$\eta\langle\phi; R_1^*[\alpha_g[R], S]R_2\rangle = 0;$$

hence $\eta_\phi(\mathfrak{R}) \subseteq \pi_\phi(\mathfrak{R})'$ indeed implies η-abelianness on ϕ. Conversely, η-abelianness on ϕ implies that $[\eta_\phi(R), \pi_\phi(S)] = 0$ for all R and S in $\mathfrak{R}$, since Φ is cyclic with respect to $\pi_\phi(\mathfrak{R})$. This concludes the proof of the theorem. ∎

From the definition of these concepts it follows trivially that η-abelianness implies G-abelianness. We now want to prove a slightly sharper version of this statement.

Corollary 1. η*-abelianness on* ϕ *implies* G*-abelianness on* ϕ.

Proof. From Theorem 7 η-abelianness on ϕ implies that $\eta_\phi(\mathfrak{R})$ is contained in $\pi_\phi(\mathfrak{R})'' \cap \pi_\phi(\mathfrak{R})' \cap U_\phi(G)'$; hence, a fortiori, $\eta_\phi(\mathfrak{R})$ is contained in the abelian von Neumann algebra $\mathfrak{N}_\phi \cap \mathfrak{N}_\phi'$. Consequently $[\pi_\phi(R), \eta_\phi(S)] = 0$ for any two elements R and S in $\mathfrak{R}$. For any two vectors Ψ' and Ψ'' in $\mathcal{H}$ we then form $(\Psi', [E_\phi\pi_\phi(R)E_\phi, E_\phi\pi_\phi(S)E_\phi]\Psi'')$. On using the lemma on p. 180 we see that this expression reduces to $(\Psi', E_\phi[\eta_\phi(R), \eta_\phi(S)]E_\phi\Psi'')$ which vanishes by virtue of the above remark. Consequently $E_\phi\pi_\phi(\mathfrak{R})E_\phi$ is abelian; by Theorem 6 this implies G-abelianness on ϕ. ∎

Corollary 2. *In the notation of Theorem* 7, η*-abelianness on* ϕ *implies* $\mathfrak{N}_\phi' = \pi_\phi(\mathfrak{R})'' \cap U_\phi(G)'$.

Proof. We decompose the proof in three steps:

(i) $\pi_\phi(\mathfrak{R})'' \cap U_\phi(G)' \subseteq \pi_\phi(\mathfrak{R})'$, hence $\pi_\phi(\mathfrak{R})'' \cap U_\phi(G)' \subseteq \mathfrak{N}'_\phi$;
(ii) $\{\pi_\phi(\mathfrak{R})'' \cap U_\phi(G)'\}E_\phi = E_\phi \pi_\phi(\mathfrak{R})'' E_\phi$, where E_ϕ is the projector on $\{\Psi \in \mathcal{H}_\phi \mid U_\phi(g)\Psi = \Psi \ \forall\, g \in G\}$;
(iii) $\mathfrak{N}'_\phi \subseteq \pi_\phi(\mathfrak{R})''$, hence $\mathfrak{N}'_\phi \subseteq \pi_\phi(\mathfrak{R})'' \cap U_\phi(G)'$.

We first notice that for every R such that $\pi_\phi(R)$ belongs to $U_\phi(G)'$ we have $\pi_\phi(R) = \eta_\phi(R)$, which by Theorem 7 is an element of $\pi_\phi(\mathfrak{R})'$; consequently $\pi_\phi(\mathfrak{R}) \cap U_\phi(G)'$ is contained in $\pi_\phi(\mathfrak{R})'$ and by continuity $\pi_\phi(\mathfrak{R})'' \cap U_\phi(G)'$ is contained in $\pi_\phi(\mathfrak{R})'$, which proves (i). Now let S be an element in $\pi_\phi(\mathfrak{R})'' \cap U_\phi(G)'$. Since $S \in U_\phi(G)'$, $SE_\phi = E_\phi S = E_\phi S E_\phi$, which belongs to $E_\phi \pi_\phi(\mathfrak{R})'' E_\phi$, since S belongs to $\pi_\phi(\mathfrak{R})''$; hence $\{\pi_\phi(\mathfrak{R})'' \cap U_\phi(G)'\}E_\phi \subseteq E_\phi \pi_\phi(\mathfrak{R})'' E_\phi$. Conversely, for every R in $\mathfrak{R}$ we know from the lemma on p. 180 that $\eta_\phi(R) \in \pi_\phi(\mathfrak{R})'' \cap U_\phi(G)'$ and $\eta_\phi(R)E_\phi = E_\phi \pi_\phi(R) E_\phi$, from which we conclude that $E_\phi \pi_\phi(\mathfrak{R}) E_\phi \subseteq \{\pi_\phi(\mathfrak{R})'' \cap U_\phi(G)'\}E_\phi$; hence by continuity $E_\phi \pi_\phi(\mathfrak{R})'' E_\phi \subseteq \{\pi_\phi(\mathfrak{R})'' \cap U_\phi(G)'\}E_\phi$, which together with the opposite inclusion proved above, establishes (ii). To prove (iii) we recall from the proof of Theorem 6 that $\mathfrak{N}_\phi E_\phi = \pi_\phi(\mathfrak{R})'' E_\phi$. From Corollary 1 we know that η-abelianness on ϕ implies G-abelianness on ϕ and then by the corollary to Theorem 6 $\mathfrak{N}'_\phi \subseteq \mathfrak{N}_\phi$. Finally, as a consequence of (i) we have that $\pi_\phi(\mathfrak{R})'' \cap U_\phi(G)' = \pi_\phi(\mathfrak{R})'' \cap \mathfrak{N}'_\phi$. Collecting the above information, we get

$$\begin{aligned}\mathfrak{N}'_\phi E_\phi &= E_\phi \mathfrak{N}'_\phi E_\phi \subseteq E_\phi \mathfrak{N}_\phi E_\phi = E_\phi \pi_\phi(\mathfrak{R})'' E_\phi \\ &= \{\pi_\phi(\mathfrak{R})'' \cap U_\phi(G)'\}E_\phi = \{\pi_\phi(\mathfrak{R})'' \cap \mathfrak{N}'_\phi\}E_\phi.\end{aligned}$$

Then, in particular, for the vector Φ corresponding to ϕ, which is obviously in $E_\phi \mathcal{H}_\phi$, we have

$$\mathfrak{N}'_\phi \Phi = \{\pi_\phi(\mathfrak{R})'' \cap \mathfrak{N}'_\phi\}\Phi,$$

which expresses that to every element T in $\mathfrak{N}'_\phi$ there corresponds at least one element S in $\pi_\phi(\mathfrak{R})'' \cap \mathfrak{N}'_\phi$ such that $S\Phi = T\Phi$. Moreover, since Φ is cyclic for $\pi_\phi(\mathfrak{R})$, hence for $\mathfrak{N}_\phi$, this vector is separating for $\mathfrak{N}'_\phi$, which means that $\{S, T \in \mathfrak{N}'_\phi, S\Phi = T\Phi\}$ implies $S = T$. Therefore we have that $T \in \mathfrak{N}'_\phi$ implies T belongs to $\pi_\phi(\mathfrak{R})'' \cap \mathfrak{N}'_\phi$, i.e., $\mathfrak{N}'_\phi$ is contained in $\pi_\phi(\mathfrak{R})'' \cap \mathfrak{N}'_\phi$. This establishes (iii). On combining (i) and (iii) we get $\mathfrak{N}'_\phi = \pi_\phi(\mathfrak{R})'' \cap U_\phi(G)'$. This concludes the proof of the corollary. ∎

We remark that since $\mathfrak{N}'_\phi \subseteq \pi_\phi(\mathfrak{R})'$, a trivial consequence of this corollary is that η-abelianness on ϕ implies that $\mathfrak{N}'_\phi$ coincides with the set of all elements in the center of $\pi_\phi(\mathfrak{R})''$ which are invariant under the mappings $Z \to U_\phi(g)\, Z U_\phi(g)^{-1}$ for all g in G.

We now prove the principal result of this subsection.

Theorem 8. *Let* G, *amenable, be a symmetry group for* $(\mathfrak{R}, \mathfrak{S}, \langle;\rangle)$ *and* η *an invariant mean on* G. *Then for every* ϕ *in* $\mathfrak{S}_{\mathrm{G}}$
(a) *the conditions*

(i) ϕ *is extremal invariant*,
(ii) $\pi_\phi(\mathfrak{R})' \cap \mathrm{U}_\phi(\mathrm{G})' = \{\lambda \mathrm{I}\}$,
(iii) $\{\pi_\phi(\mathfrak{R}), \mathrm{U}_\phi(\mathrm{G})\}'' \cap \pi_\phi(\mathfrak{R})' \cap \mathrm{U}_\phi(G)' = \{\lambda \mathrm{I}\}$,
(iv) ϕ *is* η*-clustering*,
(v) E_ϕ *is one-dimensional*,
(vi) $\pi_\phi(\mathfrak{R})'' \cap \pi_\phi(\mathfrak{R})' \cap \mathrm{U}_\phi(\mathrm{G})' = \{\lambda \mathrm{I}\}$,
(vii) $\eta_\phi(\mathrm{R}) = \langle \phi; \mathrm{R}\rangle \mathrm{I}$ *for all* R *in* $\mathfrak{R}$,
(viii) ϕ *is the only normal* G*-invariant state on* $\pi_\phi(\mathfrak{R})''$,
(ix) ϕ *is the only* G*-invariant vector state on* $\pi_\phi(\mathfrak{R})$

are in general in the following relation:

$$\begin{array}{ccccccc} (\mathrm{v}) & \Leftrightarrow & (\mathrm{iv}) & \Leftarrow & (\mathrm{vii}) & \Rightarrow & (\mathrm{viii}) \\ & & \Downarrow & & & & \Downarrow \\ (\mathrm{vi}) \Leftarrow (\mathrm{iii}) & \Leftarrow & (\mathrm{ii}) & \Leftrightarrow & (\mathrm{i}) & \Leftarrow & (\mathrm{ix}); \end{array}$$

(b) G*-abelianness on* ϕ *implies the equivalence of the first five conditions*,
(c) η*-abelianness on* ϕ *implies the equivalence of all nine conditions.*

Proof. We prove first the general implications of (a) in the theorem. We notice immediately that (ii) $\Rightarrow$ (iii) $\Rightarrow$ (vi), (vii) $\Rightarrow$ (iv) and (viii) $\Rightarrow$ (ix) are trivially true. To prove (vii) $\Rightarrow$ (viii) we recall that if ψ is a normal state on $\pi_\phi(\mathfrak{R})''$ we have for all S in $\pi_\phi(\mathfrak{R})''$

$$\langle \psi; S\rangle = \sum_i (\Psi_i, S\Psi_i) \quad \text{with} \quad \sum_i |\Psi_i|^2 = 1;$$

if, in addition, ψ is G-invariant, we have

$$\langle \psi; R\rangle = \langle \psi; \eta(R)\rangle = \sum_i (\Psi_i, \eta_\phi(R)\Psi_i) = \langle \phi; R\rangle$$

when (vii) is satisfied. This being true for all R in $\mathfrak{R}$, we have $\psi = \phi$, since $\mathfrak{R}$ is separating for the normal states of $\pi_\phi(\mathfrak{R})''$; hence (vii) $\Rightarrow$ (viii). To prove that (iv) and (v) are equivalent we recall that for all R, S in $\mathfrak{R}$

$$\eta\langle \phi; \alpha_g[R]S\rangle = (\Phi, \pi_\phi(R)\, E_\phi \pi_\phi(S)\Phi),$$

$$\langle \phi; R\rangle\langle \phi; S\rangle = (\Phi, \pi_\phi(R)\, E_\Phi \pi_\phi(S)\Phi),$$

where E_Φ is the projector on the one-dimensional subspace generated by Φ hence (iv) $\Leftrightarrow$ $\mathrm{E}_\phi = E_\Phi \Leftrightarrow$ (v), since Φ is cyclic for $\pi_\phi(\mathfrak{R})$. To prove (ix) $\Rightarrow$ (i) and (v) $\Rightarrow$ (i) $\Leftrightarrow$ (ii) suppose that ϕ is not extremal invariant, i.e., that there exists ψ in $\mathfrak{S}_G$ such that $\psi \leqslant \lambda\phi$. From the lemma on p. 174 we can write $\langle \psi; R\rangle = (B\Phi, \pi_\phi(R)B\Phi)$, with B positive in $\pi_\phi(\mathfrak{R})' \cap U_\phi(G)'$ so that

ψ is an invariant vector state and $E_\phi B\Phi = B\Phi$. If (ix) is now satisfied, ψ must coincide with ϕ and we then have (ix) $\Rightarrow$ (i); if, on the other hand, E_ϕ is one-dimensional, $B\Phi = \Phi$, since B is positive, and $\psi = \phi$, i.e., (v) $\Rightarrow$ (i). Finally (i) $\Leftrightarrow$ (ii) follows directly from the lemma on p. 174; this completes the proof of the first part of the theorem. We now consider the consequences of G-abelianness on ϕ. We already know from the corollary to Theorem 6 that G-abelianness on ϕ implies that $\mathfrak{N}_\phi' \subseteq \mathfrak{N}_\phi$ so that (ii) and (iii) are trivially equivalent; it is then sufficient for the completion of the proof of this second part to show that (ii) $\Rightarrow$ (v). To achieve this we consider the algebra of all bounded operators on $E_\phi\mathcal{H}$, namely $\mathfrak{B}(E_\phi\mathcal{H}_\phi) = E_\phi\mathfrak{B}(\mathcal{H}_\phi)E_\phi$; from (ii) this is equal to $E_\phi\mathfrak{N}_\phi E_\phi$ which is abelian; hence $\mathfrak{B}(E_\phi\mathcal{H}_\phi)$ has to be abelian, which is possible only when E_ϕ is one-dimensional. This establishes (b) of Theorem 8. To complete the proof of (c) it is sufficient to prove that (vi) $\Rightarrow$ (vii); from Theorem 7 we know that η-abelianness on ϕ implies that $\eta_\phi(\mathfrak{R})$ is contained in $\pi_\phi(\mathfrak{R})'' \cap \pi_\phi(\mathfrak{R})' \cap U_\phi(G)'$; hence (vi) implies $\eta_\phi(R) = \lambda I$ and $\langle\phi; R\rangle = \langle\phi; \eta(R)\rangle = \lambda$, so that indeed (vi) $\Rightarrow$ (vii). This concludes the proof of the theorem. ∎

We now present a few comments on this theorem; the first is directed to the structure of the theorem and its possible extensions, whereas the others are concerned with the physical meaning of the requirement that ϕ be *extremal* G-invariant.

First we want to refine our previous remarks about the fact that η-abelianness, which is weaker than our original condition of η-asymptotic abelianness, is nevertheless a stronger requirement than G-abelianness, as emphasized by the difference between parts (b) and (c) of Theorem 8. Specifically, we wonder whether there is some room in between the two concepts of η- and G-abelianness for a further notion, the strength of which would be sufficient to imply the results obtained so far. This question has been investigated by Størmer [1967a], who came up with the following concept: a group G is said to be a *large group of symmetries* for a physical system $(\mathfrak{R}, \mathfrak{S}, \langle;\rangle)$ if

$$\overline{{}^{w-op}\mathrm{co}\{\pi_\phi(\alpha_g[A]) \mid g \in G\}} \cap \pi_\phi(\mathfrak{R})' \neq \varnothing$$

for every self-adjoint element A in $\mathfrak{R}$ and every ϕ in $\mathfrak{S}_G$. This concept presents several remarkable features which we now review briefly.

1. As is the case of the substitutes for G-abelianness discussed above, this new concept does not require in its formulation that G be amenable, opening therefore the possibility of using it in relativistic theories.

2. In the case in which G is amenable and the system considered is η-abelian for some mean η on G, G is a large group of symmetries for this system.

3. If a subgroup H of G is a large group of symmetries for a given physical system, so is G; this fact leads to interesting generalizations, since it is often possible to prove η-abelianness for an amenable subgroup H of a non-amenable group G of symmetries. This is, for instance, the case in quasi-local relativistic theories in which G is the (*nonamenable*) Lorentz group and H is the (*amenable*) group T^3 of the translations in three dimensions, which always acts in a η-abelian manner in quasi-local theories (for more details see Chapter 4). Moreover, Størmer was able to prove the existence for every ϕ in $\mathfrak{S}_G$, when G is a large group of symmetries, of a unique mapping Θ_ϕ from $\pi_\phi(\mathfrak{R})''$ onto $\pi_\phi(\mathfrak{R})'' \cap \pi_\phi(\mathfrak{R})' \cap U_\phi(G)'$, mimicking exactly in its principal properties the mapping η_ϕ introduced above. Furthermore, he proved that when G is a large group of symmetries $E_\phi \mathfrak{N}_\phi E_\phi$ is abelian for each ϕ in $\mathfrak{S}_G$; we recall that this is the main consequence of G-abelianness used in the preceding pages and is the form in which this condition was originally introduced in the literature. $\mathfrak{N}_\phi'$ is then also abelian and actually turns out to be equal to $\pi_\phi(\mathfrak{R})'' \cap \pi_\phi(\mathfrak{R})' \cap U_\phi(G)'$. We should notice in this connection that this is the only consequence of η-abelianness which we used in the proof of (c) in Theorem 8. Hence, on replacing η_ϕ with Θ_ϕ, (c) in Theorem 8 holds as well when the condition of η-abelianness is replaced by the weaker requirement that G is a large group of symmetries for the system considered. Størmer [1969] (Theorem 6.2) proved the following result on the factor type of representations associated with states invariant with respect to certain "large" symmetry groups: let $\mathfrak{R}$ be a C^*-algebra and G, the additive group of real numbers. If α is a strongly continuous representation of G such that $\alpha(G)$ is a large group of symmetries of $(\mathfrak{R}, \mathfrak{S}, \langle;\rangle)$ and ϕ is a G-invariant factor-state on $\mathfrak{R}$, then π_ϕ is type I_n, I_∞, or II_1 whenever ϕ is respectively a homomorphism, a pure state and not a homomorphism, or a trace and not a homomorphism; π_ϕ is type II_∞ whenever ϕ is neither pure nor a trace and if, in addition, the vector state on $\pi_\phi(\mathfrak{R})'$ generated by the vector Φ (corresponding to the state ϕ via the GNS construction) is a trace; finally π_ϕ is type III whenever the state just defined on $\pi_\phi(\mathfrak{R})'$is not a trace.

Let us now return to the discussion of our Theorem 8. We notice that if $\mathfrak{N}_\phi' = \pi_\phi(\mathfrak{R})'' \cap \pi_\phi(\mathfrak{R})' \cap U_\phi(G)'$ (which is essentially the condition used for the proof of (c) in Theorem 8) we have in particular that $\mathfrak{N}_\phi'$ is contained in the center $\mathfrak{Z}_\phi(\mathfrak{R})$ of $\pi_\phi(\mathfrak{R})''$ for all ϕ in $\mathfrak{S}_G$; from this we conclude that η-abelianness on a primary state ϕ that is G-invariant implies that ϕ is an extremal G-invariant state and then satisfies all nine conditions of the theorem. The reciprocal implication, namely, that every extremal G-invariant state ϕ is a factor-state, would be true if $\mathfrak{N}_\phi'$ were to contain $\mathfrak{Z}_\phi(\mathfrak{R})$ for each extremal invariant state ϕ; we indeed have in such cases, by virtue of the general implication (i) $\Rightarrow$ (ii) in Theorem 8, that ϕ extremal G-invariant implies $\mathfrak{N}_\phi' = \{\lambda I\}$ and $\mathfrak{Z}_\phi(\mathfrak{R}) = \{\lambda I\}$; that is to say, $\pi_\phi(\mathfrak{R})''$ is a factor. It is interesting to know that the condition $\mathfrak{N}'_\phi \supseteq \mathfrak{Z}_\phi(\mathfrak{R})$ is actually realized in some cases relevant for physical applications; among these we want to mention here two particular examples. First, if G is a connected symmetry group for the physical system considered and α_g is norm-continuous in g (i.e., $\|\alpha_g - i\| \to 0$ as $g \to 0$), we can see from Kadison and Ringrose [1967] (Corollary 8), that $\pi_\phi(\mathfrak{R})'' \cap \pi_\phi(\mathfrak{R})'$ is contained in $\pi_\phi(\mathfrak{R})'' \cap \pi_\phi(\mathfrak{R})' \cap U_\phi(G)'$ for all ϕ in $\mathfrak{S}_G$, which obviously implies $\mathfrak{N}'_\phi \supseteq \mathfrak{Z}_\phi(\mathfrak{R})$. We should, however, realize that the norm continuity of a group of automorphisms is an extremely stringent condition; it is therefore interesting to know that $\mathfrak{N}'_\phi \supseteq \mathfrak{Z}_\phi(\mathfrak{R})$ [where $U_\phi(G)$ represents the time evolution in the representation $\pi_\phi(\mathfrak{R})$] is realized in statistical mechanical systems for which the (grand) canonical equilibrium thermodynamical states satisfy the KMS boundary conditions (see Subsection e).

We continue our discussion of the meaning of some of the conditions in Theorem 8, coming back to the case in which G is amenable, so that we know intuitively how to interpret η_ϕ as an averaging procedure. To be more specific we consider the case in which $\mathfrak{R}$ is the algebra generated by the "local observables" on a physical system and in which G is the translation group in three dimensions; the expository advantages of this particular case are (a) its immediate physical interpretation and (b) it is a η-abelian system. It should be realized, however, that the following argument is not limited to this particular case and applies as well—*mutatis mutandis*—to the general situation considered in the theorem. We start, then, from a translation invariant state ϕ, construct the representation π_ϕ, and study the properties of this representation in the light of the various conditions of the theorem; from (a) in Theorem 8 we see that (vii) is in general the strongest of the nine requirements considered. We begin our discussion with this condition, refining the arguments as we go down their logical chain. We immediately see that (vii) implies that the space-averages of the autocorrelation functions

$$\psi_A(g) = \langle \psi; \alpha_g[A]A \rangle - \langle \psi; \alpha_g[A] \rangle \langle \psi; A \rangle$$

vanish for all ψ in $\mathfrak{S}_\phi$ and all observables A in $\mathfrak{R}$; this is true in particular for any state "obtained from ϕ by a local perturbation," i.e., for any ψ of the form

$$\langle \psi; R \rangle = \langle \phi; S^*RS \rangle = (\pi_\phi(S)\Phi, \pi_\phi(R)\,\pi_\phi(S)\Phi),$$

where S is any local operation belonging to $\mathfrak{R}$. This shows that (vii) is a strong condition on ϕ, which could possibly be taken as a *sine qua non* requirement for ϕ to be a pure thermodynamical phase. This interpretation of (vii) is reinforced by the realization that this condition is equivalent to any of the following requirements:

(a) $$\eta\langle \psi; \alpha_g[A] \rangle = \langle \phi; A \rangle \quad \forall \psi \in \mathfrak{S}_\phi \qquad (\text{or } \mathfrak{B}_\phi),$$

which expresses that, on these states ψ, the space-average of the expectation-value of the translated observable $\alpha_g[A]$ is independent of ψ;

(b) $\langle\psi; (\eta_\phi(A) - \langle\psi; \eta_\phi(A)\rangle)^2\rangle = 0 \quad \forall\psi \in \mathfrak{S}_\phi \qquad (\text{or } \mathfrak{B}_\phi),$

which expresses that every such state ψ is dispersion-free for all the global observables $\eta_\phi(A)$ obtained by space averaging any local observable A;

(c) $\eta\langle\psi; \alpha_g[A]B\rangle = \eta\langle\psi; \alpha_g[A]\rangle\langle\psi; B\rangle \quad \forall\psi \in \mathfrak{S}_\phi \qquad (\text{or } \mathfrak{B}_\phi),$

which expresses that every such state is η-clustering.

As we mentioned earlier, all these conditions on ϕ are rather stringent, and we are sometimes inclined to identify a pure thermodynamical phase by requiring these conditions to hold, not necessarily for all the ψ considered above but just for the G-invariant state ϕ itself. This is precisely the meaning of condition (iv), and (c) in Theorem 8 shows that for an η-abelian system this condition is actually not weaker than (vii). In general (iv) is equivalent to (v), which means that the cyclic vector Φ canonically associated with an η-clustering (hence extremal-) G-invariant state ϕ is the only vector in $\mathcal{H}_\phi$ which is a common eigenvector of all the generators P_ϕ^k of $U_\phi(G)$. In connection with (v) we may also notice that it is the exact quantum analog of the classical condition that ϕ be *ergodic*† with respect to G.

This theorem admits the following corollary which we intend to use in Subsection e:

Corollary. *Let ϕ be an extremal* G*-invariant state on $\mathfrak{R}$ with respect to the action α of the amenable group* G*; if the vector Φ in $\mathcal{H}_\phi$, canonically associated with ϕ, is also cyclic with respect to $\pi_\phi(\mathfrak{R})'$, then* (i) *the system $\{\mathfrak{R}, \mathfrak{S}, \alpha\}$ is η-abelian on ϕ with respect to every mean η on* G *and* (ii) *ϕ satisfies the nine conditions listed in Theorem* 8.

Proof. We shall prove that ϕ satisfies property (vii) in Theorem 8 for any mean η on G; from this follow all the other properties of ϕ listed in Theorem 8, as well as η-abelianness on ϕ. Let C_ϕ^0 be the set of all vectors Ψ in $\mathcal{H}_\phi$ of the form $\Psi = S\Phi$, with S positive in $\pi_\phi(\mathfrak{R})'$, such that $|S\Phi| = 1$ and $E_\phi\Psi = \Psi$. From the lemma on p. 86 we know that the state ψ on $\mathfrak{R}$, defined by $\langle\psi; R\rangle = (\Psi, \pi_\phi(R)\Psi)$, is dominated by ϕ; furthermore ψ is G-invariant. Consequently $\psi = \phi$, since ϕ is extremal G-invariant. This means explicitly that

$$\begin{aligned}\langle\psi; R\rangle &= (S\Phi, \pi_\phi(R)S\Phi) = (S^*S\Phi, \pi_\phi(R)\Phi)\\ &= \langle\phi; R\rangle = (\Phi, \pi_\phi(R)\Phi).\end{aligned}$$

Since Φ is, by construction, cyclic in $\mathcal{H}_\phi$ with respect to $\pi_\phi(\mathfrak{R})$ and $S = S^* \geqslant 0$, we conclude from the above equality that $S\Phi = \Phi$, so that $C_\phi^0 = \{\Phi\}$.

† See, for instance, Arnold and Avez [1968], Theorem 9.7.

We now use the cyclicity of Φ with respect to $\pi_\phi(\mathfrak{R})'$ to claim that $E_\phi \mathcal{H}_\phi$ is generated, as a closed subspace of $\mathcal{H}_\phi$, by C_ϕ^0, hence by the above argument is one-dimensional [which incidentally shows that ϕ satisfies (v)]. For any mean η on G, any R in $\mathfrak{R}$, and any S in $\pi_\phi(\mathfrak{R})'$ we form, according to the lemma on p. 180,

$$\eta_\phi(R)S\Phi = S\eta_\phi(R)\Phi = SE_\phi\,\pi_\phi(R)\Phi = \langle\phi; R\rangle S\Phi.$$

The last of the above equalities follows from the fact that E_ϕ is one-dimensional; we again use the fact that Φ is cyclic with respect to $\pi_\phi(\mathfrak{R})'$ to conclude that $\eta_\phi(R) = \langle\phi; R\rangle I$. From the remark opening the proof we can now conclude that the corollary is established. ∎

Incidentally, this corollary exhibits one more situation in which, although there might be several means on G, this ambiguity does not show up in the physical properties of the system under consideration. Here neither η-abelianness nor the value of $\eta_\phi(R)$ for any R in $\mathfrak{R}$ depends on the special mean η we choose to define them.

e. The KMS Condition

The KMS condition was originally expressed by Kubo [1957] and Martin and Schwinger [1959] as a boundary condition on the analytic behavior of thermal Green functions. An algebraic formulation of this condition was proposed by Haag, Hugenholtz, and Winnink [1967], who exhibited some of its consequences on the structure of the representations associated with thermal equilibrium states in the thermodynamical limit. The algebraic treatment of the KMS condition was brought to its highest level of mathematical rigor by Kastler, Pool, and Poulsen [1969], whom we follow in part.

The organization of this subsection is the following. First we underline some remarkable properties of the representation obtained at the beginning of Section II.1.c; we next show in a semialgebraic manner how these features generalize to a wide class of finite systems; we then give the algebraic formulation of the KMS condition in the general case and derive some of its consequences. Finally we will indicate some of the recent developments that have occurred in the ever-growing literature on this subject. We should also close this outline of the present subsection by mentioning that the decomposition of a KMS state in its extremal components, and its relation to the problem of spontaneous symmetry breaking, is postponed until the next subsection.

Coming back to the first example discussed in Subsection II.1.c, we notice that the equilibrium state $\rho = \exp(\beta B\sigma^z)/\mathrm{Tr}\exp(\beta B\sigma^z)$, which corresponds to the Hamiltonian $H = -B\sigma^z$ and defined on the C^*-algebra $\mathfrak{R}$ of all 2×2 matrices with complex entries, satisfies the following property for any pair A, B of elements in $\mathfrak{R}$:

$$\langle\phi; A^*A\rangle = 0 \Leftrightarrow A = 0,$$

$$\begin{aligned}\langle\phi; \alpha_t[A]B\rangle &= \frac{\operatorname{Tr} e^{-\beta H}e^{iHt}Ae^{-iHt}B}{\operatorname{Tr} e^{-\beta H}} \\ &= \frac{\operatorname{Tr} e^{-\beta H}Be^{i(t+i\beta)H}Ae^{-i(t+i\beta)H}}{\operatorname{Tr} e^{-\beta H}} \\ &= \langle\phi; B\alpha_{t+i\beta}[A]\rangle \quad \text{for all } t \text{ in } \mathbb{R}.\end{aligned}$$

The second relation expresses the KMS condition in its most naive form. We first notice that this condition characterizes ϕ completely. Suppose, indeed, that there exists ϕ' satisfying this condition. Writing $B = I$, we see that $\langle\phi'; \alpha_t[A]\rangle = \langle\phi'; \alpha_{t+i\beta}[A]\rangle$ for all t in $\mathbb{R}$; since $\mathfrak{R} = \mathfrak{B}(\mathcal{H})$ with $\mathcal{H}$ finite-dimensional, a unique density matrix ρ' is associated with ϕ' and ρ' satisfies

$$\rho' = e^{-itH}e^{\beta H}\rho' e^{-\beta H}e^{itH},$$

from which follows that ρ' commutes with H and ϕ' is time-invariant. Incidentally, we could have seen this directly from the fact that the KMS condition implies that the entire function $\langle\phi; \alpha_t[A]\rangle$ is periodic, hence constant. We now insert again a general $B \in \mathfrak{R}$ in the KMS condition which we consider at $t = 0$ to obtain

$$\operatorname{Tr} \rho' AB = \operatorname{Tr} \rho' Be^{-\beta H}Ae^{\beta H}.$$

Since B is allowed to run over all of $\mathfrak{B}(\mathcal{H})$, this implies

$$\rho' A = e^{-\beta H}Ae^{\beta H}\rho';$$

on multiplying both sides by $e^{\beta H}$, we conclude that $e^{\beta H}\rho'$ commutes with all A in $\mathfrak{B}(\mathcal{H})$ and is then a multiple of the identity; from $\operatorname{Tr} \rho' = 1$ we then conclude that $\rho' = e^{-\beta H}/\operatorname{Tr} e^{-\beta H} = \rho$, as announced. We now want to turn to the properties of the representation associated with ϕ. At this point it would be needlessly long (although possible) to exploit the KMS condition explicitly, since we already have computed the representation π_ϕ associated with ϕ and since π_ϕ has a particularly simple form. A glimpse at this representation shows that there exists a one-to-one correspondence between the elements of $\pi_\phi(\mathfrak{R})$ and those of $\pi_\phi(\mathfrak{R})'$ (since the latter is of the form $\pi(\mathfrak{R}) \otimes I$) and that Φ is also cyclic for $\pi_\phi(\mathfrak{R})'$. A closer look at this representation shows that the mapping between $\pi_\phi(\mathfrak{R})$ and $\pi_\phi(\mathfrak{R})'$ is implemented by an operator C that satisfies the following properties:

(i) C is antiunitary, $C^2 = I$,

(ii) $C\Phi = \Phi$,

(iii) $CU_\phi(t)C = U_\phi(t)$ for all t in $\mathbb{R}$,

(iv) $C\pi_\phi(R)C = \nu_\phi(R)$ for all R in $\mathfrak{R}$, where $C = VK$, K is the complex conjugation in the basis on which the matrices $\pi_\phi(\mathfrak{R})$ are expressed on p. 90, and V is of the form

$$V = \begin{pmatrix} 1 & 0 & 0 & 0 \\ 0 & 0 & 1 & 0 \\ 0 & 1 & 0 & 0 \\ 0 & 0 & 0 & 1 \end{pmatrix}.$$

The antilinear *-representation $\nu_\phi(\mathfrak{R})$ obtained in this way coincides with $\pi_\phi(\mathfrak{R})'$ [thus showing explicitly that there is a one-to-one correspondence between $\pi_\phi(\mathfrak{R})$ and $\pi_\phi(\mathfrak{R})'$] and can alternately be obtained as follows: for every R in $\mathfrak{R}$ define the bounded operator $\nu_\phi(R)$ by

$$\nu_\phi(R)\,\Phi(T) = \Phi(TR^s)$$

where T runs over $\mathfrak{R}$ and R^s is defined from R by

$$R^s = e^{-\frac{1}{2}\beta H} R^* e^{-\frac{1}{2}\beta H}.$$

We notice, in particular, that for any pair R, T of elements of $\mathfrak{R}$ and any λ, μ in $\mathbb{C}$ we have

$$(\lambda R + \mu T)^s = \lambda^* R^s + \mu^* T^s,$$
$$(RT)^s = T^s R^s,$$

so that $\nu_\phi(\mathfrak{R})$ is indeed an antilinear *-representation of $\mathfrak{R}$.

Although we carried out these computations in the very particular case of the equilibrium state of a spin $\frac{1}{2}$-particle in a magnetic field B, the reader will suspect that the situation is more general than that. This is indeed true in every situation in which the Liouville space techniques of quantum statistical mechanics are applicable; let us follow in this regard Haag, Hugenholtz, and Winnink [1967] and consider a system in which $\mathfrak{R}$ is identifiable with $\mathfrak{B}(\mathcal{H})$ for some (separable but otherwise arbitrary) Hilbert space $\mathcal{H}$. This situation is encountered, for instance, when we consider a finite magnetic lattice or a gas of Bosons (or fermions) enclosed in a finite box.

Let H be the Hamiltonian for this system (and suppose for simplicity that the spectrum of H is discrete); let $\rho = \exp(-\beta H)/\text{Tr}\exp(-\beta H)$ be the corresponding canonical equilibrium density matrix (we might as well, for that matter, consider the grand canonical equilibrium state and interpret H as meaning $H - \mu N$; we shall not bother with it here). We then consider the Hilbert *-algebra $\mathfrak{L}(\mathcal{H})$ of all Hilbert-Schmidt operators on $\mathcal{H}$ (see p. 119), equipped with its scalar product $(K_1, K_2) = \text{Tr}\, K_1^* K_2$. We recall that $\mathfrak{L}(\mathcal{H})$ is a two-sided ideal of $\mathfrak{B}(\mathcal{H})$ so that we can define, *within* $\mathfrak{L}(\mathcal{H})$ the "representations" $\pi_\phi(\mathfrak{R})$ and $\nu_\phi(\mathfrak{R})$ of $\mathfrak{R}$ as follows:

$$\left.\begin{aligned} \pi_\phi(R)K &= RK, \\ \nu_\phi(R)K &= KR^*, \end{aligned}\right\} \quad \text{for all } R \text{ in } \mathfrak{R} \text{ and all } K \text{ in } \mathfrak{L}(\mathcal{H}).$$

[We notice that $\pi_\phi(\mathfrak{R})$ is a linear *-representation of $\mathfrak{R}$, whereas $\nu_\phi(\mathfrak{R})$ is an antilinear *-representation of $\mathfrak{R}$.] We define in a similar manner the operators $U_\phi(t)$:

$$U_\phi(t)K = \alpha_t[K] \quad \text{for all } t \text{ in } \mathbb{R} \text{ and all } k \text{ in } \mathfrak{L}(\mathcal{H}),$$

where α_t is the time-evolution of $\mathfrak{B}(\mathcal{H})$ generated by H. We finally define the operator C by

$$CK = K^* \quad \text{for all } K \text{ in } \mathfrak{L}(\mathcal{H}).$$

The relation of these objects to the state ϕ on $\mathfrak{R}$ generated by the canonical density matrix ρ is provided by considering the vector K_0 in $\mathfrak{L}(\mathcal{H})$ defined as $K_0 = \rho^{1/2}$. We have

(a) $\quad K_0$ is cyclic for $\pi_\phi(\mathfrak{R})$ and for $\nu_\phi(\mathfrak{R})$,

(b) $\quad \langle\phi; R\rangle = (K_0, \pi_\phi(R)K_0) = (K_0, \nu_\phi(R)K_0)^*$

(c) $\quad U_\phi(t)K_0 = K_0$

(d) $\quad U_\phi(t)\,\pi_\phi(R)\,U_\phi(-t) = \pi_\phi(\alpha_t[R]),$

$$U_\phi(t)\,\nu_\phi(R)\,U_\phi(-t) = \nu_\phi(\alpha_t[R]),$$

so that $\{\pi_\phi(\mathfrak{R}), U_\phi(\mathfrak{R})\}$ is the canonical covariant representation associated with ϕ; except for the replacement of "representation" with "antilinear representation," the same statement holds for $\{\nu_\phi(\mathfrak{R}), U_\phi(\mathfrak{R})\}$. Furthermore we have

(i) $\quad C$ is antiunitary, $C^2 = I$,

(ii) $\quad CK_0 = K_0,$

(iii) $\quad CU(t)C = U(t) \quad$ for all t in $\mathbb{R}$,

(iv) $\quad C\pi_\phi(R)C = \nu_\phi(R) \quad$ for all R in $\mathfrak{R}$.

To complete the extension of the results obtained with this particular model we need only to prove that $\pi_\phi(\mathfrak{R})' = \nu_\phi(\mathfrak{R})''$; this is achieved as follows. We first notice that

$$\pi_\phi(R)\,\nu_\phi(S)K = RKS^* = \nu_\phi(S)\,\pi_\phi(R)K;$$

i.e.,

$$\nu_\phi(\mathfrak{R}) \subseteq \pi_\phi(\mathfrak{R})',$$

hence

$$\nu_\phi(\mathfrak{R})'' \subseteq \pi_\phi(\mathfrak{R})'.$$

Let us denote respectively by $\pi_\phi(\mathfrak{L})$ and $\nu_\phi(\mathfrak{L})$ the restrictions of $\pi_\phi(\mathfrak{R})$ and $\nu_\phi(\mathfrak{R})$ to $\mathfrak{L}(\mathcal{H})$ considered as a subalgebra of $\mathfrak{R}$. The commutant theorem on Hilbert algebras asserts† that $\pi_\phi(\mathfrak{L})' = \nu_\phi(\mathfrak{L})''$; since $\mathfrak{L} \subseteq \mathfrak{R}$, we have:

$$\pi_\phi(\mathfrak{R})' \subseteq \pi_\phi(\mathfrak{L})' = \nu_\phi(\mathfrak{L})'' \subseteq \nu_\phi(\mathfrak{R})'';$$

† See Dixmier [1957], Theorem I.5.2.1, or Dixmier [1952].

on combining these two relations we indeed get that $\pi_\phi(\mathfrak{R})' = \nu_\phi(\mathfrak{R})''$, thus proving that a one-to-one correspondence exists between the elements of $\pi_\phi(\mathfrak{R})'$ and $\pi_\phi(\mathfrak{R})''$, hence that K_0 is also cyclic for $\pi_\phi(\mathfrak{R})'$. We have completely generalized to the present situation the remarks made before concerning the structure of the representation π_ϕ associated with the equilibrium state ϕ. We note in passing that it was not necessary to the argument considered above that $\mathfrak{R} = \mathfrak{B}(\mathcal{H})$, and that $\mathfrak{R} \supseteq \mathfrak{L}(\mathcal{H})$ would have been sufficient. For short, we refer to the existence of a conjugation C mapping $\pi_\phi(\mathfrak{R})''$ onto $\pi_\phi(\mathfrak{R})'$ in the manner prescribed above as the *commutant relation for* $\pi_\phi(\mathfrak{R})$.

We still have to formulate the KMS condition in a manner appropriate to the present situation, since the naïve form of this condition, as obtained for the initial example of this subsection, is inadequate in the general case when H is unbounded; indeed, in the latter case, $\alpha_{t+i\beta}$ maps a bounded operator R on $\mathcal{H}$ into an unbounded one. Whenever H is bounded below and such that $e^{-\gamma H}$ is of trace class for $\gamma = \beta/2$, we can compensate for this by considering

$$R_{t+i\gamma}e^{-\beta H} \equiv e^{-\gamma H}e^{itH}Re^{-itH}e^{-(\beta-\gamma)H},$$

$$e^{-\beta H}R_{t-i\gamma} \equiv e^{-(\beta-\gamma)H}e^{itH}Re^{-itH}e^{-\gamma H};$$

we notice that every term in each of these products is bounded and that at least one of them is of trace-class; consequently the operators just constructed are not only bounded but also of trace-class, since the trace-class is a two-sided ideal in $\mathfrak{B}(\mathcal{H})$ (see p. 119). As a consequence the following functions

$$F_{RS}(z) = \frac{\text{Tr } SR_z e^{-\beta H}}{\text{Tr } e^{-\beta H}},$$

$$G_{RS}(z) = \frac{\text{Tr } e^{-\beta H}R_z S}{\text{Tr } e^{-\beta H}},$$

make sense, respectively, for $z = t + i\gamma$ and $z = t - i\gamma$ with $0 \leqslant \gamma \leqslant \beta$. In the interior of the strips in which they are defined, these functions are continuous, differentiable, hence analytic; furthermore they are continuous on the boundaries and satisfy for all t in $\mathbb{R}$:

(i) $$F_{RS}(t) = \langle\phi; S\alpha_t[R]\rangle,$$

(ii) $$G_{RS}(t) = \langle\phi; \alpha_t[R]S\rangle,$$

(iii) $$F_{RS}(t + i\beta) = G_{RS}(t).$$

These properties are referred to by saying that ϕ satisfies the KMS condition. We notice that when H is bounded this condition reduces to the expression we previously gave for it in the simple model with which we opened this subsection.

Haag, Hugenholtz, and Winnink [1967] noticed that it is convenient for the passage to the thermodynamic limit to use the functions F and G for real arguments only, hence to re-express the KMS condition in the following equivalent form. We follow them in this and consider the space $\mathfrak{D}$ of infinitely differentiable functions with compact support. For every $\hat{f}$ in $\mathfrak{D}$ define

$$f(t) = \int d\omega \hat{f}(\omega) e^{i\omega t}.$$

It follows from the Paley-Wiener-Schwartz theorem† that f is extendable (in a necessarily unique way) to an entire function $f(t + i\gamma)$ and that $t^n f(t + i\gamma)$ is a bounded function of t for any fixed γ and positive n. We can therefore multiply (iii) by $f(t)$, integrate, and shift the integration path on the left-hand side to get

(iv) $$\int dt\, f_{-\beta}(t) \langle \phi; S\alpha_t[R] \rangle = \int dt\, f_0(t) \langle \phi; \alpha_t[R]S \rangle \quad \forall \hat{f} \in \mathfrak{D},$$

where we define $f_\gamma(t) = f(t + i\gamma)$. *Conversely*, denote by $\hat{F}_{RS}(\omega)$ and $\hat{G}_{RS}(\omega)$ the respective Fourier transforms of (i) and (ii); (iv) implies that $\hat{F}_{RS}(\omega) = e^{\beta\omega}\hat{G}_{RS}(\omega)$, where $\hat{F}_{RS}$ and $\hat{G}_{RS}$ are considered as distributions over $\mathfrak{D}$; $F_{RS}(t)$ and $G_{RS}(t)$ are bounded and continuous, so that $\hat{F}_{RS}$ and $\hat{G}_{RS}$ are actually distributions over the Schwartz test-function space $\mathcal{S}$ and as such still satisfy the above equality. This implies, in turn, that $F_{RS}(t)$ and $G_{RS}(t)$ extend to functions $F_{RS}(z)$ and $G_{RS}(z)$, thus satisfying all the properties mentioned in their original definition. Consequently we can refer to condition (iv) alone as the KMS condition. It might be worth mentioning that two of the properties of ϕ are readily reobtained from (iv), namely, that ϕ is time-invariant and that Φ is separating for $\pi_\phi(\mathfrak{R})''$, hence cyclic for $\pi_\phi(\mathfrak{R})'$; the other structure properties of $\pi_\phi(\mathfrak{R})$ are also reobtainable from (iv) but with more effort.

The originality of the contribution of Haag, Hugenholtz, and Winnink [1967] has been (a) to link, for the situation just discussed, the commutant relation with the fact that the equilibrium state satisfies the KMS condition and (b) to extend this link to the thermodynamical limit. Specifically, they showed that (i) under appropriate conditions on the interaction the KMS condition still holds in the thermodynamical limit for the Gibbs state ϕ and (ii) this condition (together with the condition that $\mathfrak{R}$ be simple) implies the commutant relation for $\pi_\phi(\mathfrak{R})$. They also mention in their paper the connection between the commutant relation they obtained and the work of Araki and

† This theorem is standard material in courses on the theory of distributions; see, for instance, Nachbin [1964], Section 48, or Schwartz [1951], p. 128.

Woods [1963] on the Bose gas in which these authors showed explicitly that the commutant $\pi_\phi(\mathfrak{R})'$ also provides a representation of $\mathfrak{R}$ that is unitarily equivalent to $\pi_\phi(\mathfrak{R})''$; the slight discrepancy between this result and the commutant relation (which, we recall, asserts that these two "representations" are antiunitarily equivalent) is taken care of by performing an additional antiunitary transformation on $\pi_\phi(\mathfrak{R})'$ which can actually be physically interpreted as a time-reversal operation that makes the time run in the same direction in $\pi_\phi(\mathfrak{R})$ and $\pi_\phi(\mathfrak{R})'$. We can now add that the KMS condition has since been shown to hold true for this system.† The KMS condition has also been shown to hold true, in the thermodynamic limit, for other exactly solvable models; we might mention in this connection the spin-systems treated by Robinson [1968] and the one-dimensional infinite quantum spin lattice with finite-range interaction studied by Araki [1969], in which the Gibbs state is explicitly shown to be an extremal KMS state, i.e., a state that satisfies the KMS condition and cannot be decomposed into a mixture of states also satisfying this condition.

We are now ready to attack the general case. We first recall that in the particular case just treated the essential ingredients in the formulation of the KMS condition and in the derivation of the commutant relation were, from a mathematical point of view, (a) the consideration of the test-function space $\mathfrak{D}$ which allowed the use of Fourier transform to bypass the difficulty that $\alpha_{i\beta}$ is in general not defined as a mapping from $\mathfrak{R}$ to $\mathfrak{R}$ and (b) the use, at the end of the proof of the commutant relation, of the existence of the Hilbert algebra $\mathfrak{L}(\mathcal{K})$ which is invariant under α_t and large enough in $\mathfrak{R}$. Those are the essential properties of the particular case treated before which we will have to translate to the general case. To do so we follow Kastler, Pool, and Poulsen [1969].

We first notice that the assumption, made in Subsection b, that α_t is an automorphism of $\mathfrak{R}$ such that $\langle\phi; \alpha_t[A]\rangle$ is a continuous function of t for all observables A and all states ϕ implies that α_t is continuous in the weak-topology, i.e., that $\langle f; \alpha_t[R]\rangle$ is continuous in t for all R in $\mathfrak{R}$ and all f in $\mathfrak{R}^*$. From this it follows‡ that α_t is strongly continuous in t, i.e., that $\|\alpha_t[R] - R\| \to 0$ as $t \to 0$ for all R in $\mathfrak{R}$ when $\mathfrak{R}$ is separable; when $\mathfrak{R}$ is not separable, we *assume* that α_t is strongly continuous (or restricts our attention to certain representations; see below). Now define Λ and its domain by

$$\mathfrak{D}(\Lambda) \equiv \left\{R \in \mathfrak{R} \,\middle|\, \lim_{t \to 0} \left\| \frac{\alpha_t[R] - R}{t} - \Lambda R \right\| = 0\right\};$$

† With a slight modification due to the fact that in the thermodynamical limit the time-evolution is not an automorphism of $\mathfrak{R}$ but of $\pi_\phi(\mathfrak{R})''$; the ensuing technical modifications in the proof of the commutant relation were carried out by Dubin and Sewell [1970].

‡ See Hille and Phillips [1957], Section 10.2, Corollary, p. 306.

we define similarly the domain $D(\Lambda^p)$ $(p = 1, 2, \ldots)$ of the pth power of Λ. The linear manifold $D^\infty(\Lambda)$ in $\mathfrak{R}$ is then defined as

$$\mathfrak{D}^\infty(\Lambda) = \bigcap_{p=1}^{\infty} D(\Lambda^p).$$

It results from Gårding's theorem† that $\mathfrak{D}^\infty(\Lambda)$ is norm-dense in $\mathfrak{R}$ and invariant under α_t.

We now want to define the Fourier transform of $\alpha_t[R]$. The natural way to do so is to consider the *Schwartz test-function space* $\mathcal{S}(\mathbb{R})$ of all "rapidly decreasing" infinitely differentiable functions f on $\mathbb{R}$; we recall‡ that "rapidly decreasing" means that for all positive integers n and m

$$\|\hat{f}\|_{n,m} = \sup_{t\in\mathbb{R}} \left\{ |t|^n \left| \frac{d^m \hat{f}}{dt^m}(t) \right| \right\} < \infty$$

and that an infinitely differentiable function f belongs to $\mathcal{S}(\mathbb{R})$ if and only if

$$\lim_{|t|\to\infty} t^n \frac{d^m \hat{f}}{dt^m}(t) = 0 \qquad \forall\, n, m \in \mathbb{Z}^+.$$

We now consider for each R in $\mathfrak{R}$ the function

$$X_R : \mathbb{R} \to \mathfrak{R},$$

defined by $X_R(t) = \alpha_t[R]$. From the remark made above on the continuity of α_t we see that X_R belongs to the normed *-algebra $\mathfrak{C}(\mathbb{R}, \mathfrak{R})$ of all continuous functions from $\mathbb{R}$ to $\mathfrak{R}$, where $\mathfrak{C}(\mathbb{R}, \mathfrak{R})$ is equipped with the structure

$$\begin{aligned}
(\lambda X + \mu Y)(t) &= \lambda X(t) + \mu Y(t), \\
(XY)(t) &= X(t)\, Y(t), \\
X^*(t) &= X(t)^*, \\
\|X\| &= \sup_{t\in\mathbb{R}} \|X(t)\|.
\end{aligned}$$

We can now consider X in $\mathfrak{C}(\mathbb{R}, \mathfrak{R})$ as a tempered $\mathfrak{R}$-valued distribution on $\mathbb{R}$ by defining for each f in $\mathcal{S}$

$$\langle X, f \rangle = \int dt\, f(t)\, X(t).$$

This notation has the advantage of making possible the definition of the Fourier transform $\hat{X}$ of X as a tempered distribution; namely,

$$\langle \hat{X}; f \rangle = \langle X; \hat{f} \rangle$$

† See Gårding [1947].
‡ See, for instance, Nachbin [1964] or Schwartz [1951].

where the mapping $f \to \hat{f}$ from $\mathcal{S}(\mathbb{R})$ into itself is the usual Fourier transform

$$\hat{f}(\omega) = \frac{1}{2\pi} \int dt\, f(t) e^{-i\omega t}.$$

We recall that a distribution X is said to vanish on an open set U if $\langle X; f \rangle = 0$ for all f with support in U; in analogy with the definition of the support of a function the support of a distribution is defined as the smallest closed subset outside which X vanishes.

With these preliminaries behind us we can now define the subset $\tilde{\mathfrak{R}}$ of $\mathfrak{R}$ as formed of the elements R of $\mathfrak{D}^\infty(\Lambda)$ for which $\hat{X}_R$ has compact support. This set will play, in the general case under study, the role played by $\mathfrak{L}(\mathcal{H})$ in the particular case already treated. We should notice now† that $\tilde{\mathfrak{R}}$ is a norm-dense, α_t-invariant, sub*-algebra of $\mathfrak{R}$.

We are now ready for the formulation of the KMS condition. For any R in $\tilde{\mathfrak{R}}$ and any S in $\mathfrak{R}$ consider the function

$$F_{RS}(t) = \langle \phi; S\alpha_t[R] \rangle.$$

Since $\|\alpha_t[R]\| = \|R\|$, this function is obviously bounded; furthermore, since R belongs to $\tilde{\mathfrak{R}}$, hence to $D^\infty(\Lambda)$, this function is also C^∞; we can regard it as a tempered distribution and as such define its Fourier transform by

$$\langle \hat{F}_{RS}; f \rangle = \langle F_{RS}; \hat{f} \rangle = \langle \phi; S\langle X_R; \hat{f} \rangle \rangle = \langle \phi; S\langle \hat{X}_R; f \rangle \rangle.$$

Since R belongs to $\tilde{\mathfrak{R}}$, $\hat{X}_R$ has compact support and so has the distribution $\hat{F}_{RS}$. Consequently we can use the general Paley-Wiener-Schwartz theorem‡ to conclude that $F_{RS}(t)$ is extendable (in a necessarily unique way) to an entire function $F_{RS}(t + i\gamma)$ and that $t^n f(t + i\gamma)$ is a bounded function of t for any fixed γ and positive n. The same reasoning applies to G_{RS}. We now say that a state ϕ satisfies the KMS condition for the natural temperature β whenever the corresponding functions F_{RS} and G_{RS} just defined are related by

$$F_{RS}(t + i\beta) = G_{RS}(t) \quad \text{for all } R \text{ in } \tilde{\mathfrak{R}} \text{ and all } S \text{ in } \mathfrak{R}.$$

We denote by $\mathfrak{S}_\beta$ the set of all states that satisfies this condition for the natural temperature β.

Lemma. *If $\mathfrak{R}$ possesses a unit, every ϕ in $\mathfrak{S}_\beta$ is time-invariant.*

Proof. For every R in $\tilde{\mathfrak{R}}$, $f(t) = F_{RI}(t) = G_{RI}(t)$; hence $f(z)$ is an entire periodic function: $f(t + i\beta) = f(t)$; since $f(t)$ is bounded, $f(z)$ is bounded on the strip $0 \leqslant Im z \leqslant \beta$, hence, being periodic, is bounded on $\mathbb{C}$. Consequently $f(t)$ is a constant, i.e., $\langle \phi; \alpha_t[R] \rangle = \langle \phi; R \rangle$ for all R in $\tilde{\mathfrak{R}}$; since $\tilde{\mathfrak{R}}$

† For the proof see Kastler, Pool, and Poulsen [1969], Proposition 1.

‡ See Nachbin [1964], Section 49, or Schwartz [1951], p. 128.

is norm-dense in $\mathfrak{R}$, and $\|\alpha_t[R]\| = \|R\|$, we conclude from this that the above equality extends from $\tilde{\mathfrak{R}}$ to $\mathfrak{R}$, which proves the lemma. ∎

In view of this result and for the sake of brevity, we include in the KMS condition the assumption that ϕ is time-invariant when $\mathfrak{R}$ does not possess a unit.

We now want to indicate briefly how the commutant relation follows from the KMS condition under the additional assumption that

$$\tilde{\mathfrak{K}}_\phi = \{R \in \mathfrak{R} \mid \langle \phi; R^*R \rangle = 0\} = \{0\}.$$

We first equip $\tilde{\mathfrak{R}}$ with the sesquilinear form $(R_1, R_2) = \langle \phi; R_1^* R_2 \rangle$ and then define† the group $\{J_\gamma \mid \gamma \in \mathbb{R}\}$ of algebraic automorphisms of $\tilde{\mathfrak{R}}$ by

$$\langle X_{J_\gamma(R)}; \hat{f} \rangle = \langle X_R; \hat{\gamma}(f) \rangle \quad \text{for all } R \text{ in } \mathfrak{R},$$

where $\hat{\gamma}(f)$ is the Fourier transform of the function $\exp\{\frac{1}{2}\gamma t\} f(t)$; we verify that J_γ commutes with α_t and that $J_\gamma(R^*) = J_{-\gamma}(R)^*$. Having J_γ, we then consider in particular the mappings

$$R \to R^j = J_\beta(R),$$
$$R \to R^s = J_\beta(R)^*.$$

The key to the main result of Kastler, Pool, and Poulsen [1969] is the realization that if ϕ belongs to $\mathfrak{S}_\beta$ and $\tilde{\mathfrak{K}}_\phi = \{0\}$ then $\tilde{\mathfrak{R}}$, when equipped with the scalar product $(\ldots, \ldots)$ and the mappings $R \to R^j$ and $R \to R^s$ just defined becomes a quasi-unitary algebra. For the sake of definiteness we recall that a *quasi-unitary algebra* (the word *quasi-Hilbert algebra* is also used) is an associative algebra $\mathfrak{M}$ over the complex numbers $\mathbb{C}$, equipped with the following structure:

(i) A sesquilinear form $(\ldots, \ldots)$ is given on $\mathfrak{M}$, making it a pre-Hilbert space, i.e., in particular $\{(A, A) = 0, A \in \mathfrak{M}\} \Rightarrow A = 0$.

(ii) An automorphism $A \to A^j$ of $\mathfrak{M}$ is given; in particular

$$(\lambda A + \mu B)^j = \lambda A^j + \mu B^j$$
$$(AB)^j = A^j B^j$$

for all A, B in $\mathfrak{M}$ and all λ, μ in $\mathbb{C}$.

(iii) An involuntary antiautomorphism $A \to A^s$ of $\mathfrak{M}$ is given; in particular

$$(\lambda A + \mu B)^s = \lambda^* A^s + \mu^* B^s,$$
$$(AB)^s = B^s A^s,$$
$$A^{ss} = A$$

for all A, B in $\mathfrak{M}$ and all λ, μ in $\mathbb{C}$.

† For the detailed proofs that the following properties of J_β are indeed satisfied see Kastler, Pool, and Poulsen [1969].

Furthermore we require the following relations to hold:

(iv) $(A^s, A^s) = (A, A)$ for all A in $\mathfrak{M}$,

(v) $(A^j, A) \geqslant 0$ for all A in $\mathfrak{M}$,

(vi) $(AB, C) = (B, A^{sj}C)$ for all A, B, C in $\mathfrak{M}$.

(vii) For every fixed B in $\mathfrak{M}$ the mapping $A \to BA$ is continuous with respect to the topology induced by the scalar product $(\ldots, \ldots)$.

(viii) The set $\{AB + A^jB^j \mid A, B \in \mathfrak{M}\}$ is dense in $\mathfrak{M}$ with respect to this topology.

Under these conditions the commutant theorem for quasi-Hilbert algebras asserts in general what we already saw was true for $\mathfrak{L}(\mathcal{H})$, namely, that $\pi(\mathfrak{M})'' = \nu(\mathfrak{M})'$, where $\pi(\mathfrak{M})$ [or $\nu(\mathfrak{M})$] is the left-representation of $\mathfrak{M}$ (or the right-representation of $\mathfrak{M}$), defined on the Hilbert space completion of $\mathfrak{M}$, by $\pi(A)B = AB$ [or $\nu(A)B = BA$]. Knowing this, Kastler, Pool, and Poulsen [1969] rederived the central result obtained previously by Haag, Hugenholtz, and Winnink [1967]:

Theorem 9. *Let ϕ be a KMS state on $\mathfrak{R}$ with respect to the time-evolution α_t; denote by $\{\pi_\phi(\mathfrak{R}), U_\phi(\mathbb{R})\}$ the covariant representation canonically associated with ϕ. Let $\nu_\phi(\tilde{\mathfrak{R}})$ be the antilinear *-representation of $\tilde{\mathfrak{R}}$ defined by $\nu_\phi(R)\,\Phi(S) = \Phi(SR^s)$ for all S in $\mathfrak{R}$ and all R in $\tilde{\mathfrak{R}}$; denote by $\nu_\phi(\mathfrak{R})$ the continuous extension of $\nu_\phi(\tilde{\mathfrak{R}})$ from $\tilde{\mathfrak{R}}$ to $\mathfrak{R}$. Then there exists an operator C on $\mathcal{H}_\phi$ such that*

(i) C *is antiunitary*, $C^2 = I$

(ii) $C\Phi = \Phi$,

(iii) $CU_\phi(t)C = U_\phi(t)$ *for all* t *in* $\mathbb{R}$,

(iv) $C\pi_\phi(R)C = \nu_\phi(R)$ *for all* R *in* $\mathfrak{R}$;

if, in addition,

$$\tilde{\mathfrak{R}}_\phi = \{R \in \tilde{\mathfrak{R}} \mid \langle\phi; R^*R\rangle = 0\} = \{0\},$$

then

$$\pi_\phi(\mathfrak{R})'' = \nu_\phi(\mathfrak{R})'.$$

REMARK.

(i) The first part of the theorem follows directly by construction of the representations π_ϕ and ν_ϕ: C is the extension to $\mathcal{H}_\phi$ of the mapping $R \to R^s$ defined on $\tilde{\mathfrak{R}}$. The proof of the second part of the theorem proceeds in a manner similar to that of the particular case treated before, except that the role played there by the Hilbert algebra $\mathfrak{L}(\mathcal{H})$ is now played by the quasi-unitary algebra $\tilde{\mathfrak{R}}$.

(ii) In addition to this main result, Kastler, Pool, and Poulsen pointed out that the mapping $R \rightarrow J_{-\beta}(R)$ extends to the (unbounded) self-adjoint operator $J^{-1} = e^{-\frac{1}{2}\beta H}$, thus establishing an even closer parallel with the particular case treated above.

(iii) From the fact that $\pi_\phi(\mathfrak{R})'$ is mapped onto $\pi_\phi(\mathfrak{R})''$ follows that Φ is also cyclic for $\pi_\phi(\mathfrak{R})'$, hence is separating for both $\pi_\phi(\mathfrak{R})''$ and $\pi_\phi(\mathfrak{R})'$.

Some of the consequences of this fact have already been pointed out in the preceding subsection and we shall return to it later in the present subsection.

Lemma. *For any dynamical system* $\{\mathfrak{R}, \mathfrak{S}, \alpha\}$ *the following conditions are equivalent;*

(i) $\phi \in \mathfrak{S}_\beta$

(ii) $\phi \in \mathfrak{S}$ *and*

$$\int \mathrm{dt}\, \mathrm{f}_{-\beta}(\mathrm{t})\langle\phi; \mathrm{S}\alpha_\mathrm{t}[\mathrm{R}]\rangle = \int \mathrm{dt}\, \mathrm{f}_0(\mathrm{t})\langle\phi; \alpha_\mathrm{t}[\mathrm{R}]\mathrm{S}\rangle$$

for all R *and* S *in* $\mathfrak{R}$ *and all* $\hat{\mathrm{f}}$ *infinitely differentiable with compact support; in the above equality* $\mathrm{f}_\gamma(\mathrm{t})$ (*with* $\gamma = 0$ *and* $-\beta$) *is defined by*

$$\mathrm{f}_\gamma(\mathrm{t}) \equiv \int \mathrm{d}\omega \hat{\mathrm{f}}(\omega) \mathrm{e}^{\mathrm{i}\omega(\mathrm{t}+\mathrm{i}\gamma)}$$

Proof. From the Paley-Wiener-Schwartz theorem (i) is equivalent to (ii) when in the latter R is restricted to belonging to $\tilde{\mathfrak{R}}$ (see the discussion on pp. 193–196). It is therefore sufficient to show that this restricted form of (ii) is enough to imply (ii) in the unrestricted form given in the statement of the lemma. Let R be an arbitrary element of $\tilde{\mathfrak{R}}$. Since $\tilde{\mathfrak{R}}$ is norm-dense in $\mathfrak{R}$, there exists a sequence $\{R_n\}$ of elements of $\mathfrak{R}$ such that $R_n \rightarrow R$ as $n \rightarrow \infty$ in the norm topology; we then have in particular

$$|\langle\phi; S\alpha_t[R_n - R]\rangle| \leqslant \|S\| \cdot \|R_n - R\|$$

and

$$|\langle\phi; \alpha_t[R_n - R]S\rangle| \leqslant \|S\| \cdot \|R_n - R\|.$$

For every f infinitely differentiable with compact support we have, from the Paley-Wiener-Schwartz theorem, that $\int dt\, |f_\gamma(t)| \leqslant C < \infty$ for some C (depending evidently on f and $\gamma = 0$ or $-\beta$); we therefore conclude that

$$\begin{aligned}\int dt\, f_{-\beta}(t)\langle\phi; S\alpha_t[R]\rangle &= \lim_{n\rightarrow\infty} \int dt\, f_{-\beta}(t)\langle\phi; S\alpha_t[R_n]\rangle \\ &= \lim_{n\rightarrow\infty} \int dt\, f_0(t)\langle\phi; \alpha_t[R_n]S\rangle \\ &= \int dt\, f_0(t)\langle\phi; \alpha_t[R]S\rangle,\end{aligned}$$

thus proving the lemma. ∎

Theorem 10. *Let $\mathfrak{R}$ be a C*-algebra of operators acting on some Hilbert space $\mathcal{H}$ such that $\mathcal{H} = \{R\Psi \mid R \in \mathfrak{R}, \Psi \in \mathcal{H}\}$. For any normalized vector Φ in $\mathcal{H}$ we denote by $\tilde{\phi}$ the state defined on the bicommutant $\mathfrak{R}''$ of $\mathfrak{R}$ by $\langle\tilde{\phi}; X\rangle = (\Phi, X\Phi)$ and by ϕ the restriction of $\tilde{\phi}$ to $\mathfrak{R}$. Let $\{U(t)\}$ be a weakly continuous, one-parameter group of unitary operators acting on $\mathfrak{R}$ such that $U(t)\,RU(-t)$ belongs to $\mathfrak{R}$ for every t whenever R belongs to $\mathfrak{R}$; we denote by $\tilde{\alpha}_t$ the automorphism of $\mathfrak{R}''$ defined by $\tilde{\alpha}_t[X] = U(t)\,XU(-t)$ and by α_t its restriction to $\mathfrak{R}$. If ϕ satisfies the KMS condition with respect to α_t for the natural temperature β and if the unit ball $\mathfrak{R}_1''$ of $\mathfrak{R}''$ is metrizable when equipped with the strong-operator topology, then $\tilde{\phi}$ also satisfies the KMS condition with respect to $\tilde{\alpha}_t$ for β.*

Proof. From von Neumann's density theorem (see Theorem II.1.10) we know that $\mathfrak{R}$ is dense in $\mathfrak{R}''$ for the strong-operator topology; we can therefore use Kaplanski's density theorem† to conclude that the unit ball $\mathfrak{R}_1$ of $\mathfrak{R}$ is dense in the unit ball $\mathfrak{R}_1''$ of $\mathfrak{R}''$ for the strong operator topology. Since $\mathfrak{R}_1''$ is assumed to be metrizable for this topology, we can find for any pair R, S of elements of $\mathfrak{R}_1''$ two *sequences* $\{R_n\}$ and $\{S_n\}$ of elements in $\mathfrak{R}_1$ that converge to R and S, respectively, for the strong-operator topology. We can therefore conclude that for this topology $\{\alpha_t[R_n]\}$ converges to $\tilde{\alpha}_t[R]$, hence $\{S\alpha_t[R_n]\}$ converges to $S\tilde{\alpha}_t[R]$. Consequently

$$f_n(t) \equiv f_{-\beta}(t)\langle\phi; S\alpha_t[R_n]\rangle$$

converges to

$$f(t) \equiv f_{-\beta}(t)\langle\tilde{\phi}; S\alpha_t[R]\rangle.$$

Since this convergence is *not* uniform in t, we cannot conclude so directly as in the preceding lemma that $\int dt\, f_n(t)$ converges to $\int dt\, f(t)$; we notice, however, that $|f_n(t)| \leqslant |f_{-\beta}(t)|$ and that $f_{-\beta}(t)$ is integrable by the Paley-Wiener-Schwartz theorem whenever f is infinitely differentiable with compact support. We can therefore use Lebesgue's dominated convergence theorem to conclude that

$$\lim_{n\to\infty} \int dt\, f_{-\beta}(t)\langle\phi; S\alpha_t[R_n]\rangle = \int dt\, f_{-\beta}(t)\langle\tilde{\phi}; S\tilde{\alpha}_t[R]\rangle.$$

On repeating the same process for $\langle\tilde{\phi}; \tilde{\alpha}_t[R]S\rangle$ and using condition (ii) of the lemma for $\mathfrak{R}$, we conclude that this condition is also satisfied for $\mathfrak{R}''$; hence, using the equivalence established in the lemma, we conclude that $\tilde{\phi}$ satisfies the KMS condition with respect to $\tilde{\alpha}_t$ for the natural temperature β. ∎

Corollary 1. *Let ϕ be a KMS state on $\{\mathfrak{R}, \mathfrak{S}, \alpha\}$ such that $\mathcal{H}_\phi$ is separable; then the normal extension $\tilde{\phi}$ of ϕ to $\pi_\phi(\mathfrak{R})''$ satisfies the KMS condition for the*

† See, for instance, Dixmier [1957], I.3.5, Theorem 3.

same temperature, with respect to the natural extension $\tilde{\alpha}_t$ *of* α_t *to this von Neumann algebra.*

Proof. Since $\mathcal{H}_\phi$ is separable, $\pi_\phi(\mathfrak{R})''$ is countably decomposable† and therefore‡ its unit ball is metrizable in the strong operator topology. The demonstration of the corollary is then completed directly from the theorem when we recall that $\{\pi_\phi(\mathfrak{R}), U_\phi(\mathbb{R})\}$ is the covariant representation canonically associated with ϕ, Φ is the cyclic vector corresponding to ϕ in this representation, $\langle\tilde{\phi}; X\rangle \equiv (\Phi, X\Phi)$ for all X in $\pi_\phi(\mathfrak{R})''$, and $\tilde{\alpha}_t[X] \equiv U_\phi(t)XU_\phi(-t)$. ∎

REMARKS AND DEFINITION. The proof of the theorem depends essentially on the use of the Lebesgue dominated convergence theorem, the validity of which requires the use of sequences and not only nets; this is the mathematical reason for the metrizability assumption made in Theorem 10. The proof of the corollary points out that this assumption is realized when $\mathcal{H}$ is separable. The separability of the GNS space $\mathcal{H}_\phi$, although mathematically more stringent than our metrizability assumption, is nevertheless realized § when ϕ is a locally normal state on a quasi-local algebra. In view of the physical importance of these states it seems judicious to study this case in some detail and to introduce the following definition: a KMS state is said to be *separable* when the unit ball of $\pi_\phi(\mathfrak{R})''$ is metrizable in the strong operator topology.

Corollary 2. *Let* ϕ *be a separable KMS state on* $\mathfrak{R}$; *then*

$$\mathfrak{Z}_\phi(\mathfrak{R}) \equiv \pi_\phi(\mathfrak{R})'' \cap \pi_\phi(\mathfrak{R})' \subseteq \pi_\phi(\mathfrak{R})' \cap U_\phi(\mathbb{R})' \equiv \mathfrak{N}'_\phi.$$

Proof. Consider for every X in $\pi_\phi(\mathfrak{R})^\sim$ and Z in $\mathfrak{Z}_\phi(\mathfrak{R})$ the function $F_{XZ}(t) \equiv \langle\tilde{\phi}; Z\alpha_t[X]\rangle$; X in $\pi_\phi(\mathfrak{R})^\sim$ implies, by the Paley-Wiener-Schwartz theorem that $F_{XZ}(t)$ extends uniquely to an entire function $F_{XZ}(z)$ which is bounded on a strip along the real axis. Since ϕ is a separable KMS state, the theorem applies to the present situation and

$$F_{XZ}(t + i\beta) = G_{XZ}(t) \equiv \langle\tilde{\phi}; \alpha_t[X]Z\rangle.$$

Since Z belongs to $\mathfrak{Z}_\phi(\mathfrak{R})$, hence in particular to $\pi_\phi(\mathfrak{R})'$, the right-hand side of this expression is equal to $F_{XZ}(t)$; $F_{XZ}(z)$ is therefore an entire function bounded in the entire complex plane (since it is bounded on a strip along the real axis and periodic in the imaginary part of its argument). Consequently $F_{XZ}(z)$ is a constant; hence,

$$\langle\tilde{\phi}; \{\tilde{\alpha}_t(Z) - Z\}X\rangle = 0 \;\forall\, X \in \pi_\phi(\mathfrak{R})^\sim, \quad \forall\, Z \in \mathfrak{Z}_\phi(\mathfrak{R}).$$

† See the definition on p. 146; the above assertion is then immediate.
‡ See Dixmier [1957], I.3.1, Proposition 1.
§ This fact has been pointed out by Hugenholtz and Wieringa [1969]; see our Chapter IV.

Since $\pi_\phi(\mathfrak{N})^\sim$ is norm-dense in $\pi_\phi(\mathfrak{N})$, this relation extends to all X in $\pi_\phi(\mathfrak{N})$, hence to all X of the form $X_1^* X_2$ with X_1, X_2 in $\pi_\phi(\mathfrak{N})$, so that

$$(X_1\Phi, \{\tilde{\alpha}_t(Z) - Z\}X_2\Phi) = 0.$$

Since Φ is cyclic for $\pi_\phi(\mathfrak{N})$, this implies $\tilde{\alpha}_t[Z] = Z$, which is to say $\mathfrak{Z}_\phi(\mathfrak{N}) \subseteq U_\phi(\mathbb{R})'$, hence, since $\mathfrak{Z}_\phi(\mathfrak{N}) \subseteq \pi_\phi(\mathfrak{N})'$, we get $\mathfrak{Z}_\phi(\mathfrak{N}) \subseteq \mathfrak{N}_\phi'$, which is the conclusion of the corollary. ∎

REMARK. We have already discussed briefly in the preceding subsection some of the consequences of $\mathfrak{Z}_\phi(\mathfrak{N}) \subseteq \mathfrak{N}_\phi'$ and we shall return to this point shortly. We also note in passing that an equivalent way to state the conclusion of this corollary is to say that $\tilde{\alpha}_t$ leaves the center of $\pi_\phi(\mathfrak{N})''$ *element-wise* invariant.

Lemma. *Let ϕ be a separable KMS state on the dynamical system $\{\mathfrak{N}, \mathfrak{G}, \alpha\}$; let us denote by E_ϕ the projector on $\{\Psi \in \mathcal{H}_\phi \mid U_\phi(t)\Psi = \Psi \ \forall\, t \in \mathbb{R}\}$. Then $\langle\psi; X\rangle \equiv (\Phi, X\Phi)$ is a faithful vector trace on the von Neumann algebra $E_\phi\pi_\phi(\mathfrak{N})''E_\phi$.*

Proof. Let $\mathfrak{N}_\phi$ be the von Neumann algebra generated by $\pi_\phi(\mathfrak{N})$ and $U_\phi(\mathbb{R})$ and let $\mathfrak{B}_\phi$ be the von Neumann algebra $\pi_\phi(\mathfrak{N})'' \cap U_\phi(\mathbb{R})'$. We shall need the following three preliminary results in the sequel:

(i) $E_\phi\pi_\phi(\mathfrak{N})''E_\phi$ is a von Neumann algebra,

(ii) $E_\phi\mathfrak{B}_\phi = E_\phi\pi_\phi(\mathfrak{N})''E_\phi = \mathfrak{B}_\phi E_\phi$,

(iii) $(\Phi, [X_1, X_2]\Phi) = 0 \ \forall\, X_1 \in \mathfrak{B}_\phi$ and $\forall\, X_2 \in \pi_\phi(\mathfrak{N})''$,

which we now prove. Since E_ϕ belongs to $U_\phi(\mathbb{R})''$ and then to $\mathfrak{N}_\phi$, $E_\phi\mathfrak{N}_\phi E_\phi$ is a von Neumann algebra. We recall that $E_\phi\pi_\phi(\mathfrak{N})''E_\phi = E_\phi\mathfrak{N}_\phi E_\phi$, which concludes the proof of (i). From the lemma to Theorem 7 we see that to every mean η on $\mathbb{R}$ and to every X in $\pi_\phi(\mathfrak{N})''$ corresponds an operator $\eta_\phi(X)$ in $\mathfrak{B}_\phi$ such that

$$\eta_\phi(X)E_\phi = E_\phi\eta_\phi(X) = E_\phi X E_\phi,$$

from which we see that $E_\phi\pi_\phi(\mathfrak{N})''E_\phi \subseteq \mathfrak{B}_\phi E_\phi = E_\phi\mathfrak{B}_\phi$ (the last equality follows from $E_\phi \in U_\phi(\mathbb{R})''$, hence $E_\phi \in \mathfrak{B}_\phi'$); the opposite inclusion follows trivially from $\mathfrak{B}_\phi \subseteq \pi_\phi(\mathfrak{N})''$ so that (ii) is now proved. From Corollary 1 we know that the normal extension $\tilde{\phi}$ of ϕ to $\pi_\phi(\mathfrak{N})''$ is KMS; from this condition and the fact that for every X_2 in $\mathfrak{B}_\phi$ and X_1 in $\pi_\phi(\mathfrak{N})''$ we have

$$\langle\tilde{\phi}; X_1\tilde{\alpha}_t[X_2]\rangle = \langle\tilde{\phi}; X_1X_2\rangle,$$

$$\langle\tilde{\phi}; \tilde{\alpha}_t[X_2]X_1\rangle = \langle\tilde{\phi}; X_2X_1\rangle,$$

which clearly imply the validity of (iii). We now use (ii) to conclude that for every Y in $E_\phi \pi_\phi(\mathfrak{R})'' E_\phi$ there exists at least one X in $\mathfrak{B}_\phi$ such that $Y = XE_\phi = E_\phi X$; we have therefore, on using (iii), that for every pair Y_1, Y_2 of elements of $E_\phi \pi_\phi(\mathfrak{R})'' E_\phi$

$$\begin{aligned}\langle \psi;\, Y_1 Y_2 \rangle &= (\Phi, E_\phi X_1 X_2 E_\phi \Phi) = (\Phi, X_1 X_2 \Phi) = (\Phi, X_2 X_1 \Phi)\\ &= (\Phi, E_\phi X_2 X_1 E_\phi \Phi) = \langle \psi, Y_2 Y_1 \rangle,\end{aligned}$$

so that ψ is a vector-trace on $E_\phi \pi_\phi(\mathfrak{R})'' E_\phi$. To show that it is faithful, suppose that there exists Y in $E_\phi \pi_\phi(\mathfrak{R})'' E_\phi$ such that $\langle \psi;\, Y^* Y \rangle = 0$, i.e., $Y\Phi = 0$. Now let Y_1 and Y_2 be any two elements of $E_\phi \pi_\phi(\mathfrak{R})'' E_\phi$ and let us form

$$\begin{aligned}(Y_2 \Phi, Y^* Y Y_1 \Phi) &= \langle \psi;\, Y_2^* Y^* Y Y_1 \rangle = \langle \psi;\, Y^* Y Y_1 Y_2^* \rangle\\ &= (Y^* Y \Phi, Y_1 Y_2^* \Phi) = 0.\end{aligned}$$

We now remember that Φ is cyclic in $\mathcal{H}_\phi$ with respect to $\pi_\phi(\mathfrak{R})$, hence Φ is cyclic in $E_\phi \mathcal{H}_\phi$ with respect to $E_\phi \pi_\phi(\mathfrak{R})'' E_\phi$, so that the above equality implies $Y^* Y = 0$; i.e., $Y = 0$. Hence ψ is indeed faithful on $E_\phi \pi_\phi(\mathfrak{R})'' E_\phi$. ∎

REMARK. We could also have ended the proof by remarking that Φ is separating for $\pi_\phi(\mathfrak{R})''$ as a consequence of $\tilde{\phi}$ being KMS so that Φ is also separating for $E_\phi \pi_\phi(\mathfrak{R})'' E_\phi$; aside from the fact that this ending is not markedly shorter than the one we have used, we should like to note that the end of the proof as given consists in showing that a cyclic vector trace is always faithful.

Theorem 11. *Let ϕ be a separable KMS state on the dynamical system $\{\mathfrak{R}, \mathfrak{S}, \alpha\}$ for the natural temperature β. To every KMS state ψ on $\{\mathfrak{R}, \mathfrak{S}, \alpha\}$, which is dominated by ϕ, corresponds a unique Z in $\mathfrak{B}(\mathcal{H}_\phi)$ such that*

(i) $\langle \psi;\, \mathrm{R} \rangle = (\mathrm{Z}\Phi, \pi_\phi(\mathrm{R})\mathrm{Z}\Phi) \ \forall\ \mathrm{R} \in \mathfrak{R}$,

(ii) $(\Psi, \mathrm{Z}\Psi) \geqslant 0 \ \forall\ \Psi \in \mathcal{H}_\phi$,

(iii) $\mathrm{Z} \in \mathfrak{Z}_\phi(\mathfrak{R}) \equiv \pi_\phi(\mathfrak{R})'' \cap \pi_\phi(\mathfrak{R})'$,

(iv) $|\mathrm{Z}\Phi| = 1$.

Conversely every Z satisfying (ii) *to* (iv) *generates via the relation* (i), *now taken as a definition of ψ, a separable KMS state ψ on $\{\mathfrak{R}, \mathfrak{S}, \alpha\}$ for the same natural temperature β and which is dominated by ϕ.*

Proof. We first show the converse part of the theorem. The fact that ψ is dominated by ϕ is an immediate consequence of the assumptions. From Theorem 10 we have for any R and S in $\mathfrak{R}$

$$\begin{aligned}\int dt\, f_{-\beta}(t)\langle \psi;\, S\alpha_t[R] \rangle &= \int dt\, f_{-\beta}(t)\langle \tilde{\phi};\, ZS\alpha_t[R]Z \rangle\\ &= \int dt\, f_0(t)\langle \tilde{\phi};\, Z\alpha_t[R]SZ \rangle\\ &= \int dt\, f_0(t)\langle \psi;\, \alpha_t[R]S \rangle,\end{aligned}$$

so that ψ is KMS on $\{\mathfrak{R}, \mathfrak{S}, \alpha\}$ for the natural temperature β. Now let $\mathcal{H}'_\psi = \overline{\pi_\phi(\mathfrak{R})Z\Phi}$ and $\pi'_\psi(\mathfrak{R})$ be the restriction of $\pi_\phi(\mathfrak{R})$ to $\mathcal{H}'_\psi$; $\pi'_\psi(\mathfrak{R})$ is then unitarily equivalent to the canonical representation $\pi_\psi(\mathfrak{R})$ associated with ψ. Since the mapping $(A, B) \to AB$, restricted to the unit ball, is continuous for the strong-operator topology, the unit ball of $\pi_\psi(\mathfrak{R})''$ is homeomorphic to the unit ball of $\pi'_\psi(\mathfrak{R})''$, hence metrizable, so that ψ is also a separable KMS state; this concludes the proof of the converse part of the theorem. For the direct part, we first notice that since ψ is invariant under α_t the lemma on p. 174 already asserts the existence of a unique Z that satisfies the conditions (i), (ii), (iv), and $Z \in \pi_\phi(\mathfrak{R})' \cap U_\phi(\mathbb{R})'$. We therefore conclude, as in the proof of the converse part of the theorem, that ψ is a separable KMS state. We next remark that the proof of the lemma, except for the faithfulness of the trace, would have been possible along the same lines without assuming the cyclicity of the vector Φ. We conclude that ψ is a vector-trace on $E_\phi\pi_\phi(\mathfrak{R})''E_\phi$ and is dominated by the faithful trace ϕ. There is therefore† a unique positive element Z_0 in the center $\mathfrak{Z}$ of $E_\phi\pi_\phi(\mathfrak{R})''E_\phi$ such that $\langle\psi; P\rangle = \langle\phi; Z_0^2P\rangle$ for all (positive) P in $E_\phi\pi_\phi(\mathfrak{R})''E_\phi = E_0\mathfrak{R}_\phi E_\phi$. We can now write‡

$$\mathfrak{Z} = E_\phi(\mathfrak{R}_\phi \cap \mathfrak{R}'_\phi)E_\phi = (\mathfrak{R}_\phi \cap \mathfrak{R}'_\phi)E_\phi$$
$$= (\pi_\phi(\mathfrak{R})'' \cap \pi_\phi(\mathfrak{R})' \cap U_\phi(\mathbb{R})')E_\phi.$$

From Corollary 2 the RHS of this expression reduces to $\mathfrak{Z}_\phi(\mathfrak{R})E_\phi$; there exists therefore an element Z in $\mathfrak{Z}_\phi(\mathfrak{R})$ such that $ZE_\phi = Z_0$. We now recall that Φ is cyclic for $\pi_\phi(\mathfrak{R})$, hence for $\mathfrak{R}_\phi$; consequently Φ is separating for $\mathfrak{R}'_\phi$; this is sufficient to ensure the uniqueness and positivity of Z. This Z, from its construction, then satisfies all the conditions of the theorem which is therefore proved. ∎

This theorem has the following immediate consequence:

Corollary 1. *A separable KMS state ϕ on a dynamical system $\{\mathfrak{R}, \mathfrak{S}, \alpha\}$ is extremal KMS if and only if it is primary.*

Our next problem is now to establish the existence of extremal KMS states; the solution comes as an immediate consequence of the following result:

Theorem 12. *The convex set $\mathfrak{S}_\beta$ of all KMS states on a dynamical system $\{\mathfrak{R}, \mathfrak{S}, \alpha\}$ for any fixed natural temperature β is compact in the* w**-topology.*

Proof. We first recall that ψ is KMS on $\{\mathfrak{R}, \mathfrak{S}, \alpha\}$ for β if and only if

$$\hat{F}_{RS} = e_\beta \hat{G}_{RS} \ \forall\ R \in \tilde{\mathfrak{R}} \quad \text{and} \quad S \in \mathfrak{R},$$

† See Dixmier [1957], I.6.4, Theorem 3.

‡ See Dixmier [1957], I.2.1, Corollary to Proposition 2.

where

$$\langle \hat{F}_{RS}; f \rangle \equiv \langle \psi; S\langle X_R; \hat{f} \rangle \rangle,$$

$$\langle \hat{G}_{RS}; f \rangle \equiv \langle \psi; \langle X_R; \hat{f} \rangle S \rangle,$$

and e_β is the distribution associated with $\exp(\beta\omega)$.

Now if $\{\phi_\gamma\}$ is a net of KMS states on $\{\mathfrak{R}, \mathfrak{S}, \alpha\}$ for β, converging in the w^*-topology to $\phi \in \mathfrak{S}$, we have

$$\hat{F}_{RS} \equiv \langle \phi; S\langle X_R; \hat{f} \rangle \rangle = \lim_\gamma \langle \phi_\gamma; S\langle X_R; \hat{f} \rangle \rangle$$

$$= \lim_\gamma e_\beta \langle \phi_\gamma; S\langle X_R; \hat{f} \rangle \rangle = e_\beta \hat{G}_{RS};$$

i.e., ϕ is a KMS state on $\{\mathfrak{R}, \mathfrak{S}, \alpha\}$ for β. We conclude therefore that $\mathfrak{S}_\beta$ is closed in the w^*-topology; since it is bounded (namely by 1) in the norm-topology, it is compact in the w^*-topology. ∎

REMARKS.

(i) As for all the preceding results of this subsection, this proof depends essentially on the assumed continuity of α_t; counterexamples, in which $\mathfrak{S}_\beta$ has no extreme points, are known† when α_t is not assumed to be continuous.

(ii) When $\mathfrak{R}$ is separable, $\mathfrak{S}$ is metrizable in the w^*-topology, and an alternate proof of the theorem can be obtained from the Lebesgue-dominated convergence theorem, as in the proof of Theorem 10.

From the Krein-Millman theorem we conclude immediately that $\mathfrak{S}_\beta$ is the weak*-closure of the convex hull of its extreme points or, in symbols, $\mathfrak{S}_\beta = {}^{w^*}\overline{\text{co}(\mathfrak{E}_\beta)}$; hence $\mathfrak{E}_\beta$ is nonempty as soon as $\mathfrak{S}_\beta$ itself is nonempty.

The way in which the KMS condition was introduced suggests rather strongly that a reasonable requirement on a state ϕ to be an equilibrium state for the natural temperature β could be that ϕ belong to $\mathfrak{S}_\beta$; it should be noted, however, that the opposite requirement, namely that any state in $\mathfrak{S}_\beta$ might be interpretable as an equilibrium state for the natural temperature β, must be handled with care. It has indeed been shown‡ for a class of systems which are exactly solvable by the *molecular field method* that every equilibrium state satisfies the KMS condition§; however, some other states, which also

† Takesaki and Winnink [1971].

‡ Emch and Knops [1970].

§ More precisely an appropriately generalized version of this condition, taking into account the fact that the time-evolution, in the van der Waals limit, is not an automorphism of $\mathfrak{R}$.

satisfy this KMS condition, are obtained as solutions of the usual self-consistency equations, hence are stationary points of the free energy, but do *not*, however, correspond to an absolute minimum of the free energy and consequently are not stable equilibrium states.

We now push one step further the requirement that an equilibrium state for the natural temperature β belong to $\mathfrak{S}_\beta$. We should like to assert that a *pure thermodynamical phase* for β is an equilibrium state for β that cannot be decomposed into other equilibrium states (for the same β). Suppose now that ϕ is a separable KMS state interpretable as a pure thermodynamical phase; if ϕ were not extremal KMS, there would exist a KMS state $\psi \leqslant \lambda\phi$ (for the same β by Theorem 11); hence ϕ could be decomposed into other equilibrium state for the same β, which would contradict the assertion that ϕ represents a pure thermodynamical phase. We are led to conjecture that a pure thermodynamical phase should be an extremal KMS state. This conjecture is supported by several particular results.† We first recall our earlier remark‡ that, for a finite system, $\rho_\beta = \exp(-\beta H)/\text{Tr}\exp(-\beta H)$ is the only KMS state for β, hence is extremal KMS. According to the above conjecture, ρ_β should be a pure thermodynamical phase in agreement with the theorem of Lee and Yang [1952], according to which a finite (classical) system cannot exhibit phase transitions. We next mention that in some systems§ in which it is known that no phase transitions occur it has been explicitly shown that the Gibbs state is an extremal KMS state. Finally, we notice that in the class of models studied by Emch and Knops [1970] the states that are usually interpreted as pure thermodynamical phases are indeed extremal KMS states. This tentative algebraic characterization of pure thermodynamical phases has been exploited in the formulation of a theory of crystallization,¶ and was shown to provide a rigorous basis for *Landau's argument* on the nonexistence of a critical point in the liquid-solid phase transition.

We now want to examine the question whether pure thermodynamical phases are ergodic and prove the following corollaries to Theorem 11.

Corollary 2. *Let ϕ be a separable KMS state on the dynamical system $\{\mathfrak{R}, \mathfrak{S}, \alpha\}$. Then the three conditions*

(i) *ϕ is extremal KMS,*

(ii) *ϕ is an extremal time-invariant state,*

(iii) *$\{\mathfrak{R}, \mathfrak{S}, \alpha\}$ is η-abelian on ϕ*

† See also Chapter 4. The relevance of this conjecture to the study of phase transitions has been the theme of a separate review (Emch [1971]).

‡ For an actual proof of this statement, see Wieringa [1970], Section 5.1; also Jadzyk [1969b].

§ See, for instance, Araki [1969].

¶ Emch, Knops, and Verboven [1970]; see also next subsection.

are in the following logical relation:

$$\text{(iii)} \Rightarrow \{\text{(i)} \Leftrightarrow \text{(ii)}\}$$

$$\text{(ii)} \Rightarrow \text{(iii) } \textit{and } \text{(i)}.$$

Proof. Assume (iii); by Corollary 2 to Theorem 7 we know that $\mathfrak{N}'_\phi = \pi_\phi(\mathfrak{R})'' \cap U_\phi(\mathbb{R})'$; hence $\mathfrak{N}'_\phi \subseteq \mathfrak{Z}_\phi(\mathfrak{R})$. On the other hand, we know from Corollary 2 to Theorem 10 that ϕ separable KMS implies $\mathfrak{Z}_\phi(\mathfrak{R}) \subseteq \mathfrak{N}'_\phi$. Consequently (iii) implies that $\mathfrak{N}'_\phi = \mathfrak{Z}_\phi(\mathfrak{R})$. From Corollary 1 to Theorem 11 (i) is equivalent to $\mathfrak{Z}_\phi(\mathfrak{R}) = \{\lambda I\}$; from Theorem 8 (ii) is equivalent to $\mathfrak{N}'_\phi = \{\lambda I\}$. Hence (iii) indeed implies that (i) and (ii) are equivalent. Now assume (ii); since ϕ is KMS, Φ is cyclic with respect to $\pi_\phi(\mathfrak{R})'$. We can therefore use the corollary to Theorem 8 to conclude that (iii) is satisfied; we can then use the first part of the proof to conclude that (i) is also satisfied, thus concluding the proof of the corollary. ∎

We first notice that under these circumstances E_ϕ is one-dimensional (see Theorem 8); hence ϕ is ergodic with respect to time. Should we want to conjecture that every pure thermodynamical phase ϕ for the natural temperature β is a separable KMS state for β *and* an extremal time-invariant state, we would, as the above corollary indicates, require implicitly that the dynamical system $\{\mathfrak{R}, \mathfrak{S}, \alpha\}$ be η-abelian on ϕ; we know, however, from several explicit examples† that even G-abelianness with respect to time can be violated; we therefore conclude that this alternate requirement on pure thermodynamical phases is not physically so well justified as our original one: pure thermodynamical phases are not necessarily ergodic in time.

We can, however, use part of the reasoning leading to Corollary 2 to prove the following result:

Corollary 3. *Let ϕ be a separable extremal KMS state on the dynamical system $\{\mathfrak{R}, \mathfrak{S}, \alpha\}$; let* G *be a symmetry group for $\{\mathfrak{R}, \mathfrak{S}, \langle;\rangle\}$ such that* (i) *ϕ is* G*-invariant*, (ii) G *is amenable, and* (iii) G *acts in an η-abelian manner on ϕ for some mean η on* G. *Then ϕ satisfies the nine conditions of Theorem* 8 *with respect to* G.

Proof. From Corollary 1 ϕ separable extremal KMS implies $\mathfrak{Z}_\phi(\mathfrak{R}) = \{\lambda I\}$. From Corollary 2 to Theorem 7 η-abelianness with respect to G implies that $\mathfrak{N}'_\phi \equiv \pi_\phi(\mathfrak{R})'' \cap U_\phi(G)' \subseteq \mathfrak{Z}_\phi(\mathfrak{R})$; hence $\mathfrak{N}'_\phi = \{\lambda I\}$, which, under the assumed η-abelianness with respect to G, is one of the nine equivalent conditions of Theorem 8. ∎

Physically this corollary can be used for the euclidian group G or those of its subgroups that act in a η-abelian manner on a given pure thermodynamic

† Radin [1970], Emch and Knops [1970], Narnhofer [1970].

phase of a quasi-local algebra. We discuss this point in more detail in the next subsection.

This corollary expresses in particular one of the simplest "spectrum properties" that can be derived for $U_\phi(G)$, namely, that E_ϕ is one-dimensional when ϕ is an extremal KMS state invariant under a symmetry group G. As another example of *spectrum properties* we mention the following result:

Theorem 13. *Let ϕ be a KMS state, invariant under a symmetry group* G, *which commutes with the time evolution. Then there exists an involutive antiunitary operator* C *acting on $\mathcal{H}_\phi$ such that* $\mathrm{CU}_\phi(\mathrm{g})\mathrm{C} = \mathrm{U}_\phi(\mathrm{g})$ *for all* g *in* G, *and the discrete spectrum of* $\mathrm{U}_\phi(\mathrm{G})$ *is symmetric. If, in addition,* $\mathrm{U}_\phi(\mathrm{G})$ *is an abelian, locally compact,* n-*parameter group, then its entire spectrum is symmetric.*

Proof. Let ν_ϕ be the antilinear representation associated with ϕ (see Theorem 9); from its definition we see that

$$\begin{aligned} U_\phi(g)\, \nu_\phi(R)\, \Phi(S) &= U_\phi(g)\, \Phi(SR^s) = \Phi(\alpha_g[SR^s]) \\ &= \Phi(\alpha_g[S]\, \alpha_g[R^s]) = \Phi(\alpha_g[S]\, (\alpha_g[R])^s) \\ &= \nu_\phi(\alpha_g[R])\, U_\phi(g)\, \Phi(S), \end{aligned}$$

i.e.,

$$\nu_\phi(\alpha_g[R]) = U_\phi(g)\, \nu_\phi(R)\, U_\phi(g)^{-1};$$

since, on the other hand,

$$\pi_\phi(\alpha_g[R]) = U_\phi(g)\, \pi_\phi(R)\, U_\phi(g)^{-1}$$

and

$$C\pi_\phi(R)C = \nu_\phi(R)$$

(where C is the involutive, antiunitary operator defined in Theorem 9), we conclude that

$$CU_\phi(g)\, \pi_\phi(R)\, U_\phi(g)^{-1}C = U_\phi(g)\, C\pi_\phi(R)\, CU_\phi(g)^{-1}.$$

On applying this on Φ and remembering that $U_\phi(g)\Phi = \Phi = C\Phi$, we conclude that $CU_\phi(g)C = U_\phi(g)$ for all g in G, thus proving the first part of the theorem. We now recall that a *character* χ on G (i.e., a unitary, one-dimensional, continuous representation of G on $\mathbb{C}$) is said to be a discrete point of the spectrum of $U_\phi(G)$ or equivalently to belong to the *discrete spectrum* of $U_\phi(G)$ whenever there exists a vector Ψ in $\mathcal{H}_\phi$ such that $U_\phi(g)\Psi = \chi(g)\Psi$ for all g in G. Suppose that χ belongs to the discrete spectrum of $U_\phi(G)$; we then have

$$U_\phi(g)C\Psi = CU_\phi(g)\Psi = C\chi(g)\Psi = \chi(g)^*C\Psi$$

so that χ^* is also a discrete point of the spectrum of $U_\phi(g)$, thus establishing the second part of the theorem. The third part of the theorem follows from the SNAG theorem,† since $U_\phi(g)$ is strongly continuous in g as a consequence of our definition of a symmetry group; we then have that $CU_\phi(g)C = U_\phi(g)$ for all g in G with C involutive, antiunitary implies $CE(\Delta)C = E(-\Delta)$ for every Borel set in $\mathbb{R}^n$. The remark concludes the proof of the theorem. ∎

This kind of spectrum properties has been the object of many studies; see for instance Robinson [1968], Winnink [1968], Jadczyk [1969a, b], Emch, Knops, and Verboven [1970], to mention only a few.

Having learned that pure thermodynamical phases generate primary representations, we might surmise that the determination of the types of these representations should be a more immediately tractable problem than in the case of a general equilibrium state. The first general theorem in this direction has been proved by Hugenholtz [1967], who showed that only factors of type III occur as long as β is finite. We give below a slight variation on Hugenholtz's theorem, and explain, in two remarks following the proof, the physical reasons behind the formulation chosen here.

Theorem 14. *Let ϕ be a separable, extremal KMS state on a dynamical system $\{\mathfrak{R}, \mathfrak{S}, \alpha\}$. If ϕ is invariant with respect to an amenable symmetry group* G *which acts in an η-abelian manner on ϕ, then either of the following two alternatives occurs*:

(i) $\pi_\phi(\mathfrak{R})''$ *is a type* III *factor.*
(ii) *ϕ is a trace on $\pi_\phi(\mathfrak{R})''$, $\pi_\phi(\mathfrak{R})''$ is a factor of type* II_1 *or* I_n, and $\pi_\phi(\alpha_\mathrm{t}[\mathrm{R}]) = \pi_\phi(\mathrm{R})$ *for all* R *in* $\mathfrak{R}$ *and all times* t.

Proof. From Corollary 3 to Theorem 11 ϕ satisfies the nine equivalent conditions of Theorem 8 with respect to G. Suppose now that $\pi_\phi(\mathfrak{R})''$ is not of type III. Since it is a factor (Corollary 1 to Theorem 11), $\pi_\phi(\mathfrak{R})''$ is semifinite (see the first lemma on p. 129). By definition there then exists a semifinite, faithful normal trace ψ on $\pi_\phi(\mathfrak{R})''_+$, which is unique up to a (positive) multiplicative constant,‡ since $\pi_\phi(\mathfrak{R})''$ is a factor. The fact that ψ is normal implies that there exists a family $\{\Psi_i\}$ of vectors in $\mathcal{H}_\phi$ such that $\langle\psi; X\rangle = \Sigma(\Psi_i, X\Psi_i)$ for all X in $\pi_\phi(\mathfrak{R})''_+$. We now form for each g in G and each X in $\pi_\phi(\mathfrak{R})''_+$

$$\langle\psi_g; X\rangle \equiv \langle\psi; U_\phi(g)\, XU_\phi(g)^{-1}\rangle,$$

which is again a semifinite, faithful normal trace on $\pi_\phi(\mathfrak{R})''_+$, hence is proportional to ψ. There then exists a mapping $\lambda: G \to \mathbb{R}^+$ such that $\psi_g = \lambda(g)\psi$, hence $\lambda(g_1)\,\lambda(g_2) = \lambda(g_1 g_2)$ for all g_1, g_2 in G, and $\lambda(e) = 1$, where e is

† SNAG stands for Stone [1932], Naimark [1943], Ambrose [1944], and Godement [1944]; in the most unsophisticated form, which we need here, this theorem states that if $U(G)$ is a strongly continuous unitary representation of a locally compact, abelian, n-parameter group G, then there exists a unique spectral family E on the character-group $\hat{G}$ of G such that $U(g) = \int_{\hat{G}} \chi(g)\, dE(\Delta)$ and $E(\Delta) \in \{U(G)\}''$.

‡ See Dixmier [1957], I.6.4, Corollary to Theorem 3.

the identity in G. We now want to prove that $\lambda(g) = 1$ for all g in G; suppose that this were not the case. Since $\lambda(g)\lambda(g^{-1}) = 1$, we could assume without loss of generality that $\lambda(g) < 1$. For any arbitrary finite projector E in $\pi_\phi(\mathfrak{R})''$ we form

$$\langle \psi; U_\phi(g)^n \, EU_\phi(g)^{-n} \rangle = \lambda(g)^n \langle \psi; E \rangle;$$

since E is finite, $\langle \psi; E \rangle < \infty$. Together with $\lambda(g) < 1$, the above equality would imply that $\langle \psi; U_\phi(g)^n \, EU_\phi(g)^{-n} \rangle \to 0$ as $n \to \infty$. Since ψ is normal, this would imply in turn that, for every finite projector F in $\pi_\phi(\mathfrak{R})''$, $U_\phi(g)^n \, EU_\phi(g)^{-n}F \to 0$ strongly, hence $U_\phi(g)^n \, EU_\phi(g)^{-n} \to 0$ strongly. Consequently $\langle \phi; U_\phi(g)^n \, EU_\phi(g)^{-n} \rangle \to 0$. We recall, however, that ϕ is G-invariant by assumption so that we can conclude from the above remark that $\langle \phi; E \rangle$ would vanish for every finite projector E in $\pi_\phi(\mathfrak{R})''$. We now use the KMS condition again to conclude that Φ is separating for $\pi_\phi(\mathfrak{R})''$, so that E finite would imply $E = 0$, since we have just noticed that $\langle \phi; E \rangle = 0$. This contradicts our assumption that $\pi_\phi(\mathfrak{R})''$ is a semifinite factor. Consequently $\lambda(g) = 1$ for all g in G; i.e., ψ is a G-invariant trace on $\pi_\phi(\mathfrak{R})''$. We use this fact to conclude that for every finite projector E in $\pi_\phi(\mathfrak{R})''$

$$\begin{aligned} \infty > \langle \psi; E \rangle &= \langle \psi; U_\phi(g) \, EU_\phi(g)^{-1} \rangle \\ &= \langle \psi; \eta_\phi(E) \rangle = \{\psi; \langle \phi; E \rangle I \rangle, \end{aligned}$$

where the second equality is due to the fact that ψ is normal and the third is due to (vii) in Theorem 8, which is satisfied by the remark opening the present proof. Since Φ is separating for $\pi_\phi(\mathfrak{R})''$, we have, for a nonzero finite E in $\pi_\phi(\mathfrak{R})''$, $\langle \phi; E \rangle \neq 0$, hence

$$0 < \langle \psi; \lambda I \rangle < \infty \quad \text{for some } \lambda \neq 0.$$

We can therefore normalize ψ to a *state* on $\pi_\phi(\mathfrak{R})''$. This state, which we denote again by ψ, is normal and G-invariant. By (viii) in Theorem 8 this implies that $\phi = \psi$, which is by construction a trace on $\pi_\phi(\mathfrak{R})''$. This proves the first part of the theorem. From the construction of ψ, ϕ is a faithful, normal, and evidently finite trace on $\pi_\phi(\mathfrak{R})''$. Hence $\pi_\phi(\mathfrak{R})''$ is either type II_1 or type I_n. Since ϕ is a trace, $F_{RS}(t) = G_{RS}(t)$, and we can conclude from the now familiar argument that the KMS condition $F_{RS}(t + i\beta) = G_{RS}(t)$ implies that $F_{RS}(t)$ is constant in t, i.e., explicitly,

$$\langle \phi; S\alpha_t[R] \rangle = \langle \phi; SR \rangle$$

or

$$(\Phi, \pi_\phi(S)\{U_\phi(t) - I\} \, \pi_\phi(R)\Phi) = 0.$$

Since Φ is cyclic with respect to $\pi_\phi(\mathfrak{R})$, this implies $U_\phi(t) = I$ for all times t, thus concluding the proof of the theorem. ∎

REMARKS.

(i) From Theorem 8, its corollary, and Corollary 3 to Theorem 11 we could equivalently have replaced, in the statement of the present theorem, the assumption that G acts in an η-abelian manner on ϕ by the assumption that ϕ is extremal G-invariant or that

$$E^G_\phi \mathcal{H}_\phi \equiv \{\Psi \in \mathcal{H}_\phi \mid U_\phi(g)\Psi = \Psi \ \forall\, g \in G\}$$

is one-dimensional.

(ii) Hugenholtz [1967] gave the original proof that ϕ extremal KMS implies the above alternative under the additional assumption that

$$E^{\mathbb{R}}_\phi \mathcal{H}_\phi \equiv \{\Psi \in \mathcal{H}_\phi \mid U_\phi(t)\Psi = \Psi \ \forall\, t \in \mathbb{R}\}$$

is one-dimensional. From remark (i) this amounts to assuming that the time evolution acts in a η-abelian manner on ϕ; we already expressed our reservations about this assumption in our discussion of Corollary 2 to Theorem 11. The mathematically minor difference involved in not identifying G with the time evolution has nevertheless the physical advantage of letting us choose for G the euclidian group or one of its subgroups which act in an η-abelian manner on ϕ as a consequence of locality (see Chapter 4).

(iii) To our knowledge the fact that the separation of G from the time evolution did not hamper the proof of the theorem was first published by Størmer [1967b]; the proof, as we gave it, is a slight variation on a proof given by Størmer [1969b] for the following result: "Let $\mathfrak{M}$ be a semi-finite factor acting on some Hilbert space $\mathcal{H}$. Let $\mathcal{U}$ be a group of unitary operators on $\mathcal{H}$ such that $U\mathfrak{M}U^{-1} = \mathfrak{M}$ for all $U \in \mathcal{U}$. Suppose there exists a unit vector $\Psi \in \mathcal{H}$ such that (1) Ψ is separating for $\mathfrak{M}$; (2) $\{\lambda\Psi \mid \lambda \in \mathbb{C}\}$ equals the set of vectors $\Phi \in \mathcal{H}$ such that $U\Phi = \Phi$ for all U in $\mathcal{U}$. Then $\mathfrak{M}$ is finite with trace $\mathrm{Tr}\,(A) = (\Psi, A\Psi)$ for all $A \in \mathfrak{M}$." We should notice in particular that Størmer gives his proof in such a manner that he does not have to require that his group be amenable, let alone that it act in a η-abelian manner. This added generality would be especially useful if we wanted to consider relativistic field theories in which evidently the KMS condition on ϕ has to be replaced by the assumption that Φ is a (cyclic and) separating vector for the factor $\pi_\phi(\mathfrak{R})$. For the purpose of statistical mechanics, however, the slightly simpler proof given here is still valid under assumptions that are general enough to cover the cases of interest.

(iv) It must be pointed out that the proof is strictly valid for finite temperatures only; for $\beta = 0$ (or $\beta = \infty$) the strip $0 \leqslant \mathrm{Im}\, z \leqslant \beta$ is reduced to the real line (or covers the whole complex plane), and the argument on the triviality of the time evolution miscarries so that the occurrence of primary representations of finite type cannot be eliminated for $\beta = 0$ and for $\beta = \infty$. Actually type II_1 is the only one likely to occur in a system with an infinite number of degrees of freedom in equilibrium at infinite temperature T.

(v) We might mention as a last remark to this theorem that it has been known for some time† that the Gibbs state of the infinite free Bose gas above the Bose-Einstein condensation generates a primary representation of type III.

We now want to conclude this subsection by briefly mentioning three recent developments.

First, we have assumed throughout this subsection that the time evolution of an infinite system can be described as a continuous group of automorphisms of $\mathfrak{R}$. We should realize, however, that this is by no means a trivial assumption, since the time evolution of an infinite system is defined by a limiting procedure, starting from the consideration of finite systems for which the Hamiltonian can be defined, and is directly interpretable. Stated a little bit differently, the problem then is to take the thermodynamic limit in a manner that is really consistent with the dynamics. This consistency has been recognized as being easily secured for a rather wide class of nontrivial quantum lattice systems‡ in which the time evolution has been shown to be the automorphism of the algebra $\mathfrak{R}$ we would naturally associate with the infinite system. This is not the case in general, however; for instance, it has recently been shown§ that the time evolution of an infinite, free Bose gas cannot be described as an automorphism of $\mathfrak{R}$ but can, however, be described as an automorphism of the von Neumann algebra generated by the representation associated with the Gibbs state (above the critical temperature). The same phenomena occurs for the BCS model¶ and for a class of generalized Weiss models for ferro- and antiferromagnetism;‖ in the latter case it has also been shown that the time evolution defined for each phase separately is consistent with the time evolution defined for the Gibbs state they decompose. It has been possible, for all these cases, to make the appropriate generalization of the KMS condition and to extend there most of the results of the present subsection—in particular the Commutant Theorem 9.

The second of the results we want to mention is the fact that if a classical lattice gas is embedded in the corresponding quantum lattice†† the definition of equilibrium states via the KMS condition is equivalent to the classical

† Araki [1964c] whose proof is based on different techniques.

‡ See Subsection IV.2.a.

§ Dubin and Sewell [1970], Wieringa [1970].

¶ Dubin and Sewell [1970]; see also Thirring and Wehrl [1967] and Thirring [1968].

‖ Emch and Knops [1970].

†† Brascamp [1970].

definitions of equilibrium states;† again the problem of the definition of the time evolution for the infinite system is discussed, and it is found that the time-evolution for classical interactions exists as an automorphism of the quantum algebra of observables under conditions that are weaker than those found for quantum interactions.

We finally want to mention a third development, initiated by the discovery‡ that to every faithful normal state ϕ on a von Neumann algebra $\mathfrak{N}$ there always corresponds a unique one-parameter group α_t of automorphisms of $\mathfrak{N}$ with respect to which ϕ satisfies the KMS boundary condition for $\beta = 1$. The physical significance of this remark does not yet seem to have been completely assessed, although some interesting results have recently appeared,§ based on this remark and using the mathematical techniques that lay at its foundation.

f. Decomposition Theory

Let $\mathfrak{S}_\rho$ denote the set $\mathfrak{S}$ of all states on a physical system, or the set $\mathfrak{S}_G$ of all states that are G-invariant under a symmetry group G of this system, or also the set $\mathfrak{S}_\beta$ of all KMS states on a dynamical system for a fixed natural temperature β. These convex sets have in common the feature that they are closed in the w^*-topology of $\mathfrak{R}^*$, bounded in the metric topology, hence compact in the w^*-topology. As we have already noted, this implies that $\mathfrak{S}_\rho$ is the w^*-closure of the convex hull of its extreme points; we denote by $\mathcal{E}_\rho$ the set of all extreme points of $\mathfrak{S}_\rho$. More generally, $\mathfrak{S}_\rho$ will also denote any compact, convex subset of a locally convex linear space $\mathfrak{X}$.

In this subsection we consider the question whether it is possible to decompose, in a unique way, every element ϕ in $\mathfrak{S}_\rho$ into its extremal components ψ_γ in $\mathcal{E}_\rho$.

The assumption that $\mathfrak{S}_\rho$ is a compact convex set clearly does not, by itself, suffice to decide the issue: the interval $|x| \leqslant 1$ in $\mathbb{R}$ and the disk $|\omega| \leqslant 1$ in $\mathbb{C}$ both satisfy this assumption; yet the decomposition $x = \lambda \cdot 1 + (1 - \lambda)(-1)$ (with $0 \leqslant \lambda \leqslant 1$) is unique, whereas this uniqueness is lost in the second case: every secant passing through ω_0 with $|\omega_0| < 1$ determines two points ω_1 and ω_2 on the circumference such that $\omega_0 = \lambda\omega_1 + (1 - \lambda)\omega_2$ (with $0 < \lambda < 1$).

We have therefore to appeal, if we want to establish uniqueness, to some additional property of $\mathfrak{S}_\rho$.

Let us first consider the case of $\mathfrak{S}$. Here the decomposition problem we want to discuss reduces to that of decomposing an arbitrary state ϕ into its

† Lanford and Ruelle [1969]; I am told that Lanford has now more general results on this equivalence (Gallavotti, private communication).
‡ Takesaki [1970b].
§ Takesaki [1970a], Hermann and Takesaki [1970]. Takesaki and Winnink [unpublished].

"pure state components." This decomposition is clearly *not* unique in general, as explicitly shown by the following counterexample: let $\mathcal{H}_n$ be a Hilbert space of finite dimension n and let ϕ be the state on $\mathfrak{B}(\mathcal{H}_n)$ defined by $\langle \phi; R\rangle = (1/n)\,\mathrm{Tr}\, R$; in this case any orthonormal basis $\{\Phi_i\}$ in $\mathcal{H}_n$ generates a convex decomposition:

$$\phi = \frac{1}{n}\sum_i \phi_i \quad \text{with} \quad \langle \phi_i; R\rangle = (\Phi_i, R\Phi_i)$$

of ϕ into pure states. The situation would be quite different if, instead of considering ϕ as a state on $\mathfrak{B}(\mathcal{H}_n)$, we had considered it as a state on the abelian algebra generated by all the hermitian operators in $\mathfrak{B}(\mathcal{H}_n)$ which are diagonalizable in a given fixed basis $\{\Phi_i^0\}$ in $\mathcal{H}_n$; in this new case the decomposition written above is clearly unique. Actually every state ρ on this algebra can be written as

$$\rho = \sum_{i=1}^{n} \lambda_i \phi_i,$$

thus providing a unique decomposition of ρ into its pure state components. This last feature is not an accident linked to the extreme simplicity of the particular example just discussed; the existence and uniqueness of the decomposition into pure state components of an arbitrary state is actually a characteristic of *abelian* C^*-algebras, a fact that we could surmise from Theorem I.2.9. The lesson to be learned from these preliminary remarks is that we should not in general expect a positive answer to our decomposition problem: some condition reflecting the abelianness of the trivial case discussed above will have to be imposed in order to get uniqueness. We now want to indicate that this element of "abelianness" is already present when we deal with $\mathfrak{S}_\beta$ or with $\mathfrak{S}_G$ when G acts in a G-abelian manner. The rigorous proof of this statement calls for some technical refinements which we want to postpone until after the following elementary remarks.

Let us now consider the case in which ϕ is a KMS state on the dynamical system $\{\mathfrak{R}, \mathfrak{S}, \alpha\}$ such that $\mathcal{H}_\phi$ is separable. Since $\mathfrak{Z}_\phi(\mathfrak{R})$ is an *abelian* von Neumann algebra on a separable Hilbert space, it is generated by a single hermitian operator Z_ϕ. Now suppose, in our first attempt to work our way through the theory, that Z_ϕ has a discrete spectrum; we can then write

$$\phi = \left\{\sum_i z_i P_i \mid z_i \in \mathbb{C}\right\},$$

where P_i are the eigenprojectors of Z_ϕ. Denoting by $\tilde{\phi}$ the normal extension of ϕ from $\pi_\phi(\mathfrak{R})$ to $\pi_\phi(\mathfrak{R})''$, we form

$$\langle \psi_i; R\rangle = \langle \tilde{\phi}; P_i R\rangle / \langle \tilde{\phi}; P_i\rangle,$$
$$\lambda_i = \langle \tilde{\phi}; P_i\rangle.$$

Since ϕ is a separable KMS state, $\tilde{\phi}$ is again a KMS state and Φ is separating for $\pi_\phi(\mathfrak{R})''$; on the other hand, since the projectors P_i are nonzero, we can conclude that the λ_i are strictly positive. Since $\sum_i P_i = I$, we then have $\sum_i \lambda_i = 1$. Since the P_i belong to $\mathfrak{Z}_\phi(\mathfrak{R})$ and ϕ is a separable KMS state, the ψ_i are again separable KMS states. We now want to show that they are extremal KMS states. Since P_i belongs to $\mathfrak{Z}_\phi(\mathfrak{R})$, $P_i\pi_\phi(\mathfrak{R})P_i$ is a representation of $\mathfrak{R}$ on the (separable) Hilbert space $P_i\mathcal{H}_\phi$; furthermore $\Psi_i \equiv P_i\Phi$ is cyclic in $P_i\mathcal{H}_\phi$ for this representation, and $(\Psi_i, P_i\pi_\phi(\mathfrak{R})P_i\Psi_i) = \langle\psi_i; R\rangle$. Consequently $P_i\pi_\phi(\mathfrak{R})P_i$ is unitarily equivalent to the canonical representation $\pi_{\psi_i}(\mathfrak{R})$ generated by ψ_i. To show the extremal character of the ψ_i's it is therefore sufficient to show that in $P_i\mathcal{H}_\phi$ the center of the von Neumann algebra $\{P_i\pi_\phi(\mathfrak{R})P_i\}''$ is trivial. This, however, is easy to show:

$$\begin{aligned}\{P_i\pi_\phi(\mathfrak{R})P_i\}'' \cap \{P_i\pi_\phi(\mathfrak{R})P_i\}' &= P_i\{\pi_\phi(\mathfrak{R})'' \cap \pi_\phi(\mathfrak{R})'\}P_i \\ &= P_i\mathfrak{Z}_\phi(\mathfrak{R})P_i \\ &= \{\lambda P_i \mid \lambda \in \mathbb{C}\}.\end{aligned}$$

We have therefore obtained a decomposition

$$\phi = \sum_i \lambda_i\psi_i$$

of ϕ into extremal KMS states. We further notice that if Z is any element of $\mathfrak{Z}_\phi(\mathfrak{R})$, i.e., $Z = \sum_i z_i P_i$, we have

$$\langle\tilde{\phi}; ZR\rangle = \sum_i z_i\lambda_i\langle\psi_i; R\rangle,$$

which proves from Theorem 11 that the decomposition of ϕ obtained above is unique.

This particular result evidently falls short of establishing the uniqueness of the decomposition of a general ϕ in $\mathfrak{S}_\beta$ into its extremal components. It is nevertheless worth remarking that this result has been obtained under rather weak restrictions, namely that $\mathcal{H}_\phi$ is separable and that the spectrum of Z_ϕ is discrete. The merit of this simple exploratory example is to show the direction we should take to lift these technical restrictions and solve the general case. To see this we now express the above result in the following alternate form: Let μ_ϕ be the measure defined on $\mathfrak{S}$ by $\mu_\phi = \sum_i \lambda_i\delta_{\psi_i}$; clearly μ_ϕ is concentrated on $\mathfrak{E}_\beta$; for every $Z\,(= \sum_i z_i P_i)$ in $\mathfrak{Z}_\phi(\mathfrak{R})$ we define the function φ_z in $\mathfrak{L}^\infty_{\mathbb{C}}(\mathfrak{S}, \mu_\phi)$ by $\varphi_z(\psi_i) = z_i$ and notice that $Z \to \varphi_z$ is an isomorphism from $\mathfrak{Z}_\phi(\mathfrak{R})$ to $\mathfrak{L}^\infty(\mathfrak{S}, \mu_\phi)$. We can now write

$$\langle\tilde{\phi}; ZR\rangle = \int_{\mathfrak{S}} \varphi_Z(\psi)\langle\psi; R\rangle \, d\mu_\phi(\psi),$$

which, as we shall presently see, holds under more general circumstances.

The general theory depends heavily on the following result established by Wils [1968, 1969]. For every state ϕ on a C^*-algebra $\mathfrak{R}$ there exists a unique measure μ_ϕ on $\mathfrak{S}$, called the *central measure* of ϕ such that there is a σ-continuous isomorphism φ from $\mathfrak{Z}_\phi(\mathfrak{R})$ onto $\mathfrak{L}^\infty(\mathfrak{S}, \mu_\phi)$ with the property

$$\langle \tilde{\phi}; ZR\rangle = \int_{\mathfrak{S}} \varphi_Z(\psi)\langle \psi; R\rangle \, d\mu_\phi(\psi)$$

for all R in $\mathfrak{R}$ and all Z in $\mathfrak{Z}_\phi(\mathfrak{R})$; in this expression $\tilde{\phi}$ denotes the normal extension of ϕ from $\pi_\phi(\mathfrak{R})$ to $\pi_\phi(\mathfrak{R})''$. Furthermore, this measure is concentrated in the Baire sense on the set $\mathfrak{F}$ of all primary states on $\mathfrak{R}$; i.e., $\mu_\phi(\mathfrak{C}) = 0$ for every Baire† set $\mathfrak{C}$ in $\mathfrak{S}$ with $\mathfrak{C} \cap \mathfrak{F} = \emptyset$. Finally, the connection with the usual *central disintegration of a representation*‡ is established as follows. There exists an isometric mapping U from $\mathcal{H}_\phi$ onto

$$\mathcal{H}_\mu = \int_{\mathfrak{S}} \mathcal{H}_\psi \, d\mu_\phi(\psi),$$

which establishes a unitary equivalence between π_ϕ and the representation π_μ defined by

$$\pi_\mu(R) = \int_{\mathfrak{S}} \pi_\psi(R) \, d\mu_\phi(\psi),$$

and sends the center $\mathfrak{Z}_\phi(\mathfrak{R})$ of $\pi_\phi(\mathfrak{R})''$ onto the von Neumann algebra of all *diagonal operators* on $\mathcal{H}_\mu$, i.e., the algebra of all operators $T \in \mathfrak{B}(\mathcal{H}_\mu)$ of the form

$$T = \int_{\mathfrak{S}} T_\psi \, d\mu_\phi(\psi),$$

where $T_\psi = \lambda(\psi) I_\psi$ with $\lambda(\psi) \in \mathfrak{L}^\infty(\mathfrak{S}, \mu_\phi)$ and I_ψ denotes the identity operator on $\mathcal{H}_\psi$. It should be noted, and this is the fundamental contribution of Wils, that the definition of a central measure, its existence and uniqueness, as well as those of its properties mentioned above, do not depend on the separability of $\mathfrak{R}$ for the norm-topology. This is a physically important improvement over a theory that assumes§ the separability of $\mathfrak{R}$, since the C^*-algebra of the canonical commutation relations (see Chapter 3) is not norm-separable. When $\mathfrak{R}$ is norm-separable, $\mathfrak{S}$ is metrizable in the weak*-topology; let $\{A_n\}$ be a sequence of nonzero elements, which is norm-dense in the self-adjoint portion $\mathfrak{A}$ of $\mathfrak{R}$. A metric d can then be defined, which

† For the definition of Baire sets and their relation to Borel sets the reader is referred back to p. 144 where he will also find the bibliography related to these concepts.
‡ See Dixmier [1964], Section 8.4, Sakai [1965].
§ See Sakai [1965] whose results Wils generalized to the nonseparable case.

generates the weak*-topology on $\mathfrak{S}$, namely,

$$d(\psi, \phi) = \sum_{n=1}^{\infty} \frac{|\langle \phi - \psi; A_n \rangle|}{2^n \|A_n\|};$$

Baire and Borel sets in $\mathfrak{S}$ coincide and then μ_ϕ is concentrated in the Borel sense of $\mathcal{F}$. The fact that, in general, μ_ϕ is only concentrated in the Baire sense on $\mathcal{F}$ is the price we should pay for the generalization to the case in which $\mathfrak{R}$ is not norm-separable.

The central measure μ_ϕ is the technical device that allows us to generalize to arbitrary separable KMS states our earlier particular result on the uniqueness of the decomposition of ϕ into its extremal KMS components.

Lemma. *Let ϕ be a separable KMS state; every state in the support of μ_ϕ is again a KMS state.*

Proof. Let M be any μ_ϕ-measurable subset of $\mathfrak{S}$ and let χ_M be its characteristic function; i.e., $\chi_M(\psi) = 1$ or 0, depending on whether ψ belongs to M. Since χ_M belongs to $\mathfrak{L}^\infty(\mathfrak{S}, \mu_\phi)$, we can send it back to $\mathfrak{Z}_\phi(\mathfrak{R})$ via the isomorphism φ^{-1}. Since χ_M is nonzero and positive, so then is $\varphi^{-1}(\chi_M)$; hence $\langle \tilde{\phi}; \varphi^{-1}(\chi_M) \rangle \neq 0$, since Φ is separating for $\pi_\phi(\mathfrak{R})''$. From Theorem 11 we conclude therefore that ϕ_M defined by $\langle \phi_M; R \rangle = \langle \tilde{\phi}; \varphi^{-1}(\chi_M) R \rangle$ is a positive linear functional on $\mathfrak{R}$ satisfying the KMS condition; i.e.,

$$\begin{aligned} \langle \hat{F}^M_{RS}; f \rangle &\equiv \langle \phi_M; S\langle X_R; \hat{f} \rangle \rangle \\ &= e_\beta \langle \hat{G}^M_{RS}; f \rangle \\ &= e_\beta \langle \phi_M; \langle X_R; \hat{f} \rangle S \rangle, \end{aligned}$$

or

$$\langle \tilde{\phi}; \varphi^{-1}(\chi_M)\, S\langle X_R; \hat{f} \rangle \rangle = e_\beta \langle \tilde{\phi}; \varphi^{-1}(\chi_M) \langle X_R; \hat{f} \rangle S \rangle.$$

From the definition of the central measure this implies that

$$\int_{\mathfrak{S}} \chi_M(\psi) \langle \psi; S\langle X_R; \hat{f} \rangle \rangle \, d\mu_\phi(\psi) = \int_{\mathfrak{S}} \chi_M(\psi) e_\beta \langle \psi; \langle X_R; \hat{f} \rangle S \rangle \, d\mu_\phi(\psi);$$

i.e.,

$$\int_M \{ \langle \hat{F}^\psi_{RS}; f \rangle - e_\beta \langle \hat{G}^\psi_{RS}; f \rangle \} \, d\mu_\phi(\psi) = 0$$

for all M. We can therefore conclude that $\hat{F}^\psi_{RS} = e_\beta \hat{G}^\psi_{RS}$ (since these functions are continuous in ψ for the w^*-topology), i.e., ψ belongs to $\mathfrak{S}_\beta$. ∎

We can therefore write for every separable KMS state

$$\langle \tilde{\phi}; ZR \rangle = \int_{\mathfrak{S}_\beta} \varphi_Z(\psi) \langle \psi; R \rangle \, d\mu_\phi(\psi).$$

To prove that μ_ϕ is concentrated on $\mathfrak{S}_\beta$ we need some rudiments of Choquet theory, which we now survey† briefly.

Let $\mathfrak{S}_\rho$ be a compact convex subset of a Hausdorff locally convex linear space $\mathfrak{X}$. A real-valued continuous function f on $\mathfrak{S}_\rho$ is said to be *convex* if

$$f(\lambda\phi_1 + (1 - \lambda)\phi_2) \leqslant \lambda f(\phi_1) + (1 - \lambda)f(\phi_2)$$

for all ϕ_1, ϕ_2 in $\mathfrak{S}_\rho$ and all λ in $[0, 1]$; the set of all convex functions on $\mathfrak{S}_\rho$ is denoted by $\mathcal{C}(\mathfrak{S}_\rho)$. We denote by $\mathcal{A}(\mathfrak{S}_\rho)$ the set $\mathcal{C}(\mathfrak{S}_\rho) \cap -\mathcal{C}(\mathfrak{S}_\rho)$ of all *affine* functions; we notice that $f \in \mathcal{A}(\mathfrak{S}_\rho)$ if and only if

$$f(\lambda\phi_1 + (1 - \lambda)\phi_2) = \lambda f(\phi_1) + (1 - \lambda)f(\phi_2)$$

for all ϕ_1, ϕ_2 in $\mathfrak{S}_\rho$ and all λ in $[0, 1]$. We denote by $\mathcal{M}(\mathfrak{S}_\rho)$ the set of all positive *Radon measures* μ on $\mathfrak{S}_\rho$ and by $\langle\mu; f\rangle$ the integral $\int f(\psi)\, d\mu(\psi)$ with f in $\mathcal{C}(\mathfrak{S}_\rho)$. We say that ν in $\mathcal{M}(\mathfrak{S}_\rho)$ is *majorized* by μ in $\mathcal{M}(\mathfrak{S}_\rho)$, which we denote by $\nu \prec \mu$, if $\langle\nu; f\rangle \leqslant \langle\mu; f\rangle$ for all f in $\mathcal{C}(\mathfrak{S}_\rho)$. Clearly, $\nu \prec \mu$ implies that $\langle\nu; f\rangle = \langle\mu; f\rangle$ for all f in $\mathcal{A}(\mathfrak{S}_\rho)$; whenever the latter situation occurs, we say that ν and μ have the same *resultant*, which we denote $\nu \sim \mu$. In particular, we say that μ in $\mathcal{M}(\mathfrak{S}_\rho)$ *represents* a point ϕ in $\mathfrak{S}_\rho$ whenever $\mu \sim \delta_\phi$, i.e., $\int f(\psi)\, d\mu(\psi) = f(\phi)$ for all f in $\mathcal{A}(\mathfrak{S}_\rho)$. The relation $\nu \prec \mu$ defines a *partial ordering* in $\mathcal{M}(\mathfrak{S}_\rho)$, and a measure is said to be *maximal* if it is maximal with respect to this partial ordering.

Our interest in maximal measures arises from the following heuristic argument. A convex function assumes its maximum values at the extreme points of $\mathfrak{S}_\rho$, so that $\nu \prec \mu$ means that the "mass" of μ is concentrated closer to the extreme points than that of ν; another way to verbalize this is to say that μ is "more diffuse" than ν. Hence maximal measures are the closest we can get to having a measure concentrated on the extreme points. The reader unfamiliar with these notions is invited to consider again the interval $|x| \leqslant 1$ in $\mathbb{R}$ and the circle $|z| \leqslant 1$ in $\mathbb{C}$.

The first result we should know is that every measure ν in $\mathcal{M}(\mathfrak{S}_\rho)$ is majorized by some maximal measure in $\mathcal{M}(\mathfrak{S}_\rho)$; in particular, every point in $\mathfrak{S}_\rho$ can be represented at least by one maximal measure.‡ Since $\mathfrak{S}_\beta$ is a compact convex subset of a Hausdorff, locally convex linear space (namely $\mathfrak{R}^*$), we know that there is at least a solution, in terms of maximal measures, to our decomposition problem. We now want to show that there are important physical cases in which this solution is unique.

† See Phelps [1966] or Choquet [1969], Chapter 6, for textbook expositions; the reader interested in the original papers should read Choquet and Meyer [1963] and Bishop and deLeeuw [1959].

‡ For a proof, see for instance Choquet [1969], Proposition 26.9.

Theorem 15. *The central measure* μ_ϕ *of a separable KMS state* ϕ *is the unique measure on* $\mathfrak{S}_\beta$ *which is maximal and represents* ϕ.

Proof. Let ν be a measure representing ϕ on $\mathfrak{S}_\beta$. For each f in $\mathcal{C}(\mathfrak{S}_\beta)$ and each $\varepsilon > 0$ there exists† a discrete measure

$$\nu_{f,\varepsilon} = \sum_{i=1}^{n} \lambda_i \delta_{\phi i},$$

with $\lambda_i > 0$, $\sum_{i=1}^{n} \lambda_i = 1$, and $\phi_i \in \mathfrak{S}_\beta$, which satisfies

$$\langle \nu; f\rangle - \langle \nu_{f,\varepsilon}; f\rangle \leqslant \varepsilon$$

and represents ϕ so that we have, in particular, for every R in $\mathfrak{R}$

$$\langle \phi; R\rangle = \sum_{i=1}^{n} \lambda_i \langle \phi_i; R\rangle.$$

Since ϕ_i belongs to $\mathfrak{S}_\beta$, it determines uniquely by Theorem 11 a positive element Z_i in $\mathfrak{Z}_\phi(\mathfrak{R})$ such that

$$\langle \phi_i; R\rangle = \langle \tilde{\phi}; Z_i R\rangle = \int_{\mathfrak{S}_\beta} \varphi_{Z_i}(\psi)\langle \psi; R\rangle \, d\mu_\phi(\psi)$$
$$= \int_{\mathfrak{S}_\beta} \langle \psi; R\rangle \, d\mu_i(\psi),$$

where $d\mu_i(\psi) \equiv \varphi_{Z_i}(\psi)\, d\mu_\phi(\psi)$. We use this relation, which we can write

$$\phi = \int_{\mathfrak{S}_\beta} \psi \, d\mu_i$$

to conclude that the second inequality below holds:

$$\langle \nu; f\rangle - \varepsilon \leqslant \langle \nu_{f,\varepsilon}; f\rangle = \sum_{i=1}^{n} \lambda_i f(\phi_i) \leqslant \sum_{i=1}^{n} \lambda_i \int_{\mathfrak{S}_\beta} f(\psi)\, d\mu_i(\psi)$$
$$= \int_{\mathfrak{S}_\beta} \sum_{i=1}^{n} \lambda_i \, \varphi_{Z_i}(\psi) f(\psi)\, d\mu_\phi(\psi);$$

since Φ is separating on $\pi_\phi(\mathfrak{R})''$, we have

$$\sum_{i=1}^{n} \lambda_i Z_i = I.$$

Furthermore since φ is an isomorphism from $\mathfrak{Z}_\phi(\mathfrak{R})$ to $\mathcal{L}^\infty(\mathfrak{S}_\beta, \mu_\phi)$, we have

$$\sum_{i=1}^{n} \lambda_i \varphi_{Z_i}(\psi) = 1;$$

† See Choquet [1969], Lemma 26.14.

hence our previous majorization becomes

$$\langle \nu; f\rangle - \varepsilon \leqslant \int_{\mathfrak{S}_\beta} f(\psi)\, d\mu_\phi(\psi) \equiv \langle \mu_\phi; f\rangle.$$

Since $\varepsilon > 0$ and f in $\mathcal{C}(\mathfrak{S}_\beta)$ were chosen arbitrarily, we conclude that $\nu < \mu_\phi$. Since ν was only assumed to be a measure representing ϕ on $\mathfrak{S}_\beta$, we conclude that μ_ϕ majorizes every other measure representing ϕ on $\mathfrak{S}_\beta$, thus proving the theorem.

REMARKS.

(i) The conclusion of the theorem has been reached under the assumption that ϕ be separable. As we mentioned when we introduced this notion, we can bring forward some physical justification for this assumption. From the mathematical point of view, however, we notice that this assumption might become redundant; for instance, although the C^*-algebra of the canonical commutation relations is not norm-separable, the C^*-algebra of the canonical anticommutation relations is norm-separable. The latter condition is sufficient to ensure that $\mathcal{H}_\phi$ will be separable, hence that every state is separable, since the unit ball of a von Neumann algebra acting on a separable Hilbert space is metrizable for the strong-operator topology. In this case the conclusion of the theorem holds for all ϕ in $\mathfrak{S}_\beta$. This situation is characterized by saying that $\mathfrak{S}_\beta$ is a *simplex*. We might ask whether this feature persists under more general circumstances. In this respect we mention that Takesaki's paper [1970b] ends with the following remark: "The set of all states satisfying the KMS boundary condition with respect to a fixed one-parameter automorphism group and a fixed β forms a Choquet simplex." With the tools used here we did not reach a conclusion of such generality: we had to require that ϕ be separable to get that the central measure was the unique maximal measure representing ϕ on $\mathfrak{S}_\beta$. We should note, however, that the role played by this assumption is a rather far-fetched one: it ensures that $\tilde{\phi}$ is KMS on the *von Neumann algebra* $\pi_\phi(\mathfrak{R})''$. Two consequences of this fact were used in the proof: first Φ is separating for $\pi_\phi(\mathfrak{R})''$ and second $\mathfrak{Z}_\phi(\mathfrak{R})''$ is pointwise invariant under the time evolution, a fact used in the proof of Theorem 11 on which the present theorem depends.

(ii) Our second remark on Theorem 15 emphasizes that this theorem establishes the existence and uniqueness of a decomposition of ϕ, separable KMS, in terms *only* of maximal measures. Now, to assert the existence of a strict decomposition of ϕ into "its extremal components" is to assert the existence of a measure that represents ϕ *and* such that its support is actually contained in $\mathfrak{E}_\beta$. If it exists, this measure is then† maximal and the above theorem ensures its uniqueness. It should be noticed, however, that the *existence* of a measure decomposing ϕ in this stricter sense is not established by Theorem 15; μ_ϕ maximal only implies that μ_ϕ is concentrated on $\mathfrak{E}_\beta$ in the Baire sense. When $\mathfrak{R}$ is norm-separable, $\mathfrak{S}$ is metrizable and μ_ϕ is concentrated on $\mathfrak{E}_\beta$ in the Borel sense. This refinement can be obtained under less restrictive assumptions; in particular, we can show that μ_ϕ is concentrated on $\mathfrak{E}_\beta$ in the Borel sense whenever ϕ is locally normal.‡ This, however, does not yet ensure that μ_ϕ will actually be concentrated on $\mathfrak{E}_\beta$. We can, however, force this last circumstance under certain reasonably weak assumptions,§ thus asserting that under the appropriate physical circumstances an equilibrium state can be decomposed in a *unique* way into pure thermodynamical phases.

If ϕ is invariant under a symmetry group G, different from the time evolution, its pure thermodynamical phase components do not necessarily inherit the full symmetry of G, thus allowing for a spontaneous symmetry breaking. In particular, the case in which G is the euclidian group in three dimensions has been studied in detail¶; the possible subsymmetries inherited by the pure thermodynamical phases entering into the composition of the (euclidian invariant) Gibbs state have been classified and brought into one-to-one correspondence with certain clustering and spectrum properties characteristic of the states to be considered. Crystalline phases, in particular, have been shown to be accommodated in this classification.

We now turn our attention to the set $\mathfrak{S}_G$ of all states on the C^*-algebra $\mathfrak{R}$ which are invariant under a symmetry group G acting in a G-abelian manner. We know from the corollary to Theorem 6 that G-abelianness implies that the von Neumann algebra $\mathfrak{N}'_\phi = \pi_\phi(\mathfrak{R})' \cap U_\phi(G)'$ is abelian. Lanford and Ruelle [1967] noticed that the latter condition provides the "abelianness" necessary to ensure the uniqueness of a decomposition of $\phi \in \mathfrak{S}_G$ into its extremal G-invariant components. A hint to the fact that it is indeed the case is gained by consideration of the simplified example discussed on

† See Phelps [1966], Corollary 9.8, or Choquet [1969], Proposition 27.2 and Theorem 27.4.
‡ See Chapter 4; we only mention here in anticipation that the proof will rest on the existence of a countable family $\{\mathfrak{R}_\gamma \mid \gamma \in \Gamma\}$ of C^*-subalgebras of $\mathfrak{R}$ such that (a) $\bigcup_{\gamma\in\Gamma} \mathfrak{R}_\gamma$ is dense in $\mathfrak{R}$ and (b) each $\mathfrak{R}_\gamma$ possesses a closed two-sided ideal $\mathfrak{T}_\gamma$ with the property that $\|\phi\,|_{\mathfrak{T}_\gamma}\| = 1$ for all γ in Γ. Hence the lack of separability of $\mathfrak{R}$ is made up by an adequate "separability" assumption on ϕ.
§ See, for instance, Emch, Knops, and Verboven [1970].
¶ Emch, Knops, and Verboven [1970]; the original idea to use the central decomposition of the Gibbs state to get a theory accounting for the symmetry breaking which occurs in the crystalline process is due to Kastler, Haag, and Michel [1968] who, however, did not give to their exploratory study the motivation linked to the KMS condition given here.

pp. 214–215 for $\phi \in \mathfrak{S}_\beta$, where we substitute $\phi \in \mathfrak{S}_G$ for $\phi \in \mathfrak{S}_\beta$ and $\mathfrak{N}'_\phi$ for $\mathfrak{Z}_\phi(\mathfrak{R})$. We now want to indicate that under the assumption of G-abelianness $\mathfrak{N}'_\phi$ plays in the decomposition of $\phi \in \mathfrak{S}_G$ into its extremal G-invariant components the role played by $\mathfrak{Z}_\phi(\mathfrak{R})$ in the decomposition of $\phi \in \mathfrak{S}_\beta$; further, in the present case we have the added simplification that in contrast to Theorem 11 the lemma on p. 174 does not require any separability assumption on ϕ.

Let ϕ be a G-invariant state on $\mathfrak{R}$ and let us denote by $\approx\!\!\phi$ the natural extension of ϕ from $\mathfrak{R}$ to $\mathfrak{N}_\phi \equiv \{\pi_\phi(\mathfrak{R}) \cup U_\phi(G)\}''$, defined by

$$\langle \overset{\approx}{\phi}; X\rangle = (\Phi, X\Phi) \qquad X \in \mathfrak{N}_\phi;$$

we denote, as previously, by $\tilde{\phi}$ the restriction of $\overset{\approx}{\phi}$ to $\pi_\phi(\mathfrak{R})''$. Let $\tilde{\mu}_\phi$ be the central measure of $\overset{\approx}{\phi}$ and let $\tilde{\varphi}$ be the associated isomorphism between

$$\mathfrak{Z}_\phi(\mathfrak{N}_\phi) \equiv \mathfrak{N}_\phi \cap \mathfrak{N}'_\phi = \mathfrak{N}'_\phi \quad \text{and} \quad \mathfrak{L}^\infty(\mathfrak{S}(\mathfrak{N}_\phi), \tilde{\mu}_\phi).$$

We have, in particular, for every B in $\mathfrak{N}'_\phi$ and any R in $\mathfrak{R}$

$$\langle \tilde{\phi}; BR\rangle = \int_{\mathfrak{S}(\mathfrak{N}_\phi)} \tilde{\varphi}_B(\psi)\langle \psi; \pi_\phi(R)\rangle \, d\tilde{\mu}_\phi(\psi).$$

We prove, as for the lemma to Theorem 15, that this reduces to

$$\langle \tilde{\phi}; BR\rangle = \int_{\mathfrak{S}_G} \tilde{\varphi}_B(\psi)\langle \psi; R\rangle \, d\tilde{\mu}_\phi(\psi).$$

We can now adapt, in a straightforward manner, the proof of Theorem 15 to yield the following result:

Theorem 16. *Let* G *be a symmetry group on* $\{\mathfrak{R}, \mathfrak{S}, \langle;\rangle\}$. *If this system is* G-*abelian, then* $\mathfrak{S}_G$ *is a simplex.*

REMARKS.

(i) If G acts in a G-abelian manner on ϕ alone, then there still exists a unique maximal measure $\tilde{\mu}_\phi$ representing this ϕ on $\mathfrak{S}_G$; this remark reflects the fact that the proof of the theorem is made for each ϕ in $\mathfrak{S}_G$ separately.

(ii) The theorem could have been proved directly by adapting the proof of Wils's theorem to this case.

(iii) From the general theory of Choquet-Meyer [1963] the assertion that $\mathfrak{S}_G$ is a simplex is equivalent to the assertion that the cone $\mathfrak{T}_G$ of all positive G-invariant elements of $\mathfrak{R}^*$ is a lattice when equipped with its natural partial-ordering relation; this fact has been used by Doplicher, Kastler, and Størmer [1969] to show that G-abelianness implies that $\mathfrak{S}_G$ is a simplex.

(iv) The second of the two remarks made about Theorem 15 can be transposed to the present case; in particular, the fact that ϕ is locally normal implies that $\tilde{\mu}_\phi$ is concentrated in the Borel sense on $\mathfrak{S}_G$ has been established by Ruelle [1966], whose reasoning has subsequently been adapted to the KMS case to get the result mentioned in the discussion of Theorem 15.

We now want to mention only a few of the applications of Theorem 16. Suppose that ϕ is an extremal G-invariant state and H is a subgroup of G. Is ϕ still extremal invariant with respect to H and, if not, what is the physical meaning of a decomposition of ϕ into its extremal H-invariant components? This problem has been analyzed by Kastler and Robinson [1966] and Robinson and Ruelle [1967a] (see also Ruelle [1969a] Section 6.5.2), in particular in connection with the occurrence of spontaneous symmetry breaking in phase transitions. Several aspects of their results have been (later) incorporated into the decomposition of a KMS ϕ into its extremal KMS components, the latter procedure having the advantage of specifying H more tightly. A definitive choice between these two speculative approaches, however, seems hard to make in the absence of any exactly soluble (three-dimensional) model for crystallization. In the present stage it appears that these approaches should be regarded rather as possible guides toward the building of such models.

Another particular application of Theorem 16 has been the decomposition of G-invariant partial states into extremal G-invariant partial states and its connection with the two-dimensional Ising model spontaneous magnetization.† Finally, we want to mention Kastler's [1967] review article on broken symmetries and the Goldstone theorem in axiomatic field theory; we shall come back to this point in Chapter 4.

† Emch, Knops, and Verboven [1968a,b].

CHAPTER 3

Canonical Commutation and Anticommutation Relations

It was the study of quantum field theory that pointed most emphatically to the necessity for a new approach transcending Fock-space techniques. In the early thirties the idea of an algebraic approach seemed to have occurred to physicists like Jordan, von Neumann, and Wigner, as we pointed out in Chapter 1. The mathematical techniques of the time, however, were not ready to support their ideas; even Segal's proposal after World War II did not immediately command the attention it deserved, and finally received after the publication of the papers by Araki and Woods and Haag and Kastler, some 15 years later. An explanation (but certainly no excuse) for this lack of receptivity on the part of physicists might perhaps be that theorems were established in the thirties by von Neumann (and Jordan and Wigner) asserting the uniqueness, up to unitary equivalence, of the irreducible representation of the CCR (or the CAR) for every finite number of degrees of freedom, thus giving a status of universality to the Schrödinger representation. The Fock-space formulation of quantum field theory then seemed to be such a natural generalization of the Schrödinger representation that for quite a while the problem of its uniqueness did not seem to have bothered too many physicists. In the meantime, the field-theoretical investigations of Friedrichs, van Hove, Schweber, and Wightman, in particular, indicated that von Neumann's uniqueness theorem could not extend to quantum field theories. Not only was the existence of inequivalent representations firmly established, but so many of them were actually pouring out of the mill that such names as "strange" and "myriotic" were attached to them; this semantic remark might help in conveying the feeling that this discovery, instead of improving the situation, apparently made it worse: how, indeed, could the physicist choose one particular representation from such a profusion of possibilities? This chapter is devoted mainly to that question and breaks naturally into two sections: the CCR and the CAR. On purely didactic grounds one might think of treating the CAR first (since they are simpler) and enter the discussion of the CCR only afterwards, with all the complications due to the proper choice of a test function space; for reasons of economy and continuity we elected, however, to take the other road: the CCR are discussed in Section 1 and the treatment of the CAR is to be found in Section 2.

SECTION 1 CANONICAL COMMUTATION RELATIONS

OUTLINE. We first describe the various formulations of the canonical commutation relations for a finite number of degrees of freedom and analyze the physical meaning of the Weyl form of the CCR; von Neumann's theorem is stated but its proof is postponed to a later place in this section. We then define the general C^*-algebra of the CCR; in doing so we introduce the mathematical concept of the C^*-inductive limit of C^*-algebras which also play a central role in the next chapter. A general structure theorem for the representations of this algebra is then derived from the GNS construction; as a particular case we then prove von Neumann's theorem. The two parts of Haag's theorem are analyzed separately. We next give a further illustration of the general structure theorem by constructing some special representations. The section also includes some remarks on the proper test-function space to be associated with a given representation. We conclude with some indications on the limitations of some of representations which we used for illustrative purposes in this section.

a. Properties of the Schrödinger Representation

In this subsection we define the Schrödinger representation of the canonical commutation relation for one degree of freedom; we then collect some of its properties with the aim of delineating the sense in which this representation can be expected to be essentially unique.

The usual Hilbert space used to describe the quantum mechanics of a single particle with no other degree of freedom than that resulting from the assignment to move along the real line $\mathbb{R}$ is the (separable!) Hilbert space $\mathfrak{L}^2(\mathbb{R})$ of all square integrable (with respect to the Lebesgue measure) functions from $\mathbb{R}$ to $\mathbb{C}$. This is the space of the Schrödinger representation that we want to consider.

We consider the self-adjoint operator Q defined on

$$\mathfrak{D}(Q) = \{\Psi \in \mathfrak{L}^2(\mathbb{R}) \mid x\Psi \in \mathfrak{L}^2(\mathbb{R})\}$$

by $(Q\Psi)(x) = x\Psi(x)$ and interpret it as the *position operator*. We further consider the weakly continuous one-parameter group $\{U(a)\}$ of the unitary operators defined on $\mathfrak{L}^2(\mathbb{R})$ by $(U(a)\Psi)(x) = \Psi(x - a)$. We notice that the dense linear manifold $\mathfrak{D}(Q)$ is stable with respect to $U(a)$ for all a in $\mathbb{R}$ and that

$$U(a)\, QU(-a) = Q - aI \quad \text{on} \quad \mathfrak{D}(Q),$$

so that we can indeed interpret $U(a)$ as providing a description of the action of the *translations* along $\mathbb{R}$. This is our first form of the Schrödinger representation of the canonical commutation relation for one degree of freedom. Two other forms are usually encountered in the literature and we now want to describe them briefly.

From Stone's theorem we know that there exists a self-adjoint operator P with domain $\mathfrak{D}(P)$ such that

$$s\text{-}\lim_{a \to 0} \frac{1}{a}\, [U(a) - I] = -iP \quad \text{on} \quad \mathfrak{D}(P);$$

P is referred to as the generator of $\{U(a)\}$ and we interpret it as the *momentum operator*. From this definition we see that P is equal to $-i(d/dx)$ and that $\mathfrak{D}(P)$ is the dense linear manifold in $\mathfrak{L}^2(\mathbb{R})$ consisting of all absolutely continuous functions on $\mathbb{R}$ with derivative in $\mathfrak{L}^2(\mathbb{R})$.

We recall† that the Fourier transform

$$(V\Psi)(x) = \frac{1}{\sqrt{2\pi}} \int_{-\infty}^{+\infty} e^{ixy}\Psi(y)\, dy,$$

mapping $\mathfrak{L}^2(\mathbb{R})$ isometrically onto itself, induces a unitary equivalence $P = VQV^{-1}$ between the unbounded, self-adjoint operators P and Q just defined; both P and Q have a simple continuous spectrum extending from $-\infty$ to $+\infty$.

Foias, Geher, and Sz.-Nagy [1960] noticed that the dense linear manifold $\mathfrak{S}(\mathbb{R})$ of all infinitely differentiable functions Ψ from $\mathbb{R}$ to $\mathbb{C}$ for which

$$\lim_{|x| \to \infty} x^n \frac{d^m}{dx^m} \Psi(x) = 0 \quad \text{for all } m, n = 0, 1, 2 \cdots$$

enjoys the following properties pertinent to the study of the Schrödinger representation of the canonical commutation relations; $\mathfrak{S}(\mathbb{R})$, contained in $\mathfrak{D}(PQ - QP) = \mathfrak{D}(PQ) \cap \mathfrak{D}(QP) \subseteq \mathfrak{D}(P) \cap \mathfrak{D}(Q)$, is stable with respect to P and Q and furthermore is mapped onto itself by the operators $(P \pm iI)$ and $(Q \pm iI)$. Moreover, the restrictions to $\mathfrak{S}(\mathbb{R})$ of P, Q and $P^2 + Q^2$ are essentially self-adjoint. Finally we clearly have

$$PQ - QP = -iI \quad \text{on} \quad \mathfrak{S}(\mathbb{R}).$$

The fact that $PQ - QP$ coincides with $-iI$ on some dense linear manifold in $\mathfrak{L}^2(\mathbb{R})$ is the second among the forms of the Schrödinger representation of

† See, for instance, Stone [1932a], Theorem 10.9.

the canonical commutation relation we want to consider here. Referred to as the *Heisenberg uncertainty principle*, this relation has a direct (and historically primordial) interpretation that we can formulate here as follows:

$$\langle\psi; (P - \langle\psi; P\rangle)^2\rangle\langle\psi; (Q - \langle\psi; Q\rangle)^2\rangle \geqslant \tfrac{1}{4},$$

where ψ denotes the state on the algebra $\mathfrak{P}$ of all polynomials in P and Q, defined by

$$\langle\psi; X\rangle = (\Psi, X\Psi) \quad \text{for all } X \text{ in } \mathfrak{P}$$

with Ψ any element in $\mathcal{S}(\mathbb{R})$ with $|\Psi| = 1$.

Since Q is self-adjoint, it generates a weakly continuous, one-parameter group $\{V(b)\}$ of unitary operators defined on $\mathcal{L}^2(\mathbb{R})$ with

$$(V(b)\Psi)(x) = (e^{-iQb}\Psi)(x) = e^{-ixb}\Psi(x)$$

[notice that this relation, clearly true on $\mathcal{S}(\mathbb{R})$, can be extended by continuity from $\mathcal{S}(\mathbb{R})$ to $\mathcal{L}^2(\mathbb{R})$]. We then have, on using the explicit form of $U(a)$ and $V(b)$ given above,

$$U(a)\, V(b) = e^{iab}\, V(b)\, U(a) \quad \text{for all } a, b \text{ in } \mathbb{R}.$$

This is the *Weyl form* of the Schrödinger representation of the canonical commutation relation for one degree of freedom; we have already encountered some of its generalizations in Chapter 2. The main advantage of this form is that it involves only unitary operators and thus avoids the somewhat cumbersome domain questions arising when we deal directly with P and Q. On the other hand, it has the inconvenience of being rather removed from the dynamics as it occurs in the canonical formalism of quantum mechanics which is inherited from the Hamilton relations of classical mechanics.

Finally, from the fact that a function in $\mathcal{L}^2(\mathbb{R})$ with compact support cannot have a Fourier transform with compact support unless it is identically zero we conclude that the only subspace of $\mathcal{L}^2(\mathbb{R})$ stable with respect to P and Q is $\{0\}$, so that the Schrödinger representation studied in this subsection is *irreducible*.

b. Uniqueness Theorems

The problem investigated in this subsection is whether the various forms of the canonical commutation relation discussed in the preceding subsection determine, at least up to unitary equivalence, the representation in which P and Q form an irreducible set.

More precisely, we say that a pair (P, Q) of self-adjoint operators acting on a separable Hilbert space $\mathcal{H}$ form a Heisenberg representation of the canonical commutation relations if there exists a dense linear manifold $\mathcal{D}$ in $\mathcal{H}$ with

$\mathfrak{D} \subseteq \mathfrak{D}(P) \cap \mathfrak{D}(Q)$ such that

$$PQ - QP = -iI \quad \text{on} \quad \mathfrak{D};$$

the representation is said to be irreducible if the only subspaces of $\mathcal{H}$ stable with respect to P and Q are $\{0\}$ and $\mathcal{H}$. We first want to prove that $\mathfrak{D}$ has to be strictly contained in $\mathcal{H}$ by showing that at least one of the operators P or Q must be unbounded.

Theorem 1. *There is no Heisenberg representation of the canonical commutation relations by bounded operators.*†

Proof. Suppose on the contrary that there exist two bounded operators P and Q acting on $\mathcal{H}$ and such that $PQ - QP = -iI$; we can assume without loss of generality that P is nonsingular, i.e., that P possesses a unique two-sided inverse, denoted P^{-1}, for if this were not the case we could replace P with $P_\lambda = P - \lambda I$ with $\lambda > \|P\|$ without changing the commutation relation and then have it hold with P_λ nonsingular and bounded. This being the case, we notice that for any complex number μ: $(PQ - \mu I) = P(QP - \mu I)P^{-1}$, so that the spectrum Sp (PQ) of PQ coincides with the spectrum Sp (QP) of QP. We further notice, as a result of the commutation relation $PQ = QP - iI$, that for any complex number ν we have $(PQ - i(\nu - 1)I) = (QP - i\nu I)$ so that $i\nu$ belongs to Sp (QP) if and only if $i(\nu - 1)$ belongs to Sp (PQ). Now using the previously established identity of Sp (PQ) and Sp (QP) and the fact that Sp (PQ) is not empty, we conclude that Sp (PQ) contains together with some $i\nu$ the whole sequence $\{i(\nu + n) \mid n = 0, 1, 2, \ldots\}$ so that Sp (PQ) is unbounded. This contradicts $\|PQ\| \leqslant \|P\| \cdot \|Q\| < \infty$; hence P and Q cannot both be bounded. ∎

REMARK. The result obtained in this theorem is maximal under the assumptions imposed; in particular, the case in which only one of the two operators P and Q is unbounded cannot be ruled out, even on physical grounds, as the following counterexample will show. Consider, indeed, the case of a (free) particle moving on the interval $(0, 1)$ with periodic boundary conditions (the latter even including a phase-change Θ with $0 \leqslant \Theta < 2\pi$); in this case an analysis paralleling that conducted in the preceding subsection leads to the operators Q and P defined as follows:

$(Q\Psi)(x) = x\,\Psi(x)$ on $\{\Psi \in \mathfrak{L}^2(0, 1) \mid x\,\Psi(x) \in \mathfrak{L}^2(0, 1)\}$,

$(P\Psi)(x) = -i\dfrac{d}{dx}\Psi(x)$ on the linear manifold of all absolutely continuous functions such that $\Psi'(x) \in \mathfrak{L}^2(0, 1)$ and $\Psi(1) = \Psi(0) \exp i\Theta$.

† As evidence that this theorem is an integral part of the folklore of mathematical physics, we might mention that a sketch of its proof can be found even in an article in the *Encylcopaedia Britannica* written by Kadison [1969]; original credit might be given to Wintner [1947] and Wielandt [1949]; see also Wightman and Schweber [1955], Note No. 34.

In this case Q is bounded; moreover, P has a discrete spectrum. Both facts indicate that this representation of the canonical commutation relation is not unitarily equivalent to the Schrödinger representation studied in Subsection a.

Incidentally, the fact that P has a discrete spectrum in the above realization of $[P, Q] \subseteq -iI$ allows us to exhibit very simply vector states [namely $\langle\psi; X\rangle = (\Psi, X\Psi)$ with $\Psi(x) = \exp(i\lambda x)$ and $\lambda = 2\pi n + \Theta$] for which

$$\langle\psi; (P - \langle\psi; P\rangle)^2\rangle\langle\psi; (Q - \langle\psi; Q\rangle)^2\rangle = 0.$$

This evidently does not contradict $[P, Q] \subseteq -iI$ but rather illustrates the fact that the proof that the left-hand side admits $\frac{1}{4}$ for its lower bound requires more conditions than just $\Psi \in \mathfrak{D}(P) \cap \mathfrak{D}(Q)$.

We conclude from this counterexample that there exist Heisenberg representations of the canonical commutation relation, even for one degree of freedom, which are *not* unitarily equivalent to the Schrödinger representation studied in the preceding subsection; this is not too surprising in view of the fact that a Heisenberg representation, as defined in the beginning of the present subsection, concentrates its requirement essentially on the local aspect of the canonical commutation relations, neglecting the physics contained in the boundary conditions of the problem at hand. Consequently, if we want any uniqueness theorem at all, we need to impose some conditions that are more stringent than those required in the definition of a Heisenberg representation; one way to achieve this is to impose conditions on the domains and properties of the self-adjoint operators P and Q. Several results have been reported in this direction; among them we might first mention that of Dixmier [1958]:

Theorem 2. *In order that a pair* (P, Q) *of closed symmetric operators, acting within a (separable) Hilbert space* $\mathcal{H}$, *be unitarily equivalent to a direct sum of Schrödinger representations it is necessary and sufficient that there exists a linear manifold* $\mathfrak{D}$, *contained in* $\mathfrak{D}(\mathrm{P}) \cap \mathfrak{D}(\mathrm{Q})$ *and dense in* $\mathcal{H}$, *such that*

(i) $\mathfrak{D}$ *is stable with respect to* P *and* Q,
(ii) *the restriction of* $(\mathrm{P}^2 + \mathrm{Q}^2)$ *to* $\mathfrak{D}$ *is essentially self-adjoint,*
(iii) $\mathrm{PQ} - \mathrm{QP} = -i\mathrm{I}$ *on* $\mathfrak{D}$.

Furthermore these conditions are sufficient to ensure that P *and* Q *will be self-adjoint and even that the restrictions of* P *and* Q *to* $\mathfrak{D}$ *will be essentially self-adjoint.*

This theorem is an improvement on an earlier result of Rellich [1946] who assumed in addition that $P^2 + Q^2$ is "decomposable in $\mathfrak{D}$," i.e., that this operator, defined on $\mathfrak{D}$, admits a self-adjoint closure $\int \lambda \, dE_\lambda$ such that for any finite interval (a, b) and subspace $(E_b - E_a)$ is contained in $\mathfrak{D}$.

In the same vein we might mention the following theorem due to Foias, Geher, and Sz.-Nagy [1960] (corollary to their Theorem II):

Theorem 3. *In order that a pair* (P, Q) *of closed symmetric operators, acting within a (separable) Hilbert space* $\mathcal{H}$, *be unitarily equivalent to a direct sum of Schrödinger representations, it is necessary and sufficient that there exists a linear manifold* $\mathfrak{D}$, *contained in* $\mathfrak{D}(PQ - QP)$ *and dense in* $\mathcal{H}$, *such that*

(i) $(P \pm iI)\mathfrak{D} \subseteq \mathfrak{D}$; $(Q \pm iI)\mathfrak{D} \subseteq \mathfrak{D}$,
(ii) $PQ - QP = -iI$ *on* $\mathfrak{D}$.

In this theorem (i) can be replaced by the condition (see Foias, Geher, and Sz.-Nagy [1960], Theorem II):

(i′) $(P + iI)(Q + iI)\mathfrak{D}$ or $(Q + iI)(P + iI)\mathfrak{D}$ be dense in $\mathcal{H}$.

Finally we want to mention the result obtained by Tillman [1963]:

Theorem 4. *Let* P *and* Q *be two closed symmetric operators acting within a (separable) Hilbert space* $\mathcal{H}$, *with* $\mathfrak{D}(P) \cap \mathfrak{D}(Q)$ *dense in* $\mathcal{H}$, *and such that*

(i) $(Q\Psi, P\Phi) - (P\Psi, Q\Phi) = -i(\Psi, \Phi) \qquad \forall \Psi, \Phi \in \mathfrak{D}(P) \cap \mathfrak{D}(Q)$,
(ii) $(Q + iP)^* = (Q - iP)^{\sim}$;

then P *and* Q *are self-adjoint and the pair* (P, Q) *is unitarily equivalent to a direct sum of Schrödinger representations.*

The preceding three theorems were mentioned here without proofs, with the main intention of directing the reader's attention to the quite extensive body of information available on this subject in the mathematical literature; for more details and for proofs the reader is referred to the monograph by Putnam [1967] (see, in particular, his Chapter IV).

We now come back to the point mentioned earlier and try to take into account, as simply as possible, the boundary conditions relevant to the fact that we want to determinate the most general representation capable of describing the motion of a single quantum mechanical particle on the infinitely extended real line.

Theorem 5. *Let* Q *be a self-adjoint operator with domain* $\mathfrak{D}(Q)$ *dense in some Hilbert space* $\mathcal{H}$; *let* $\{U(a)\}$ *be a weakly continuous, one-parameter group of unitary operators acting on* $\mathcal{H}$ *and such that*

$$U(a)\,QU(-a) = Q - aI \quad \text{on} \quad \mathfrak{D}(Q),\ \forall\, a \in \mathbb{R}.$$

Then the weakly continuous, one-parameter group $\{V(b)\}$ *generated by* Q *satisfies the relation*

$$U(a)\,V(b) = e^{iab}V(b)\,U(a) \qquad \forall\, a, b \in \mathbb{R};$$

furthermore, if $\{U(a), V(b) \mid a, b \in \mathbb{R}\}$ *is irreducible,* $\mathcal{H}$ *is separable.*

Proof. We first remark that the hypothesis of the theorem implicitly contains the fact that $\mathfrak{D}(Q)$ is stable with respect to $U(a)$ for all a in $\mathbb{R}$. Since the self-adjoint operators Q and $Q - aI$ are unitarily equivalent, they have the same spectrum; we therefore conclude that the spectrum of Q extends from $-\infty$ to $+\infty$ and that if $\{E_\lambda\}$ denotes the spectral family of Q then†

$$U(a)\, E_\lambda U(-a) = E_{\lambda+a} \quad \text{for all } \lambda, a \text{ in } \mathbb{R}.$$

From this assertion follows the first part of the theorem. Let us now consider the von Neumann algebra $\mathfrak{W}''$ generated by $\{U(a), V(b) \mid a, b \in \mathbb{R}\}$. We denote by $\mathfrak{W}_0$ the set of all finite sums of the form $\sum_i z_i U(a_i)\, V(b_i)$, where $z_i = c_i + id_i$ and a_i, b_i, c_i, and d_i are restricted to the rationals; $\mathfrak{W}_0$ is then countable. Since $U(a)$ and $V(b)$ are *both* weakly (and then strongly) continuous, one-parameter groups of unitary operators on $\mathcal{H}$, $U(a)\, V(b)$ is jointly continuous in a and b, so that $\mathfrak{W}_0$ is dense with respect to the strong-operator topology in the set $\mathfrak{W}_1$ of all finite sums of the form $\sum_i z_i U(a_i)\, V(b_i)$ (without restriction now on a_i, b_i, and z_i). Because of the form of the canonical commutation relation, we have that every finite product of terms of the form $U(a_i)\, V(b_i)$ can still be written in the form $U(a)\, V(b)$; so $\mathfrak{W}_1$ is a *-algebra with unit, the strong closure of which is $\mathfrak{W}''$. We can conclude therefore that $\mathfrak{W}''$ possesses a countable, everywhere dense family of elements, namely $\mathfrak{W}_0$. Hence $\mathfrak{W}''$ is separable in the strong-operator topology. Now if $\{U(a), V(b) \mid a, b \in \mathbb{R}\}$ is irreducible, every vector Φ in $\mathcal{H}$ is cyclic with respect to $\mathfrak{W}''$ and then with respect to $\mathfrak{W}_0$; since $\mathfrak{W}_0\Phi$ is countable, $\mathcal{H}$ is indeed separable. This concludes the proof of the theorem. ∎

REMARK. It is clear from the proof that the separability of $\mathcal{H}$ follows from the existence of *one* vector Φ in $\mathcal{H}$, cyclic with respect to $\mathfrak{W}''$.

The following example will show that the assumption that *both* $\{U(a)\}$ and $\{V(b)\}$ are weakly continuous in a and b, respectively, is essential for the proof of the separability of the representation space. Consider, in the Schrödinger representation studied in Subsection a, the abelian C^*-algebra $\mathfrak{V}$ generated by $\{V(b) \mid b \in \mathbb{R}\}$ and the automorphisms $\{\alpha_a \mid a \in \mathbb{R}\}$ of $\mathfrak{V}$ defined by

$$\alpha_a[V(b)] = U(a)\, V(b)\, U(-a) = e^{iab}\, V(b).$$

For any element R in $\mathfrak{V}$ $\alpha_a[R]$ is then defined by $U(a)\, RU(-a)$. Furthermore, for any such R and any $\varepsilon > 0$ there exists a finite sum $V = \sum_i c_i V(b_i)$ such that $\|R - V\| \leqslant \varepsilon$. For this V and the same ε there exists an a_0 in $\mathbb{R}$ such that $\|\alpha_a[V] - V\| \leqslant \varepsilon$ for all $|a| < a_0$. For these a we then have

$$\|\alpha_a[R] - R\| \leqslant \|\alpha_a[R - V]\| + \|\alpha_a[V] - V\| + \|V - R\| \leqslant 4\varepsilon,$$

† See, for instance, Stone [1932a], Theorem 7.1.

so that in particular $\langle\phi; \alpha_a[R]\rangle$ is continuous in a for all R in $\mathfrak{V}$ and all states ϕ on $\mathfrak{V}$; hence $\mathbb{R}$ is a symmetry group for $\mathfrak{V}$, which we interpret naturally as describing the action of the translations along $\mathbb{R}$. Since $\mathfrak{V}$ possesses a unit and $\mathbb{R}$ is abelian (hence amenable), we know from the lemma on p. 172 that there exists at least one state on $\mathfrak{V}$ invariant with respect to the group $\{\alpha_a \mid a \in \mathbb{R}\}$. Furthermore, from the discussion on p. 174 we know that there then exists at least one extremal translation-invariant state ϕ on $\mathfrak{V}$. Let $\{\pi_\phi(\mathfrak{V}), U_\phi(\mathbb{R})\}$ be the irreducible (see Theorem II.2.8) covariant representation associated with ϕ by Theorem II.2.5. Writing for simplicity $\pi_\phi(V(b))$ as $V_\phi(b)$, we have

$$U_\phi(a)\, V_\phi(b)\, U_\phi(-a) = e^{iab}\, V_\phi(b) \;\forall\, a, b \in \mathbb{R};$$

we have therefore obtained an irreducible representation of the canonical commutation relation in Weyl's form. We now want to analyze this representation. We notice first that the translation invariance of ϕ implies

$$\langle\phi; V(b)\rangle = \langle\phi; \alpha_a[V(b)]\rangle = e^{iab}\langle\phi; V(b)\rangle \;\forall\, a, b \in \mathbb{R},$$

so that $\langle\phi; V(b)\rangle = 0$ unless $b = 0$, in which case $\langle\phi; V(0)\rangle = 1$. Now consider for each b in $\mathbb{R}$ the vector $\Phi(b) = V_\phi(b)\Phi$, where Φ is the cyclic vector associated with the state ϕ; we then have $|\Phi(b)|^2 = 1$ and $(\Phi(b_1), \Phi(b_2)) = 0$ whenever $b_1 \neq b_2$; hence the closed linear manifold spanned by $\{\Phi(b) \mid b \in \mathbb{R}\}$, which coincides with $\mathcal{H}_\phi$, contains a *continuous* orthonormal basis; that is to say $\mathcal{H}_\phi$ is *not* separable. We therefore have constructed explicitly a representation which contradicts the conclusion of the second part of Theorem 5; there must then be at least one of the assumptions of the theorem that is not satisfied in this representation. This is indeed the case: $\{V_\phi(b)\}$ is *not* weakly continuous in b, since, for example, $(\Phi, V_\phi(b)\Phi) = \langle\phi; V(b)\rangle$ is not continuous at $b = 0$ as we have already noted. On the other hand, $U_\phi(a)$ is continuous in a in the weak-operator topology, according to Theorem II.2.5. Since Φ is invariant under $\{U_\phi(a)\}$ by Theorem II.2.5, we have

$$U_\phi(a)\, \Phi(b) = U_\phi(a)\, V_\phi(b)\, U_\phi(-a)\Phi = e^{iab}\Phi(b)$$

for all a, b in $\mathbb{R}$, so that the momentum-operator P_ϕ, defined as usual as the generator of $\{U_\phi(a) \mid a \in \mathbb{R}\}$, satisfies the relation $P_\phi\Phi(b) = -b\Phi(b)$. From the physical point of view we remark then that the representation just constructed accommodates the set of all plane waves as an orthonormal basis for the nonseparable Hilbert space $\mathcal{H}_\phi$. With this interpretation we understand on physical grounds why $V_\phi(b)$ is not weakly continuous: if it were, a position operator Q_ϕ would exist, which clearly does not make sense when the representation space $\mathcal{H}_\phi$ is spanned by plane waves! Incidentally, this representation might possibly provide a framework for a rigorous potential-scattering theory in which plane waves could be used directly.

The assumptions in Theorem 5 seem to be the closest possible to the requirements associated with the physical interpretation of the operators P and Q as the momentum and position of a particle, the motion of which is restricted to the infinitely extended real line. This being admitted, the uniqueness problem reduces to the uniqueness of the representation of the canonical commutation relation in Weyl's form; this question is answered by the following theorem, due to von Neumann [1931]:

Theorem 6. *Let* $\mathcal{H}$ *be a separable Hilbert space,* $\{U(a) \mid a \in \mathbb{R}\}$ *and* $\{V(b) \mid b \in \mathbb{R}\}$ *be two weakly continuous, one-parameter groups of unitary operators acting on* $\mathcal{H}$*, such that*

$$U(a)\,V(b) = e^{iab}\,V(b)\,U(a) \qquad \forall a, b \in \mathbb{R};$$

then $\{U(a), V(b) \mid a, b \in \mathbb{R}\}$ *is unitarily equivalent to a direct sum of Schrödinger representations.*

The proof of this theorem is given† in subsection c as an illustration of the forthcoming Theorem 7.

It might be worth mentioning here that Theorem 3 has been established in two steps; in the first, Foias, Geher, and Sz.-Nagy proved that the assumptions of their theorem ensure that P and Q will generate $\{U(a)\}$ and $\{V(b)\}$ to satisfy the Weyl form of the commutation relation; the second step consisted in using the von Neumann theorem just mentioned.

In closing we want to mention that the passage from one to a *finite number* of degrees of freedom does not involve any essential modification of the results reviewed in this subsection.

c. The C^*-Algebra of the Canonical Commutation Relations

The aim of this subsection is to set up a formalism in which we can discuss systematically the canonical commutation relations in the typical case of the scalar Bose field.

We first present a formulation of the Weyl representation encompassing any (finite or infinite) number of degrees of freedom. Following Kastler and his school,‡ we define a C^*-algebra carrying some of the main features attached to the concept of the Weyl representations. As an elementary illustration we finally proceed to a proof of von Neumann's theorem (Theorem 6) which emphasizes its essential limitation to the case of a finite number of degrees of freedom.

† Another proof of this theorem, based on the theory of induced group representations, can be found in Mackey [1968].

‡ See for instance Kastler [1965a], Loupias and Miracle-Sole [1966], or Manuceau [1968]; we cannot consider here the interesting formulations of Segal [1963], Mackey [1957], and Glimm [1961] which would carry us farther than we can afford in this book.

We say that $\mathfrak{W}_{\mathcal{H}}(\mathcal{T})$ is a *Weyl representation* of the canonical commutation relations for the test function space $\mathcal{T}$ if

(Ia) $\mathcal{T}$ is a *real* pre-Hilbert space (the completion of which with respect to the scalar product $(\cdots,\cdots)$, we denote $\overline{\mathcal{T}}$);
(Ib) to each f in $\mathcal{T}$ corresponds a pair of unitary operators $U(f)$ and $V(f)$ acting on a Hilbert space $\mathcal{H}$, common to all f in $\mathcal{T}$, and called the "representation space";
(Ic) these operators satisfy the following algebraic conditions:

$$
\begin{aligned}
&U(f_1)\, U(f_2) = U(f_1 + f_2), \\
&V(f_1)\, V(f_2) = V(f_1 + f_2), \qquad\qquad \forall\, f_1, f_2 \in \mathcal{T}. \\
&U(f_1)\, V(f_2) = V(f_2)\, U(f_1) e^{i(f_1, f_2)},
\end{aligned}
$$

(Id) $U(\lambda f)$ and $V(\lambda f)$ are continuous in $\lambda \in \mathbb{R}$, in the weak-operator topology of $\mathfrak{B}(\mathcal{H})$, for all f in $\mathcal{T}$.

With these four assumptions only we can already walk some of the path toward a field theory; in particular, for each f in $\mathcal{T}$, we define the field operator $F(f)$ and its canonical conjugate $P(f)$ as the self-adjoint generators of $V(\lambda f)$ and $U(\lambda f)$, respectively:

$$
\begin{aligned}
V(\lambda f) &= \exp\{-i\lambda F(f)\}, \\
U(\lambda f) &= \exp\{-i\lambda P(f)\}.
\end{aligned}
$$

From Gårding's theorem [1947] we know that for any *finite* sequence $\mathcal{F} = f_1, f_2, \cdots, f_n$ of elements in $\mathcal{T}$ there exists in $\mathcal{H}$ a dense linear manifold $\mathfrak{D}_{\mathcal{F}}$ stable with respect to $F(f_j)$ and $P(f_j)$ for all f_j in $\mathcal{F}$. On $\mathfrak{D}_{\mathcal{F}}$ the above composition laws imply for all f_j, f_k in $\mathcal{F}$

$$
\begin{aligned}
[F(f_j), F(f_k)] &= 0 = [P(f_j), P(f_k)] \\
[F(f_j), P(f_k)] &= i(f_j, f_k).
\end{aligned}
$$

We further define on $\mathfrak{D}_{\mathcal{F}}$ the annihilation and creation operators

$$
a(f_j) = \frac{\{F(f_j) + iP(f_j)\}}{\sqrt{2}},
$$

$$
a^*(f_j) = \frac{\{F(f_j - iP(f_j)\}}{\sqrt{2}},
$$

which then satisfy the CCR relations

$$
\begin{aligned}
[a(f_j), a(f_k)] &= 0 = [a^*(f_j), a^*(f_k)], \\
[a(f_j), a^*(f_k)] &= (f_j, f_k).
\end{aligned}
$$

The case in which $\mathcal{T} = \overline{\mathcal{T}}$ is $\mathbb{R}$ with $(a, b)_{\mathcal{T}} = ab$ reduces to the Weyl form of the canonical commutation relation for a system with one degree of freedom; similarly, if $\mathcal{T} = \overline{\mathcal{T}}$ is n-dimensional (with $n < \infty$), this formulation corresponds to the description of a system with n degrees of freedom. These situations were covered by the results obtained in Subsections a and b.

The case in which $\overline{\mathcal{T}}$ is infinite-dimensional is typical of field theories, and more generally of the "many-body" problems, of which we have already seen concrete examples, in particular in the illustrations given in Chapter 2. In this case, which we shall now study systematically, we need some more technical assumptions if we want to establish some kind of reasonable contact with field theories of, say, Wightman's type. We should be able to compute expectation values for all (finite) products of $F(f)$ and $P(f)$ with f running over the whole of $\mathcal{T}$. To achieve this aim, we impose the following condition:

(II) There exists in the representation space $\mathcal{H}$ a dense linear manifold $\mathfrak{D}$, stable with respect to all $F(f)$ and all $P(f)$ with f in $\mathcal{T}$, i.e., $\mathfrak{D} \subseteq \mathfrak{D}(F(f))$, $\mathfrak{D} \subseteq \mathfrak{D}(P(f))$, and $F(f)\mathfrak{D} \subseteq \mathfrak{D}$, $P(f)\mathfrak{D} \subseteq \mathfrak{D}$. Furthermore, we might want $F(f)$ and $P(f)$ to be essentially self-adjoint on $\mathfrak{D}$.

We denote by $\mathfrak{P}_{\mathcal{H}}(\mathcal{T})$ the *-algebra of all polynomials in the $F(f)$'s and $P(f)$'s, as defined over $\mathfrak{D}$, when f runs over $\mathcal{T}$; by $\mathfrak{Q}_{\mathcal{H}}(\mathcal{T})$, the corresponding algebra in which only the $F(f)$ enter; and by $\overline{\mathfrak{W}_{\mathcal{H}}(\mathcal{T})}$, the C^*-algebra generated by $\{U(f), V(f) \mid f \in \mathcal{T}\}$.

Our last assumption is a very mild restriction on $\mathcal{T}$:

(III) If $\mathcal{T}$ is infinite-dimensional there exists an amenable group $G = \{g\}$ of unitary transformations $\{f \to g[f] \mid f \in \mathcal{T}\}$ of $\overline{\mathcal{T}}$ such that

(a) $g[\mathcal{T}] \subseteq \mathcal{T}$ for all g in G,

(b) the resulting group of automorphisms $\{g \to \alpha_g \mid g \in G\}$ of $\overline{\mathfrak{W}_{\mathcal{H}}(\mathcal{T})}$ satisfies the condition of η-abelianness.

A natural, and apparently innocuous, question to ask at this point could be the following: given $\mathcal{T}$ find all (classes of unitary equivalence of) the Weyl representations satisfying conditions (I), (II), and (III) stated above. In cases in which $\mathcal{T}$ is finite-dimensional, condition (II) becomes redundant, whereas condition (III) is irrelevant, as is pointed out on an heuristic but physical basis at the end of the next subsection. Only condition (I) is then left in this case, and the answer is given by von Neumann's theorem (stated as Theorem 6 in the preceding subsection for the one-dimensional case and proved at the end of the present subsection). In cases in which $\mathcal{T}$ is infinite-dimensional a word of caution is necessary: it is generally recognized that there exist uncountably many inequivalent representations which could possibly be relevant for physical purposes, and we should therefore be equipped with some clean tools before attempting to account for the situation. A first step in this direction is provided by the construction of an appropriate C^*-algebra which incorporates the requirements of condition (I).

Essentially for notational sake we condense our condition (I) by the following trick: let $\mathcal{T}_{\mathbb{C}}$ be the natural complexification of $\mathcal{T}$ with the scalar product

$$(f_1 + ig_1, f_2 + ig_2) = (f_1, f_2)_{\mathcal{T}} + (g_1, g_2)_{\mathcal{T}} + i\{(f_1, g_2)_{\mathcal{T}} - (g_1, f_2)_{\mathcal{T}}\}.$$

For each pair f, g of elements of $\mathcal{T}$ we then define the operator

$$W(f + ig) = U(f)\, V(g) e^{-i(f,g)_{\mathcal{T}}/2};$$

we finally denote by $\mathfrak{W}_{\mathcal{H}}(\mathcal{T}_{\mathbb{C}})$ the set $\{W(f) \mid f \in \mathcal{T}_{\mathbb{C}}\}$. We obtain in this fashion

(a) a prehilbert space $\mathcal{T}_{\mathbb{C}}$ over $\mathbb{C}$;
(b) for each f in $\mathcal{T}_{\mathbb{C}}$, a unitary operator $W(f)$ acting on $\mathcal{H}$, with $W(f)^* = W(-f)$;
(c) the composition law: $W(f_1)W(f_2) = W(f_1 + f_2)e^{i\,\mathrm{Im}(f_1, f_2)/2}$
(d) $W(\lambda f)$ are continuous in $\lambda \in \mathbb{R}$, in the weak-operator topology on $\mathfrak{B}(\mathcal{H})$ for all f in $\mathcal{T}_{\mathbb{C}}$.

Furthermore, writing for each f in $\mathcal{T}$: $U(f) = W(f)$ and $V(f) = W(if)$, we recover all four conditions (I), to which the four conditions above are then equivalent.

We are now ready for the construction of the C^*-algebra of the canonical commutation relations.

We first define the normed $*$-algebra $\Delta(\mathcal{T}_{\mathbb{C}})$ as follows:

(i) Its elements are the mappings R from $\mathcal{T}_{\mathbb{C}}$ to $\mathbb{C}$ which vanish everywhere, except for a finite subset $\mathcal{F}_R$ of $\mathcal{T}$.
(ii) This set of mappings is equipped with the natural laws of addition and multiplication by scalars, namely,

$$(R_1 + R_2)(f) = R_1(f) + R_2(f),$$
$$(\lambda R)(f) = \lambda R(f),$$

which equip $\Delta(\mathcal{T}_{\mathbb{C}})$ with the structure of a complex linear space.
(iii) A ("convolution" or "twisted") product is then defined on $\Delta(\mathcal{T}_{\mathbb{C}})$ as follows:

$$(R_1R_2)(f) = \sum_{g \in \mathcal{T}_{\mathbb{C}}} R_1(g)\, R_2(f - g) e^{i\,\mathrm{Im}(g,f)/2}.$$

We first notice that this sum contains only a finite number of terms, due to the definition of the R's; furthermore we verify that this product is associative and distributive with respect to the addition, so that equipped with this product $\Delta(\mathcal{C}_{\mathbb{C}})$ is now an algebra.

(iv) An involution is defined on $\Delta(\mathcal{C}_{\mathbb{C}})$ by $R^*(f) = R(-f)^*$; we verify easily that this involution satisfies the property $(R_1R_2)^* = R_2^*R_1^*$, so that $\Delta(\mathcal{C}_{\mathbb{C}})$ becomes an involutive algebra.

(v) We finally observe that the mapping $\|\cdots\|_1$ from $\Delta(\mathcal{C}_{\mathbb{C}})$ to $\mathbb{R}$, defined by $\|R\|_1 = \sum_{f\in\mathcal{C}_{\mathbb{C}}} |R(f)|$, is well-defined (since the sums contain only a finite number of nonvanishing terms) and satisfies the axioms of a norm; furthermore, since $\|R^*\|_1 = \|R\|_1$ for all R in $\Delta(\mathcal{C}_{\mathbb{C}})$, the latter is indeed a normed *-algebra.

We denote by $\Delta_1(\mathcal{C}_{\mathbb{C}})$ the involutive Banach algebra obtained by completing $\Delta(\mathcal{C}_{\mathbb{C}})$ with respect to this norm; this is, however, not a C^*-algebra, as the relation $\|R^*R\|_1 = \|R\|_1^2$ would fail to hold in general.

To get an idea of what the proper norm on $\Delta(\mathcal{C}_{\mathbb{C}})$ might be, which would allow us to generate from it a C^*-algebra, we first establish the connection between $\Delta(\mathcal{C}_{\mathbb{C}})$ and the Weyl form of the canonical commutation relations as follows. For each f in $\mathcal{C}_{\mathbb{C}}$ we form the element R_f of $\Delta(\mathcal{C}_{\mathbb{C}})$ defined by $R_f(g) = 0$ for all $g \neq f$ and $R_f(f) = 1$. From definition (iv) of the involution we have $R_f^* = R_{-f}$ and from definition (iii) of the twisted product we see that

$$R_{f_1}R_{f_2} = R_{f_1+f_2}e^{i\,\mathrm{Im}\,(f_1,f_2)/2},$$

so that the elements R_f are unitary ($R_0 = I$) and then satisfy the algebraic requirements of conditions (b) and (c). It is worth observing that the set $\Delta_0(\mathcal{C}_{\mathbb{C}}) = \{R_f \mid f \in \mathcal{C}_{\mathbb{C}}\}$ is a free system, generating $\Delta(\mathcal{C}_{\mathbb{C}})$ by linearity; i.e., $\Delta_0(\mathcal{C}_{\mathbb{C}})$ is a "basis" for $\Delta(\mathcal{C}_{\mathbb{C}})$. Furthermore, since $\|R_f\|_1 = 1$, we have

$$\left\|\sum_{j=1}^{n} \lambda_j R_{f_j}\right\|_1 = \sum_{f\in\mathcal{C}_{\mathbb{C}}} \left|\sum_{j=1}^{n} \lambda_j R_{f_j}(g)\right| = \sum_{j=1}^{n} |\lambda_j|.$$

We further verify that every nondegenerate representation $\pi: \Delta(\mathcal{C}_{\mathbb{C}}) \to \mathfrak{B}(\mathcal{H})$ is faithful, so that the set $\pi(\Delta_0(\mathcal{C}_{\mathbb{C}}))$ of unitary operators acting on $\mathcal{H}$ satisfies the algebraic conditions imposed on a Weyl representation. Furthermore, since $\|\pi(R_f)\| = 1$, we have

$$\left\|\pi\left(\sum_{j=1}^{n} \lambda_j R_{f_j}\right)\right\| \leqslant \sum_{j=1}^{n} |\lambda_j| = \left\|\sum_{j=1}^{n} \lambda_j R_{f_j}\right\|_1.$$

so that π is continuous on $\Delta(\mathcal{C}_{\mathbb{C}})$ and then extends by continuity to a representation of $\Delta_1(\mathcal{C}_{\mathbb{C}})$; and we have for every R in $\Delta_1(\mathcal{C}_{\mathbb{C}})$: $\|\pi(R)\| \leqslant \|R\|_1$ [which is trivial anyhow, since $\Delta_1(\mathcal{C}_{\mathbb{C}})$ is an involutive Banach algebra]. To take into account the continuity condition (d) we restrict our attention to the set $P(\mathcal{C}_{\mathbb{C}})$ of all nondegenerate representations of $\Delta_1(\mathcal{C}_{\mathbb{C}})$ for which $\pi(R_{\lambda f})$ is continuous in $\lambda \in \mathbb{R}$ with respect to the weak-operator topology of $\mathfrak{B}(\mathcal{H})$ for all f in $\mathcal{C}_{\mathbb{C}}$.

We now define a new norm on $\Delta(\mathcal{C}_{\mathbb{C}})$ by

$$\|R\| = \sup_{\pi \in P(\mathcal{C}_{\mathbb{C}})} \|\pi(R)\| \leqslant \|R\|_1$$

and verify trivially that $\|R^*\| = \|R\|$ and $\|R^*R\| = \|R\|^2$, so that the completion $\overline{\Delta(\mathcal{C}_{\mathbb{C}})}$ of $\Delta(\mathcal{C}_{\mathbb{C}})$ with respect to this norm is a C^*-algebra, which we refer to as the *C^*-algebra of the canonical commutation relations.* Incidentally, for every π in $P(\mathcal{C}_{\mathbb{C}})$ we have trivially $\|\pi(R)\| \leqslant \|R\|$ for all R in $\Delta(\mathcal{C}_{\mathbb{C}})$, so that π can be extended by continuity to $\overline{\Delta(\mathcal{C}_{\mathbb{C}})}$; we will not need to distinguish by a special notation in the sequel between π in $P(\mathcal{C}_{\mathbb{C}})$ considered as either a representation of $\Delta(\mathcal{C}_{\mathbb{C}})$, $\Delta_1(\mathcal{C}_{\mathbb{C}})$, or $\overline{\Delta(\mathcal{C}_{\mathbb{C}})}$. We might also mention in passing that the C^*-algebra $\overline{\Delta(\mathcal{C}_{\mathbb{C}})}$ is not "too big" for the purpose of studying the set of all Weyl representations that satisfy only the defining conditions (I); we have indeed that the set $\Delta_0(\mathcal{C}_{\mathbb{C}})$—of fundamental interest to us—generates $\overline{\Delta(\mathcal{C}_{\mathbb{C}})}$ in the sense of normed linear spaces.

The interest of this formulation is that it will allow us to attack the problem of constructing representations of the Weyl canonical commutation relations by a procedure analogous to the GNS construction of representations of a C^*-algebra.

We might note in this connection that the "twisted" product (iii) could be used to equip $\mathcal{C}_{\mathbb{C}}$ with a group structure. The procedure followed here appears to be similar in spirit to the technique described briefly in Subsection II.2.c for constructing weakly continuous unitary representations of locally compact groups.

We denote by $\mathfrak{W}_{\mathrm{I}}(\mathcal{C}_{\mathbb{C}})$ the set of all Weyl representations, defined only by condition (I)

Lemma. *For any element* π *of* $\mathrm{P}(\mathcal{C}_{\mathbb{C}})$ *we define* $\mathfrak{W}_\pi(\mathcal{C}_{\mathbb{C}}) = \{\mathrm{W}_\pi(\mathrm{f}) = \pi(\mathrm{R}_\mathrm{f}) \mid \mathrm{f} \in \mathcal{C}_{\mathbb{C}}\}$; *then the mapping* $\pi \in \mathrm{P}(\mathcal{C}_{\mathbb{C}}) \to \mathfrak{W}_\pi \in \mathfrak{W}_{\mathrm{I}}(\mathcal{C}_{\mathbb{C}})$ *is a bijection; furthermore* π *is cyclic (resp. irreducible) if and only if* $\mathfrak{W}_\pi$ *is cyclic (resp. irreducible).*

Proof. From the definition of $P(\mathcal{C}_{\mathbb{C}})$ it is clear that $\mathfrak{W}_\pi(\mathcal{C}_{\mathbb{C}})$ belongs to $\mathfrak{W}_{\mathrm{I}}(\mathcal{C}_{\mathbb{C}})$. Suppose now that π_1 and π_2 belong to $P(\mathcal{C}_{\mathbb{C}})$ and $W_{\pi_1}(f) = W_{\pi_2}(f)$ for all f in $\mathcal{C}_{\mathbb{C}}$. We then have

$$\pi_1\left(\sum_{j=1}^{n} \lambda_j R_{f_j}\right) = \sum_{j=1}^{n} \lambda_j W_{\pi_1}(f) = \sum_{j=1}^{n} \lambda_j W_{\pi_2}(f) = \pi_2\left(\sum_{j=1}^{n} \lambda_j R_{f_j}\right),$$

so that π_1 and π_2 coincide on $\Delta(\mathcal{C}_{\mathbb{C}})$, since $\Delta_0(\mathcal{C}_{\mathbb{C}})$ linearly generates $\Delta(\mathcal{C}_{\mathbb{C}})$, hence $\pi_1 = \pi_2$; consequently the mapping $\pi \to \mathfrak{W}_\pi$ is injective. To show that it is surjective we start from an arbitrary $\mathfrak{W}_{\mathcal{H}}(\mathcal{C}_{\mathbb{C}})$ in $\mathfrak{W}_{\mathrm{I}}(\mathcal{C}_{\mathbb{C}})$ and form

$$\pi\left(\sum_{j=1}^{n} \lambda_j R_{f_j}\right) \equiv \sum_{j=1}^{n} \lambda_j W_\pi(f_j),$$

which is possible without contradiction, since the R_f are linearly independent. Since, moreover, $\Delta_0(\mathcal{T}_{\mathbb{C}})$ spans $\Delta(\mathcal{T}_{\mathbb{C}})$, this defines a representation π of $\Delta(\mathcal{T}_{\mathbb{C}})$ in $\mathfrak{B}(\mathcal{H})$; we verify then that the continuity condition is satisfied as a result of condition (d) on $\mathfrak{W}_{\mathcal{H}}(\mathcal{T}_{\mathbb{C}})$, so that π actually belongs to $P(\mathcal{T}_{\mathbb{C}})$; hence the mapping $\pi \to \mathfrak{W}_\pi$ is surjective and we have completed the proof of the first part of the lemma. The proof of the second part is trivial with the two natural definitions: $\mathfrak{W}_{\mathcal{H}}(\mathcal{T}_{\mathbb{C}})$ is said to be cyclic if there exists a vector Φ in $\mathcal{H}$ such that $(\Psi, W(f)\Phi) = 0$ for all f in $\mathcal{T}_{\mathbb{C}}$ implies $\Psi = 0$; $\mathfrak{W}_{\mathcal{H}}(\mathcal{T}_{\mathbb{C}})$ is said to be irreducible if $\mathfrak{W}_{\mathcal{H}}(\mathcal{T}_{\mathbb{C}})'' = \mathfrak{B}(\mathcal{H})$. ∎

Since every nondegenerate representation of an involutive algebra is a direct sum of cyclic representations, the above lemma reduces the study of Weyl representations to that of the *cyclic* Weyl representations.

Furthermore, the preceding lemma is instrumental, in conjunction with the GNS construction, in establishing the following theorem, which is the central result of the present subsection:

Theorem 7. *To every mapping $\hat{\phi}$ from $\mathcal{T}_{\mathbb{C}}$ to $\mathbb{C}$ such that* (i) $\hat{\phi}(0) = 1$, (ii) $\hat{\phi}(\lambda f + g)$ *is continuous in $\lambda \in \mathbb{R}$ for all* f, g *in $\mathcal{T}_{\mathbb{C}}$, and* (iii) *for every finite sequence* $\{(\lambda_j, f_j) \mid j = 1, 2, \ldots, n\}$ *of pairs* (λ_j, f_j) *with λ_j in $\mathbb{R}$ and f_j in $\mathcal{T}_{\mathbb{C}}$:*

$$\sum_{j,k=1}^{n} \lambda_k^* \lambda_j \hat{\phi}(f_j - f_k) \exp\left\{\frac{-i\,\mathrm{Im}\,(f_k, f_j)}{2}\right\} \geqslant 0,$$

corresponds a unique (up to unitary equivalence) representation $\mathfrak{W}_{\hat{\phi}}(\mathcal{T}_{\mathbb{C}})$ in $\mathfrak{W}_I(\mathcal{T}_{\mathbb{C}})$ with cyclic vector Φ such that $(\Phi, W_{\hat{\phi}}(f)\Phi) = \hat{\phi}(f)$ for all f in $\mathcal{T}_{\mathbb{C}}$; conversely, every cyclic representation in $\mathfrak{W}_I(\mathcal{T}_{\mathbb{C}})$ can be obtained in this way.

Proof. The idea of the proof is to show that $\hat{\phi}$ defines canonically a state ϕ on $\Delta_1(\mathcal{T}_{\mathbb{C}})$, then to construct the GNS representation π_ϕ in $P(\mathcal{T}_{\mathbb{C}})$, and finally to infer from the preceding lemma the existence of the representation $\mathfrak{W}_{\hat{\phi}}$; the proof of the converse part of the theorem is obtained by retracing these steps in the reverse order. Specifically, let $\hat{\phi}: \mathcal{T}_{\mathbb{C}} \to \mathbb{C}$ satisfy the assumptions of the theorem; since the R_f's are linearly independent and generate linearly $\Delta(\mathcal{T}_{\mathbb{C}})$, we can define without possible contradictions for any element

$$R = \sum_{j=1}^{n} \lambda_j R_{f_j} \quad \text{of } \Delta(\mathcal{T}_{\mathbb{C}}),$$

$$\langle \phi; R \rangle = \sum_{j=1}^{n} \lambda_j \hat{\phi}(f_j),$$

which is then clearly a linear mapping from $\Delta(\mathcal{T}_{\mathbb{C}})$ to $\mathbb{C}$. We notice on writing $f_1 = f$ and $f_2 = 0$ in (iii) and on using (i), that we get $\hat{\phi}(-f) = \hat{\phi}(f)^*$ for all f in $\mathcal{T}_{\mathbb{C}}$, so that ϕ is hermitian on $\Delta(\mathcal{T}_{\mathbb{C}})$; (iii) actually has a stronger consequence on ϕ. Indeed from the definition of the involution and of the "twisted" product on $\Delta(\mathfrak{T}_{\mathbb{C}})$ we get

$$\left\langle \phi; \left(\sum_{j=1}^{n} \lambda_j R_{f_j}\right)^* \left(\sum_{j=1}^{n} \lambda_j R_{f_j}\right)\right\rangle = \sum_{j,k=1}^{n} \lambda_k^* \lambda_j \hat{\phi}(f_j - f_k) e^{-i\,\mathrm{Im}\,(f_k, f_j)/2},$$

which is positive by condition (iii); we then have obtained a positive linear form ϕ on the involutive algebra $\Delta(\mathfrak{T}_{\mathbb{C}})$. From Schwartz's inequality we have

$$|\langle \phi; R^*S\rangle|^2 \leqslant \langle \phi; R^*R\rangle\langle \phi; S^*S\rangle$$

and in particular, with $R = I$ and $S = R_f$,

$$|\langle \phi; R_f\rangle|^2 \leqslant \langle \phi; I\rangle = \hat{\phi}(0) = 1.$$

Consequently,

$$\left|\left\langle \phi; \sum_{j=1}^{n} \lambda_j R_{f_j}\right\rangle\right| \leqslant \sum_{j=1}^{n} |\lambda_j|\, |\langle \phi; R_{f_j}\rangle| \leqslant \sum_{j=1}^{n} |\lambda_j| = \left\|\sum_{j=1}^{n} \lambda_j R_{f_j}\right\|_1;$$

hence ϕ is continuous on $\Delta(\mathfrak{T}_{\mathbb{C}})$ and can be extended by continuity to a positive (continuous) linear form on the involutive Banach algebra $\Delta_1(\mathfrak{T}_{\mathbb{C}})$, with $\langle \phi; I\rangle = 1$; i.e., ϕ is a state on $\Delta_1(\mathfrak{T}_{\mathbb{C}})$. Looking back to the proof of Theorem I.3.14, we see that it goes through when $\mathfrak{R}$ is only an involutive Banach algebra with unit. Let π_ϕ be the GNS representation of $\Delta_1(\mathfrak{T}_{\mathbb{C}})$ associated with ϕ. Using the cyclicity of π_ϕ, we can verify by a straightforward computation that the weak-operator continuity of $\pi_\phi(R_{\lambda f})$ in λ, $\forall\, \lambda \in \mathbb{R}$, $f \in \mathfrak{T}_{\mathbb{C}}$ follows from condition (ii); hence π_ϕ belongs to $P(\mathfrak{T}_{\mathbb{C}})$. On using the preceding lemma, we get a representation $\mathfrak{W}_{\hat{\phi}} = \mathfrak{W}_{\pi_\phi}$ in $\mathfrak{W}_{\mathrm{I}}(\mathfrak{T}_{\mathbb{C}})$ with $(\Phi, W(f)\Phi) = \hat{\phi}(f)$ for all f in $\mathfrak{T}_{\mathbb{C}}$. Incidentally, since π_ϕ belongs to $P(\mathfrak{T}_{\mathbb{C}})$, we know already that it can be extended by continuity to a unique representation of $\overline{\Delta(\mathfrak{T}_{\mathbb{C}})}$, so that ϕ becomes a vector-state on $\pi(\overline{\Delta(\mathfrak{T}_{\mathbb{C}})})$ and then a state on $\overline{\Delta(\mathfrak{T}_{\mathbb{C}})}$. This establishes the existence of a representation $\mathfrak{W}_{\hat{\phi}}(\mathfrak{T}_{\mathbb{C}})$ that satisfies the conditions of the theorem. We now want to prove the uniqueness; we first notice that if ψ is a state on $\Delta_1(\mathfrak{T}_{\mathbb{C}})$ [or for that matter on $\overline{\Delta(\mathfrak{T}_{\mathbb{C}})}$] which coincides with $\hat{\phi}$ on $\Delta_0(\mathfrak{T}_{\mathbb{C}})$ the very construction of ϕ, the linearity of ψ and its continuity, imply that $\psi = \phi$. Hence to each $\hat{\phi}$ corresponds a *unique* state ϕ on $\Delta_1(\mathfrak{T}_{\mathbb{C}})$ [or $\overline{\Delta(\mathfrak{T}_{\mathbb{C}})}$], the restriction of which coincides with $\hat{\phi}$ on $\Delta_0(\mathfrak{T}_{\mathbb{C}})$. Now from Theorem II.1.1, we get that ϕ determines π_ϕ up to a unitary equivalence, so that indeed $\hat{\phi}$ determines $\mathfrak{W}_{\hat{\phi}}$ up to a unitary equivalence. To prove the second part of the theorem it is sufficient to prove that if $\mathfrak{W}_{\mathcal{H}}(\mathfrak{T}_{\mathbb{C}})$ is a representation in $\mathfrak{W}_{\mathrm{I}}(\mathfrak{T}_{\mathbb{C}})$, with cyclic vector Φ, then $\hat{\phi}(f) \equiv (\Phi, W(f)\Phi)$ satisfies the assumptions of the theorem. A straightforward computation again shows that this is indeed the case. This completes the proof of the theorem. ∎

As an illustration of this theorem we can now give a brief proof of von Neumann's uniqueness theorem. Let us consider for notational sake the simplest case of one degree of freedom; then $\mathcal{C}_{\mathbb{C}}$ is one-dimensional, i.e., $\mathcal{C}_{\mathbb{C}} = \mathbb{C}$ and $f = z = x + iy$; condition (II) is redundant and condition (III) is ignored. The unique normalized measure

$$d\mu(z) = \frac{1}{\sqrt{2\pi}} e^{-|z|^2/4}\, dz = \frac{1}{\sqrt{2\pi}} e^{-(x^2+y^2)/4}\, dx\, dy,$$

invariant with respect to all transformations $z \to z'$ with $|z| = |z'|$, is used to form for *any* irreducible representation $\mathfrak{W}_{\mathcal{H}}(\mathbb{C})$ of the Weyl commutation relations, the operator $A = \int W(z)\, d\mu(z)$ defined by

$$(\Phi, A\Psi) = \int (\Phi, W(z)\Psi)\, d\mu(z)$$

for all pairs Φ, Ψ of vectors in $\mathcal{H}$. A is clearly bounded [with norm at most equal to 1, since the $W(z)$ are unitary] and self-adjoint, since $W(z)^* = W(-z)$ and $d\mu(z) = d\mu(-z)$. Moreover, von Neumann proved in his 1931 paper, in which he introduced this operator, that the canonical commutation relations provide, after some straightforward computations, that $AW(z)A = Ae^{-|z|^2/4}$ for each z in $\mathbb{C}$ and ensure that $A \neq 0$; from this we get that A is a nonzero projector by writing $z = 0$ in the above result; consequently there exists at least one normalized vector Φ in $\mathcal{H}$ such that $A\Phi = \Phi$. Since $\mathfrak{W}_{\mathcal{H}}(\mathbb{C})$ is supposed to be irreducible, Φ is cyclic (see the above lemma); for this vector we form $\hat{\phi}(z) = (\Phi, W(z)\Phi) = (\Phi, AW(z)A\Phi) = (\Phi, A\Phi)e^{-|z|^2/4} = e^{-|z|^2/4}$. We now notice that in the Schrödinger representation in $\mathfrak{L}^2(\mathbb{R})$, $(W(z)\Psi)(\xi) = e^{-iy(\xi - x/2)}\Psi(\xi - x)$ for all Ψ in $\mathfrak{L}^2(\mathbb{R})$; on computing $\hat{\phi}(z) = (\Phi, W(z)\Phi)$ on the "vacuum" vector $\Phi(\xi) = \pi^{-1/4}e^{-\xi^2/2}$ we find $\hat{\phi}(z) = e^{-|z|^2/4}$. We can therefore conclude from the theorem that every irreducible representation of the canonical commutation relations in Weyl form, for one degree of freedom, is unitarily equivalent to the Schrödinger representation. Had the original representation not been irreducible, we would have obtained that each subspace of $\mathcal{H}$ spanned by $\{W(z)\Phi \mid z \in \mathbb{C}\}$ with Φ such that $A\Phi = \Phi$ is a stable subspace for the representation considered, supporting an irreducible representation unitarily equivalent to the Schrödinger representation. On considering this construction for an orthonormal basis $\{\Phi_j\}$ in the subspace of $\mathcal{H}$ formed by all vectors stable under A we get a complete proof of Theorem 6. The generalization to the case of n ($< \infty$) degrees of freedom is indeed trivial, since it is sufficient to replace in the beginning of the proof the measure $d\mu(z)$ by the gaussian measure $d\mu(f) = (2\pi)^{-n/2}e^{-|f|^2/4}\, df$ which incidentally is the only normalized product measure invariant under the group of all unitary transformations of $\mathcal{C}_{\mathbb{C}}$.

The proof of uniqueness can clearly not be extended to the case of an infinite number of degrees of freedom, since no such measure exists. We emphasize that Theorem 7 nevertheless holds true irrespective of the dimension (finite or infinite) of $\mathcal{C}_{\mathbb{C}}$. Let us, for instance, characterize the Fock representation by the fact that there exists a cyclic vector Φ such that $a(f)\Phi = 0$ for all f in $\mathcal{C}$. To compute the function $\hat{\phi}(f)$ characteristic of this representation we first notice that $a(f)\Phi = 0$ implies that $\{P(f)^2 + Q(f)^2\}\Phi = \Phi$ for every f in $\mathcal{C}$ with $|f|^2 = 1$; we use this fact to verify that the function

$$\psi(a, b) \equiv (\Phi, U(af)\, V(bf)\Phi)$$

satisfies the differential equations

$$\left(\frac{\partial^2}{\partial a^2} + \frac{\partial^2}{\partial b^2}\right)\psi(a, b) = \left(b^2 - 1 + 2ib\frac{\partial}{\partial a}\right)\psi(a, b)$$

$$= \left(a^2 - 1 + 2ia\frac{\partial}{\partial b}\right)\psi(a, b).$$

From these features it follows that

$$\psi(a, b) = \exp\left\{-\tfrac{1}{4}(a^2 + b^2) + \frac{i}{2}ab\right\};$$

hence for each degree of freedom $f \in \mathcal{C}$ with $|f|^2 = 1$

$$(\Phi, W(af + ibf)\Phi) = \exp\{-\tfrac{1}{4}(a^2 + b^2)\},$$

from which we conclude that for each f in $\mathcal{C}_{\mathbb{C}}$

$$\hat{\phi}(f) = (\Phi, W(f)\Phi) = \exp\{-\tfrac{1}{4}|f|^2\},$$

which we have already encountered at the beginning of our study of the free Bose gas (Subsection II.1.c). We should also take note of the fact that inequivalent norms can be put on $\mathcal{C}_{\mathbb{C}}$, leading to different $\hat{\phi}(f)$, hence to inequivalent Fock-space representations which depend on the norm chosen on $\mathcal{C}_{\mathbb{C}}$. This remark should be linked to some of the "strange phenomena" occurring in our discussion of the van Hove model in Section I.1.

This remark calls for some elaboration. Looking back at the definition of our algebra, we see that, although it is essential that $\mathcal{C}$ be a real pre-Hilbert space (equipped as such with an inner product), the fact that $\mathcal{C}_{\mathbb{C}}$ is equipped with an inner product can be dispensed with. The essential object of the theory is the *symplectic form* $\sigma(f, g)$ defined on $\mathcal{C}_{\mathbb{C}}$ by

$$\sigma(f, g) = \frac{i\{(f_1, g_2) - (f_2, g_1)\}}{2},$$

where the elements f_1, f_2, g_1, g_2 of $\mathcal{C}$ are related to the elements f, g of $\mathcal{C}_{\mathbb{C}}$ by

$$f = f_1 + if_2 \quad \text{and} \quad g = g_1 + ig_2.$$

If we insist on having an inner product on $\mathcal{T}_{\mathbb{C}}$, we are evidently still free to define it in many inequivalent ways, all with the same $\sigma(f, g) = i \operatorname{Im}(f, g)/2$; for instance, let $\hat{f}$ denote the Fourier transform of an element f in $\mathcal{T}$ and $\alpha(k)$, $\beta(k)$ be such that the following relations hold (almost everywhere in k):

$$\alpha(k)^* = \alpha(-k) \quad \text{and} \quad \beta(k)^* = \beta(-k),$$

(e.g., $\alpha(k) = \beta(k)^{-1} = (k^2 + m^2)^{-1/8}$); we can then define

$$(f, g)_{\alpha,\beta} = \int dk\{\alpha(k)^* \alpha(k) \hat{f}_1(k)^* \hat{g}_1(k) + \beta(k)^* \beta(k) \hat{f}_2(k)^* \hat{g}_2(k)\}$$
$$+ i \int dk\{\alpha(k)^* \beta(k) \hat{f}_1(k)^* \hat{g}_2(k) - \beta(k)^* \alpha(k) \hat{f}_2(k)^* \hat{g}_1(k)\}.$$

As long as $\alpha(k)^* \beta(k) = 1$ (almost everywhere in k) we have indeed $\sigma(f, g) = i \operatorname{Im} \{(f, g)_{\alpha\beta}\}/2$. We can then define $\hat{\phi}(f) = \exp \{-\frac{1}{4} |f|^2_{\alpha,\beta}\}$ and check that this functional gives rise to an irreducible "Fock" representation of the canonical commutation relations and that this representation is inequivalent to the representation with $\alpha = \beta = 1$. It is worth noticing in this respect that the norms $|\cdots|_{1,1}$ and $|\cdots|_{\alpha,\beta}$ defined by

$$|f|^2_{\alpha,\beta} = d\int k\{\alpha^*(k) \alpha(k) \hat{f}_1(k)^* \hat{f}_1(k) + \beta^*(k) \beta(k) \hat{f}_2(k)^* \hat{f}_2(k)\}$$

lead to different completions of the space $\mathcal{T}_{\mathbb{C}}$, thus touching the general problem of determining the "proper" test function space relative to a given representation. We should emphasize that no special significance has to be attached to the inner product $(f, g)_{\alpha,\beta}$ itself; we repeat that the only significant ingredients are the form $\sigma(f, g)$ which defines the CCR algebra and the norm $|\cdots|_{\alpha,\beta}$ which defines the "quadratic representation" considered. These particular representations of the CCR algebra (which are not irreducible when measurable deviations from $\alpha^*(k) \beta(k) = 1$ occur) have been studied by Klauder and Streit [1969], who illustrate their general theory of "quadratic" representations with examples chosen from quantum field theory (relativistic free and generalized free fields, rotationally symmetric model, the latter accommodating interactions, etc.) and quantum statistical mechanics (free and generalized free fields in thermal equilibrium, etc.); the reader is referred to this paper for further details (including a bibliography) on this aspect of the theory of representations of the CCR.

From Theorem 7 and the examples just mentioned we conclude that the aim fixed in the beginning of this subsection has now been attained: we have a general algebraic framework in which we can understand some of the particular results obtained earlier. Aside from its didactic value, such a generalized framework is badly needed in many-body theory; this last statement is elaborated on in the next subsection for the case of field-theoretical scattering theory on the basis of a general argument due originally to Haag.

From their construction the cyclic representations $\mathfrak{W}_{\hat{\phi}}$ obtained with Theorem 7 are such that the mappings $\lambda \in \mathbb{R} \to U(\lambda f) \in \mathfrak{B}(\mathcal{H}_{\hat{\phi}})$ and $\lambda \in \mathbb{R} \to V(\lambda g) \in \mathfrak{B}(\mathcal{H}_{\hat{\phi}})$ are continuous in the strong operator topology for each fixed f and g in $\mathcal{T}$. As we just remarked, we are still at liberty to equip $\mathcal{T}$ with any topology we please; the question is whether there is one topology that is more natural than the others, and if this is the case whether anything is to be gained by considering $\mathcal{T}$ as equipped with this topology. In this connection Hegerfeldt and Klauder [1970]† proved some interesting results which we want to mention. The metric defined by

$$\|f\|^2_{P,\hat{\phi}} = \int_{-\infty}^{+\infty} d\lambda e^{-\lambda^2} \,|(U_{\hat{\phi}}(\lambda f) - I)\Phi_0|^2$$

equips $\mathcal{T}$ with the weakest topology under which it becomes a linear topological space such that the mapping $f \in \mathcal{T} \to U_{\hat{\phi}}(f) \in \mathfrak{B}(\mathcal{H}_{\hat{\phi}})$ is continuous. Let $\mathcal{T}_{p,\hat{\phi}}$ be the completion of $\mathcal{T}$ with respect to this metric. For every sequence $\{f_n\}$ converging to $\tilde{f}$ in $\mathcal{T}_{P,\hat{\phi}}$, we define the unitary operator:

$$\tilde{U}_{\hat{\phi}}(\tilde{f}) = s - \lim_{n\to\infty} U_{\hat{\phi}}(f_n).$$

We define similarly

$$\|g\|_{F,\hat{\phi}} = \int_{-\infty}^{+\infty} d\lambda e^{-\lambda^2} \,|(V(\lambda g) - I)\Phi_0|^2,$$

which equips $\mathcal{T}$ with the weakest topology, under which it becomes a linear topological space such that the mapping $g \in \mathcal{T} \to V_{\hat{\phi}}(g) \in \mathfrak{B}(\mathcal{H}_{\hat{\phi}})$ is continuous. We define analogously $\mathcal{T}_{F,\hat{\phi}}$ and $\tilde{V}_{\hat{\phi}}(\tilde{f})$. The set

$$\tilde{W}_{\hat{\phi}}(\mathcal{T}_{\mathbb{C}}) \equiv \{\tilde{U}_{\hat{\phi}}(\tilde{f}), \tilde{V}_{\hat{\phi}}(\tilde{g}) \,|\, \tilde{f} \in \mathcal{T}_{P,\hat{\phi}}, \tilde{g} \in \mathcal{T}_{F,\hat{\phi}}\}$$

defines a generalized representation of the canonical commutation relations:

$$\tilde{U}_{\hat{\phi}}(\tilde{f}_1)\, \tilde{U}_{\hat{\phi}}(\tilde{f}_2) = \tilde{U}_{\hat{\phi}}(\tilde{f}_1 + \tilde{f}_2),$$
$$\tilde{V}_{\hat{\phi}}(\tilde{g}_1)\, \tilde{V}_{\hat{\phi}}(\tilde{g}_2) = \tilde{V}_{\hat{\phi}}(\tilde{g}_1 + \tilde{g}_2),$$
$$\tilde{U}_{\hat{\phi}}(\tilde{f})\, \tilde{V}_{\hat{\phi}}(\tilde{g}) = \tilde{V}_{\hat{\phi}}(\tilde{g})\, \tilde{U}_{\hat{\phi}}(\tilde{f}) e^{i[\tilde{f},\tilde{g}]}$$

for all $\tilde{f}, \tilde{f}_1, \tilde{f}_2$ in $\mathcal{T}_{P,\hat{\phi}}$ and all $\tilde{g}, \tilde{g}_1, \tilde{g}_2$ in $\mathcal{T}_{F,\hat{\phi}}$; clearly $\tilde{U}_{\hat{\phi}}(f) = U_{\hat{\phi}}(f)$ for all f in $\mathcal{T}$ and $\tilde{V}_{\hat{\phi}}(g) = V_{\hat{\phi}}(g)$ for all g in $\mathcal{T}$. This representation is a generalization of our usual representations in two respects: first the bilinear form $[\tilde{f}, \tilde{g}]$ obtained by continuity from (f, g) might be degenerate; second, the test

† See also Woods [1970], Araki and Woods [1971].

function spaces for the field and its "canonical conjugate" are not necessarily the same. The gain is that we now have the maximal extensions of the test function space compatible with the conditions that $\tilde{f} \to \tilde{U}_{\hat{\phi}}(\tilde{f})$ and $\tilde{g} \to \tilde{V}_{\hat{\phi}}(\tilde{g})$ are continuous. It should be noted that although these conditions determine the topology uniquely, they do not determine the metric itself, and many equivalent metrics can be constructed; up to this point the choice of the metric associated with the Gaussian measure $d\mu_\lambda = \exp(-\lambda^2)\, d\lambda$ is to be understood on the basis of its simplicity and mathematical convenience in the proofs.

We want to emphasize that up to this point the general theory—culminating in Theorem 7—has been developed without reference to any particular choice of an orthonormal basis in $\mathcal{T}$. This is in sharp contrast to the approach of Gårding and Wightman [1954b].† Although the approach of these authors belongs more to the realm of functional analysis than to the algebraic methods that are our primary interest in this book, we want to describe briefly their results without, however, entering into their proofs.

They define a representation of the canonical commutation relations (for a system with an infinite number of degrees of freedom) as follows: for each k in $\mathbb{Z}^+$ there are two continuous, one-parameter groups $\{U_k(s), V_k(t) \mid s, t \in \mathbb{R}\}$ of unitary operators acting on some separable Hilbert space $\mathcal{H}$ and such that

$$[U_k(s), U_l(t)] = 0,$$
$$[V_k(s), V_l(t)] = 0,$$
$$U_k(s)\, V_l(t) = V_l(t)\, U_k(s) e^{its\delta_{kl}}.$$

The contact with our formulation is evidently given as follows: let $\overline{\mathcal{T}}$ be a real Hilbert space, $\{e_k \mid k \in \mathbb{Z}^+\}$, an orthonormal basis in $\overline{\mathcal{T}}$, and $\mathcal{T}$, the linear manifold spanned by $\{e_k\}$; then $U_k(t) \equiv U(te_k)$, $V_l(s) \equiv V(se_l)$; conversely $W(\mathcal{T}_{\mathbb{C}})$ can be immediately reconstructed from

$$\{U_k(s), V_l(t) \mid k, l \in \mathbb{Z}^+;\, s, t \in \mathbb{R}\}.$$

Having thus made contact with their particular definition, we can proceed to describe their construction. Let $\mathfrak{X}$ be the set of all infinite sequences $\{n_1, n_2, \ldots, n_k, \ldots\} \equiv n$ of non-negative integers. Physically n characterizes the number n_k of particles in the state e_k. Mathematically, we notice that $\mathfrak{X}$ is a semigroup under componentwise addition. We then consider for every pair (k, j) of elements in $\mathbb{Z}^+$ the subset S_{kj} of $\mathfrak{X}$ of all sequences such that $n_k = j$; let $\mathcal{S}$ be the σ-ring generated by these subsets. We denote by Δ the set of all "rational" sequences, i.e., $\delta \in \Delta \subset \mathfrak{X}$ iff $\delta_k = 0$ for all but a finite

† See also Wightman and Schweber [1955].

number of k in $\mathbb{Z}^+$. For any measure μ on $(\mathfrak{X}, \mathcal{S})$ and any $\delta \in \Delta$ we define μ_δ as the measure obtained from μ by $\mu_\delta(S) = \mu(S + \delta)$; a measure μ on $(\mathfrak{X}, \mathcal{S})$ is said to be *quasi-invariant* whenever μ and μ_δ are absolutely continuous with respect to each other for every δ in Δ.

Let μ be a quasi-invariant measure on $(\mathfrak{X}, \mathcal{S})$, normalized to 1, let ν be a quasi-invariant, positive integer-valued function on $\mathfrak{X}$, and let $\mathcal{H}_n$ be a Hilbert space of dimension $\nu(n)$ for each $n \in \mathfrak{X}$. Then let $\mathcal{L}^2(\mathfrak{X}, \mu, \mathcal{H}_n)$ denote the direct integral $\int \mathcal{H}_n \, d\mu(n)$. For every k in $\mathbb{Z}^+$ we define the domain

$$\mathcal{D}_k = \left\{ f \in \mathcal{L}^2(\mathfrak{X}, \mu, \mathcal{H}_n) \,\middle|\, \int n_k \, |f(n)|^2 \, d\mu(n) < \infty \right\},$$

where $|f(n)|$ denotes the norm of $f(n)$ in $\mathcal{H}_n$. We then define for every f in $\mathcal{D}_k$

$$(a_k f)(n) = -i\sqrt{n_k + 1}\, C_k(n) f(n + \delta_k) \left[\frac{d\mu(n + \delta_k)}{d\mu(n)} \right]^{1/2},$$

where δ_k in Δ is the sequence $\{n_j = \delta_{kj}\}$ and $C_k(n)$ are measurable unitary operators from $\mathcal{H}_{n+\delta_k}$ to $\mathcal{H}_n$ which satisfy

$$C_k(n)\, C_l(n + \delta_k) = C_l(n)\, C_k(n + \delta_l).$$

We finally form

$$P_k = \frac{i(a_k^* - a_k)}{\sqrt{2}},$$

$$Q_k = \frac{a_k + a_k^*}{\sqrt{2}},$$

and verify that the closures of these operators are self-adjoint and generate $\{U_k(t), V_k(s)\}$, which form a representation of the canonical commutation relations in the sense just defined. Furthermore, we prove that every representation of the CCR is unitarily equivalent to one of these representations.

This is the original form in which our Theorem 7 was proved. The representations obtained in this way do, however, depend on the basis $\{e_k\}$ chosen to construct them; consequently, if we use this approach, we must check whether the results so obtained are themselves basis independent. As a positive example, we want to mention, in closing this subsection, the following result due to Reed [1969a]:

Reed's Theorem on Garding Domains: *Let* $\{U_k(s), V_l(t)\}$ *with* $k, l \in \mathbb{Z}^+$ *be a representation of the canonical commutation relations in the sense of Gårding and Wightman; in particular* $\mathcal{H}$ *is assumed to be separable and* P_k, Q_k *denote, respectively, the generators of* $U_k(s)$ *and* $V_k(t)$. *Then there exists a*

Banach space $\mathfrak{B}$ *of sequences of real numbers and a domain* $\mathfrak{D}$, *dense in* $\mathfrak{H}$, *such that for all* $\{c_k \mid k \in \mathbb{Z}^+\}$ *in* $\mathfrak{B}$

(i) $\sum_{k\in\mathbb{Z}^+} c_k Q_k$, $\sum_{k\in\mathbb{Z}^+} c_k P_k$ *are well defined and essentially self-adjoint on* $\mathfrak{D}$;

(ii) $(\sum_{k\in\mathbb{Z}^+} c_k Q_k)\mathfrak{D} \subseteq \mathfrak{D}$, $(\sum_{k\in\mathbb{Z}^+} c_k P_k)\mathfrak{D} \subseteq \mathfrak{D}$;

(iii) *if* $\{c_k^n\} \to \{c_k\} \in \mathfrak{B}$ *and* $\Psi \in \mathfrak{D}$, *then* $\sum_{k\in\mathbb{Z}^+} c_k^n Q_k \Psi \to \sum_{k\in\mathbb{Z}^+} c_k Q_k \Psi$ *and* $\sum_{k\in\mathbb{Z}^+} c_k^n P_k \Psi \to \sum_{k\in\mathbb{Z}^+} c_k P_k \Psi$.

Furthermore, Reed showed how far his result extends when $\mathfrak{T}$ is either the Schwartz test-function space $\mathcal{S}(\mathbb{R}^n)$ or some Hilbert space $\overline{\mathfrak{T}}$, thus establishing contact with the approaches of Gelfand and Vilenkin [1964], Segal [1963], and that followed in the main body of this subsection.

d. Haag's Theorem

The ideas developed in this subsection originated in a paper by Haag [1955], who pointed explicitly to some of the serious difficulties associated with the conventional formulation of a quantum field theory which attempts to be both relativistic and nontrivial. Haag's central idea was refined and reworked by many authors and standard proofs are available in the textbook literature (see, for instance, Chapter IV, Section 5 in Streater and Wightman [1963], Section 14 in Barton [1963]); most of these proofs, however, rely rather heavily on the analytic properties of the Wightman functions, which themselves reflect the locality and spectrum conditions, and tend to obscure the simple algebraic and group-theoretical facts actually responsible for the results obtained. These facts have recently been brought to light in a very elegant note by Streit [1968]; the account of Haag's theorem presented in this subsection rests mainly on that paper. In the sequel we consider Weyl representations of the canonical commutation relations that satisfy all conditions of subsection c; condition (II) is necessary to ensure contact with ordinary quantum field theory and is used in the subsequent argument. It might be noted that condition (III), in particular, the amenability of G and η-abelianness, is not the weakest possible assumption sufficient for the proof of the essential results of this subsection; we nevertheless keep these two conditions as stated, as they allow some simplifications and are satisfied anyhow in most cases of physical interest. Consider, for instance, the following special but typical case. Let $\overline{\mathfrak{T}}$ be the Hilbert space of all square-integrable real functions on $\mathbb{R}^3$ and $\mathfrak{T}$, the manifold of all real-valued C^∞-functions with compact support; F is then the scalar neutral field. For G we take the standard representation of the translation group T^3: $(g[f])(x) = f(g^{-1}[x])$; G is then amenable and the condition of η-abelianness is satisfied (incidentally, it is only through this mild consequence that the usual "locality assumption" will manifest itself in the proof presented here). We systematically suppose in the sequel that $\mathfrak{T}$ is infinite-dimensional.

Theorem 8 (Haag's Theorem, Part I). *Let* $\mathfrak{W}_{\mathcal{H}}(\mathfrak{T}_{\mathbb{C}})$ *be a Weyl representation of the canonical commutation relations, with cyclic vector* Φ. *Suppose that the state* $\hat{\phi}$ *on* $\mathfrak{W}_{\mathcal{H}}(\mathfrak{T}_{\mathbb{C}})$ *corresponding to* Φ *is* G-*invariant and* η-*clustering and that there exists a normalized vector* Ω *in* $\mathcal{H}$ *such that* $a(f)\Omega = 0$ *for all* f *in* $\mathfrak{T}$. *Then* $\Phi = \lambda\Omega$ *with* $\lambda \in \mathbb{C}$ and $|\lambda| = 1$.

Proof. From Theorem 7 we know that $\mathfrak{W}_{\mathcal{H}}(\mathfrak{T}_{\mathbb{C}})$ is unitarily equivalent to $\mathfrak{W}_{\hat{\phi}}(\mathfrak{T}_{\mathbb{C}})$ with $\hat{\phi}(f) = (\Phi, W(f)\Phi)$. From the lemma on p. 238 $\mathfrak{W}_{\hat{\phi}}(\mathfrak{T}_{\mathbb{C}})$ extends canonically to the representation π_ϕ of the C^*-algebra $\overline{\Delta(\mathfrak{T}_{\mathbb{C}})}$, where ϕ is the natural extension of $\hat{\phi}$ from $\mathfrak{W}_{\mathcal{H}}(\mathfrak{T}_{\mathbb{C}})$ to $\overline{\Delta(\mathfrak{T}_{\mathbb{C}})}$; ϕ is then clearly G-invariant and η-clustering, so that we can infer from Theorem II.2.8 that ϕ is the only G-invariant vector-state on the cyclic representation $\pi(\overline{\Delta(\mathfrak{T}_{\mathbb{C}})})$. On the other hand, consider the closed subspace $\mathcal{H}_\Omega$ spanned by $\mathfrak{W}_{\mathcal{H}}(\mathfrak{T}_{\mathbb{C}})\Omega$; $\mathcal{H}_\Omega$ is obviously stable under $\mathfrak{W}_{\mathcal{H}}(\mathfrak{T}_{\mathbb{C}})$. Let $\mathfrak{W}_\Omega(\mathfrak{T}_{\mathbb{C}})$ be the Weyl representation obtained by restricting $\mathfrak{W}_{\mathcal{H}}(\mathfrak{T}_{\mathbb{C}})$ to $\mathcal{H}_\Omega$; Ω is then cyclic for this representation and $\hat{\omega}(f) = (\Omega, W(f)\Omega)$. Since $a(f)\Omega = 0$ for all f in $\mathfrak{T}$, we know from the end of the preceding subsection that $\hat{\omega}(f) = \exp\{-\frac{1}{4}\,|f|^2\}$. We then have $\hat{\omega}(f) = \hat{\omega}(g[f])$ for all f in $\mathfrak{T}_{\mathbb{C}}$ and all g in G; hence the state $\langle\omega; R\rangle = (\Omega, \pi(R)\Omega)$, which coincides with the canonical extension of $\hat{\omega}$ from $\mathfrak{W}_{\mathcal{H}}(\mathfrak{T}_{\mathbb{C}})$ to $\pi(\overline{\Delta(\mathfrak{T}_{\mathbb{C}})})$, is a G-invariant vector-state on the representation $\pi(\overline{\Delta(\mathfrak{T}_{\mathbb{C}})})$. We now recall from the beginning of the proof that ϕ is the only such state, so that $\phi = \omega$ and, in particular, $\langle\phi; a^*(f)\,a(f)\rangle = 0$; i.e., $a(f)\Phi = 0$ for all f in $\mathfrak{T}$. Suppose now that Φ and Ω are not collinear; whenever this is the case, there exists Ω_1 in $\mathcal{H}$ such that $(\Omega_1, \Phi) = 0$ and $a(f)\Omega_1 = 0$ for all f in $\mathcal{H}$. A straightforward computation shows that $(\Omega_1, W(f)\Phi) = 0$ for all f in $\mathfrak{T}_{\mathbb{C}}$. Since Φ is cyclic by hypothesis, this implies that $\Omega_1 = 0$ and therefore Φ and Ω must be collinear; since they are both normalized to 1 we have indeed $\Omega = \lambda\Phi$ with $|\lambda| = 1$. ∎

REMARKS. We can paraphrase the result of Haag's theorem (Part I) by saying that it *prevents vacuum polarization* from occurring. More explicitly, in conventional quantum field theory we identify the "physical vacuum" ϕ as a G-invariant, η-clustering† vector state from which the representation space can be reconstructed by successive applications of the creation operators (i.e., Φ is supposed to be cyclic); in contradistinction, the *no-particle state* (or "bare vacuum") ω is identified as a vector state (for the same

† Even stronger clustering properties are usually expected from the "physical vacuum," but η-clustering is already sufficient for the argument (see Chapter 4).

representation!) that satisfies $\langle\omega; a^*(f)\, a(f)\rangle = 0$ for all f in $\mathfrak{T}$. We say that a given representation exhibits *vacuum polarization* when the physical vacuum and the no-particle state are distinguishable from one another; we sometimes even mean by this expression that the vector Φ generating ϕ and the vector Ω generating ω are not collinear. The theorem says precisely that even the first (hence the second) statement cannot hold. Our second remark emphasizes that the assumptions of the theorem do *not* presuppose that Ω is cyclic, nor that it is G-invariant, nor even that it is unique; all these properties come rather as consequences of the other assumptions which seem a priori less restrictive. Our last remark is that the condition of η-abelianness can be replaced by the weaker condition of G-abelianness (the formulation of which does not require the amenability of G; see Theorem II.2.6) if we require that G possess no other one-dimensional representation than the identity, and if, *in addition*, we suppose that Ω is unique up to a phase. To see this we first notice that we can assume without loss of generality that $\mathcal{H}$ hosts a representation $U(G)$ of G such that $U(g)\, a(f)\, U(g)^{-1} = a(g[f])$ for all f in $\mathfrak{T}$ and all g in G. This implies that for any g in G we have $a(f)\, U(g)\Omega = 0$ for all f in $\mathfrak{T}$, so that, by the assumed uniqueness of Ω, $U(g)\Omega = \lambda(g)\Omega$, where $\lambda(g)$ is a one-dimensional representation of G and must then be 1. Thus Ω is G-invariant and the rest of the proof goes as before, with the added simplification that G-abelianness is sufficient to ensure that E_0 is one-dimensional and Ω must be collinear with Φ. This way to proceed is closer to the standard presentation. Physically, however, the theorem, as stated, seems to be more general, since η-abelianness with respect to T^3 is expected to hold for most of the physically reasonable choices of $\mathfrak{T}$ we can think of, and since we did not need to assume the uniqueness of Ω. In the same vein we might notice that the proof would hold without modifications if η-abelianness were to be replaced by the condition that G be a large group of automorphisms for the representation $\mathfrak{W}_{\mathcal{H}}(\mathfrak{T}_{\mathbb{C}})$.

Haag's theorem (Part II) reinforces the results just stated, and their physical interpretation, by taking the dynamics explicitly into account. Loosely speaking, this theorem points out the inability of conventional field theories to describe scattering situations, using the interaction picture, in which the S-matrix is different from the identity.

We proceed in the following discussion of the second part of Haag's theorem in two steps: first we prove an abstract and elementary lemma and then prove and state the physically relevant theorem.

Lemma. *Let* G *be an amenable group of symmetries for* $(\mathfrak{R}, \mathfrak{S}, \langle;\rangle)$ *and assume η-abelianness in* G; *let* $\phi_{\rm j}$ ($\rm j = 1, 2$) *be* G*-invariant states on* $\mathfrak{R}$, *with* ϕ_1 *η-clustering; let* $\{\pi_{\rm j}(\mathfrak{R}), \rm U_j(G)\}$ *be the covariant representation associated with* $\phi_{\rm j}$, $\mathcal{H}_{\rm j}$ *the corresponding representation space, and* $\Phi_{\rm j}$ *the corresponding* *cyclic vector; let* $\{U_j(t) \mid t \in \mathbb{R}\}$ *be a weakly continuous, one-parameter group of unitary operators acting within* $\mathcal{H}_j$, *and such that* $U_j(t)\Phi_j = \Phi_j \ \forall\, t \in \mathbb{R}$; *let* $\{\pi_j^{(t)}(\mathfrak{R}), U_j^{(t)}(G)\}$ *be the covariant representation defined for each* t *in* $\mathbb{R}$ *by*

$$\pi_j^{(t)}(R) = U_j(t)\,\pi_j(R)\,U_j(-t),$$

$$U_j^{(t)}(g) = U_j(t)\,U_j(g)\,U_j(-t).$$

We suppose that for some $t = t_0$ *there exists a unitary mapping* $V(t_0)$ *from* $\mathcal{H}_2$ *onto* $\mathcal{H}_1$ *such that*

$$V(t_0)^*\,\pi_1^{(t_0)}(R)\,V(t_0) = \pi_2^{(t_0)}(R) \quad \textit{for all } R \textit{ in } \mathfrak{R}.$$

Then, $\phi_1 = \phi_2$, *and for each* t *in* $\mathbb{R}$ *there exists a unitary mapping* $U_0(t)$ *from* $\mathcal{H}_2$ *onto* $\mathcal{H}_1$ *such that*

(a) $$\Phi_2 = U_0^*(t)\Phi_1,$$

(b) $$\pi_2^{(t)}(R) = U_0^*(t)\pi_1^{(t)}(R)\,U_0(t)$$

(c) $$U_2^{(t)}(g) = U_0^*(t)\,U_1^{(t)}(g)\,U_0(t),$$

(d) $$U_0(t) = U_1(t)\,U_0(0)\,U_2(-t).$$

Proof. We first notice that under the assumptions of the lemma $\pi_1^{(t)}(\mathfrak{R})$ and $\pi_2^{(t)}(\mathfrak{R})$ are unitarily equivalent for all t in $\mathbb{R}$, the unitary mapping being given by $V^*(t) = U_2(t - t_0)\,V^*(t_0)\,U_1(t_0 - t)$. Writing $V = V(0)$, we can then assume without loss of generality that $\pi_2(R) = V^*\pi_1(R)V$ for all R in $\mathfrak{R}$. Now let ψ_1 be the vector state on $\pi_1(\mathfrak{R})$ generated by the vector $\Psi_1 \equiv V\Phi_2$: $\langle\psi_1; R\rangle = (\Psi_1, \pi_1(R)\Psi_1) = (V\Phi_2, \pi_1(R)V\Phi_2) = (\Phi_2, \pi_2(R)\Phi_2) = \langle\phi_2; R\rangle$, i.e., $\psi_1 = \phi_2$. Since ϕ_2 is G-invariant, so then is ψ_1; on the other hand, since ϕ_1 is G-invariant and η-clustering, it is the only G-invariant vector state on the representation $\pi_1(\mathfrak{R})$. We then have $\phi_1 = \psi_1 = \phi_2$ which proves the first part of the lemma. Since Φ_1 and Φ_2 are two cyclic vectors generating the same G-invariant state, there exists a unitary mapping U_0 from $\mathcal{H}_2$ onto $\mathcal{H}_1$ such that

(a) $$\Phi_2 = U_0^*\Phi_1,$$

(b) $$\pi_2(R) = U_0^*\pi_1(R)U_0,$$

(c) $$U_2(g) = U_0^*U_1(g)U_0;$$

defining $U_0(t)$ from $U_0(0) = U_0$ via the property (d) in the conclusion of the lemma, we see immediately that the latter is true. ∎

We now consider the following application of this lemma, which leads directly to the second part of Haag's theorem: let $\pi_j(\mathfrak{R})$ be two Weyl representations of the canonical commutation relations for the test function space $\mathcal{T}$; let G be a group of unitary transformations of $\mathcal{T}_{\mathbb{C}}$ that satisfies all the assumptions imposed above. We further denote by H_j the generator of the time evolution: $U_j(t) = \exp\{iH_j t\}\ \forall\ t \in \mathbb{R}$. From $U_j(t)\Phi_j = \Phi_j$ for all t in $\mathbb{R}$ we have $H_j\Phi_j = 0$. We further suppose that $\mathfrak{D}(H_j) \supseteq \mathfrak{D}_j$ and $H_j\mathfrak{D}_j \subseteq \mathfrak{D}_j$ (for the definition of $\mathfrak{D}_j$ see condition (II), p. 235). Moreover, we impose the fundamental assumption of the canonical formalism, namely that $P_j(f)$ is the canonical conjugate of $F_j(f)$, i.e.,

$$P_j(f) = i[H_j, F_j(f)] \quad \text{on} \quad \mathfrak{D}_j \qquad \forall f \in \mathcal{C}.$$

We now consider the time derivative at $t = 0$ of the finite product $F_1(f_1, t) \cdots F_1(f_n, t)\Phi_1$; since $F_1(f, t) = U_1(t)\, F_1(f)\, U_1(-t)$ and $U_1(t)\Phi_1 = \Phi_1$, we find

$$iH_1F_1(f_1) \cdots F_1(f_n)\Phi_1 = \sum_{\nu=1}^{n} F_1(f_1) \cdots F_1(f_{\nu-1})\, P_1(f_\nu)\, F_1(f_{\nu+1}) \cdots F_1(f_n)\Phi_1.$$

Under the assumptions of the lemma we find that this expression is equal to

$$U \sum_{\nu=1}^{n} F_2(f_1) \cdots F_2(f_{\nu-1})\, P_2(f)\, F_2(f_{\nu+1}) \cdots F_2(f_n)\Phi_2 = iUH_2U^*F_1(f_1) \cdots F_1(f_n)\Phi_1$$

so that

$$H_1 = UH_2U^* \quad \text{on} \quad \mathfrak{Q}_1(\mathcal{C})\Phi_1.$$

If we now suppose that Φ_1 is cyclic, not only with respect to $\mathfrak{W}_1(\mathcal{C}_{\mathbb{C}})$ but also with respect to $\mathfrak{Q}_1(\mathcal{C})$, i.e., $\mathfrak{Q}_1(\mathcal{C})\Phi_1$ is dense in $\mathcal{H}_1$, it is then a nontrivial but technical step to prove that the restriction of H_1 to this domain has a unique self-adjoint extension, namely H_1, which is unitarily equivalent to H_2. This has been proved by Streit [1968] on the basis of Nelson's theorem [1959] on analytic vectors, but we do not want to make this account unnecessarily heavy by reproducing the last part of this proof here. We then have that $U_2(t) = U^*U_1(t)U$ for all t in $\mathbb{R}$ and consequently $\pi_2(R) = U^*\pi_1(R)U$ for all R in $\mathfrak{R}$, with the *same* U. We can summarize the results so far obtained as follows:

Theorem 9 (Haag's Theorem, Part II). *Let* $\{U_j(f), V_j(f) \mid f \in \mathcal{C}\}$ $(j = 1, 2)$ *be two Weyl representations of the canonical commutation relations satisfying the assumptions of the lemma. Suppose that the respective generators* $F_j(f)$, $P_j(f)$, *and* H_j *of* $V(_jf)$, $U(_jf)$, *and* $U_j(t)$ *satisfy the following conditions:*

(i) *there exists a linear manifold* $\mathfrak{D}_j$, *dense in* $\mathcal{H}_j$, *and stable with respect to* $F_j(f)$, $P_j(f)$, *and* H_j;
(ii) *on* $\mathfrak{D}_j$, $P_j(f) = i[H_j, F_j(f)]$;
(iii) $\mathfrak{Q}_1(\mathcal{C})\Phi_1$ *is dense in* $\mathcal{H}_1$.

Then $H_2 = U^*H_1U$, *where* U *is the unitary operator, the existence and properties of which have been established in the lemma.*

The assumptions of the theorems obtained in this subsection are satisfied in conventional field theories; in particular, it was tacitly assumed, before Haag's paper that all "physically relevant" representations of the canonical relations were unitarily equivalent; this was one of the reasons why so much attention was devoted to Fock-space techniques in the first phase of the development of field theory; other representations were referred to as "strange." In this conventional framework Haag's theorem stirred up much trouble, since it asserts that two quantum fields (the time evolutions of which are supposed to be unitarily implemented) that are unitarily equivalent at any given time are both free if one of them is supposed to be free; in particular, the theorem shows that the "interaction picture" used in conventional field theory for scattering purposes can describe only free fields, so that the S-matrix can only be trivial unless some drastic mutilation is performed on the formalism.

As already indicated by the concrete van Hove model discussed in Section I.1 and as stated in Haag's original paper, the only possible way to get out of this difficulty is to stand ready to accept the physical significance of unitarily inequivalent (irreducible) representations of the canonical commutation relations. To show that such representations do (fortunately!) exist in great multiplicity is the object of Subsection f. Once again we want to recall that this result is in sharp contradistinction to what we learned in Subsection b, where we saw that all irreducible representations of the canonical commutation relations corresponding to a system with a *finite* number of degrees of freedom are unitarily equivalent. We might wonder, incidentally, why for such systems some theorem similar to Haag's has not hampered the development of most usual scattering theory; the reason for this is, expectably, that at least one of the assumptions in Haag's theorem is so typical of field theories that it cannot be true for a system with a finite number of degrees of freedom. This is the case in all proofs given of Haag's theorem, and we want to close this subsection by pointing out what this assumption is in the presentation made here. Suppose, indeed, that $\mathfrak{T}$ is finite-dimensional and let $\{f_j \mid j = 1, 2, \ldots, n\}$ be an orthonormal basis in $\mathfrak{T}$; we then form

$$A = \frac{\sum_{j=1}^{n} a^*(f_j)^2}{2n}$$

and see immediately that if g is a unitary transformation $f \to g[f]$ of $\mathfrak{T}$

$$\frac{\sum_{j=1}^{n} a^*(g[f_j])^2}{2n} = A.$$

If ϕ is now a G-invariant state, Φ, the corresponding cyclic vector, and $U_\phi(G)$, the corresponding representation of G, then $A\Phi$ is also invariant under $U_\phi(G)$, so that E_ϕ is not one-dimensional; furthermore, the G-invariant vector state ϕ' generated by $A\Phi$ is different from ϕ, so that ϕ cannot be the only G-invariant vector state on the representation considered, a condition that was central in the proofs, as given here, and in the physical interpretation of the results. We then have indeed that Haag's theorem is characteristic of conventional field theory and does not apply to systems with a finite number of degrees of freedom.

e. C^*-Inductive Limit and IDPS

This subsection is devoted to a rapid survey of the theory of inductive limits and infinite direct products of C^*-algebras. These notions can be traced back in the mathematical literature to the papers of Murray and von Neumann [1936] and von Neumann [1938]; the extension to abstract C^*- and W^*-algebras has been worked out by Turumaru [1952, 1953, 1954, 1956] and by Takeda [1955]. Although the theory is developed here with the CCR-algebra in mind, its physical applications extend further than the study of Boson-systems and cover similar aspects of the study of Fermi-systems (see, for instance, Guichardet [1966], Størmer [1969a], Balsev, Manuceau, and Verbeure [1968]) and more generally any quasi-local system (see Chapter 4).

We recall that a set $\mathcal{F}$ is said to be a *directed set* if $\mathcal{F}$ is a partially ordered set such that for any two elements F_1 and F_2 in $\mathcal{F}$ there exists an element F in $\mathcal{F}$ with $F_1 \leqslant F$ and $F_2 \leqslant F$; for example, $\mathcal{F}$ could be the set of all compact subsets of $\mathbb{R}^3$ or the set of all finite-dimensional subspaces of an infinite-dimensional Hilbert space, the partial ordering being, in both cases, provided by the natural set inclusion. We now suppose that with any element F in $\mathcal{F}$ we associate some C^*-algebra $\mathfrak{R}_F$ (with unit I_F); if there exists a C^*-algebra $\mathfrak{R}$ (with unit I) such that for every F in $\mathcal{F}$ there exists an injective *-homomorphism i_F from $\mathfrak{R}_F$ into $\mathfrak{R}$ such that

$$i_F(I_F) = I,$$

$$i_{F_1}(\mathfrak{R}_{F_1}) \subseteq i_{F_2}(\mathfrak{R}_{F_2}), \quad \text{whenever } F_1 \leqslant F_2,$$

$$\overline{\bigcup_{F\in\mathcal{F}} i_F(\mathfrak{R}_F)} = \mathfrak{R},$$

then $\mathfrak{R}$ is called the *C^*-inductive limit* of $\{\mathfrak{R}_F \mid F \in \mathcal{F}\}$. Takeda [1955] proved that a sufficient condition for a family $\{\mathfrak{R}_F \mid F \in \mathcal{F}\}$ (where $\mathcal{F}$ is a directed set) to admit a C^*-inductive limit is that there exists, for every pair F_1, F_2 of elements in $\mathcal{F}$ with $F_1 \leqslant F_2$, an injective *-homomorphism $i_{2,1}$ from $\mathfrak{R}_{F_1}$ into $\mathfrak{R}_{F_2}$ such that

$$i_{2,1}(I_{F_1}) = I_{F_2}$$

and

$$i_{3,1} = i_{3,2}i_{2,1}, \quad \text{whenever} \quad F_1 \leqslant F_2 \leqslant F_3.$$

We now proceed with the example relevant to the forthcoming discussion of the Weyl representations of the canonical commutation relations. Let $\overline{\mathcal{E}}$ be a separable real Hilbert space, let $\{e_j \mid j \in \mathbb{Z}^+\}$, where $\mathbb{Z}^+$ is the set of all non-negative integers, be an orthonormal basis in $\overline{\mathcal{E}}$, and let $\mathcal{E}$ be the linear manifold spanned by $\{e_j\}$, i.e., the set of all finite linear combinations (with real coefficients) of the e_j's. We further introduce the natural complexification $\mathcal{E}_{\mathbb{C}}$ of $\mathcal{E}$. If, for instance, $\overline{\mathcal{E}}_{\mathbb{C}} = \mathfrak{L}^2(\mathbb{R})$, we could take for $\{e_j\}$ the set of eigenfunctions

$$e_j(x) = (\sqrt{\pi}2^j j!)^{-1/2} e^{-1/2 x^2} H_j(x)$$

of the harmonic oscillator (H_j is the Hermite polynomial of degree j). We now construct $\mathcal{F}$ as the family of all linear subspaces $F_{\mathbb{C}}$ of $\mathcal{E}_{\mathbb{C}}$ spanned by $\{e_j \mid j \in J\}$ (where J runs over all *finite* subsets of $\mathbb{Z}^+$). The natural inclusion of subspaces provides a partial ordering of $\mathcal{F}$ which thus becomes clearly a directed set. We notice incidentally that $\mathcal{F}$ is *absorbing* in $\mathcal{E}_{\mathbb{C}}$, i.e., $\mathcal{E}_{\mathbb{C}} = \bigcup_{\mathcal{F}} F_{\mathbb{C}}$. We then associate with every $F_{\mathbb{C}}$ in $\mathcal{F}$ the C^*-algebra $\overline{\Delta(F_{\mathbb{C}})}$.

Theorem 10. $\overline{\Delta(\mathcal{E}_{\mathbb{C}})}$ *is the* C^**-inductive limit of* $\{\overline{\Delta(F_{\mathbb{C}})} \mid F_{\mathbb{C}} \in \mathcal{F}\}$.

Proof. Since $\mathcal{F}$ is absorbing for $\mathcal{E}_{\mathbb{C}}$, we clearly have $\Delta(\mathcal{E}_{\mathbb{C}}) = \bigcup_{F_{\mathbb{C}} \in \mathcal{F}} \Delta(F_{\mathbb{C}})$ and we immediately have a natural embedding of $\Delta(F_{\mathbb{C}})$ into $\Delta(\mathcal{E}_{\mathbb{C}})$, which is an injective *-homomorphism that sends the identity of $\Delta(F_{\mathbb{C}})$ to the identity of $\Delta(\mathcal{E}_{\mathbb{C}})$; the remaining part of the proof consists in showing that this embedding preserves the norms and can be extended by continuity to the C^*-algebras $\overline{\Delta(F_{\mathbb{C}})}$ and $\overline{\Delta(\mathcal{E}_{\mathbb{C}})}$, respectively. To see this we first remark that the restriction π_F to $\Delta(F_{\mathbb{C}})$ of any π in $P(\mathcal{E}_{\mathbb{C}})$ belongs to $P(F_{\mathbb{C}})$; furthermore, since F is finite-dimensional, we know from von Neumann's theorem that every representation in $P(F)$ is a direct sum of Schrödinger representations, so that we have

$$\|R\|_F \equiv \sup_{\rho \in P(F_{\mathbb{C}})} \|\rho(R)\| = \|\rho(R)\|$$

for any R in $\Delta(F_{\mathbb{C}})$ and any representation ρ in $P(F_{\mathbb{C}})$; putting together these two facts, we have

$$\|R\| = \sup_{\pi \in P(\mathcal{E}_{\mathbb{C}})} \|\pi(R)\| = \|\pi_F(R)\| = \|R\|_F.$$

The embedding considered above can then be extended to an embedding of $\overline{\Delta(F_{\mathbb{C}})}$ into $\overline{\Delta(\mathcal{E}_{\mathbb{C}})}$. To complete the proof of the theorem it is now sufficient to notice that

$$\Delta(\mathcal{E}_{\mathbb{C}}) = \bigcup_{F_{\mathbb{C}} \in \mathfrak{F}} \Delta(F_{\mathbb{C}}) \subset \bigcup_{F_{\mathbb{C}} \in \mathfrak{F}} \overline{\Delta(F_{\mathbb{C}})} \subset \overline{\Delta(\mathcal{E}_{\mathbb{C}})}$$

so that $\bigcup_{F_{\mathbb{C}} \in \mathfrak{F}} \overline{\Delta(F_{\mathbb{C}})}$ is dense in $\overline{\Delta(\mathcal{T}_{\mathbb{C}})}$, since the latter is the completion, in the norm just considered, of $\Delta(\mathcal{T}_{\mathbb{C}})$. ▮

Manuceau [1968b] shows that the following result is a consequence of Theorem 10:

Corollary. $\overline{\Delta(\mathcal{T}_{\mathbb{C}})}$ *is* not *separable.*

We now want to introduce the concept of the infinite direct product of C^*-algebras; to do so we define successively the infinite direct products of vector spaces, involutive algebras, and C^*-algebras; in the process we make a digression on the related notion of the infinite direct product of Hilbert spaces which also play an important role in the sequel.

We consider first† an indexed family $\{E_\gamma \mid \gamma \in \Gamma\}$ of complex vector spaces E_γ and denote by $\times_{\gamma\in\Gamma} E_\gamma$ the vector space obtained as the quotient of the vector space of all formal linear combinations of families $\{x_\gamma \mid \gamma \in \Gamma;\, x_\gamma \in E_\gamma\}$ by the vector space generated by elements of the form (a) $\{x_\gamma\} + \{y_\gamma\} + \{z_\gamma\}$ with $x_\delta + \gamma_\delta = z_\delta$ for one index $\delta \in \Gamma$ and $x_\gamma = y_\gamma = z_\gamma$ for all $\gamma \neq \delta$ or (b) $\{x_\gamma\} - \lambda\{y_\gamma\}$ with $x_\delta = \lambda y_\delta$ for one index $\delta \in \Gamma$ and $x_\gamma = y_\gamma$ for all $\gamma \neq \delta$.

The *direct product* (or *tensor product*) $\times_{\gamma\in\Gamma} E_\gamma$ enjoys the following three fundamental properties:

(i) For every family $\{x_\gamma \mid \gamma \in \Gamma;\, x_\gamma \in E_\gamma\}$ there exists an element $\times_{\gamma\in\Gamma} x_\gamma$ of $\times_{\gamma\in\Gamma} E_\gamma$, linear in each of its components; furthermore, every element in $\times_{\gamma\in\Gamma} E_\gamma$ is a linear combination of such elements.

(ii) For every multilinear mapping $f : \Pi_{\gamma\in\Gamma} E_\gamma \to F$, where F is an arbitrary vector space, there exists a unique linear mapping $\psi : \times_{\gamma\in\Gamma} E_\gamma \to F$ such that $f(\{x_\gamma\}) = \psi(\times\, x_\gamma)$.

(iii) For every partition $\{\Gamma_M\} = \bigcup_{\mu\in M} \Gamma_\mu$ of Γ there exists a unique isomorphism from $\times_{\gamma\in\Gamma} E_\gamma$ onto $\times_{\mu\in M}(\times_{\gamma\in\Gamma_\mu} E_\gamma)$ mapping $\times_{\gamma\in\Gamma} x_\gamma$ to $\times_{\mu\in M} (\times_{\gamma\in\Gamma_\mu} x_\gamma)$.

For an arbitrary family $\{a_\gamma \mid \gamma \in \Gamma;\, a_\gamma \in E_\gamma;\, a_\gamma \neq 0\}$ we denote by $\times_{\gamma\in\Gamma} {}^a E_\gamma$ the subspace of $\times_{\gamma\in\Gamma} E_\gamma$ spanned by all elements of $\times_{\gamma\in\Gamma} E_\gamma$ of the form $\times_{\gamma\in\Gamma} x_\gamma$ with $x_\gamma = a_\gamma$ for all but a finite number of γ in Γ.

Suppose now that $\{\mathcal{H}_\gamma \mid \gamma \in \Gamma\}$ is a family of Hilbert spaces and let $a = \{a_\gamma \mid \gamma \in \Gamma;\, a_\gamma \in \mathcal{H}_\gamma;\, |a_\gamma| = 1\}$ we form $\times_{\gamma\in\Gamma} {}^a\mathcal{H}_\gamma$ in the way just indicated, and we equip this vector space with the scalar product

$$\Big(\underset{\gamma\in\Gamma}{\times} x_\gamma, \underset{\gamma\in\Gamma}{\times} y_\gamma\Big) = \prod_{\gamma\in\Gamma} (x_\gamma, y_\gamma);$$

† For more details see Bourbaki [1948], Chapter III, Appendix I. Bourbaki, and many other authors, denote $\times_{\gamma\in\Gamma} E_\gamma$ by the symbol $\otimes_{\gamma\in\Gamma} E_\gamma$; we prefer to keep the symbol $\otimes$ for the cases in which a topological completion has been performed. This allows a somewhat lighter notation, more adapted to our specific needs here.

$\times_{\gamma\in\Gamma}{}^a\mathcal{H}_\gamma$ then becomes a prehilbert space, the completion of which (with respect to the norm associated with this scalar product) we denote by $\bigotimes_{\gamma\in\Gamma}{}^a\mathcal{H}_\gamma$. This is the von Neumann [1938] incomplete direct product space (IDPS) relative to the family $a = \{a_\gamma \mid \gamma \in \Gamma;\, a_\gamma \in \mathcal{H}_\gamma;\, |a_\gamma| = 1\}$; since the definition of this object is sometimes given in another form, we first notice that, given a, to any family $x = \{x_\gamma \mid \gamma \in \Gamma;\, x_\gamma \in \mathcal{H}_\gamma\}$ such that

(i) $$\sum_{\gamma\in\Gamma} \big||x_\gamma|^2 - 1\big| < \infty,$$

(ii) $$\sum_{\gamma\in\Gamma} |(x_\gamma, a_\gamma) - 1| < \infty$$

corresponds a vector $\bigotimes_{\gamma\in\Gamma} x_\gamma$ in $\bigotimes_{\gamma\in\Gamma}{}^a\mathcal{H}_\gamma$, obtained as the limit of $\times_{\gamma\in\Gamma'} x_\gamma$, where Γ' runs over the finite subsets of Γ, and such that

(i) $$\Big|\bigotimes_{\gamma\in\Gamma} x_\gamma\Big| = \prod_{\gamma\in\Gamma} |x_\gamma| \,(<\infty),$$

(ii) $$\Big|\bigotimes_{\gamma\in\Gamma} x_\gamma\Big| = 0 \quad \text{iff at least one } x_\gamma \text{ is } 0.$$

This remark suggests the definition of two equivalence relations between the families $\{a_\gamma \mid \gamma \in \Gamma;\, a_\gamma \in \mathcal{H}_\gamma;\, |a_\gamma| = 1\}$; we say that two such families are in the relation $a \approx b$ (resp. $a \sim b$) whenever $\sum_{\gamma\in\Gamma} |(a_\gamma, b_\gamma) - 1| < \infty$ (resp. $\sum_{\gamma\in\Gamma} \big||(a_\gamma, b_\gamma)| - 1\big| < \infty$). These two relations are clearly equivalence relations; furthermore $a \sim b$ if and only if there exists a family $\alpha = \{\alpha_\gamma \mid \gamma \in \Gamma;\, \alpha_\gamma \in \mathbb{C};\, |\alpha_\gamma| = 1\}$ such that $b \approx \alpha a = \{\alpha_\gamma a_\gamma \mid \gamma \in \Gamma\}$. The interesting feature about these relations is that if a, b, and α are defined as above and satisfy $b \approx \alpha a$, there exists a unique isomorphism from $\bigotimes_{\gamma\in\Gamma}{}^a\mathcal{H}_\gamma$ onto $\bigotimes_{\gamma\in\Gamma}{}^b\mathcal{H}_\gamma$, which sends $\bigotimes_{\gamma\in\Gamma} x_\gamma$ ($\in \bigotimes_{\gamma\in\Gamma}{}^a\mathcal{H}_\gamma$ with $x_\gamma = a_\gamma$ for all but a finite number of γ in Γ) to $\bigotimes_{\gamma\in\Gamma} \alpha_\gamma x_\gamma$ ($\in \bigotimes_{\gamma\in\Gamma}{}^a\mathcal{H}_\gamma$). This establishes the connection with von Neumann's [1938] notation: $\bigotimes_{\gamma\in\Gamma}{}^a\mathcal{H}_\gamma$ is the space denoted $\Pi\bigotimes_{\gamma\in\Gamma}{}^{\mathfrak{C}}\mathcal{H}_\gamma$ (where $\mathfrak{C}$ stands for the $\approx$-equivalence class of a); the "complete" product is then obtained as follows: pick one representative a in each $\approx$-equivalence class and form the direct sum of the corresponding $\bigotimes_{\gamma\in\Gamma}{}^a\mathcal{H}_\gamma$; this space is denoted $\Pi\bigotimes_{\gamma\in\Gamma}\mathcal{H}_\gamma$ and its connection with the universal "van Hove receptacle" is discussed later on. Finally the space denoted $\Pi\bigotimes_{\gamma\in\Gamma}{}^{\mathfrak{C}}w\mathcal{H}_\gamma$ is obtained in the same manner except that we restrict the a's to the same $\sim$-equivalence class.

After this digression on the infinite direct products of Hilbert space let us come back to the point at which, on p. 255, we defined the restricted infinite product $\times_{\gamma\in\Gamma}{}^aE_\gamma$ of $\{E_\gamma \mid \gamma \in \Gamma\}$ with respect to $a = \{a_\gamma \mid \gamma \in \Gamma\}$. We now consider a family $\{\mathfrak{R}_\gamma \mid \gamma \in \Gamma\}$ of involutive algebras with unit I_γ and repeat the above construct with respect to $I = \{I_\gamma \mid \gamma \in \Gamma\}$ to get the vector space $\times_{\gamma\in\Gamma}{}^I\mathfrak{R}_\gamma$, which we simply denote as $\times_{\gamma\in\Gamma}\mathfrak{R}_\gamma$ (by an abuse of language that might anyhow lead to confusion only in the case in which the index set Γ is infinite). We can now transfer to $\mathsf{X}_{\gamma\in\Gamma}\,\mathfrak{R}_\gamma$ the involutive algebraic structure of the $\mathfrak{R}_\gamma$'s by defining the product

$$\left(\mathop{\mathsf{X}}_{\gamma\in\Gamma} R_\gamma\right)\left(\mathop{\mathsf{X}}_{\gamma\in\Gamma} S_\gamma\right) = \mathop{\mathsf{X}}_{\gamma\in\Gamma}(R_\gamma S_\gamma)$$

and the involution

$$\left(\mathop{\mathsf{X}}_{\gamma\in\Gamma} R_\gamma\right)^* = \mathop{\mathsf{X}}_{\gamma\in\Gamma} R_\gamma^*,$$

which naturally extend by linearity to $\mathsf{X}_{\gamma\in\Gamma}\,\mathfrak{R}_\gamma$. The last step in our brief review is to reach a satisfactory definition for the infinite direct product of C^*-algebras. To achieve this we first consider the case of a family $\{\mathfrak{R}_\gamma \mid \gamma\in\Gamma\}$ of C^*-algebras with units over a *finite* index set Γ; we then define the involutive algebra $\mathsf{X}_{\gamma\in\Gamma}\,\mathfrak{R}_\gamma$ as just indicated. To define on it a norm we choose a family $\{\pi_\gamma : \mathfrak{R}_\gamma \to \mathfrak{B}(\mathcal{H}_\gamma) \mid \gamma\in\Gamma\}$ of *faithful* representations π_γ; since Γ is finite, we can form without problems the representation $\pi = \mathsf{X}_{\gamma\in\Gamma}\,\pi_\gamma$ of $\mathsf{X}_{\gamma\in\Gamma}\,\mathfrak{R}_\gamma$ into $\mathfrak{B}(\bigotimes_{\gamma\in\Gamma}\mathcal{H}_\gamma)$ defined by

$$\pi\left(\mathop{\mathsf{X}}_{\gamma\in\Gamma} R_\gamma\right)\left(\bigotimes_{\gamma\in\Gamma} x_\gamma\right) = \bigotimes_{\gamma\in\Gamma}(\pi_\gamma(R_\gamma)x_\gamma).$$

We now introduce the norm

$$\left\| \sum_{j=1}^m \mathop{\mathsf{X}}_{\gamma\in\Gamma} R_{\gamma,j} \right\| = \left\| \sum_{j=1}^m \mathop{\mathsf{X}}_{\gamma\in\Gamma} \pi_\gamma(R_{\gamma,j}) \right\|;$$

it is a nontrivial result that this norm actually does *not* depend on the choice of the representations π_γ, provided that the latter are faithful; for a proof in the case in which $\Gamma = \{1, 2\}$, see Theorem 1 in Wulfsohn [1963]; the proof in the case of a finite index set proceeds along the same lines or by induction. We now define $\bigotimes_{\gamma\in\Gamma}\mathfrak{R}_\gamma$ as the completion, with respect to this norm, of the involutive algebra $\mathsf{X}_{\gamma\in\Gamma}\,\mathfrak{R}_\gamma$ and obtain in this way a C^*-algebra called the *direct product* of $\{\mathfrak{R}_\gamma \mid \gamma\in\Gamma\}$. This definition was worked out for the case in which the index set Γ is finite; some care should be exercised when Γ is of arbitrary cardinality. In the latter case we consider first the set $\mathcal{F}$, the elements of which are the finite subsets $\mathcal{F}$ of the index set Γ; equipped with the partial ordering provided by the inclusion of sets, $\mathcal{F}$ clearly becomes a directed set. For each $F = \{\gamma_1, \gamma_2, \ldots, \gamma_n\}$ we then form the C^*-algebra

$$\mathfrak{R}_F = \mathfrak{R}_{\gamma_1} \otimes \mathfrak{R}_{\gamma_2} \otimes \cdots \otimes \mathfrak{R}_{\gamma_n};$$

the natural embedding of $\mathfrak{R}_{F_1}$ into $\mathfrak{R}_{F_2}$, when $F_1 \subseteq F_2$, satisfies Takeda's condition (p. 253), so that the C^*-inductive limit of $\{\mathfrak{R}_F \mid F\in\mathcal{F}\}$ exists; we denote† it by $\bigotimes_{\gamma\in\Gamma}\mathfrak{R}_\gamma$ and refer to it as the *infinite direct product of the C^*-algebras* $\mathfrak{R}_\gamma(\gamma\in\Gamma)$.

† This product is often denoted in the literature by $\bigotimes^*_{\gamma\in\Gamma}\mathfrak{R}_\gamma$.

With these definitions cleared up, we now come back to the C^*-algebra of the canonical commutation relations for the test function space $\mathcal{T}$ considered above. We denote by F_j the one-dimensional complex subspace of $\mathcal{T}_{\mathbb{C}}$ generated by the basis vector e_j; again using von Neumann's theorem, we see that for any subspace $F_{\mathbb{C}}$ of $\mathcal{T}_{\mathbb{C}}$, spanned by $\{e_j \mid j \in J\}$, where J is any finite subset of $\mathbb{Z}^+$, we have

$$\overline{\Delta(F_{\mathbb{C}})} = \bigotimes_{j \in J} \overline{\Delta(F_j)}.$$

From the very definition of the infinite direct product, and from Theorem 10, we then get the following:

Theorem 11. $$\overline{\Delta(\mathcal{T}_{\mathbb{C}})} = \bigotimes_{j \in \mathbb{Z}^+} \overline{\Delta(F_j)}.$$

Various generalizations of this theorem can be thought of; in particular, given any test function space $\mathcal{T}$ and any symplectic basis† $\{e_j, f_j \mid j \in \mathbb{Z}^+\}$, we can repeat the above construction and get a similar result (see, for instance, Manuceau [1968b]). Theorem 11, as it stands, will nevertheless be sufficient for the discussion in Subsection f.

f. Representations Associated with Product States

We now come back to the general problem of constructing representations of the canonical commutation relations which could be of interest in a description of physical systems exhibiting an infinite number of degrees of freedom.

The special application which guides the development of this subsection is the study of certain Weyl representations for the particular case in which $\mathcal{T}$ is the linear manifold spanned by the orthonormal basis $\{e_j\}$ of $\overline{\mathcal{T}}$. The object of this detailed study is to emphasize those aspects of the theory that make it a genuine and nontrivial generalization of the case in which we have only to consider a finite number of degrees of freedom. This approach, developed to a great extent as a sequel to a paper by Kastler [1965], also allows us to tie up: (i) von Neumann's [1931] theorem for the case in which dim $\mathcal{T} < \infty$ (our Theorem 6); (ii) the classification of the Weyl representations in the case in which dim $\mathcal{T} = \infty$, obtained by Gårding and Wightman [1954 b], Wightman and Schweber [1955] (see the end of our Subsection c); and (iii) the applications of the general theory due to Araki and

† A subset $\{e_j, f_j \mid j \in J\}$ of $\mathcal{T} \times \mathcal{T}$ is said to be a *symplectic basis* if given a symplectic form σ: (i) $\sigma(e_j, f_j) = 1 \; \forall j \in J$, (ii) $\sigma(e_i, e_j) = \sigma(f_i, f_j) = 0$ if $i \neq j$, (iii) $\bigcup_{j \in J} \{e_j, f_j\}$ is a basis in $\mathcal{T}$.

Woods [1963], Streit [1967], Hugenholtz [1967], Størmer [1969a, b], Manuceau and Verbeure [1968], Manuceau, Rocca, and Testard [1968], to mention only a few papers among many.

The structure of $\Delta(\mathcal{C}_{\mathbb{C}})$, as given by Theorem 11, is now used to construct a wealth of inequivalent representations of the canonical commutation relations by a procedure that provides an immediate identification of the physical meaning of the representations obtained. We proceed in the immediate sequel from the general to the particular, establishing first some general results.

Suppose that $\bigotimes_{\gamma\in\Gamma} \mathfrak{R}_\gamma$ is the infinite direct product of the family $\{\mathfrak{R}_\gamma \mid \gamma \in \Gamma\}$ of C^*-algebras $\mathfrak{R}_\gamma$; suppose, further, that we are given a family $\{\phi_\gamma \mid \gamma \in \Gamma\}$ of states ϕ_γ respectively defined on $\mathfrak{R}_\gamma$. Let $\pi_\gamma: \mathfrak{R}_\gamma \to \mathfrak{B}(\mathcal{H}_\gamma)$ be the representation of $\mathfrak{R}_\gamma$, obtained for each γ, from the state ϕ_γ by the GNS construction. Let Φ_γ be the normalized cyclic vector corresponding to ϕ_γ. We then form the incomplete direct product space $\bigotimes_{\gamma\in\Gamma} {}^{\Phi}\mathcal{H}_\gamma$ associated with the family $\Phi = \{\Phi_\gamma \mid \gamma \in \Gamma\}$. For every family $\{R_\gamma \mid \gamma \in \Gamma;\ R_\gamma \in \mathfrak{R}_\gamma;\ R_\gamma = I_\gamma$ for all but a finite number of indices $\gamma \in \Gamma\}$ and for every family $\{\Psi_\gamma \mid \gamma \in \Gamma; \Psi_\gamma \in \mathcal{H}_\gamma;$ $\Psi_\gamma = \Phi_\gamma$ for all but a finite number of indices $\gamma \in \Gamma\}$ we form†

$$\pi(\times R_\gamma)(\otimes \Psi_\gamma) = \otimes\, \pi_\gamma(R_\gamma)\Psi_\gamma,$$

which then extends by linearity to a bounded mapping of $\times\, {}^{\Phi}\mathcal{H}_\gamma$ into itself; we can then extend by continuity this mapping to an operator $\pi(\times R_\gamma)$ in $\mathfrak{B}(\otimes\, {}^{\Phi}\mathcal{H}_\gamma)$; thus we obtain a representation π of $\times\, \mathfrak{R}_\gamma$ into $\mathfrak{B}(\otimes\, {}^{\Phi}\mathfrak{R}_\gamma)$ which we then extend by continuity to a representation, denoted by $\pi = \otimes\, {}^{\Phi}\pi_\gamma$, of $\otimes\, \mathfrak{R}_\gamma$, acting in the space $\otimes\, {}^{\Phi}\mathcal{H}_\gamma$. We notice in particular that the vector $\Phi = \otimes\, \Phi_\gamma$ in $\otimes\, {}^{\Phi}\mathcal{H}_\gamma$ is cyclic for this representation and generates a state ϕ on $\otimes\, \mathfrak{R}_\gamma$ such that

$$\langle \phi; \times\, \mathfrak{R}_\gamma \rangle = \Pi \langle \phi_\gamma; R_\gamma \rangle$$

for every family $\{R_\gamma \mid \gamma \in \Gamma;\ R_\gamma \in \mathfrak{R}_\gamma;\ R_\gamma = I_\gamma$ for all but a finite number of indices $\gamma \in \Gamma\}$; for this reason ϕ is denoted $\otimes\, \phi_\gamma$ and is referred to as the *product state* of the family $\{\phi_\gamma \mid \gamma \in \Gamma;\ \phi_\gamma \in \mathfrak{S}_\gamma\}$. Since Φ is cyclic for the representation $\otimes\, {}^{\Phi}\pi_\gamma$, the latter is determined up to a unitary equivalence by $\{\phi_\gamma \mid \gamma \in \Gamma;\ \phi_\gamma \in \mathfrak{S}_\gamma\}$.

In connection with this construction it is interesting to recall here the definition of the incomplete direct product of a family $\{\mathfrak{R}_\gamma \mid \gamma \in \Gamma;\ \mathfrak{R}_\gamma \subseteq \mathfrak{B}(\mathcal{H}_\gamma)\}$ of von Neumann algebras; we start again with a family $\Phi = \{\Phi_\gamma \mid \gamma \in \Gamma;\ \Phi_\gamma \in \mathcal{H}_\gamma;\ |\Phi_\gamma| = 1\}$ and form $\otimes\, {}^{\Phi}\mathcal{H}_\gamma$. For each R_{γ_0} in any $\mathfrak{R}_{\gamma_0}$

† When no confusion is likely to occur, we lighten an otherwise rather heavy notation and drop the index sets on which the products are performed; clearly the expression considered here should read $\pi(\times_{\gamma\in\Gamma} R_\gamma)(\bigotimes_{\gamma\in\Gamma} \Psi_\gamma) = \bigotimes_{\gamma\in\Gamma} \pi_\gamma(R_\gamma)\Psi_\gamma$.

we define the bounded linear operator $\bar{R}_{\gamma_0}$, acting on $\otimes\,^{\Phi}\mathfrak{R}_\gamma$, by

$$\bar{R}_{\gamma_0}(\otimes \Psi_\gamma) = \otimes \Psi'_\gamma$$

where

$$\Psi'_\gamma = \begin{cases} R_{\gamma_0}\Psi_{\gamma_0} & \text{if} \quad \gamma = \gamma_0 \\ \Psi_{\gamma_0} & \text{if} \quad \gamma \neq \gamma_0 \end{cases}$$

the *incomplete direct product* $\otimes\,^{\Phi}\mathfrak{N}_\gamma$ is now defined as the von Neumann algebra generated in $\mathfrak{B}(\otimes \mathcal{H}_\gamma)$ by the operators $\bar{R}_\gamma$ just defined, with R_γ running over $\mathfrak{R}_\gamma$ and γ running over Γ. We clearly have $\otimes\,^{\Phi}\mathfrak{N}'_\gamma \subseteq (\otimes\, \mathfrak{N}_\gamma)'$; we can prove† furthermore that $\otimes\,^{\Phi}\mathfrak{N}_\gamma = \mathfrak{B}(\otimes\,^{\Phi}\mathcal{H}_\gamma)$ if and only if $\mathfrak{N}_\gamma = \mathfrak{B}(\mathcal{H}_\gamma)$ for *all* $\gamma \in \Gamma$ and that $(\otimes\,^{\Phi}\mathfrak{N}_\gamma) \cap (\otimes\,^{\Phi}\mathfrak{N}_\gamma)' = \{\lambda I\}$ if and only if $\mathfrak{N}_\gamma \cap \mathfrak{N}'_\gamma = \{\lambda I_\gamma\}$ for *all* $\gamma \in \Gamma$. Coming back to the product state $\phi = \otimes_{\gamma\in\Gamma} \phi_\gamma$ on the C^*-algebra $\otimes\, \mathfrak{R}_\gamma$, we conclude from the above result that ϕ is primary (resp. pure) if and only if ϕ_γ is primary (resp. pure) for *each* $\gamma \in \Gamma$.

We notice further that in the above construction of the representation $\pi = \otimes\,^{\Phi}\pi_\gamma$ of the C^*-algebra $\otimes\, \mathfrak{R}_\gamma$ we could have started from the representations π_γ and the family Φ, without assuming that the Φ_γ were cyclic for π_γ, and still proceed as above in the construction of $\otimes_{\gamma\in\Gamma}\,^{\Phi}\pi_\gamma$; this raises the question of the relation that might then exist between the representations so obtained from a fixed family $\{\pi_\gamma \mid \gamma \in \Gamma\}$ of representations. The answer can be formulated as follows:

Theorem 12. *Let* $\{\mathfrak{R}_\gamma \mid \gamma \in \Gamma\}$ *be a family of* C*-*algebras, let* $\{\pi_\gamma : \mathfrak{R}_\gamma \to \mathfrak{B}(\mathcal{H}_\gamma) \mid \gamma \in \Gamma\}$ *be a family of representations of the* $\mathfrak{R}_\gamma$; *with* k = 1, 2, *let* $\Phi^{(k)} = \{\Phi_\gamma^{(k)} \mid \gamma \in \Gamma;\ \Phi_\gamma^{(k)} \in \mathcal{H}_\gamma;\ |\Phi_\gamma^{(k)}| = 1\}$, *and let* $\pi^{(k)}$ *be the direct product representations of* $\otimes\, \mathfrak{R}_\gamma$ *corresponding respectively to* $\Phi^{(k)}$. *If* $\Phi^{(1)} \sim \Phi^{(2)}$, *then* $\pi^{(1)}$ *is unitarily equivalent to* $\pi^{(2)}$. *Conversely, if all the* π_γ *are irreducible and* $\Phi^{(1)}$ *not* $\sim$ *equivalent* $\Phi^{(2)}$, *then* $\pi^{(1)}$ *is not unitarily equivalent to* $\pi^{(2)}$.

Proof. Suppose that $\Phi^{(1)} \sim \Phi^{(2)}$; i.e.,

$$\sum_{\gamma\in\Gamma} \big|\,|(\Phi_\gamma^{(2)}, \Phi_\gamma^{(1)})| - 1\big| < \infty.$$

There exists then $\alpha = \{\alpha_\gamma \mid \gamma \in \Gamma;\ \alpha_\gamma \in \mathbb{C};\ |\alpha_\gamma| = 1\}$ such that

$$\sum |(\Phi_\gamma^{(2)}, \alpha_\gamma \Phi_\gamma^{(1)}) - 1| < \infty;$$

i.e., $\Phi^{(2)} \approx \alpha\Phi^{(1)}$. We know (see p. 256) that there exists a unique isomorphism from $\mathcal{H}^{(1)} = \otimes\,^{\Phi^{(1)}}\mathcal{H}_\gamma$ onto $\mathcal{H}^{(2)} = \otimes\,^{\Phi^{(2)}}\mathcal{H}_\gamma$, sending $\otimes \Psi_\gamma$ $(\in \mathcal{H}^{(1)})$, with $\Psi_\gamma = \Phi_\gamma^{(1)}$ for all but a finite number of indices γ in Γ, to $\otimes\, \alpha_\gamma\Psi_\gamma$ $(\in \mathcal{H}^{(2)})$. Let

† See von Neumann [1938], Bures [1963], and Guichardet [1966].

U be this isomorphism. For each family $\{R_\gamma \mid \gamma \in \Gamma;\ R_\gamma \in \mathfrak{R}_\gamma;\ R_\gamma = I_\gamma$ for all but a finite number of indices$\}$ we form

$$\pi^{(2)}(\textsf{X}\, R_\gamma)\, U(\otimes \Psi_\gamma) = \pi^{(2)}(\textsf{X}\, R_\gamma)(\otimes\, \alpha_\gamma \Psi_\gamma) = \otimes\, \alpha_\gamma \pi_\gamma(R)\Psi_\gamma$$
$$= U\pi^{(1)}(\textsf{X}\, R_\gamma)(\otimes \Psi_\gamma).$$

By linearity and continuity we conclude that $\pi^{(2)}(R)U = U\pi^{(1)}(R)$ for all R in $\otimes\, \mathfrak{R}_\gamma$, which proves the first part of the theorem. We merely sketch the proof of the second part. Let $\{\Gamma_n\}$ be any strictly increasing, and absorbing, sequence of finite subsets Γ_n of Γ. For each n we define $\Psi_\gamma^{(n)} = \Phi_\gamma^{(2)}$ for all $\gamma \in \Gamma_n$ and $\Psi_\gamma^{(n)} = \Phi_\gamma^{(1)}$ for all $\gamma \notin \Gamma_n$; the normalized vectors $\Psi^{(n)} = \textsf{X}_{\gamma\in\Gamma} \Psi_\gamma^{(n)}$ belong to $\mathcal{H}^{(1)}$ and then define naturally a sequence $\{\psi^{(n)}\}$ of vector states on the representation $\pi^{(1)}$. We verify that $\psi^{(n)}$ converges in the w^*-topology to $\phi^{(2)}$, the state on $\pi^{(2)}$ associated with the vector $\Phi^{(2)}$ in $\mathcal{H}^{(2)}$; furthermore, it can be shown that $\Psi^{(n)}$ converges weakly to zero. This implies that $\phi^{(2)}$ cannot be a vector state on the irreducible representation $\pi^{(1)}$; by Theorem II.1.1 this implies in turn that $\pi^{(1)}$ and $\pi^{(2)}$ cannot be unitarily equivalent. ∎

The interest of this general theorem goes far beyond the special application to the canonical commutation relations we are about to consider.† Because of Theorem 11, we can now use Theorem 12 to construct inequivalent representations of the canonical commutation relations; in this case we know from Theorem 6 that the irreducible representations π_γ are all unitarily equivalent to the Schrödinger representation studied in Subsection a, so that we can assume, without loss of generality, that $\mathcal{H}_\gamma = \mathfrak{L}^2(\mathbb{R})$ $(\gamma = 1, 2, \ldots)$.‡

Corollary. *Every IDPS constructed from an infinite sequence of copies of* $\mathfrak{L}^2(\mathbb{R})$ *supports an irreducible representation of the canonical commutation relations; two representations obtained in this way are unitarily equivalent if and only if the corresponding IDPS's are constructed from two* $\sim$*-equivalent families* $\{\Phi_\gamma^{(k)} \mid \gamma \in \mathbb{Z}^+\}$ $(k = 1, 2)$ *of normalized vectors in* $\mathfrak{L}^2(\mathbb{R})$.

The proof of this corollary is immediate from the observations preceding its statement. As an application of this corollary, we can, for instance, recover the results of Wightman and Schweber [1955] concerning the irreducible, "discrete" representations of the canonical commutation relations. Indeed, let $\{\Psi_n\}$ be an orthonormal basis in $\mathfrak{L}^2(\mathbb{R})$ consisting of eigenvectors of the (one-dimensional) harmonic oscillator. Now take for Γ the set $\mathbb{Z}^+$ of all

† This theorem has been stated in the above form and proved in detail by Guichardet [1966], Propositions 2.7 and 2.8.

‡ This particular application of Guichardet's theorem has been recognized by Klauder, McKenna, and Woods [1966], Theorem 4.3; they actually proved it independently; for still another proof see Streit [1967].

positive integers and consider all families of the form $\Phi = \{\Phi_\gamma = \Psi_{n(\gamma)} \mid \gamma \in \mathbb{Z}^+\}$; for each such family form the irreducible representation $\otimes^\Phi \pi_\gamma$ constructed in the IDPS $\otimes^\Phi \mathcal{H}_\gamma$. From the corollary two representations $\pi^{(k)} = \otimes^{\Phi(k)} \pi_\gamma$ obtained in this way are unitarily equivalent if and only if

$$\sum \big| |(\Phi_\gamma^{(1)}, \Phi_\gamma^{(2)})| - 1 \big| = \sum \big| |(\Psi_{n_1(\gamma)}, \Psi_{n_2(\gamma)})| - 1 \big|$$
$$= \sum (\delta_{n_1(\gamma), n_2(\gamma)} - 1)$$

is finite, that is to say, if and only if the two sequences $\{n_1(\gamma) \mid \gamma \in \mathbb{Z}^+\}$ and $\{n_2(\gamma) \mid \gamma \in \mathbb{Z}^+\}$ differ only in a finite number of entries. Two such sequences $\{n^{(1)}(\gamma)\}$ and $\{n^{(2)}(\gamma)\}$ are said to be equivalent; but, on the other hand, $\{n^{(k)}(\gamma)\}$ characterizes $\{\Phi_\gamma^{(k)}\}$, and we see that two representations $\pi^{(1)}$ and $\pi^{(2)}$ are unitarily equivalent if and only if the corresponding sequences $\{n^{(k)}(\gamma)\}$ of integers are equivalent in the above sense. The physical meaning of this condition is made clear if we notice that in $\mathcal{L}^2(\mathbb{R})$ we have $a^*a\Psi_n = n\Psi_n$, so that $n(\gamma)$ is the occupation number of the state corresponding to e_γ in $\mathcal{T}$ in the state corresponding to the vector Φ in $\otimes^\Phi \mathcal{H}_\gamma$. Hence two representations obtained as just described are unitarily equivalent if and only if they give the same occupation numbers $n(\gamma)$ for all but at most a finite number of the basis vectors e_γ in the test function space $\mathcal{T}$. Since we can obviously construct an infinity of nonequivalent sequences $\{n(\gamma) \mid \gamma \in \mathbb{Z}^+\}$, *this result establishes the existence of an infinity of inequivalent irreducible representations of the canonical commutation relations;* furthermore, it gives, together with an explicit construction of these representations, their physical interpretation. The Fock representation appears naturally as one of these representations, namely, that corresponding to the equivalence class of the sequence $\{n(\gamma) = 0 \mid \gamma \in \mathbb{Z}^+\}$; the vector Φ corresponding to the sequence $\{\Psi_{n(\gamma)} = \Psi_0\}$ is the familiar vacuum, generating the no-particle state ϕ_0. We notice finally that the Fock representation is, among all the irreducible representations of the canonical commutation relations just constructed, the only one (up to a unitary equivalence) that admits either a no-particle vector state ϕ_0 or a total number operator $N = \sum_{\gamma \in \mathbb{Z}^+} a_\gamma^* a_\gamma$. This intuitively clear result appeared as an immediate consequence of the work of Gårding and Wightman [1954b] and Wightman and Schweber [1955]. It has been generalized to representations of the CCR, which are not necessarily irreducible, by Dell'Antonio, Doplicher, and Ruelle [1966] and Dell'Antonio and Doplicher [1967];† these authors showed in particular that the following three conditions are equivalent: (a) π_ϕ is a multiple of the Fock representation, (b) ϕ can be written as a density matrix in the Fock representation, and (c) a total number operator exists in $\mathcal{H}_\phi$; the latter condition hinges on the proper definition of a total number operator, i.e., a self-adjoint (unbounded), densely defined

† Similar results had been obtained previously by Chaiken [1966].

operator, independent of the choice of the orthonormal basis chosen in $\mathfrak{T}$ to define it. With this generality this result has an important physical consequence: the state ϕ describing a system in thermodynamical equilibrium with nonzero density cannot be written as a density-matrix in the Fock representation; if, however, a hard core is present in the interaction, which prevents an infinite number of particles from coexisting simultaneously in a finite region Ω of space, then ϕ when restricted to the local algebra $\mathfrak{R}(\Omega)$ associated with Ω can be described as a density matrix in the Fock space $\mathcal{H}(\Omega)$.

We mentioned, at the end of Subsection c, that two problems remained unsolved after the general result of Theorem 7 and had to be approached with tools that are definitely more analytical than the purely algebraic tools that made the proof of Theorem 7 possible. These two problems are (a) to determine the maximal test function space proper to a given representation, and (b) to establish the existence of a Gårding domain. As is to be expected, these problems have received earlier answers in the particular case of the representations discussed in the subsection. We now want to review briefly some of the results obtained in this connection.

Streit [1967] established the following result, valid for an arbitrary IDPS $\otimes_{\gamma\in\mathbb{Z}^+}\mathcal{H}_\gamma$ (with $\mathcal{H}_\gamma = \mathfrak{L}^2(\mathbb{R})\ \forall\ \gamma \in \mathbb{Z}^+$): the unitary operators $U(\lambda f)$ and $V(\lambda g)$ (with $f = \sum f_\gamma e_\gamma$ and $g = \sum g_\gamma e_\gamma$), acting on the IDPS constructed on Φ, are weakly continuous in λ real and satisfy the algebraic canonical commutation relations *if and only if* there exists a family $\Psi \approx \Phi$ such that

$$\sum_{\gamma\in\mathbb{Z}^+} |f_\gamma^k(\Psi_\gamma, Q_\gamma^k\Psi_\gamma)| < \infty,$$
$$\sum_{\gamma\in\mathbb{Z}^+} |g_\gamma^k(\Psi_\gamma, P_\gamma^k\Psi_\gamma)| < \infty, \qquad k = 1, 2,$$

where Q and P are the Schrödinger position and momentum operators defined in Subsection a. Two remarks should be made concerning this remarkable result; first, it appears from Streit's analysis that Ψ might depend in general on f or g; second, we might notice that this result passes over in silence the domain questions discussed in Subsection c, which are of serious importance if we wish to establish reasonable contact with ordinary field theory.

We now turn our attention to the latter problem. Let $\otimes^\Phi \pi_\gamma$ with $\{\Phi_\gamma = \Psi_{n(\gamma)} \mid \gamma \in \mathbb{Z}^+\}$ be an irreducible "discrete" representation of the canonical commutation relations, as defined above; let $[n]$ denote the equivalence class of the sequence $\{n(\gamma) \mid \gamma \in \mathbb{Z}^+\}$. We denote by m the elements in the equivalence class $[n]$: any $m \in [n]$ is then a sequence $\{m(\gamma) \mid \gamma \in \mathbb{Z}^+\}$ of positive integers, differing from $\{n(\gamma) \mid \gamma \in \mathbb{Z}^+\}$ by at most a finite number of entries. We denote by Φ_m the vector in $\otimes^\Phi \mathcal{H}_\gamma$ associated with the family $\{\Phi_\gamma = \Psi_{m(\gamma)} \mid \gamma \in \mathbb{Z}^+\}$ with $m \in [n]$; the set $\{\Phi_m \mid m \in [n]\}$ constitutes clearly an orthonormal basis in $\otimes^\Phi \mathcal{H}_\gamma$ so that

$$\otimes^\Phi \mathcal{H}_\gamma = \left\{ \sum_{m\in[n]} \lambda_m \Phi_m \,\middle|\, \sum_{m\in[n]} |\lambda_m|^2 < \infty \right\}.$$

We notice that $W(\lambda f)$ is unitary on $\otimes^\Phi \mathcal{H}_\gamma$, and continuous in λ real for all f obtained as finite, real, linear combinations of our basis vectors e_γ. We can then, in particular, define $F_\gamma = F(e_\gamma)$, $P_\gamma = P(e_\gamma)$, $a_\gamma = a(e_\gamma)$ and $a_\gamma^* = a^*(e_\gamma)$ for all e_γ, $\gamma \in \mathbb{Z}^+$; we further have for all γ in $\mathbb{Z}^+$

$$a_\gamma \Phi_m = \sqrt{m(\gamma)}\, \Phi_{m-\delta\gamma},$$

$$a_\gamma^* \Phi_m = \sqrt{m(\gamma)+1}\, \Phi_{m+\delta\gamma},$$

where $m - \delta_\gamma$ (or $m + \delta_\gamma$) is the sequence identical to m except for the fact that its γth entry has been decreased (or increased) by unity. The natural domain of a^* is then

$$\mathfrak{D}(a_\gamma^*) = \left\{ \Psi \in \otimes^\Phi \mathcal{H}_\gamma \,\middle|\, \sum_{m\in[n]} (m(\gamma)+1)\,|(\Phi_m, \Psi)|^2 < \infty \right\};$$

we define in a similar manner the natural domain of a_γ. We notice further that the linear manifold $\mathfrak{D}$ spanned by $\{\Phi_m \mid m \in [n]\}$, dense in $\otimes^\Phi \mathcal{H}_\gamma$, is contained in

$$\bigcap_{\gamma\in\mathbb{Z}^+} (\mathfrak{D}(a_\gamma) \cap \mathfrak{D}(a_\gamma^*))$$

and *stable* under these operators, hence under all F_γ and P_γ. Consequently $\mathfrak{D}$ is stable under all $F(f)$ and $P(f)$ where

$$f = \sum_{\gamma=1}^{k} f_\gamma e_\gamma, \qquad k < \infty, \qquad f_\gamma \in \mathbb{R}.$$

The linear manifold of all of these f's is precisely the original test function space $\mathcal{T}$ from which we started our present investigation. We now want to see whether $\mathcal{T}$ can be extended without losing the domain properties just established. This is, indeed, possible, as shown by the following construction.† Let

$$\mathcal{T}_{[n]} = \{f = \sum f_\gamma e_\gamma \text{ with } \sum (n(\gamma)+1)\,|f_\gamma|^2 < \infty\};$$

for each f in $\mathcal{T}_{[n]}$ we form

$$f^{(k)} = \sum_{\gamma=1}^{k} f_\gamma e_\gamma \in \mathcal{T}_{\mathbb{C}}$$

and notice that the sequence $W(\lambda f^{(k)})$ converges strongly to a unitary operator, which we denote by $W(\lambda f)$; furthermore, the set

$$\{W(f) \mid f \in \mathcal{T}_{[n]}\}$$

† See Shelupski [1966].

constructed in this manner again satisfies the canonical commutation relations. In connection with the aforementioned result of Streit we have

$$\sum |(\mathbb{R}\mathrm{e}\, f_\gamma)(\Psi_{n(\gamma)}, Q_\gamma \Psi_{n(\gamma)})| = 0,$$
$$\sum |(\mathrm{Im}\, f_\gamma)(\Psi_{n(\gamma)}, P_\gamma \Psi_{n(\gamma)})| = 0,$$

and

$$\sum |(\mathbb{R}\mathrm{e}\, f_\gamma)^2(\Psi_{n(\gamma)}, Q_\gamma^2 \Psi_{n(\gamma)})| = \sum (\mathbb{R}\mathrm{e}\, f_\gamma)^2 (n(\gamma) + \tfrac{1}{2})$$
$$\leqslant \sum |f_\gamma|^2 (n(\gamma) + 1) < \infty \;\forall f \in \mathcal{T}_{[n]};$$

similarly

$$\sum |(\mathrm{Im}\, f_\gamma)^2(\Psi_{n(\gamma)}, P_\gamma^2 \Psi_{n(\gamma)})| \leqslant \sum |f_\gamma|^2 (n(\gamma) + 1) < \infty \;\forall f \in \mathcal{T}_{[n]}.$$

This confirms that $U(\lambda f)$ and $V(\lambda f)$ are continuous in λ real for all real f in $\mathcal{T}_{[n]}$. From the generators of these unitary, weakly continuous one-parameter groups we define as usual $a^*(f)$ and $a(f)$ and notice that

$$|a^*(f)\Phi_m|^2 = \sum (m(\gamma) + 1) \, |f_\gamma|^2,$$

which is finite for all f in $\mathcal{T}_{[n]}$ and all $m \in [n]$; a similar expression is obtained for $|a(f)\Phi_m|^2$ so that $\mathfrak{D}$ is actually contained in the domain of an $a^*(f)$ and $a(f)$ for all f in $\mathcal{T}_{[n]}$. We further conclude from the above computation that $\mathcal{T}$ cannot be extended more than to $\mathcal{T}_{[n]}$ if we want to preserve the latter domain property.† In the elementary example of the Fock-space representation, for which $n(\gamma) = 0 \;\forall\; \gamma \in \mathbb{Z}^+$, we determine from these remarks that the test function space $\mathcal{T}_0$ attached to this represenation is identical to $\overline{\mathcal{T}} = \mathfrak{L}^2(\mathbb{R})$; similarly, we could conclude that $\mathcal{T}_0 = \mathfrak{L}^2_\mu(\mathbb{R}^3)$ for a free, relativistic scalar boson field of mass $m \neq 0$.

Aside from its meaning in connection with the determination of a common dense domain for the operators $a^*(f)$ and $a(f)$ with f in $\mathcal{T}_{[n]}$, the condition defining $\mathcal{T}_{[n]}$ has the following interesting consequence which we shall presently exploit to gain some further understanding of the situation encountered during our discussion of the van Hove model (Chapter 1, Section 1): the mapping $f \to W(f)$ from $\mathcal{T}_{[n]}$ into $\mathfrak{B}(\otimes^{\Phi}\mathcal{H}_\gamma)$ is continuous when $\mathcal{T}_{[n]}$ is equipped with the topology associated with $\|f\|_n^2 = \sum (n(\gamma) + 1)|f_\gamma|^2$, and $\mathfrak{B}(\otimes^{\Phi}\mathcal{H}_\gamma)$ is equipped with its strong operator topology.

Indeed let $\chi(f)$ be a complex-valued linear function defined on $\mathcal{T}_{[n]}$, and let us form

$$R_f \to R_f e^{-i\,\mathrm{Im}\chi(f)};$$

this transformation clearly extends, in a unique way, to an automorphism of the C^*-algebra $\overline{\Delta(\mathcal{T}_{[n]})}$ of the canonical commutation relations associated

† The existence of a common dense domain D of essential self-adjointness of $F(f)$ and $P(f)$ has been established by Reed [1968, 1969b]: D is the finite linear span of $\{\Psi \in \otimes^{\Phi}\mathcal{H}_\gamma \mid \Psi = \otimes \Psi_\gamma; \Psi_\gamma = \Phi_\gamma \quad \text{for } \gamma \geqslant N, N \text{ arbitrary}\}$.

to our test function space $\mathcal{T}_{[n]}$; let us denote this automorphism by α_χ. Its physical interest is that in any representation of $\overline{\Delta(\mathcal{T}_{[n]})}$, satisfying the continuity condition that $\pi(R_{\lambda f})$ be strongly continuous in λ real for all f in $\mathcal{T}_{[n]}$, we have for the corresponding field operators and their canonical conjugates

$$F(f) \to \hat{F}(f) = F(f) + \operatorname{Im} \chi(f),$$

$$P(f) \to \hat{P}(f) = P(f) + \mathbb{R}\mathrm{e}\, \chi(f),$$

and

$$a^*(f) \to \hat{a}^*(f) = a^*(f) - \frac{i\chi(f)}{\sqrt{2}},$$

$$a(f) \to \hat{a}(f) = a(f) + \frac{i\chi(f)^*}{\sqrt{2}},$$

which is the pseudo-canonical transformation that links the free and interpolating fields of the van Hove model. The present transformation is naturally of more general interest, since we have not yet specified the form of $\chi(f)$ nor have we restricted the original representation to be the Fock representation. We then obtain in this fashion a new irreducible representation $\{\hat{W}(f) \mid f \in \mathcal{T}_{[n]}\}$ of the canonical commutation relations. The question now is whether $\hat{\mathfrak{W}}(\mathcal{T}_{[n]})$ is equivalent to $\mathfrak{W}(\mathcal{T}_{[n]})$; the answer is provided by the following result:† the automorphism α_χ is unitarily implemented in a representation $f \to W(f)$ which is strongly continuous in f if and only if χ is continuous in f. In our case this means (Riesz theorem) that $\chi(f)$ can be written in the form $\chi(f) = (g_\chi, f)_{[n]}$ with g_χ in $\mathcal{T}_{[n]}$, i.e., $\sum (n(\gamma) + 1) |\chi_\gamma|^2 < \infty$, where χ_γ is defined for each γ in $\mathbb{Z}^+$ by $\chi_\gamma = \chi(e_\gamma)$. Whenever this condition is realized, we have $\hat{W}(f) = \alpha_\chi[W(f)] = W(g_\chi)\, W(f)\, W(-g_\chi)$, i.e., the automorphism of the image of $\overline{\Delta(\mathcal{T}_{[n]})}$ by the representation considered is inner. The last two results have already been pointed out by Shelupski [1966], who proved in addition that when the condition

$$\sum (n(\gamma) + 1) |\chi_\gamma|^2 < \infty$$

is *not* satisfied, the representation $\hat{\mathfrak{W}}(\mathcal{T}_{[n]})$ is unitarily inequivalent to every irreducible discrete representation of the canonical commutation relations, although $\langle \hat{\phi}; \hat{W}(f) \rangle = \langle \phi; W(f) \rangle e^{-i \operatorname{Im} \chi(f)}$ is still a product of pure states. The idea in Shelupski's proof is that if $\{\hat{a}(f) = a(f) + i\, \chi(f)^*/\sqrt{2}\}$ were unitarily equivalent to some discrete representation, constructed, say, on the IDPS associated with the equivalence class $[n']$, the inverse image Ψ in

† See Manuceau [1968a].

the original representation of the vector $\Phi_{[n']} = \otimes \Psi_{n'(\gamma)}$ would satisfy the relation

$$\left(a_\gamma^* - \frac{i\chi_\gamma}{\sqrt{2}}\right)\left(a_\gamma + \frac{i\chi_\gamma^*}{\sqrt{2}}\right)\Psi = n'(\gamma)\Psi;$$

this condition, together with the condition that $\Psi = \sum_{m\in[n]} (\Phi_m, \Psi)\Phi_m$ belong to the original IDPS, i.e., that $\sum_{m\in[n]} |(\Phi_m, \Psi)|^2 < \infty$, can then be seen to be equivalent to $\sum_{\gamma\in\mathbb{Z}^+} (n(\gamma) + 1)\, |\chi_\gamma|^2 < \infty$, which is impossible by assumption.

At this point two comments are in order. *First*, the facts just established should now make clear the origin of the mathematical difficulties encountered in the solution of the van Hove model (consequences of the "ultraviolet divergence"); in particular, the statements made on pp. 27–28 should now appear natural. Incidentally, the "myriotic" representations produced by Friedrichs [1952] in connection with the occurrence of infrared divergences have the same origin: namely the divergence of $\sum_{\gamma\in\mathbb{Z}^+} (n(\gamma) + 1)\, |\chi_\gamma|^2$. *Second*, the above discussion exhibits explicitly a family of irreducible representations which are outside the class of the discrete representations of Wightman and Schweber [1955], who actually pointed out this circumstance in the particular case in which the representation $\mathfrak{W}(\mathcal{C}_{[n]})$ is the Fock representation; incidentally, their physical motivation for this remark was also the study of the van Hove model, from which they concluded that "Configuration space methods based on interaction or Heisenberg representation fields are *not* of general applicability in field theory. The assumption that a field theory possesses such a configuration space representation may contradict the equations of motion of the theory." This remark has been confirmed repeatedly, starting with Haag's general theorem (Subsection d) and continuing with the recent investigations of exactly soluble models (see Chapter 4 for some bibliographical references).

Aside from the above applications to the mathematical consequences of ultraviolet and infrared divergences in field theory, representations associated with product states have also been found in some simple statistical mechanical systems. In addition to the already mentioned paper of Araki and Woods [1963] on the free Bose gas (see also Robinson [1965a]), we might mention here the papers of Verboven [1966a, b] and Verbeure and Verboven [1966, 1967a, b], to cite only those that are directly connected with the treatment of physical systems satisfying Bose statistics. An important class of representations generated by product states has been studied by Størmer [1969a], whose general results apply in particular to the above-mentioned papers. These results are described in more detail in Chapter 4, together with some other applications.

We want to close this subsection on the representations associated with product states on the C^*-algebra of the canonical commutation relations by noting that, in spite of the great wealth of representations obtained in this way, we have nevertheless some reason to believe that more general representations might be involved in the description of properly interacting fields (i.e., $S \neq I$)† as well as in the description of thermodynamical systems that exhibit phase transitions with high-order singularities. On this account it is suggested that the reader study the original articles of Powers [1967], Klauder [1965, 1967], Sinha and Emch [1969], Sinha [1969], Manuceau and Verbeure [1968], Manuceau, Rocca, and Testard [1969]; to various degrees of generality the purpose of these papers is to analyze the conditions under which a representation is bound to describe only quasi-free systems. The hope is then that, once these conditions are properly identified, we might be able to go beyond the restriction they impose. We first mention the result due to Robinson [1965a]. Let ϕ be a state on a C^*-algebra $\mathfrak{R}$; for any sequence $R_1, R_2, \ldots, R_n$ of elements in $\mathfrak{R}$ we define $\langle \phi_T; R_1 R_2 \cdots R_n \rangle$ by the recursion formula

$$\langle \phi; R_1 R_2 \cdots R_n \rangle = \sum \langle \phi_T; R_{i_k} \cdots \rangle \cdots \langle \phi_T; \cdots R_{i_j} \rangle,$$

where the sum is carried over all possible clusterings of the sequence R_1, $R_2, \ldots, R_n$, and where the order within each cluster is that inherited from the order in the original sequence; for instance,

$$\langle \phi_T; R \rangle = \langle \phi; R \rangle,$$
$$\langle \phi_T; R_1 R_2 \rangle = \langle \phi; R_1 R_2 \rangle - \langle \phi; R_1 \rangle \langle \phi; R_2 \rangle,$$

etc.

Depending on the specific applications in mind, we refer to the functional ϕ_T obtained in this way as either the *n-point truncated functions* for ϕ or the *n-fold correlation functions* relative to ϕ. Robinson proved then that if the truncated functions

$$\langle \phi_T; f_1 \cdots f_n \rangle = (\Psi_0, a(f_1) \cdots a(f_k)\, a^*(f_{k+1}) \cdots a^*(f_n) \Psi_0)_T$$

vanish for all $n \geqslant N > 2$ they all vanish whenever $n > 2$; he also proved a similar result for the truncated functions

$$\langle \phi_T; f_1 \cdots f_n \rangle = (\Psi_0, F(f_1) \cdots F(f_n) \Psi_0)_T$$

relative to any local Wightman relativistic (neutral, scalar) Bose field that satisfies the condition of positivity of the energy. In other words, either the theory considered is "trivial" (i.e., quasi-free) or its exact solution involves an infinity of nonvanishing n-fold correlation functions. Powers proved that if ϕ is a translational-invariant state on the C^*-algebra of the canonical anticommutation relations (for a definition of the latter see Section 2 of this

† It is the former fact that is sometimes referred to by the statement: "The universal von Hove receptacle is empty." See Wightman [1967].

chapter) and if ϕ can be approximated in the norm-topology by states $\phi' = \bigotimes_{\gamma\in\Gamma}\phi_\gamma$, where $\mathcal{T} = \bigoplus_{\gamma\in\Gamma}\mathcal{T}_\gamma$ is a decomposition of the test function space $\mathcal{T}$ into a direct sum of finite dimensional orthogonal subspaces, the truncated n-point functions for ϕ vanish for all $n > 2$ (i.e., ϕ is a quasi-free state).† In the Boson case Sinha's principal result can be summarized as follows: let $F(\boldsymbol{x}, t)$ be a local Wightman field (scalar and neutral for simplicity), i.e., in particular $F(\boldsymbol{x}, t)$ is an operator-valued distribution in $\boldsymbol{x}$ over a suitable test function space on $\mathbb{R}^N$; suppose that there exists a dense linear manifold $\mathfrak{D}$ in the irreducible representation space of the Weyl algebra generated by the field F and its canonical conjugate $P(f, t) = \dot{F}(f, t)$ at $t = 0$ such that $\mathfrak{D}$ is stable under the action of $F(f)$, $P(g)$, and $U(t)$; then for space dimension $N \geqslant 3$ we have

$$\partial_t^2 F(f, t) = F(Tf, t) + c(f)I,$$

where T is a linear operator acting on the test function space; hence $F(\boldsymbol{x}, t)$ is a quasi-free field. As representative of the results obtained by the Marseille School, we might mention that Manuceau and Verbeure gave a complete classification of the quasi-free states on the C^*-algebra of the canonical commutation relations and analyzed in this connection the role of the Bogoliubov transformations.

SECTION 2 CANONICAL ANTICOMMUTATION RELATIONS

The problem of the canonical anticommutation relations is much easier to handle than that of the canonical commutation relations on at least one account: no recourse to indirect techniques like the Weyl form of the canonical commutation relations is necessary, for all the operators of interest are bounded; furthermore, they are bounded in such a way that all questions about the proper test function space attached to a particular representation become trivially simple. This is what we show first in this section; we then indicate how the theory can be made to run parallel with the theory of the representations of the canonical commutation relations discussed in Section 1. Since this is possible with the simplifications already mentioned, this section is much shorter than the preceding one; we mainly survey the field and give references to the literature which the reader might then want to consult for specific details. In this spirit we also present toward the end of the section a few introductory remarks to establish the connection between this line of approach and that of several authors who used Clifford algebras as a tool for the study of the representations of the canonical anticommutation relations.

† Powers' proof involves a chain of rather tricky estimates, the carrying out of which is possible only when the dimension of the underlying physical space is at least equal to 2.

We say that $\mathfrak{A}_{\mathcal{H}}(\mathcal{T})$ is a *representation of the anticommutation relations* for the test function space $\mathcal{T}$ if

(Ia) $\mathcal{T}$ is a real pre-Hilbert space (the completion of which with respect to the scalar product $(\ldots,\ldots)$ we denote by $\overline{\mathcal{T}}$);

(Ib) to each f in $\mathcal{T}$ corresponds a pair of operators $a(f)$ and $a^*(f)$ densely defined on a Hilbert space $\mathcal{H}$, common to all f in $\mathcal{T}$ and called the "representation space";

(Ic) these operators satisfy the following algebraic conditions:

$$\begin{aligned}[][a(f), a(g)]_+ &= a(f)\,a(g) + a(g)\,a(f) = 0,\\ [a^*(f), a^*(g)]_+ &= 0,\\ [a(f), a^*(g)]_+ &= (f, g)I\end{aligned}$$

for all f and g in $\mathcal{T}$;

(Id) $a^*(f)$ is the hermitian adjoint of $a(f)$, i.e.,

$$a^*(f) = a(f)^* \quad \text{for all } f \text{ in } \mathcal{T},$$

and $a(f)$ is linear in f, i.e.,

$$a(\lambda f + \mu g) = \lambda a(f) + \mu a(g)$$

for all f and g in $\mathcal{T}$ and all λ, μ, in $\mathbb{R}$;

(II) There exists in the representation space $\mathcal{H}$ a dense linear manifold $\mathfrak{D}$, stable with respect to all $a(f)$ and $a^*(f)$ with f in $\mathcal{T}$.

As an immediate consequence of these definitions, we prove the following result:

Theorem 1. *For all* f *in* $\mathcal{T}$ $\|a^\natural(f)\| \leqslant |f|_{\mathcal{T}}$, *where* $a^\natural$ *denotes either* a *or* a*.

Proof. For every f in $\mathcal{T}$ and every Ψ in the domain $\mathfrak{D}$ of definition of these operators we form

$$\begin{aligned}|a(f)\Psi|^2 + |a^*(f)\Psi|^2 &= (\Psi, a^*(f)\,a(f)\Psi) + (\Psi, a(f)\,a^*(f)\Psi)\\ &= |f|_{\mathcal{T}}^2\,|\Psi|^2;\end{aligned}$$

hence $\|a^\natural(f)\| \leqslant |f|_{\mathcal{T}}$ for all f in $\mathcal{T}$, which proves the theorem. ∎

REMARK. For each f in $\mathcal{T}$, $a^\natural(f)$, being bounded on the dense linear manifold $\mathfrak{D}$, can be extended by continuity in a unique manner to $\mathcal{H}$ itself, so that the domain of assumption (II) can be assumed without loss of generality to extend to the whole of $\mathcal{H}$. Furthermore, since $a^\natural(f)$ is bounded by $|f|_{\mathcal{T}}$, $a^\natural$ can be extended by continuity in a unique manner to $\overline{\mathcal{T}}$ itself, so that we can assume without loss of generality that $\mathcal{T} = \overline{\mathcal{T}}$; i.e., $\mathcal{T}$ is complete with respect to the original norm $|\cdots|_{\mathcal{T}}$. These two facts, in sharp contrast with the situation encountered in the case of the canonical commutation relations, are responsible for some substantial simplifications of the formalism developed in Section 1; in particular, we do not limit the generality of the forthcoming considerations by picking an arbitrary orthonormal basis $\{e_\gamma \mid \gamma \in \mathbb{Z}^+\}$ in $\mathfrak{G}$ and considering only the operators $a^\natural_\gamma = a^\natural(e_\gamma)$, since every $a^\natural(f)$ can be reconstructed from these particular operators by linearity and continuity (the latter is evidently to be invoked only when $\mathfrak{G}$ is infinite-dimensional).

If $\mathfrak{G}$ is *finite-dimensional*, the equivalent of von Neumann's uniqueness theorem (recorded in this book as Theorem III.1.6), namely, that there exists, up to a unitary equivalence, only one irreducible representation of the canonical anticommutation relations, has been proved by Jordan and Wigner [1928].†

From now on we shall concentrate on the nontrivial case in which $\mathfrak{G}$ is *infinite-dimensional* with countable basis (i.e., $\mathfrak{G}$ is a separable real Hilbert space).

Following Guichardet [1966], we now define the *C^*-algebra $\mathfrak{A}$ of the canonical anticommutation relations* as $\mathfrak{A} = \bigotimes_{\gamma\in\mathbb{Z}^+} \mathfrak{R}_\gamma$, where $\mathfrak{R}_\gamma$ is, for each γ in $\mathbb{Z}^+$, a copy of the C^*-algebra $\mathfrak{M}_2$ of all matrices of rank 2 with complex entries. We recall that every element M in $\mathfrak{M}_2$ can be written in the form

$$M = \sum_{\mu=0}^{3} \lambda_\mu \sigma^\mu,$$

where λ_μ are complex numbers, σ^0 is the identity matrix $\begin{pmatrix}1 & 0\\ 0 & 1\end{pmatrix}$ and σ^i ($i = x, y, z$) are the three Pauli matrices:

$$\sigma^x = \begin{pmatrix}0 & 1\\ 1 & 0\end{pmatrix}, \quad \sigma^y = \begin{pmatrix}0 & -i\\ +i & 0\end{pmatrix}, \quad \sigma^z = \begin{pmatrix}1 & 0\\ 0 & -1\end{pmatrix}.$$

We denote by i_γ the canonical embedding of $\mathfrak{R}_\gamma$ into $\mathfrak{A}$ and define σ^μ_γ as $i_\gamma(\sigma^\mu)$. For each γ in $\mathbb{Z}^+$ we now form

$$a_\gamma = \sigma^z_1\sigma^z_2 \cdots \sigma^z_{\gamma-1}(\tfrac{1}{2}(\sigma^x_\gamma + i\sigma^y_\gamma)),$$

$$a^*_\gamma = \sigma^z_1\sigma^z_2 \cdots \sigma^z_{\gamma-1}(\tfrac{1}{2}(\sigma^x_\gamma - i\sigma^y_\gamma)).$$

The elements a_γ and a^*_γ of $\mathfrak{A}$ obtained in this fashion clearly satisfy the canonical anticommutation relations:

$$[a_\gamma, a_{\gamma'}]_+ = 0 = [a^*_\gamma, a^*_{\gamma'}],$$

$$[a_\gamma, a^*_{\gamma'}]_+ = \delta_{\gamma,\gamma'}.$$

† Incidentally, this is the same paper in which the canonical anticommutation relations were introduced.

Theorem 2. *The algebra of the canonical anticommutation relations is simple.*

Proof. To prove that $\mathfrak{A}$ is simple it is sufficient to prove that every nonzero morphism π from $\mathfrak{A}$ to some C^*-algebra $\mathcal{B}$ is isometric. We first notice that each $\mathfrak{R}_\gamma$, being a copy of $\mathfrak{M}_2$, is simple, so that for every finite subset Γ of $\mathbb{Z}^+$ $\mathfrak{A}_\Gamma = \bigotimes_{\gamma\in\Gamma} \mathfrak{R}_\gamma$ is simple (actually $\mathfrak{A}_\Gamma$ is isomorphic to the algebra of all bounded linear operators on a Hilbert space of dimension 2^n, where n is the number of points in Γ). Now let π be a nonzero morphism from $\mathfrak{A}$ to some C^*-algebra $\mathcal{B}$; since $\mathfrak{A}$ is the C^*-inductive limit of the $\mathfrak{A}_\Gamma$ just constructed, there exists at least one finite subset Γ_0 of $\mathbb{Z}^+$ such that $\pi(\mathfrak{A}_{\Gamma_0}) \neq 0$. Otherwise $\pi(\mathfrak{A})$ would be zero in contradiction to our assumption. For every finite subset Γ of $\mathbb{Z}^+$ we therefore have $\pi(\mathfrak{A}_{\Gamma\cup\Gamma_0}) \neq 0$. We now recall that, $\Gamma \cup \Gamma_0$ being a finite subset of $\mathbb{Z}^+$, $\mathfrak{A}_{\Gamma\cup\Gamma_0}$ is simple. Hence π restricted to $\mathfrak{A}_{\Gamma\cup\Gamma_0}$, being nonzero, is isometric. Consequently π restricted to $\mathfrak{A}_\Gamma$ is isometric. Since Γ is an arbitrary finite subset of $\mathbb{Z}^+$ and $\mathfrak{A}$ is the C^*-inductive limit of these $\mathfrak{A}_\Gamma$, we conclude that π is isometric on $\mathfrak{A}$ itself. As no restrictions were imposed on π except that it be a nonzero morphism from $\mathfrak{A}$ to some C^*-algebra $\mathfrak{B}$, we conclude that $\mathfrak{A}$ is simple. ∎

We now denote by $\mathfrak{A}_0$ the subset $\{a_\gamma, a_\gamma^* \mid \gamma \in \mathbb{Z}^+\}$ of $\mathfrak{A}$. For the purpose of this section we are interested primarily in $\mathfrak{A}_0$; the interest in $\mathfrak{A}$ is based on the following result:

Theorem 3. *For any representation π of $\mathfrak{A}$ we define π_0 as the restriction of π to $\mathfrak{A}_0$; then the mapping $\pi \to \pi_0$ is a bijection from the set* P *of all representations of $\mathfrak{A}$ to the set* P_0 *of all representations of the canonical anticommutation relations. Furthermore, π is cyclic* (*or irreducible*) *if and only if π_0 is cyclic* (*or irreducible*). *Finally, two representations π and ρ of $\mathfrak{A}$ are equivalent* (*or quasi-equivalent*) *if and only if π_0 and ρ_0 are equivalent* (*or quasi-equivalent*).

Proof. We first notice that from the above definition of $\mathfrak{A}_0$ it follows immediately that $\pi(\mathfrak{A}_0)$ is a representation of the canonical anticommutation relations. Conversely, if we have a representation of the canonical anticommutation relations, we can reconstruct uniquely from it a representation of $\mathfrak{A}$ by defining

$$\sigma_\gamma^z = 2a_\gamma^* a_\gamma - 1,$$

$$\sigma_\gamma^x = \sigma_1^z\sigma_2^z \cdots \sigma_{\gamma-1}^z(a_\gamma^* + a_\gamma),$$

$$\sigma_\gamma^y = i\sigma_1^z\sigma_2^z \cdots \sigma_{\gamma-1}^z(a_\gamma^* - a_\gamma);$$

the latter clearly satisfy the standard commutation relations of the Pauli-matrices:

$$(\sigma_\gamma^i)^2 = I,$$

$$[\sigma_\gamma^j, \sigma_{\gamma'}^k] = 2i\delta_{\gamma',\gamma}\sigma_\gamma^l,$$

where j, k, l is a cyclic permutation of the indices x, y, z. From the σ_γ^i ($i = x, y, z$) we construct $\mathfrak{R}_\gamma$ and then $\mathfrak{A}$, so that we indeed obtain a representation of $\mathfrak{A}$ by operators acting in the representation space of the representation of the canonical anticommutation relations we started with. Hence the mapping $\pi \rightarrow \pi_0$ is surjective; the other assertions of the theorem follow in the same way from the fact that $\mathfrak{A}$ is generated by $\{\sigma_\gamma^i \mid i = x, y, z; \gamma \in \mathbb{Z}^+\}$. ▌

REMARK. The interest in this theorem is that is reduces the study of the representations of the canonical anticommutation relations to the study of the representations of the C^*-algebra $\mathfrak{A}$: we recall that $\mathfrak{A}$ is defined as the infinite direct product of identical copies of the four-dimensional C^*-algebra $\mathfrak{M}_2$, so that we can use all the tools developed in Section 1 in connection with the representations of a direct product of C^*-algebras. We even have here the further simplification that $\mathfrak{R}_\gamma$ is now finite-dimensional. We have already seen in Theorem 2 one consequence of this fact; the algebra $\mathfrak{A}$ will also be studied in another connection in Chapter 4. In particular, the various consequences of Theorem III.1.11 have their analogs here as consequences of Theorem 3. Just to mention one example, the discrete representations obtained in Section 1 as particular cases of IDPS representations appear here in the same manner and are characterized, up to a unitary equivalence, by an equivalence class $[n]$ of infinite sequences $m = \{m_\gamma \mid \gamma \in \mathbb{Z}^+; m_\gamma = 0 \text{ or } 1\}$. Among the discrete representations the representation associated to the sequence $\{m_\gamma = 0 \mid \gamma \in \mathbb{Z}^+\}$ is singled out as the standard Fock-space representation constructed on the vacuum.†

Corollary. *For any representation of the canonical anticommutation relations* $\|a^\natural(f)\| = |f|_{\mathcal{T}}$ *for all* f *in* $\mathcal{T}$.

Proof. From Theorem 2 we know that $\mathfrak{A}$ is simple, hence that every representation π of $\mathfrak{A}$ is faithful and therefore for every A in $\mathfrak{A}$ $\|\pi(A)\| = \|A\|$. From Theorem 3 we know that every representation of the canonical anticommutation relations can be obtained as the restriction π_0 of a representation π of $\mathfrak{A}$; hence $a^\natural(f)$ has the same norm in every representation and therefore we can calculate it from any one of them. This is especially easy for the Fock representation for which we immediately find $\|a^\natural(f)\| = |f|_{\mathcal{T}}$; this proves the corollary. ▌

As an orientation for further reading, we indicate briefly how Clifford algebras enter rather naturally into the study of the representations of the canonical anticommutation relations.

† We discussed in detail the Fock-space representation for the boson case in Chapter 1, Section 1; for a mathematical discussion of the fermion case, carried in complete parallel, see again the definitive article by Cook [1953].

The following definition of $\mathfrak{C}_0(\mathcal{C})$ is admittedly sketchy; the mathematically inclined reader will find a more serious definition in Bourbaki [1959].†

The *Clifford algebra* $\mathfrak{C}_0(\mathcal{C})$ of the real Hilbert space $\mathcal{C}$ is the complex algebra generated by an identity, denoted I, and the monomials $[f_1][f_2]\cdots[f_n]$ with f_i in $\mathcal{C}$; we impose in the definition of the composition laws the following restrictions:

$$c_1[f_1] + c_2[f_2] = [c_1 f_1 + c_2 f_2],$$
$$[f]^2 = |f|^2 I,$$

and we define an involution on $\mathfrak{C}_0(\mathcal{C})$ by

$$I^* = I,$$

$$\left(\sum_\lambda c_\lambda [f_1][f_2]\cdots[f_{n(\lambda)}]\right)^* = \sum_\lambda c_\lambda^* [f_{n(\lambda)}]\cdots[f_2][f_1].$$

We notice that $\mathfrak{C}_0(\mathcal{C})$ contains $\mathcal{C}$ as a real subspace (via the canonical embedding $f \to [f]$) and that $\mathfrak{C}_0(\mathcal{C})$ is generated, as a complex algebra, by $[\mathcal{C}]$; also, as a direct consequence of the above composition laws,

$$[f_1][f_2] + [f_2][f_1] = 2(f_1, f_2)I.$$

The Clifford algebra $\mathfrak{C}_0(\mathcal{C})$ and its states have been studied in detail by Shale and Stinespring [1964], who proved that $\mathfrak{C}_0(\mathcal{C})$ is simple and discussed its connection with the representations of the canonical anticommutation relations. A norm can be defined abstractly on $\mathfrak{C}_0(\mathcal{C})$ and the norm closure $\mathfrak{C}(\mathcal{C})$ of $\mathfrak{C}_0(\mathcal{C})$ can then be defined.‡ The connection between $\mathfrak{C}(\mathcal{C})$ and $\mathfrak{A}$ is established by the following remark:§ there exists a unique isomorphism T from $\mathfrak{C}(\mathcal{C})$ onto $\mathfrak{A}$ such that if $\{e_1, e_2, \ldots; e_1', e_2', \ldots\}$ is an orthonormal basis in $\mathcal{C}$ then

$$Te_\gamma = a_\gamma + a_\gamma^*$$
$$Te_\gamma' = -i(a_\gamma - a_\gamma^*);$$

furthermore, for each f in $\mathcal{C}$, $\|Tf\| = |f|_{\mathcal{C}}$.

The work of Shale and Stinespring [1964] on states of the Clifford algebra $\mathfrak{C}_0(\mathcal{C})$ and certain automorphisms of $\mathfrak{A}$ associated with unitary transformations of $\mathcal{C}_{\mathbb{C}}$ has been developed by several authors, among whom we might mention Guichardet [1966], Balsev and Verbeure [1968], and Størmer [1969a].

Among the more traditional references to the representations of the canonical anticommutation relations we ought to mention the complete

† Specifically, Chapter 9, Section 9.1.

‡ In this connection see the early work of Segal [1956], to which this approach might be traced.

§ See Guichardet [1966], Proposition 3.7.

classification obtained by Gårding and Wightman [1954a] (see also Wightman and Schweber [1955]). Among the physical applications of the general theory we mention the study of the free fermion gas by Araki and Wyss [1964] and the many papers on the BCS-model (see Chapter 1, Section 1 for bibliographical references; see also Fano and Loupias [1971]); we also recall the result by Powers [1967c] mentioned at the end of the preceding section.

The study of quasi-free states on the algebra of the canonical anticommutation relations and of their connection with Bogoliubov automorphisms has been carried out in several papers, among which are Balsev, Manuceau, and Verbeure [1968], Rocca, Sirugue, and Testard [1969], Manuceau, Rocca, and Testard [1969], Powers and Størmer [1970], and Araki [1970a]; see also Rideau [1968] and dell'Antonio [1968].

CHAPTER 4

Quasi-Local Theories

In Chapter 2 we tried to develop the consequences of the following assumption: a C^*-algebra can be constructed in such a manner that its self-adjoint elements represent the bounded observables of a physical system. This approach was motivated by the hope that the many-body character of the systems we want to consider (in field theory and in statistical mechanics) is best exhibited when we pass to the limit in which these systems extend to an infinite region of space and have an infinite number of degrees of freedom.

The aim of this chapter is to take into account the following fact: the measurements that a physicist actually performs are limited in space and in time. As a consequence we often have enough information on the local structure of the system under consideration to be able to build up the structure of the idealized infinite system we want to consider. The first section deals with the general postulates of this formulation, and the second presents two types of simple illustrative example borrowed from statistical mechanics.

SECTION 1 GENERAL THEORY OF LOCAL SYSTEMS

OUTLINE. The C^*-algebra of all quasi-local observables on a physical system is defined in the beginning of the first subsection. The postulates of isotony, covariance, and local commutativity, which are common to all theories of local observables, are presented in this subsection. We then discuss the concept of locally normal states and its relevence for statistical mechanics.

The second subsection deals mainly with Lorentz covariant, quasi-local theories. Cluster properties and uniqueness of the vacuum are analyzed, first on the basis of locality alone, then in conjunction with the spectrum condition. Some of the consequences of the latter and of the weak additivity postulate (Reeh-Schlieder theorem) are then studied in connection with the factor types of local algebra and the simplicity of the quasi-local algebra associated with certain quantum fields. Some areas of research in which the algebraic approach has proved itself to be a useful tool are briefly indicated toward the end of this subsection. We finally come back once again to the study of cluster properties of certain states, this time in connection with the "decomposition at infinity," and we indicate the role of the latter in the theory of symmetry breaking.

a. Quasi-Local Algebras and Locally Normal States

The consideration of "physical systems" which are infinitely extended in space is prompted in statistical mechanics by the simplifications we expect to occur when the thermodynamical limit is taken, such as the appearance of sharp phase transitions and the elimination of Poincaré cycles. In quantum field theory the consideration of infinitely extended systems is motivated by the hope of exploiting fully the principle of special relativity as summarized in the covariance of the theory under Lorentz transformations. Hence in all three cases just mentioned, namely equilibrium and nonequilibrium statistical mechanics and quantum field theory, a limit is taken as a mathematical convenience to eliminate from the description of the theory what we would like to call unwanted or subsidiary boundary effects.

Taking these limits, however, involves some technical problems which have to be solved. The first of these problems is to decide what form the description of the infinite system should take. The basic element entering into this decision is the idea that although the system itself is *very* large (actually infinite) the information we have on it is gained from experiments that are essentially localized. Consequently we shall attempt to construct the global description as a summary, or compound, of all the local descriptions that are pertinent to the physical problem we are interested in. Let us now express this idea in mathematical language.

First let $\mathbb{R}^3$ (the three-dimensional euclidian space), or $\mathfrak{M}^4$ (the $(1 + 3)$-dimensional Minkowski space) be the *configuration space* in which we have decided to operate. In *both* cases we equip the configuration space with the topology it inherits from the euclidian distance. In line with the postulates agreed on (see Chapter 1, Section 2) we associate a C^*-algebra $\mathfrak{R}(\Omega)$ with each bounded open region Ω of our configuration space, and we interpret the self-adjoint elements of $\mathfrak{R}(\Omega)$ as the *local observables* pertinent to the region Ω. The set $\mathcal{F}$ of all bounded open regions Ω, equipped with the partial ordering of the set-theoretical inclusion, is a directed set. We now assume that the postulate of *isotony* will hold; that is, we assume that for every pair Ω_1, Ω_2 of elements of $\mathcal{F}$ with $\Omega_1 \subseteq \Omega_2$ there exists an *injective* *-homomorphism $i_{2,1}$ from $\mathfrak{R}(\Omega_1)$ into $\mathfrak{R}(\Omega_2)$ such that

$$i_{2,1}(I_1) = I_2,$$
$$i_{3,2}i_{2,1} = i_{3,1} \quad \text{whenever } \Omega_1 \subseteq \Omega_2 \subseteq \Omega_3.$$

We then define the *quasi-local algebra* $\mathfrak{R}$ as the C^*-inductive limit† of $\{\mathfrak{R}(\Omega) \mid \Omega \in \mathcal{F}\}$ and we write symbolically

$$\mathfrak{R} = \overline{\bigcup_{\Omega \in \mathcal{F}} \mathfrak{R}(\Omega)}.$$

At this point the reader should realize that a good deal of the physics of the problem is captured in the choice of the local algebras $\mathfrak{R}(\Omega)$. In some simple (yet nontrivial) systems, such as those discussed in Subsection 2a, the choice of the $\mathfrak{R}(\Omega)$ presents no problems. This, however, is not the case in general, and in the absence of better physical intuition we often have to make this choice on the basis of mathematical convenience. As a result of different choices of $\mathfrak{R}(\Omega)$, different algebras $\mathfrak{R}$ might occur, singling out different sets of states as "naturally more physical." The choice of the $\mathfrak{R}(\Omega)$ then becomes a rather delicate matter; a general discussion of this situation and its consequences has been started recently by Haag, Kadison, and Kastler [1970]. For the time being we shall take the pragmatic attitude that a choice has been made and we shall examine what can be said from there on; the knowledge of the latter point anyhow will help in understanding, at a later stage, the more subtle problem of the choice of the $\mathfrak{R}(\Omega)$.

We might also mention at this point that in the early days of the algebraic approach to relativistic quantum field theories the assumption was often made‡ that the $\mathfrak{R}(\Omega)$ are von Neumann algebras acting on some Hilbert space $\mathcal{H}$ (independent of Ω) and that the appropriate global algebra is

$$\mathfrak{N} = \{\mathfrak{R}(\Omega) \mid \Omega \in \mathcal{F}\}''.$$

Now let us come back to our general scheme and let G denote either the euclidian group $\mathbb{E}^3$ or the inhomogeneous proper Lorentz group $L_+^\uparrow$, depending on whether the configuration space is chosen to be $\mathbb{R}^3$ or $\mathfrak{M}^4$. As every g in G maps every Ω in $\mathcal{F}$ into $g[\Omega]$ in $\mathcal{F}$, we can identify for each R in $\mathfrak{R}(\Omega)$ the element $\alpha_g[R]$ in $\mathfrak{R}(g[\Omega])$ and we assume that α_g is a $*$-isomorphism from $\mathfrak{R}(\Omega)$ onto $\mathfrak{R}(g[\Omega])$. Then let $\{\mathfrak{R}_n \in \mathfrak{R}(\Omega_n)\}$ be a Cauchy sequence converging to R in $\mathfrak{R}$; $\alpha_g[R_n]$ is again a Cauchy sequence in $\mathfrak{R}$. Let us denote its limit point by $\alpha_g[R]$. Hence we arrived at the formulation of the postulate of *covariance* of the theory; that is, we assume that there exists a homomorphism α from G into Aut $(\mathfrak{R})$ such that $\alpha_g[\mathfrak{R}(\Omega)] = \mathfrak{R}(g[\Omega])$ for every Ω in $\mathcal{F}$ and every g in G.

In order to formulate in a concise manner the postulate of *local commutativity* (also called *Einstein causality* when the configuration space is $\mathfrak{M}^4$), we introduce the following notations. Depending on whether the configuration space is $\mathbb{R}^3$ or $\mathfrak{M}^4$, we write $\Omega_1 \,\dot{\cap}\, \Omega_2$ to mean either $\Omega_1 \cap \Omega_2 = \varnothing$ or

† For the definition see Section III.1.e.

‡ See, for instance, the remarkable paper of Araki [1964b].

$g_{\mu\nu}(x-y)^{\mu}(x-y)^{\nu} < 0$ for all x in Ω_1 and all y in Ω_2; $g_{\mu\nu}$ is the metrics tensor $g_{00} = +1$, $g_{ii} = -1$ for $i = 1, 2, 3$, and $g_{\mu\nu} = 0$ for $\mu \neq \nu$. Since $\mathcal{F}$ is a directed set, we have for each Ω_1 and Ω_2 in $\mathcal{F}$ an element Ω_0 in $\mathcal{F}$ such that $\Omega_1 \subseteq \Omega_0$ and $\Omega_2 \subseteq \Omega_0$; we then identify $\mathfrak{R}(\Omega_1)$ [or $\mathfrak{R}(\Omega_2)$], with its image in $\mathfrak{R}(\Omega_0)$ via the isotony mapping, the existence of which we assumed earlier. We then write $\mathfrak{R}(\Omega_1) \smile \mathfrak{R}(\Omega_2)$ to mean that $R_1R_2 = R_2R_1$ for all R_1 in $\mathfrak{R}(\Omega_1)$ and all R_2 in $\mathfrak{R}(\Omega_2)$. With these notations we now assume (*local commutativity*) that $\Omega_1 \between \Omega_2$ implies $\mathfrak{R}(\Omega_1) \smile \mathfrak{R}(\Omega_2)$.

These three postulates—isotony, covariance, and local commutativity—form the core of every global theory based on the algebras of local observables; the formulation of local commutativity given here applies also to local fields that obey Bose statistics. Fermi fields require some adaptations that we shall study later. We shall show some of the general consequences of these postulates in the subsections that follow and discuss some of their particular realizations in Section 2.

We should now like to particularize our preliminary remarks to non-relativistic statistical mechanics; it is customary in this case to take for Ω bounded open regions in $\mathbb{R}^3$. Two problems which are actually linked have to be handled: the definition of the time-evolution and the definition and properties of the Gibbs state.

The current recipe for defining the time-evolution for a system that is infinitely extended in space can be formulated as follows: let $\mathcal{F}_0$ be a subset of $\mathcal{F}$ consisting of an increasing sequence $\{\Omega_n\}$ such that for every Ω in $\mathcal{F}$ there is some finite positive integer $N(\Omega)$ with the following property: $\Omega \subseteq \Omega_n$ for all $n \geqslant N(\Omega)$; for instance $\{\Omega_n\}$ could be a sequence of cubes of edge L^n, centered at the origin. Furthermore, suppose that the time evolution $\alpha_n(t)$ is well-defined in Ω_n; this amounts in particular to assuming certain boundary conditions in Ω_n. For each R in some Ω of $\mathcal{F}$ we then study the limit of $\alpha_n(t)[R]$ as $n \to \infty$ "in the appropriate topology"; if the latter exists and defines an element $\alpha(t)[R]$ of $\mathfrak{R}$ and furthermore if α_t is isometric, we can then extend it by continuity from $\bigcup \mathfrak{R}(\Omega)$ to $\mathfrak{R}$. We should then check whether α_t is continuous in t. This program evidently involves some convergence proofs. As we shall see in Section 2, the latter work all the way for some class of interactions on a quantum spin lattice. Within this class, however, there are still interactions that are sufficiently long range so that local observables at time $t = 0$ lose their local character in the course of the time evolution. In the extreme case of the van der Waals limit α_t can no longer be defined as an automorphism of $\mathfrak{R}$ but can still be defined, for the representations π of interest, as an automorphism of $\pi(\mathfrak{R})''$. As we shall see in Section 2, a similar situation occurs in the case of the free Bose gas.

Let us now turn to the definition of the Gibbs state. We assume again that a choice has been made of an increasing sequence $\{\Omega_n\}$ that satisfies the property mentioned above. Suppose that for each Ω_n, $\mathfrak{R}(\Omega_n)$ is represented faithfully on some Hilbert space $\mathcal{H}_n$ and that the time evolution is unitarily implemented in this representation; let H_n be the corresponding Hamiltonian which is supposed to be such that the canonical density matrix

$$\rho_n = \frac{\exp(-\beta H_n)}{\operatorname{Tr}\exp(-\beta H_n)}$$

is well-defined. For every R in $\mathfrak{R}(\Omega)$ and every $n > N(\Omega)$ we can then define

$$\langle\phi_n; R\rangle = \operatorname{Tr}_{\mathcal{H}_n} \rho_n R,$$

where R is identified with its canonical injection into $\mathfrak{R}(\Omega_n)$. We then study the limit, as $n \to \infty$, of $\langle\phi_n; R\rangle$; if it exists as a state ϕ on $\mathfrak{R}$, we identify ϕ with the canonical equilibrium state (or Gibbs state) for the natural temperature β. We define in an analogous manner the grand canonical equilibrium state by replacing throughout H by $(H - \mu N)$. As for the time evolution, these definitions involve limiting procedures which have to be proved to converge for each individual physical situation. We should remark at this point that these limiting procedures involve a choice of boundary conditions; we assume this choice to be consistent with the intrinsic symmetry of the problem; for instance, periodic boundary conditions preserve the flip-flop symmetry of the Ising model and the corresponding Gibbs state inherits this symmetry. A subsequent decomposition of this Gibbs state as a mixture of pure thermodynamical phases might then produce states which individually do not necessarily inherit the full symmetry of the Gibbs state itself. When the symmetry of the pure thermodynamical phases obtained in this manner is lower than the symmetry of the original problem (i.e., the Gibbs state), we speak of *spontaneous symmetry breaking*. We could then raise the question whether it might be possible to reproduce these pure thermodynamical phases directly by a limiting procedure in which we would start from boundary conditions that do not preserve the symmetry of the problem; for instance, we could compute the canonical equilibrium state of a finite Ising lattice by embedding it into an infinite one in which all the sites outside the finite region under consideration are occupied by a spin pointing down. Incidentally, the reader who thinks that this condition is unnatural should consider its meaning for the lattice-gas equivalent of the Ising model. The finite-volume equilibrium states obtained in the manner just described are evidently not flip-flop invariant. The Gibbs state defined as the limit of such states is then, in general, not flip-flop invariant. Playing with the boundary conditions can thus turn out to be a subtle way of breaking some of the symmetry of the theory and to favor one thermodynamical phase over the others, thus providing an alternate scheme for the study of spontaneous symmetry breaking. This approach has been successfully exploited by

Dobrushin [1968a, b] and some other authors (see, for instance, Ginibre [1969a, b] and Robinson [1969]).

Dell'Antonio [1966] proved the following theorem to the effect that the Gibbs state is a locally normal state under certain conditions often realized in physical applications:

Theorem 1. *Let* $\mathfrak{R} = \overline{\bigcup_{\Omega\in\mathcal{F}} \mathfrak{R}(\Omega)}$ *where* $\mathfrak{R}(\Omega)$ *are von Neumann algebras; let* $\{\Omega_n\} \subset \mathcal{F}$ *be an increasing sequence such that for every* Ω *in* $\mathcal{F}$ *there exists a positive integer* $\mathrm{N}(\Omega) < \infty$ *for which* $\Omega \subseteq \Omega_n$ *for all* $\mathrm{n} \geqslant \mathrm{N}(\Omega)$. *Suppose further that for each* n *a normal state* ϕ_n *on* $\mathfrak{R}(\Omega_n)$ *is given and that* $\langle\phi;\, \mathrm{R}\rangle \equiv \lim_{n\to\infty}\langle\phi_n;\, \mathrm{R}\rangle$ *exists for every* R *in* $\bigcup_{\Omega\in\mathcal{F}}\mathfrak{R}(\Omega)$; *then for every* Ω *in* $\mathcal{F}$ *the restriction* $\phi(\Omega)$ *of* ϕ *to* $\mathfrak{R}(\Omega)$ *is a normal state.*

Proof. We recall that a state ψ on a von Neumann algebra $\mathfrak{N}$ is normal if and only if $\langle\psi;\, \sum E_i\rangle = \sum\langle\psi;\, E_i\rangle$ for every family $\{E_i \mid i \in I\}$ of pairwise orthogonal projectors in $\mathfrak{N}$. From the definition of $\mathfrak{R}$ there exists for each Ω in $\mathcal{F}$ an *injective* *-homomorphism from $\mathfrak{R}(\Omega)$ into $\mathfrak{R}(\Omega_n)$ whenever $\Omega_n \subseteq \Omega$; consequently the restriction $\phi_n(\Omega)$ of ϕ_n to $\mathfrak{R}(\Omega)$ [for all $n \geqslant N(\Omega)$] is a normal state on $\mathfrak{R}(\Omega)$. The theorem is then reduced to proving that a *sequence* ψ_n [here $\phi_n(\Omega)$] of normal states on a von Neumann algebra $\mathfrak{N}$ [here $\mathfrak{R}(\Omega)$] which converges pointwise on $\mathfrak{N}$ converges to a normal state ψ [here $\phi(\Omega)$] on $\mathfrak{N}$. The proof of the latter assertion can be done in the following manner.† Let $\mathcal{E} = \{E_i \mid i \in I\}$ be any family of pairwise orthogonal projectors in $\mathfrak{N}$ and denote by $\mathfrak{N}(\mathcal{E})$ the abelian von Neumann subalgebra of $\mathfrak{N}$ generated by $\mathcal{E}$: $\mathfrak{N}(\mathcal{E}) \equiv \{E_i \mid i \in I\}''$; ψ_n restricted to $\mathfrak{N}(\mathcal{E})$ are then evidently normal. Since $\mathfrak{N}(\mathcal{E})$ is abelian, there exists‡ a locally compact space $\mathcal{X}$, a positive measure μ on $\mathcal{X}$, and an isometric isomorphism from $\mathfrak{N}(\mathcal{E})$ (considered as a normed involutive algebra) onto the normed involutive algebra $\mathcal{L}^\infty_{\mathbb{C}}(\mathcal{X}, \mu)$ of all complex-valued, μ-essentially bounded functions on $\mathcal{X}$. The set of all positive normal linear forms on $\mathfrak{N}$, which is the positive cone in the predual of $\mathfrak{N}$, is then isomorphic to the positive cone in the predual of $\mathcal{L}^\infty_{\mathbb{C}}(\mathcal{X}, \mu)$. On the other hand, we know§ that the dual of $\mathcal{L}^1_{\mathbb{C}}(\mathcal{X}, \mu)$ is $\mathcal{L}^\infty_{\mathbb{C}}(\mathcal{X}, \mu)$ and that $\mathcal{L}^1_{\mathbb{C}}(\mathcal{X}, \mu)$ is weakly sequentially complete. Hence the images of the ψ_n in $\mathcal{L}^1_{\mathbb{C}}(\mathcal{X}, \mu)$ converge in $\mathcal{L}^1_{\mathbb{C}}(\mathcal{X}, \mu)$ and $\{\psi_n\}$ converge to ψ normal on $\mathfrak{N}(\mathcal{E})$, i.e., $\langle\psi;\, \sum E_i\rangle = \sum \langle\psi;\, E_i\rangle$. Since $\mathcal{E}$ is an arbitrary family of pairwise orthogonal projectors in $\mathfrak{N}$, we conclude that ψ is normal and the proof of the theorem is achieved. ∎

REMARKS.

(i) It is essential to the completion of the proof, and the validity of the theorem, that $\{\phi_n\}$ be a *sequence* of normal states, since the set of all

† See dell'Antonio [1967]; see also Sakai [1957].

‡ See Dixmier [1957], Theorem I.7.3.1.

§ Dunford and Schwartz [1957], IV.8.5 and IV.8.6.

normal states is weakly dense in the space of all states on $\mathfrak{N}$. We have here one of the cases in which it is important to distinguish between convergence of sequences and convergence of nets.

(ii) If, in addition to the assumptions of the theorem, the $\mathfrak{N}(\Omega)$ are assumed to be type I factors (or more generally that they are von Neumann algebras of type I with totally atomic center), dell'Antonio proved that the convergence $\phi_n(\Omega) \to \phi(\Omega)$ is uniform, i.e., for every $\varepsilon > 0$ there exists n_0 such that $n, m > n_0$ implies

$$\sup_{\substack{R \in \mathfrak{N}(\Omega) \\ \|R\| \leqslant 1}} |\langle \phi_n ; R \rangle - \langle \phi_m ; R \rangle| \leqslant \varepsilon .$$

We now want to show that under certain conditions the representation of $\mathfrak{R}$ canonically associated to the Gibbs state is defined on a separable Hilbert space.

Theorem 2. *Let* $\mathfrak{R} = \overline{\bigcup \mathfrak{R}(\Omega_m)}$, *where* $\{\mathfrak{R}(\Omega_m)\}$ *is a sequence of von Neumann algebras respectively defined on some separable Hilbert spaces* $\mathfrak{H}_m$. *Let* ϕ *be a state on* $\mathfrak{R}$ *such that its restriction* ϕ_m *to each* $\mathfrak{R}(\Omega_m)$ *is normal. Then the space* $\mathcal{H}_\phi$ *of the representation* π_ϕ *of* $\mathfrak{R}$ *canonically associated with* ϕ *is separable.*

Proof. Porta and Schwartz [1967] showed that if $\mathfrak{B}(\mathfrak{H})$ denotes the algebra of all bounded operators acting on a separable Hilbert space $\mathfrak{H}$ then a state ψ on $\mathfrak{B}(\mathfrak{H})$ is normal if and only if the Hilbert space $\mathcal{H}_\psi$ of the representation π_ψ of $\mathfrak{B}(\mathfrak{H})$ canonically associated with ψ is separable. Now suppose that $\mathfrak{N}$ is a von Neumann subalgebra of $\mathfrak{B}(\mathfrak{H})$ and that ψ is a normal state on $\mathfrak{N}$. This implies that there exists $\{\Psi_i\} \subset \mathfrak{H}$ such that $\sum_i |\Psi_i|^2 = 1$ and $\langle \psi ; N \rangle = \sum_i (\Psi_i, N\Psi_i)$ for all N in $\mathfrak{N}$; we then define the extension $\tilde{\psi}$ of ψ from $\mathfrak{N}$ to $\mathfrak{B}(\mathfrak{H})$ by $\langle \tilde{\psi} ; B \rangle = \sum_i (\Psi_i, B\Psi_i)$ for all B in $\mathfrak{B}(\mathfrak{H})$, so that $\tilde{\psi}$ is a normal form on $\mathfrak{B}(\mathfrak{H})$. There exists therefore a separable Hilbert space $\tilde{\mathcal{H}}_\psi$ accommodating a cyclic representation $\tilde{\pi}_\psi$ of $\mathfrak{B}(\mathfrak{H})$ with cyclic vector Ψ such that $\langle \tilde{\psi} ; B \rangle = (\Psi, \tilde{\pi}_\psi(B)\Psi)$ for all B in $\mathfrak{B}(\mathfrak{H})$. We denote by $\mathcal{H}_\psi$ the closure of the pre-Hilbert space $\{\tilde{\pi}_\psi(N)\Psi \mid N \in \mathfrak{N}\}$; $\mathcal{H}_\psi$ is clearly invariant under $\tilde{\pi}_\psi(N)$ for all N in $\mathfrak{N}$. We can therefore define the representation π_ψ of $\mathfrak{N}$, acting on $\mathcal{H}_\psi$, as the restriction of $\tilde{\pi}_\psi$ from $\mathfrak{B}(\mathfrak{H})$ to $\mathfrak{N}$; we clearly have Ψ cyclic in $\mathcal{H}_\psi$ for $\pi_\psi(\mathfrak{N})$ and $\langle \psi ; N \rangle = (\Psi, \pi_\psi(N)\Psi)$ for all N in $\mathfrak{N}$. Hence $\pi_\psi(\mathfrak{N})$ is unitarily equivalent to the representation of $\mathfrak{N}$ canonically associated with ψ by the GNS construction. Since $\mathcal{H}_\psi$ is a subspace of the separable Hilbert space $\tilde{\mathcal{H}}_\psi$, it is also separable (since† $\tilde{\mathcal{H}}_\psi$ is a metric space). Consequently the Hilbert space $\mathcal{H}_m$ of the representation π_m of $\mathfrak{R}(\Omega_m)$ canonically associated with ϕ_m is separable, since ϕ_m is normal by assumption. Let us then consider in $\mathcal{H}_\phi$ the Hilbert space $\mathcal{H}_m'$ obtained as the completion of $\{\pi_\phi(R)\Phi \mid R \in \mathfrak{R}(\Omega_m)\}$,

† See Dunford and Schwartz [1957], I.6.12.

where Φ is the cyclic vector associated with ϕ. The restriction $\pi'_m(\mathfrak{R}(\Omega_m))$ of $\pi(\mathfrak{R}(\Omega_m))$, from $\mathcal{H}_\phi$ to $\mathcal{H}'_m$, is then unitarily equivalent to $\pi_m(\mathfrak{R}(\Omega_m))$. This implies that $\mathcal{H}'_m$ is separable, so that the countable union $\bigcup_m \mathcal{H}'_m$ is separable; the latter is dense in $\mathcal{H}_\phi$, hence $\mathcal{H}_\phi$ is separable, which is what we wanted to prove. ∎

REMARKS. In the case in which $\mathfrak{R}(\Omega_m)$ is $\mathfrak{B}(\mathfrak{H}_m)$ this result has been found by Hugenholtz and Wieringa [1969] (see also Ruelle [1967]); furthermore it is clear from the beginning of the proof of the theorem that in this particular case the converse of the theorem is also true, namely, that ϕ is locally normal (with respect to $\{\Omega_m\}$) if $\mathcal{H}_\phi$ is separable; the latter result, proved by Wieringa [1970], has been generalized to the case in which the $\mathfrak{R}(\Omega_m)$ are factors by Haag, Kadison, and Kastler [1970], who used the fact that every representation of a factor on a separable Hilbert space is normal.†

We have already noticed that any positive linear form ϕ on $\mathfrak{B}(\mathfrak{H})$ can be decomposed uniquely as $\phi = \phi_1 + \phi_2$ with ϕ_1 normal and $\langle \phi_2; K\rangle = 0$ for all K in $\mathfrak{C}(\mathfrak{H})$, the closed two-sided ideal of $\mathfrak{B}(\mathfrak{H})$ formed by all compact operators on $\mathfrak{H}$. When $\mathfrak{H}$ is separable, $\mathfrak{C}(\mathfrak{H})$ is also separable, since it is the norm-closure of the algebra of all finite-rank operators on $\mathfrak{H}$. We can then transpose these properties to an abstract setting and say that a state ϕ on a C^*-algebra $\mathfrak{R}$ is "locally normal" if there exists a countable family $\{\mathfrak{R}_n\}$ of sub-C^*-algebras of $\mathfrak{R}$ such that (a) $\bigcup \mathfrak{R}_n$ is dense in $\mathfrak{R}$, (b) each $\mathfrak{R}_n$ possesses a separable closed two-sided ideal $\mathcal{C}_n$, and (c) the restriction ϕ_n of ϕ to $\mathcal{C}_n$ still has norm 1. Ruelle [1966] and Lanford and Ruelle [1967], who introduced this characterization of "locally normal" states, proved that it is sufficient to secure the following consequences:

1. The representation π_ϕ canonically associated with ϕ acts in a separable Hilbert space.
2. The set $\mathcal{N}(\{\mathfrak{R}_n\})$ of all states which are "locally normal" with respect to a fixed sequence $\{\mathfrak{R}_n\}$ is separated from $\mathfrak{S}$ by a sequence $\{A_i\}$ of self-adjoint elements of $\mathfrak{R}$; i.e., there exists at least one sequence $\{A_i\}$ of self-adjoint elements of $\mathfrak{R}$ such that $\langle \phi; A_i\rangle \neq \langle \psi; A_i\rangle$ for at least one member of this sequence whenever $\phi \in \mathcal{N}(\{\mathfrak{R}_n\})$ and $\psi \in \mathfrak{S}$.
3. The unique maximal measure μ_ϕ associated (see II.2.f) with a locally normal G-invariant state on a G-abelian system is concentrated in the Borel sense on $\mathfrak{S}_G$; as we have already mentioned in Section II.2.f, the same result can be transposed to $\mathfrak{S}_\beta$. The key to the proof of the latter fact is first to define a continuous mapping σ from $\mathfrak{S}_\rho$ (meaning either $\mathfrak{S}_G$ or $\mathfrak{S}_\beta$) to $\mathbb{R}^\infty$ by

$$\sigma(\phi) = \{\langle \phi; A_i\rangle \mid A_i \in \{A_i\}\}.$$

† For this see Kaplansky [1952], Feldman and Fell [1957], and Takesaki [1960].

We then associate with each ϕ in $\mathcal{N}(\{\mathfrak{R}_n\}) \cap \mathfrak{S}_\rho$ the measure ν_ϕ on $\sigma(\mathfrak{S}_\rho)$ defined by

$$\langle \nu_\phi; f \rangle = \langle \mu_\phi; f \circ \sigma \rangle$$

for all convex f on $\sigma(\mathfrak{S}_\rho)$. Since μ_ϕ is maximal on $\mathfrak{S}_\rho$, ν_ϕ is maximal on the compact, convex, *metrizable* set $\sigma(\mathfrak{S}_\rho)$, hence is concentrated *in the Borel sense* on the extremal points of $\sigma(\mathfrak{S}_\rho)$; μ_ϕ is then concentrated in the Borel sense on the inverse image $\sigma^{-1}(\mathcal{E}[\sigma(\mathfrak{S}_\rho)])$ of this set. We can easily check that $\mathcal{N}(\{\mathfrak{R}_n\})$ is a Borel set of μ_ϕ-measure 1 so that μ_ϕ is a maximal measure, concentrated in the Borel sense on $\mathcal{N}(\{\mathfrak{R}_n\}) \cap \sigma^{-1}(\mathcal{E}[\sigma(\mathfrak{S}_\rho)])$. We now use the fact that $\{A_i\}$ separates $\mathcal{N}(\{\mathfrak{R}_n\})$ from $\mathfrak{S}$ to conclude that μ_ϕ is concentrated in the Borel sense on $\mathcal{N}(\{\mathfrak{R}_n\}) \cap \mathcal{E}_\rho$.

b. First Consequences of the Postulates

In this subsection we systematically use the notation appropriate to the case in which the configuration space is $\mathfrak{M}^4$ unless explicit mention is made to the contrary; the case in which the configuration space is $\mathbb{R}^3$ can be followed trivially by the thread of the following arguments.

We recall that $\mathcal{F}$ denotes the directed set of all bounded open regions Ω in $\mathfrak{M}^4$ (where "bounded" and "open" are to be understood in the topology induced by the euclidian metric). We denote by a an arbitrary space-time translation and by $\mathbf{a}$ any space like translation. (When $\mathbb{R}^3$ takes the place of $\mathfrak{M}^4$, this means that $\mathbf{a}$ is a translation in $\mathbb{R}^3$.)

Theorem 3. *For any two elements* R *and* S *in* $\mathfrak{R} = \overline{\bigcup_{\Omega \in \mathcal{F}} \mathfrak{R}(\Omega)}$ *and any* $\varepsilon_0 > 0$ *there exists a finite positive number* $\mathrm{K}(\mathrm{R}, \mathrm{S}; \varepsilon_0)$ *such that* $\| [\mathrm{R}, \alpha_{\mathbf{a}}[\mathrm{S}]]\| \leqslant \varepsilon_0$ *for all* $\mathbf{a}$ *such that* $|\mathbf{a}| \geqslant \mathrm{K}(\mathrm{R}, \mathrm{S}; \varepsilon_0)$.

Proof. We first notice, as a consequence of the isotony postulate, that for every R and S in $\mathfrak{R}$ and any $\varepsilon > 0$ there exists Ω in $\mathcal{F}$ and R_0, S_0 in $\mathfrak{R}(\Omega)$ such that

$$\|R - R_0\| \leqslant \varepsilon \quad \text{and} \quad \|S - S_0\| \leqslant \varepsilon.$$

Since Ω belongs to $\mathcal{F}$, there exists a finite number $K(R, S; \varepsilon)$ such that $\Omega \,\check{o}\, \mathbf{a}[\Omega]$ for all $\mathbf{a}$ with $|\mathbf{a}| \geqslant K(R, S; \varepsilon)$. As a consequence of local commutativity and translational covariance, $[R_0, \alpha_{\mathbf{a}}[S_0]] = 0$ for these translations. We now form

$$\begin{aligned} \| [R, \alpha_{\mathbf{a}}[S]]\| &= \| [R - R_0 + R_0, \alpha_{\mathbf{a}}[S - S_0 + S_0]]\| \\ &\leqslant \| [R - R_0, \alpha_{\mathbf{a}}[S - S_0]]\| \\ &\quad + \| [R - R_0, \alpha_{\mathbf{a}}[S_0]]\| + \| [R_0, \alpha_{\mathbf{a}}[S - S_0]]\| \\ &\leqslant 2\varepsilon(\varepsilon + \|S_0\| + \|R_0\|) \\ &\leqslant 2\varepsilon(\|R\| + \|S\| + 3\varepsilon). \end{aligned}$$

For R, S, and ε_0 fixed, we can choose ε so that the right-hand side of this inequality becomes smaller than ε_0, thus proving our theorem. ∎

This assertion has several interesting consequences; the first formalizes the observation that a Lorentz transformation can hardly be conceived as a quasi-local operation, so that we should be surprised to see it implemented by an element of the quasi-local algebra. We can indeed support this almost semantic remark by a mathematical proof which we now present.†

Corollary 1. *A nontrivial Lorentz transformation cannot be an inner automorphism of* $\mathfrak{R}$.

Proof. We first prove this result for the translation part of $L_+^\uparrow$. Suppose that there exists for some space-time translation b a unitary element U_b of $\mathfrak{R}$ such that $\alpha_b[R] = U_b R U_b^{-1}$ for all R in $\mathfrak{R}$; from the theorem we know that for every R in $\mathfrak{R}$ and $\varepsilon > 0$ there exists $K(U_b, R, \varepsilon)$ such that $\|[U_b, \alpha_{\mathbf{a}}[R]]\| \leqslant \varepsilon$ for all $|\mathbf{a}| \geqslant K(U_b, R, \varepsilon)$; on the other hand, we have

$$\begin{aligned}\|[U_b, \alpha_{\mathbf{a}}[R]]\| &= \|[U_b, \alpha_{\mathbf{a}}[R]]\| \cdot \|U_b^{-1}\| \\ &= \|U_b\alpha_{\mathbf{a}}[R]U_b^{-1} - \alpha_{\mathbf{a}}[R]\| \\ &= \|\alpha_b\alpha_{\mathbf{a}}[R] - \alpha_{\mathbf{a}}[R]\| \\ &= \|\alpha_{\mathbf{a}}[\alpha_b[R] - R]\| \\ &= \|\alpha_b[R] - R\|.\end{aligned}$$

Since ε is independent of R and b, hence as small as we want, we conclude that $\alpha_b[R] = R$ for all R in $\mathfrak{R}$, which is absurd (unless $b = 0$). We therefore have that no space-time translation can be implemented by a unitary element of $\mathfrak{R}$. To prove this we used $\alpha_{\mathbf{a}}\alpha_b = \alpha_b\alpha_{\mathbf{a}}$, i.e., the translations commute among themselves; on noticing, however, that the homogeneous Lorentz transformations act as automorphisms of the set of space-like translations, the above proof can be extended to cover the homogeneous Lorentz transformations as well, thus using the semidirect product property of the inhomogeneous Lorentz group. ∎

REMARKS.

(i) When the configuration space is taken to be $\mathbb{R}^3$, read Euclidian transformations instead of Lorentz transformations throughout.

(ii) This corollary shows, for instance that the total energy-momentum is not going to exist as an element of the algebra $\mathfrak{R}$ of the quasi-local observables. In the representations of $\mathfrak{R}$ in which α_b is implemented by a strongly continuous group U_b of unitary operators, the generators of U_b

†J. Dollard pointed out to us that Haag and Kastler [1964] gave a somewhat incomplete proof of the following result. The proof here is due to C. Radin and K. Sinha.

are defined. We should, however, be wary in general of any physical interpretation of these quantities that goes further than that.

(iii) We should also insist on the fact that the corollary rules out only the possibility of a global inner implementation of α_g but does *not* rule out the existence of local implementation.

The result of the theorem, namely that

$$\lim_{|\mathbf{a}|\to\infty} \|[R, \alpha_{\mathbf{a}}[S]]\| = 0 \qquad \forall\, R, S \in \mathfrak{R},$$

is referred to by saying that the spacelike translations act in a *norm-asymptotically abelian* manner on $\mathfrak{R}$. This is true in particular of the group of all space translations on $\mathbb{R}^3$. As we have already mentioned on p. 177, this is a much stronger condition than η-abelianness and a fortiori than G-abelianness. As we saw in Theorem II.2.6, the latter property is equivalent to the assertion that the von Neumann algebra generated by

$$E_\phi(\mathbb{R}^3)\, \pi_\phi(\mathfrak{R})\, E_\phi(\mathbb{R}^3)$$

is abelian for every $\mathbb{R}^3$-invariant state on $\mathfrak{R}$; we recall that $E_\phi(\mathbb{R}^3)$ is the subspace consisting of all vectors of $\mathcal{H}_\phi$ that are invariant under $U_\phi(\mathbb{R}^3)$. Furthermore, by the remark on p. 179, if ϕ is $\mathbb{R}^4$-invariant [or $L_+^\uparrow$-invariant], we have that the von Neumann algebras generated by $E_\phi(\mathbb{R}^4)\, \pi_\phi(\mathfrak{R})\, E_\phi(\mathbb{R}^4)$† [or $E_\phi(L_+^\uparrow)\, \pi_\phi(\mathfrak{R})\, E_\phi(L_+^\uparrow)$] are abelian. Therefore we can apply the results of Subsection II.2.d; in particular, we see that $\mathbb{R}^3$ acts in an η-abelian manner on $\mathfrak{R}$ and the equivalence of the nine conditions in Theorem II.2.8 holds true.

Since the condition of norm-asymptotic abelianness is significantly stronger than those studied in Subsection II.2.d, we should expect to receive more information from it. This is indeed the case, as we shall see presently.

Corollary 2. *For any representation* π *of* $\mathfrak{R}$ $\mathrm{w}-\lim_{|\mathbf{a}|\to\infty} [\mathrm{B}, \pi(\alpha_{\mathbf{a}}[\mathrm{R}])] = 0$ *for all* R *in* $\mathfrak{R}$ *and all* B *in* $\mathfrak{Z}_\pi(\mathfrak{R})'$, *where* $\mathfrak{Z}_\pi(\mathfrak{R}) = \pi(\mathfrak{R})'' \cap \pi(\mathfrak{R})'$.

Proof. We first use the fact that $\mathfrak{Z}_\pi(\mathfrak{R})'$ is the closure in the strong-operator topology of the set $\mathfrak{Y}$ of all finite linear combinations of the form

$$Y = \sum_i \lambda_i\, \pi(R_i) X_i,$$

with λ_i in $\mathbb{C}$, R_i in $\mathfrak{R}$, and X_i in $\pi(\mathfrak{R})'$. Hence for any B in $\mathfrak{Z}_\pi(\mathfrak{R})'$, any Φ and Ψ in $\mathcal{H}_\pi$, and any $\varepsilon > 0$ there exists B_0 in $\mathfrak{Y}$ such that

$$\|(B - B_0)\Phi\| \leqslant \varepsilon \quad \text{and} \quad \|(B - B_0)\Psi\| \leqslant \varepsilon.$$

On the other hand, we conclude from the condition of norm-asymptotic abelianness that for each B_0 in $\mathfrak{Y}$, each R in $\mathfrak{R}$, and each $\varepsilon > 0$ there exists a

† Compare with Araki [1964b] and in conventional field theory Borchers [1962] and Reeh and Schlieder [1962].

finite number $K(B_0, R; \varepsilon)$ such that

$$\|[B_0, \pi(\alpha_{\mathbf{a}}[R]]\| \leqslant \varepsilon \quad \text{for all} \quad |\mathbf{a}| \geqslant K(B_0, R; \varepsilon).$$

Joining these two facts we get

$$\begin{aligned}|(\Phi, [B, \pi(\alpha_{\mathbf{a}}[R])]\Psi)| &\leqslant |(\Phi, [B - B_0, \pi(\alpha_{\mathbf{a}}[R])]\Psi)| + |(\Phi, [B_0, \pi(\alpha_{\mathbf{a}}[R])]\Psi)| \\ &\leqslant \varepsilon(2\,\|R\| + 1)\,\|\Phi\| \cdot \|\Psi\|,\end{aligned}$$

hence proving the weak-operator convergence we were looking for. ∎

This corollary is a preliminary to the following result.

Corollary 3. *For any primary representation π of $\mathfrak{R}$, any two vector states ϕ and ψ on $\pi(\mathfrak{R})$, and any* R *in* $\mathfrak{R}$

$$\lim_{|\mathbf{a}| \to \infty} \langle \phi; \alpha_{\mathbf{a}}[\mathrm{R}]\rangle - \langle \psi; \alpha_{\mathbf{a}}[\mathrm{R}]\rangle = 0.$$

Proof. Let Φ and Ψ be any two normalized vectors in $\mathcal{H}_\pi$; then there exists at least one unitary operator U on $\mathcal{H}_\pi$ such that $U\Phi = \Psi$. Since π is primary, $\mathfrak{Z}_\pi(\mathfrak{R}) = \{\lambda I\}$; hence U belongs to $\mathfrak{Z}_\pi(\mathfrak{R})' = \mathfrak{B}(\mathcal{H}_\pi)$. We can then use Corollary 2 to assert that

$$\begin{aligned}\lim_{|\mathbf{a}| \to \infty} \langle \phi; \alpha_{\mathbf{a}}[R]\rangle - \langle \psi; \alpha_{\mathbf{a}}[R]\rangle &= \lim_{|\mathbf{a}| \to \infty} (\Psi, \{U^*\pi(\alpha_{\mathbf{a}}[R])U - \pi(\alpha_{\mathbf{a}}[R])\}\Psi) \\ &= \lim_{|\mathbf{a}| \to \infty} (\Phi, [\pi(\alpha_{\mathbf{a}}[R]), U]\Psi) = 0. \quad ∎\end{aligned}$$

Our next result, which relates physically to the question of uniqueness and cluster properties of the "vacuum," strengthens the conclusions of Theorem II.2.8 in replacing space-averages with pointwise limits:†

Theorem 4. *Let ϕ be an $\mathbb{R}^3$-invariant, primary state on $\mathfrak{R}$. Then*

(i) *ϕ is extremal $\mathbb{R}^3$-invariant;*
(ii) *$\mathrm{E}_\phi(\mathbb{R}^3)$ is one-dimensional;*
(iii) *ϕ is the only $\mathbb{R}^3$-invariant vector state on $\pi_\phi(\mathfrak{R})$;*
(iv) *$\tilde{\phi}$ is the only $\mathbb{R}^3$-invariant normal state on $\pi_\phi(\mathfrak{R})''$;*
(v) *for every space-like translation* **a** *and every* R *in* $\mathfrak{R}$

$$w - lim_{\lambda \to \infty}\, \pi_\phi(\alpha_{\lambda \mathbf{a}}[\mathrm{R}]) = \langle \phi; \mathrm{R}\rangle \mathrm{I};$$

(vi) *for every spacelike translation* **a** *and every* R, R_1, R_2 *in* $\mathfrak{R}$

$$\lim_{\lambda \to \infty} \langle \phi; \mathrm{R}_1^*\alpha_{\lambda \mathbf{a}}[\mathrm{R}]\mathrm{R}_2\rangle = \langle \phi; \mathrm{R}\rangle\langle \phi; \mathrm{R}_1^*\mathrm{R}_2\rangle.$$

† Compare with Araki [1964b].

Proof. The properties (i) to (iv) follow directly from Theorem II.2.8 and from the fact that $\mathbb{R}^3$ acts in a η-abelian manner on $\mathfrak{R}$. From Corollary 3 we have for every normalized vector Ψ in $\mathcal{H}_\phi$

$$\lim_{|\mathbf{a}|\to\infty} (\Psi, \pi_\phi(\alpha_{\mathbf{a}}[R])\Psi) = \langle\phi; R\rangle,$$

hence, by polarization, for every pair Ψ_1, Ψ_2 of vectors in $\mathcal{H}_\phi$

$$\lim_{|\mathbf{a}|\to\infty} (\Psi_1, \pi_\phi(\alpha_{\mathbf{a}}[R])\Psi_2) = \langle\phi; R\rangle(\Psi_1, \Psi_2),$$

of which (v) and (vi) are particular cases. This concludes the proof of the theorem. ∎

We now want to investigate some of the consequences of the *spectrum condition* imposed on *relativistic* quantum field theories. By the usual abuse of language we denote by $\mathbb{R}^4$ the group of all translations in $\mathfrak{M}^4$; we also use (x, y) to denote the pseudo-scalar product $g_{\mu\nu}x^\mu x^\nu$.

Theorem 5. *Let ϕ be an $\mathbb{R}^4$-invariant state on $\mathfrak{R}$, let $(\pi_\phi(\mathfrak{R}), \mathrm{U}_\phi(\mathbb{R}^4))$ be the covariant representation associated canonically with ϕ, and let $\{\mathrm{E}(\Delta)\}$ be the projection-valued measure associated with $\mathrm{U}_\phi(\mathbb{R}^4)$ by the SNAG theorem. Let $\bar{\mathrm{V}}_+ \equiv \{\mathrm{p} \in \mathfrak{M}^4 \mid (\mathrm{p}, \mathrm{p}) \geqslant 0\ \mathrm{p}^0 \geqslant 0\}$. If $\Delta \cap \bar{\mathrm{V}}_+ = \varnothing$ implies $\mathrm{E}(\Delta) = 0$, then $\mathrm{U}_\phi(\mathbb{R}^4) \subseteq \pi_\phi(\mathfrak{R})''$, and ϕ is extremal $\mathbb{R}^4$-invariant if and only if $\pi_\phi(\mathfrak{R})$ is irreducible.*

Proof. For any R in $\mathfrak{R}$ and any X in $\pi_\phi(\mathfrak{R})'$ we form

$$\begin{aligned}(\Phi, \pi_\phi(R)\, U_\phi(a)X\Phi) &= (\Phi, U_\phi(-a)\, \pi_\phi(R)\, U_\phi(a)X\Phi)\\ &= (\Phi, \pi_\phi(\alpha_{-a}[R])X\Phi) = (\Phi, X\pi_\phi(\alpha_{-a}[R])\Phi)\\ &= (\Phi, XU_\phi(-a)\, \pi_\phi(R)\Phi).\end{aligned}$$

On using the SNAG theorem and the spectrum condition imposed in the statement of our theorem, we see that the LHS of the above equality is the Fourier transform of a complex measure, the support of which lays in $\bar{V}_+$, whereas the RHS is the Fourier transform of a complex measure, the support of which lays in $-\bar{V}_+$. This is true for all a, and we can therefore conclude that $(\Phi, \pi_\phi(R)\, U_\phi(a)X\Phi)$ is a constant in a, since $\bar{V}_+ \cap -\bar{V}_+ = \{0\}$. Therefore by writing $R = S^*T$ with S and T in $\mathfrak{R}$ we have

$$(\pi_\phi(S)\,\Phi, \{U_\phi(a)\, XU_\phi(-a) - X\}\, \pi_\phi(T)\Phi) = 0.$$

Since Φ is cyclic for $\pi_\phi(\mathfrak{R})$, this implies that for any a in $\mathbb{R}^4$ and any X in $\pi_\phi(\mathfrak{R})'$, $[U_\phi(a), X] = 0$, i.e., $U_\phi(\mathbb{R}^4) \subseteq \pi_\phi(\mathfrak{R})''$ and $\pi_\phi(\mathfrak{R})' \subseteq U_\phi(\mathbb{R}^4)''$ so that $\pi_\phi(\mathfrak{R})' \cap U_\phi(\mathbb{R}^4)' = \pi_\phi(\mathfrak{R})'$. The condition that ϕ be extremal $\mathbb{R}^4$-invariant is equivalent to the condition that the LHS of this equality be $\{\lambda I\}$, hence to the condition that $\pi_\phi(\mathfrak{R})' = \{\lambda I\}$. This concludes the proof of the theorem. ∎

REMARKS.

(i) From the proof it is clear that the first part of the theorem still holds true for any covariant representation of $(\mathfrak{R}, \mathbb{R}^4)$ satisfying the spectrum condition, provided that $E_\phi(\Delta = 0)$ is cyclic under $\pi_\phi(\mathfrak{R})$. If, in addition, $E_\phi(\Delta = 0)$ is one-dimensional, $\pi_\phi(\mathfrak{R})$ is then irreducible on the subspace of $\mathcal{H}$ generated by $\pi_\phi(\mathfrak{R}) E_\phi(\Delta = 0)\mathcal{H}$; it is in this form that the theorem has been proven by Araki [1964b] (see also Ruelle [1962]).

(ii) The second part of the theorem is sometimes referred to by saying that in a local relativistic quantum field theory satisfying the spectrum condition (namely, that the "spectrum of the energy-momentum lies in the forward light cone") the uniqueness of the vacuum (i.e., $E_\phi(\Delta = 0) = \{\lambda \Phi \mid \lambda \in \mathbb{C}\}$) is equivalent to the "completeness of the theory" (i.e.,† $\pi_\phi(\mathfrak{R})'' = \mathfrak{B}(\mathcal{H}_\phi)$).

(iii) We noticed in the proof of the theorem that

$$\pi_\phi(\mathfrak{R})' = \mathfrak{N}'_\phi \equiv \pi_\phi(\mathfrak{R})' \cap U_\phi(\mathbb{R}^4)'.$$

We have already noted (p. 286) that $E_\phi(\mathbb{R}^4)\mathfrak{N}_\phi E_\phi(\mathbb{R}^4)$ is abelian; from the corollary to Theorem II.2.6 this implies that $\mathfrak{N}'_\phi$ [hence $\pi_\phi(\mathfrak{R})'$] is abelian. We have therefore that

$$\mathfrak{Z}_\phi(\mathfrak{R}) \equiv \pi_\phi(\mathfrak{R})'' \cap \pi_\phi(\mathfrak{R})' = \mathfrak{N}'_\phi$$

so that the central decomposition of ϕ coincides with its decomposition into $\mathbb{R}^4$-extremal invariant states. The latter again satisfy the spectrum condition. We also notice that the central decomposition in this case reduces to a decomposition into pure states. Consequently the decomposition of ϕ into states that are extremal invariant with respect to any subgroup of $\mathbb{R}^4$ would provide the same components as the central decomposition.

(iv) The proof of the theorem can be extended‡ to show that if Lorentz-covariance is substituted for $\mathbb{R}^4$-covariance we again have $U_\phi(L_+^\uparrow) \subseteq \pi_\phi(\mathfrak{R})''$ so that the Lorentz invariance of ϕ is not broken in the above-mentioned central decomposition.

(v) Several algebraic conditions equivalent to the *existence* of a covariant representation of $(\mathfrak{R}, \mathbb{R}^4)$ satisfying the spectrum condition have been formulated§ and proved¶ to be independent of the other axioms (i.e., isotony, local commutativity, and $\mathbb{R}^4$-covariance) of the theory, even

† See, for instance, Haag and Schroer [1962].
‡ In the conventional approach to quantum field theory in which the fields are described in terms of their Wightman functions the following result, as well as several of the features mentioned in these remarks, has been proved by Borchers [1962]; see also Reeh and Schlieder [1962].
§ See Doplicher [1965], Montvay [1965], and Borchers [1969].
¶ See Doplicher, Regge, and Singer [1968].

if we restrict the latter by further imposing that $\mathfrak{R}$ be simple and that it satisfy the "time-slice axiom"; that is, if Δ is the region of $\mathfrak{M}^4$ included between two space hyperplanes $(t_1, \mathbb{R}^3)$ and $(t_2, \mathbb{R}^3)$ then $\{\mathfrak{R}(\Omega) \mid \Omega \in \mathcal{F}, \Omega \subset \Delta\}$ already generates $\mathfrak{R}$. This axiom is also referred to as the axiom of "weak primitive causality."†

Theorem 6. *With the assumptions of Theorem 5, suppose further that* $Sp_a(U_\phi(\mathbb{R}^4)) = \{0\}$ *and that for any bounded open* Ω *in* $\mathfrak{M}^4$, $\mathfrak{R} = \overline{\bigcup_{a \in \mathbb{R}^4} \mathfrak{R}(a[\Omega])}$. *The cyclic vector* Φ *canonically associated with* ϕ *is then cyclic (in* $\mathcal{H}_\phi$*) and separating for every* $\pi_\phi(\mathfrak{R}(\Omega))$ *with* Ω *bounded open in* $\mathfrak{M}^4$.

Proof. Let Ω_0 be a bounded open region in $\mathfrak{M}^4$ such that its closure is contained in Ω. The first assertion of the theorem will be proved if we show that for each Ψ in $\mathcal{H}_\phi$, such that $(\Psi, \pi_\phi(R)\Phi) = 0$ for all R in $\mathfrak{R}(\Omega)$, we also have $(\Psi, \pi_\phi(R)\Phi) = 0$ for all R in $\mathfrak{R}$ [since Φ is cyclic by construction for $\pi_\phi(\mathfrak{R})$]. Let $R_1, R_2, \ldots, R_n$ be an arbitrary sequence of elements of $\mathfrak{R}(\Omega_0)$ and let $a_1, \ldots, a_n$ run over $\mathbb{R}^4$. We form

$$F(a_1, a_2 - a_1, \ldots, a_n - a_{n-1}) = (\Psi, \alpha(a_1)[R_1]\, \alpha(a_2)[R_2] \cdots \alpha(a_n)[R_n]\Phi).$$

Since $\bar{\Omega}_0$ is closed and contained in Ω, which is open, there exists a neighborhood $N_1 \times N_2 \times \cdots \times N_n$ in $(\mathbb{R}^4)^n$ such that $a_i[\Omega_0] \subset \Omega$ for $a_i \in N_i$ $(i = 1, 2, \ldots, n)$. By construction F vanishes for these values of a_i. On the other hand, the strengthened spectrum condition implies that $F(a_1, a_2 - a_1, \ldots, a_n - a_{n-1})$ is the boundary value of a function $F(z_1, z_2 - z_1, \ldots, z_n - z_{n-1})$ analytic for Im (z_1), Im $(z_2 - z_1), \ldots,$ Im $(z_n - z_{n-1})$ in the interior V_+ of the positive light cone $\bar{V}_+$. Consequently‡ $F(a_1, a_2 - a_1, \ldots, a_n - a_{n-1})$ is identically zero. Letting then $R_1, R_2, \ldots, R_n$ run over $\mathfrak{R}(\Omega_0)$ and $a_1, a_2, \ldots, a_n$ run over $\mathbb{R}^4$, we generate $\mathfrak{R}$, which implies that $(\Psi, R\Phi) = 0$ for all R in $\mathfrak{R}$. Hence the first part of the theorem is proved. Now let $\Omega_1 \circ\!\!\!\! \mid \Omega$; we have as a consequence of local commutativity that $\pi_\phi(\mathfrak{R}(\Omega)) \subseteq \pi_\phi(\mathfrak{R}(\Omega_1))'$. On using the first part of the theorem we get that Φ is cyclic for $\pi_\phi(\mathfrak{R}(\Omega_1))$, hence separating for $\pi_\phi(\mathfrak{R}(\Omega_1))'$; hence Φ is separating for $\pi_\phi(\mathfrak{R}(\Omega))$ which is contained in $\pi_\phi(\mathfrak{R}(\Omega_1))'$. This completes the proof of the theorem. ▮

REMARKS

(i) This result is known as the Reeh and Schlieder theorem.§

(ii) Note that local commutativity has not been used in the proof of the first part of the theorem.

† See Haag and Schroer [1962].

‡ See, for instance, Theorem 2.17 in Streater and Wightman [1964].

§ See Reeh and Schlieder [1961] who proved it in the conventional framework of quantum field theory; for a proof in English see Streater and Wightman [1964] or Jost [1965]; the algebraic version of the theorem is due to Araki [1964a] (see also Borchers [1965a]).

(iii) If ϕ is invariant under the inhomogeneous Lorentz group, the condition $Sp_d\{U_\phi(\mathbb{R}^4)\} = \{0\}$ is redundant; suppose, indeed, that there existed Ψ in $\mathcal{H}_\phi$ such that $U_\phi(1, a)\Psi = e^{i(a,p)}\Psi$ for all a in $\mathbb{R}^4$. We would then have for every homogeneous Lorentz transformation

$$(\Psi, U_\phi(1, a)\, U_\phi(\Lambda, 0)\Psi) = (\Psi, U_\phi(\Lambda, 0)\, U_\phi(1, \Lambda^{-1}a)\Psi),$$

hence

$$(e^{i(a,p)} - e^{i(\Lambda^{-1}a,p)})(\Psi, U_\phi(\Lambda, 0)\Psi) = 0.$$

Now for each $p \neq 0$ there exists Λ arbitrary small such that $(a, p) \not\equiv (\Lambda^{-1}a, p) \pmod{2\pi}$ for some a. This implies that $(\Psi, U_\phi(\Lambda, 0)\Psi) = 0$, hence $|\{U_\phi(\Lambda, 0) - I\}\Psi| = 2$, which contradicts the continuity of $U_\phi(\Lambda, 0)$.

(iv) The additional condition imposed on the structure of $\mathfrak{R}$ can be slightly modified (without modifying the proof and the conclusion of the theorem) by requiring it to hold only for the representation π_ϕ or to hold in the sense of von Neumann algebras (it is then called *weak additivity*) instead of the uniform closure characteristic of a purely C^*-algebraic approach. If we strengthen it to hold for **a** running over $\mathbb{R}^3$ only, we obtain a version of the weak primitive causality of Haag and Borchers.

We might mention in this connection that, on using their condition of weak primitive causality and a result obtained by Borchers [1961] in the conventional approach to quantum field theory, Haag and Schroer [1962] argued that in an irreducible representation π of $\mathfrak{R}$, $\pi(\mathfrak{R}(\Omega))''$ are factors. Working on this assumption, Kadison [1963] proved that $\pi(\mathfrak{R}(\Omega))$ cannot be of finite type; it seems also that they cannot be type I in general, and it has actually been proved by Araki [1964a] that type I is certainly excluded in particular cases. Apparently the only solid result in this direction has been obtained by Størmer [1967b] who assumed that (a) to each bounded region Ω we associate a C^*-algebra $\mathfrak{R}(\Omega)$ of operators acting on some infinite dimensional Hilbert space $\mathcal{H}$; (b) a strongly continuous representation $a \to U(a)$ of $\mathbb{R}^4$ is given on $\mathcal{H}$, which satisfies the covariance condition on the weak closure $\mathfrak{N}(\Omega)$ of $\mathfrak{R}(\Omega)$ and the condition that its spectrum lies in the forward light cone; (c) the usual isotony and local commutativity hold for the $\mathfrak{R}(\Omega)$; (d) the following version of weak additivity holds: if $\{\Omega_n\}$ is a covering of the unbounded region Ω in $\mathfrak{M}^4$ by bounded regions Ω_n, then $\{\mathfrak{R}(\Omega_n) \mid \Omega_n \in \{\Omega_n\}\}''$ is independent of the covering and is denoted by $\mathfrak{N}(\Omega)$; (e) there exists, up to a scalar multiple, a unique vector Ψ in $\mathcal{H}$ (the "vacuum") cyclic under $\mathfrak{N}(\mathfrak{M}^4)$ and such that $U(a)\Psi = \Psi$ for all a in $\mathbb{R}^4$. Under these conditions he proved that $\mathfrak{N}(\Omega)$ is a factor of type III for all unbounded regions

Ω in $\mathfrak{M}^4$ such that there exist a spacelike vector $\mathbf{a}$ with $\Omega = \Omega + \mathbb{R}\mathbf{a}$ and an open nonvoid region, spacelike to Ω. This result generalizes a previous result by Araki [1963], who proved that $\mathfrak{N}(\Omega)$ is not type I when Ω is chosen to be $\{(p_0, p_1, p_2, p_3) \mid p_1 > 0 \text{ and } |p_0| < |p_1|\}$. Hence it appears that in quantum field theory, as in statistical mechanics, type III factors come into play; the reasons for their appearance, however, are physically quite different.

We also mention that Misra [1965] proved that $\mathfrak{R}$ itself is simple if (a) it is the uniform closure of von Neumann algebras $\mathfrak{N}(\Omega)$ defined for each bounded open region Ω of $\mathfrak{M}^4$ in a separable Hilbert space, (b) $\mathfrak{R}$ satisfies the usual isotony, translation-covariance, and local commutativity requirements, and (c) for every bounded open Ω in $\mathfrak{M}^4$ there exists a bounded open Ω_1 such that $\Omega \subseteq \Omega_1$ and $\mathfrak{N}(\Omega_1)$ is a factor.† He further proved that if all $\mathfrak{N}(\Omega)$ are infinite factors then for any two representations π_1 and π_2 of $\mathfrak{R}$ there exists a *-isomorphism α from $\pi_1(\mathfrak{R})$ onto $\pi_2(\mathfrak{R})$ such that $\alpha \circ \pi_1(\mathfrak{N}(\Omega)) = \pi_2(\mathfrak{N}(\Omega))$ for every bounded open Ω; for each Ω separately α is then unitarily implemented.

The last few results are admittedly somewhat disconnected; they are mentioned merely to indicate the kind of general results obtained so far on the algebraic structure of local theories such as the local relativistic quantum field theories. To proceed for a while in the same vein let us mention briefly three types of problem in which algebraic methods are the relevant tools to be used. The first one is the *Goldstone theorem*. Basically it seems to be understandable as an application of the decomposition theory expounded in Subsection II.2.f and to which we came back in the discussion in Theorem 3; we have to complete the picture by the introduction of an "internal symmetry" group under which ϕ is also invariant. Aside from the fact that this symmetry group G commutes with time evolution and tentatively with the other Lorentz transformations, its characteristic as internal is that $\alpha_g[\mathfrak{R}(\Omega)] = \mathfrak{R}(\Omega)$ for all g in G and all Ω in $\mathcal{F}$. The purpose is to determine the circumstances under which the central decomposition of ϕ leads to states in which only a lower internal symmetry persists and which can be obtained from one another by the action of G to restore the G-invariance of ϕ. Some evidence indicates that such spontaneous symmetry breaking is linked to the fact that the energy-spectrum extends continuously to zero, which in a relativistic theory should in turn be linked to the presence in the theory of particles of zero mass. For a review the reader is directed to Kastler [1967b] or Swieca [1969]; original papers of special interest in connection with algebraic methods

† A comparable result has been obtained recently by Haag, Kadison, and Kastler [1970], who showed that if there exists in $\mathfrak{R}$ a net $\{\mathfrak{N}_\alpha\}$ of type I factors possessing the following properties (a) $\mathfrak{R}$ in the C^*-inductive limit of $\{\mathfrak{N}_\alpha\}$, (b) each $\mathfrak{N}_\alpha$ has a representation on a separable Hilbert space, (c) each $\mathfrak{N}_\alpha$ is included in a $\mathfrak{N}_\beta$ in which it has infinite relative commutant, then $\mathfrak{R}$ is simple.

are, for instance, Kastler, Robinson, and Swieca [1966], Ezawa and Swieca [1967]; see also Symanzik [1967] and Wagner [1966].

Another development has been the proof by Borchers [1960] that (a) if two local fields $A(x)$ and $B(x)$ are such that $A_{\text{in}}(x) = B_{\text{in}}(x)$ then $\{A(x)$ is local with respect to $B(x)\}$ is equivalent to $\{A(x)$ and $B(x)$ give the same S-matrix$\}$, and (b) there is more than one interpolating field to a given causal S-matrix. In algebraic language the relative locality of two fields is expressed by $\mathfrak{R}_A(\Omega_1) \subseteq \mathfrak{R}_B(\Omega_2)'$ for all Ω_1 and Ω_2 in $\mathcal{F}$ such that $\Omega_1 \,\delta\, \Omega_2$.

The third area of application has been the recent development of constructive field theories due to Segal, Glimm and Jaffe, and Guenin; see their reviews in Jost [1969].

Algebraic methods have provided general "soft" guidelines in each of these fields of research; these guidelines, however, have had to be supplemented by a considerable amount of "hard analysis", the exposition of which does not fall within the scope of this book. For a review of the situation in rigorous quantum field theory the reader is invited to study the following proceedings: Jost [1969], Hagen, Guralnik, and Mathur [1967], and Lurçat [1967].

We should like to conclude this section with the proof that the local structure of $\mathfrak{R}$ can be used to provide a criterion for the occurrence of a cluster property significantly stronger than the cluster properties we have already discussed. Since the applications of the following results have been in the realm of statistical mechanics, the reader is invited to think of the "configuration space" for the remainder of this section as being either $\mathbb{R}^3$ or any array of points such as a regular lattice in $\mathbb{R}^n$.

We say that a state ϕ on $\mathfrak{R} = \overline{\bigcup_{\Omega \in \mathcal{F}} \mathfrak{R}(\Omega)}$ is *uniformly clustering* if, given $\varepsilon > 0$ and R in $\mathfrak{R}$, there exists Ω in $\mathcal{F}$ such that

$$|\langle \phi; RS \rangle - \langle \phi; R \rangle \langle \phi; S \rangle| \leqslant \varepsilon \, \|S\|$$

for all S in

$$\mathfrak{R}(\Omega_\delta) \equiv \overline{\bigcup_{\substack{\Omega_1 \in \mathcal{F} \\ \Omega_1 \delta \Omega}} \mathfrak{R}(\Omega_1)}.$$

In the same manner as $\mathfrak{R}(\Omega)$ is interpreted as the algebra of observables associated with the open bounded region Ω in $\mathcal{F}$, $\mathfrak{R}(\Omega_\delta)$ is interpreted as the algebra of observables associated with the region of space Ω_δ, which is (causally) disjoint from Ω. For any representation π of $\mathfrak{R}$ we define the von Neumann algebras

$$\mathfrak{B}_\pi(\Omega) \equiv \pi(\mathfrak{R}(\Omega_\delta))'' \quad \text{and} \quad \mathfrak{B}_\pi = \bigcap_{\Omega \in \mathcal{F}} \mathfrak{B}_\pi(\Omega),$$

which are interpreted respectively as the algebra of global observables associated to Ω_δ and the algebra of global observables at infinity. When π_ϕ is the representation canonically associated with the state ϕ, we simply denote the corresponding algebras by $\mathfrak{B}_\phi(\Omega)$ and $\mathfrak{B}_\phi$. We notice immediately that $\mathfrak{B}_\pi \subseteq \pi(\mathfrak{R})'' \cap \pi(\mathfrak{R})'$; indeed, on the one hand, $\mathfrak{B}_\pi \subseteq \pi(\mathfrak{R})''$ by construction; on the other hand, local commutativity implies that for each Ω_1 in $\mathcal{F}$ with $\Omega_1 \,\text{ó}\, \Omega$ we have $\pi(\mathfrak{R}(\Omega_1)) \subseteq \pi(\mathfrak{R}(\Omega))'$, hence $\pi(\mathfrak{R}(\Omega_\delta))'' \subseteq \pi(\mathfrak{R}(\Omega))'$ which is therefore valid for all Ω in $\mathcal{F}$. Taking the intersection over $\mathcal{F}$, we get

$$\mathfrak{B}_\pi \subseteq \bigcap_{\Omega \in \mathcal{F}} \pi(\mathfrak{R}(\Omega))' = \{\pi(\mathfrak{R}(\Omega)) \mid \Omega \in \mathcal{F}\}' = \pi(\mathfrak{R})',$$

thus proving our assertion.

Lanford and Ruelle [1969] proved the following result:†

Theorem 7. *Let ϕ be a state on the quasi-local algebra $\mathfrak{R}$; then ϕ is uniformly clustering if and only if $\mathfrak{B}_\phi = \{\lambda I\}$.*

Proof. Suppose $\mathfrak{B}_\phi = \{\lambda I\}$ and that ϕ is not uniformly clustering. There then exist R in $\mathfrak{R}$, an increasing net $\{\Omega^\alpha\}$ of bounded open regions such that $\bigcup_\alpha \Omega^\alpha$ covers the whole configuration space, and a family

$$\{S^\alpha \in \mathfrak{R}(\Omega_\delta^\alpha) \mid \|S^\alpha\| \leqslant 1\}$$

such that $\lim_\alpha \{\langle\phi; RS^\alpha\rangle - \langle\phi; R\rangle\langle\phi; S^\alpha\rangle\} \neq 0$. We can assume without loss of generality that there exists a subnet such that $\pi_\phi(S^\alpha)$ converges in the weak operator topology to an element in $\mathfrak{B}_\phi = \{\lambda I\}$; this implies immediately that the LHS of the above inequality is zero, thus providing the looked for contradiction with our tentative assumption that ϕ is not uniformly clustering when $\mathfrak{B}_\phi$ is trivial. Conversely, suppose that ϕ is uniformly clustering. Let B be in $\mathfrak{B}_\phi$; i.e., B belongs to $\mathfrak{B}_\phi(\Omega_\delta)$ for all Ω in $\mathcal{F}$. By Kaplanski's density theorem there exists C in $\mathfrak{R}(\Omega_\delta)$ such that, given R in $\mathfrak{R}$ and $\varepsilon > 0$,

(i) $\|C\| \leqslant \|B\|$,

(ii) $|(\Phi, \pi_\phi(R)\, \pi_\phi(C)\Phi) - (\Phi, \pi_\phi(R)B\Phi)| \leqslant \varepsilon$,

(iii) $|(\Phi, \pi_\phi(C)\Phi) - (\Phi, B\Phi)| \cdot |(\Phi, \pi_\phi(R)\Phi)| \leqslant \varepsilon$.

This implies

$$|(\Phi, \pi_\phi(R)B\Phi) - (\Phi, \pi_\phi(R)\Phi)(\Phi, B\Phi)| \leqslant 2\varepsilon + |\langle\phi; RC\rangle - \langle\phi; R\rangle\langle\phi; C\rangle|.$$

From the uniform clustering property we conclude that the last term on the right-hand side can be made smaller than $\varepsilon\, \|C\|$, hence smaller then $\varepsilon\, \|B\|$. Since ε can be made arbitrarily small, we conclude that

$$(\Phi, \pi_\phi(R)B\Phi) = (\Phi, \pi_\phi(R)\Phi)(\Phi, B\Phi).$$

On (a) writing $R = S^*T$ with S and T in $\mathfrak{R}$, (b) using the remark made earlier that $\mathfrak{B}_\phi$ is contained in $\mathfrak{Z}_\phi(\mathfrak{R})$, and (c) that Φ is cyclic in $\mathfrak{H}_\phi$ for $\pi_\phi(\mathfrak{R})$, we conclude indeed that $B = (\Phi, B\Phi)I$, thus completing the proof of the theorem. ∎

† See also Ruelle [1969b, 1970].

REMARKS.

(i) The "if" part of the theorem strengthens the remark made on p. 185 to the effect that if ϕ is translation-invariant and $\mathfrak{B}_\phi = \{\lambda I\}$ (the latter always being realized when ϕ is primary) we clearly have that ϕ is extremal translation invariant and not only satisfies the cluster properties mentioned before but also the present uniform clustering property.

(ii) As a consequence of the strong abelian character of the action of the translations on a quasi-local algebra (see Theorem 3), another result, stronger than the consequenc of remark (i), can be proved,† namely, that not only $\mathfrak{N}'_\phi \subseteq \mathfrak{Z}_\phi(\mathfrak{R})$ but also $\mathfrak{N}'_\phi \subseteq \mathfrak{B}_\phi$; we shall see presently an interesting consequence of this fact in the theory of symmetry breaking. However, we ought to point out that the above remarks (but not the theorem!) become trivial when $\mathfrak{B}_\phi$ and $\mathfrak{Z}_\phi(\mathfrak{R})$ coincide, which actually occurs in quantum lattice systems or when ϕ is a locally normal state on the quasi-local algebra of a Bose gas.

We noticed earlier that $\mathfrak{B}_\phi$ is contained in $\mathfrak{Z}_\phi(\mathfrak{R})$; this implies trivially that

$$\mathfrak{B}_\phi \subseteq \pi_\phi(\mathfrak{R})' \subseteq \mathfrak{B}'_\phi,$$
$$\mathfrak{B}_\phi \subseteq \pi_\phi(\mathfrak{R})'' \subseteq \mathfrak{B}'_\phi.$$

Substituting $\mathfrak{B}_\phi$ for $\mathfrak{N}'_\phi$ in the treatment of the decomposition of an invariant state into its extremal invariant components, as carried out in Subsection II.2.f, we extend ϕ to $\mathfrak{B}'_\phi$ and make its central decomposition with respect to this algebra. This provides not only a disintegration of $\mathfrak{B}_\phi$, but also a disintegration of $\pi_\phi(\mathfrak{R})''$ and $\pi_\phi(\mathfrak{R})'$ which, as we just saw, are contained in $\mathfrak{B}'_\phi$. It is then easy to check that the measure corresponding to this disintegration is concentrated (first in the Baire sense and then in the Borel sense when the usual appropriate separability assumptions are made) on the set of states ψ such that $\mathfrak{B}_\psi = \{\lambda I\}$. This decomposition therefore is a decomposition of ϕ into its uniform clustering components. It is called the *decomposition at infinity* of ϕ. If ϕ is an equilibrium state, we might tentatively interpret these components as pure thermodynamical phases. When ϕ [or its extension to $\pi_\phi(\mathfrak{R})''$] is a locally normal KMS state [and is such that $\mathfrak{B}_\phi = \mathfrak{Z}_\phi(\mathfrak{R})$], this decomposition evidently coincides with our previous interpretation of the central decomposition of ϕ as the decomposition of an "equilibrium state" into its "pure thermodynamical components." Should there be physical cases in which these particular conditions are *not* satisfied, the above decomposition [which is coarser than the central decomposition whenever $\mathfrak{B}_\phi \neq \mathfrak{Z}_\phi(\mathfrak{R})$] would constitute another approach to the problem of symmetry breaking in phase transitions.

† See Ruelle [1969b].

SECTION 2 SOME SIMPLE MODELS OF STATISTICAL MECHANICS

OUTLINE. This section has been placed at the end of the book to be used as the basis for illustrations of the general theory developed earlier. It is purposely written in a rather loose manner, thus inviting the reader to exercise.

One of the class of physical systems in which the algebraic approach to statistical mechanics has been most successful in deriving specific results consists of the quantum lattice systems. The simplicity of these models is such that most aspects of the general theory developed earlier can be applied directly to them. For orientation we first define their quasi-local algebra. Under general assumptions on the interaction we show how to prove the existence of the thermodynamical limit and of the time evolution for the infinitely extended lattice. We finally mention some applications to equilibrium and nonequilibrium statistical mechanics.

The second subsection deals with the simplest continuous quantum systems: the free Fermi gas and free Bose gas. Their extreme simplicity naturally limits their physical interest and they are certainly best used for illustrative purposes. The Fermi gas poses no problems except that of defining properly the algebra of observables as opposed to the algebra of fields, and, as we show, this problem can easily be solved. The Bose gas was the first model ever analyzed by these methods and actually served the role of a prototype in the algebraic approach to the many-body problem. In spite of this, its time-evolution presents some interesting features which have been recognized only recently and which point out, among other things, to an interesting generalization of the use of the KMS condition studied earlier.

a. Quantum Lattice Systems

The physical systems considered in this subsection consist of an array of spins located at the vertices of an infinite lattice, interacting among themselves and with an external magnetic field.

Specifically, we denote the vertices (or "sites") of the lattice by a parameter ω whose range is $\mathbb{Z}^\nu$ (where $\nu = 1, 2, 3, \ldots$ is the dimensionality of the lattice); we assume that at each site sits a spin, and we shall take for simplicity $s = \frac{1}{2}$. We denote the spin at ω by $\boldsymbol{\sigma}_\omega = \{\sigma_\omega^x, \sigma_\omega^y, \sigma_\omega^z\}$ with the usual commutation relations

$$[\sigma_\omega^k, \sigma_{\omega'}^l] = 2i\sigma_\omega^m \delta_{\omega,\omega'}$$

(where k, l, m is any cyclic permutation of the indices x, y, z). The algebra $\mathfrak{R}_\omega$ of observables at the site ω is therefore a copy of $\mathfrak{M}_2$, the algebra of all 2×2 matrices with complex entries. For future reference we denote simply by ω the isomorphic mapping $\omega:\mathfrak{M}_2 \to \mathfrak{R}_\omega$. To each finite collection Ω of sites corresponds the local algebra

$$\mathfrak{R}(\Omega) = \bigotimes_{\omega\in\Omega} \mathfrak{R}_\omega;$$

whenever $\Omega_1 \subseteq \Omega_2$, we define

$$i_{2,1}\Big(\bigotimes_{\omega\in\Omega_1} R_\omega\Big) = \Big(\bigotimes_{\omega\in\Omega_1} R_\omega\Big) \otimes \Big(\bigotimes_{\omega\in\Omega_2-\Omega_1} I_\omega\Big),$$

which extends trivially to an injective mapping from $\mathfrak{R}(\Omega_1)$ into $\mathfrak{R}(\Omega_2)$. Denoting by $\mathcal{F}$ the set of all finite collections Ω of sites in $\mathbb{Z}^\nu$, we define

$$\mathfrak{R} = \overline{\bigcup_{\Omega\in\mathcal{F}} \mathfrak{R}(\Omega)} = \bigotimes_{\omega\in\mathbb{Z}^\nu} \mathfrak{R}_\omega.$$

The collection of injections $i_{2,1}$ defined above provides a relation of *isotony* in $\mathfrak{R}$. The *local commutativity* of Section 1 is to be read here with $\Omega_1 \,\delta\, \Omega_2$ meaning $\Omega_1 \cap \Omega_2 = \varnothing$. Although we could take the group of all symmetries of the lattice for the *covariance* group, it is sufficient for our purpose to consider the group $\mathbb{Z}^\nu$ of translations of our lattice; for each translation a in $\mathbb{Z}^\nu$ and each site ω in $\mathbb{Z}^\nu$ we denote by $\omega + a$ the translated site and define for each R in $\mathfrak{M}_2$

$$\alpha_a o\omega[R] = (\omega + a)[R],$$

i.e.,

$$\alpha_a[R_\omega] = R_{\omega+a}.$$

This clearly defines an isometric isomorphism from $\mathfrak{R}_\omega$ to $\mathfrak{R}_{\omega+a}$ which trivially extends first to an isometric isomorphism from $\mathfrak{R}(\Omega)$ to $\mathfrak{R}(a[\Omega])$ (where $a[\Omega] = \{\omega + a \mid \omega \in \Omega\}$) and then to an automorphism of $\mathfrak{R}$.

Equipped with this structure, $\mathfrak{R}$ becomes a *quasi-local algebra* in the sense of Section 1.

Except for the notational substitution of $\mathbb{Z}^+$ by $\mathbb{Z}^\nu$, this algebra is the algebra of the canonical anticommutation relations; in particular, we have (Theorem III.2.2) that $\mathfrak{R}$ is *simple*. Furthermore, let Ω_n ($n = 1, 2, \ldots$) be the collection of all n^ν nearest neighboring sites to a fixed site (say, for instance, $\omega = 0$) in $\mathbb{Z}^\nu$. Clearly every Ω in $\mathcal{F}$ is contained in some Ω_n and $\mathfrak{R} = \overline{\bigcup_n \mathfrak{R}(\Omega_n)}$; i.e., $\mathfrak{R}$ is the closure in the norm topology of the union of an increasing sequence of separable algebras (the $\mathfrak{R}(\Omega_n)$). Hence $\mathfrak{R}$ is *separable*. Moreover, since $\mathfrak{R}(\Omega_n)$ is the algebra of all $2^{n^\nu} \times 2^{n^\nu}$ matrices, $\mathfrak{R}(\Omega_n)$ is a factor of type I_{n^ν} and $\mathfrak{R}$ is the "closure in the norm of the union of an increasing sequence of factors of type I_{p_n} with $p_n \to \infty$ as $n \to \infty$"; i.e., $\mathfrak{R}$ is a *uniformly hyperfine* (*UHF*) *algebra*. The UHF algebras have been defined and studied by Glimm [1960], and some of their representations have been analyzed by Powers [1967a, b] we shall come back to this aspect of our systems later in this subsection.

The thermodynamics of these systems has been studied by Robinson [1967 and 1968] who established the next two theorems.† Let $H(\Omega) \in \mathfrak{R}(\Omega)$ be the Hamiltonian defined for the finite system Ω (considered independently of the rest of the lattice); for instance, in the Ising model

$$H(\Omega) = -\sum_{\omega,\omega' \in \Omega} \varepsilon_{\omega,\omega'} \sigma^z_\omega \sigma^z_{\omega'} - \sum_{\omega} B(\omega)\sigma^z_\omega.$$

We assume that $H(\Omega)$ is self-adjoint and translation-"invariant", i.e.,

$$H(a[\Omega]) = \alpha_a[H(\Omega)] \;\forall\; a \in \mathbb{Z}^\nu, \qquad \forall\; \Omega \in \mathcal{F}.$$

To single out the individual contributions of n-body potentials, we define recursively the potentials $V(\Omega)$ by

$$H(\Omega) = \sum_{\Omega_1 \subseteq \Omega} V(\Omega_1);$$

for instance, for the Ising model described above

$$\begin{aligned} V(\omega_1) &= -B(\omega_1), \\ V(\omega_1, \omega_2) &= -\varepsilon_{\omega_1,\omega_2}\sigma^z_{\omega_1}\sigma^z_{\omega_2}, \qquad \forall\; \omega_1, \omega_2, \ldots, \omega_n \in \mathbb{Z}^\nu, \\ V(\omega_1, \omega_2, \ldots, \omega_{n>2}) &= 0, \end{aligned}$$

the last equality expressing that the interactions in the Ising model are at most two-body potentials. Since $V(\Omega)$ is defined recursively from $H(\Omega)$, it inherits from it the translational invariance

$$V(a[\Omega]) = \alpha_a[V(\Omega)].$$

For any fixed ω in $\mathbb{Z}^\nu$ let us denote by $\mathcal{F}_\omega$ the set of all elements Ω in $\mathcal{F}$ such that $\omega \in \Omega$. We say that $V:\mathcal{F} \to \mathfrak{R}$ is of *finite range* if, for a fixed ω in $\mathbb{Z}^\nu$, $V(\Omega) = 0$ for all but a finite number of Ω in $\mathcal{F}_\omega$. Clearly, the translational invariance of V implies that it is sufficient to require this with one ω in $\mathbb{Z}^\nu$ in order to have it for all ω in $\mathbb{Z}^\nu$. Let $\mathfrak{B}_0$ denote the set of all potentials with finite range and by $\mathfrak{B}$ the real Banach space of all potentials $V:\mathcal{F} \to \mathfrak{R}$ such that

$$|||V||| \equiv \sum_{\Omega \in \mathcal{F}_0} \frac{\|V(\Omega)\|}{N(\Omega)} < \infty,$$

where $N(\Omega)$ is the number of sites in Ω. This condition is clearly a *stability* condition, since $V \in \mathfrak{B}$ implies that the average energy per site $H(\Omega)/N(\Omega)$ is uniformly (i.e., as Ω run over $\mathcal{F}$) bounded in norm, that is, by $|||V|||$, hence is not allowed to blow up as the system is made to become larger and larger.

† See also Ruelle [1969a].

This condition is sufficient to establish the classical results of van Hove on the existence of the thermodynamical limit.†

Theorem 1. *Let* $\mathfrak{R}$ *be the quasi-local algebra of a quantum lattice system and* V *be a potential in* $\mathfrak{B}$. *Then*

(i) $\mathrm{P}_\Omega(\mathrm{V}) \equiv \mathrm{N}(\Omega)^{-1} \ln \mathrm{Tr}_\Omega \exp \{-\mathrm{H}(\Omega)\}$ *converges to a finite result* P(V) *as* Ω *tends to infinity in the sense of van Hove;*
(ii) *if* V_1 *and* V_2 *are in* $\mathfrak{B}$, $|\mathrm{P}(\mathrm{V}_1) - \mathrm{P}(\mathrm{V}_2)| \leqslant |||\mathrm{V}_1 - \mathrm{V}_2|||$;
(iii) P, *considered as a function on* $\mathfrak{B}$, *is convex.*

Proof. We merely sketch the proof, since it is available from many sources.‡ From the classical inequality

$$|\ln \mathrm{Tr} \exp A - \ln \mathrm{Tr} \exp B| \leqslant \|A - B\|,$$

valid for any pair A, B of self-adjoint operators on a (complex) finite dimensional Hilbert space $\mathcal{H}_n$, we get immediately

$$|P_\Omega(V_1) - P_\Omega(V_2)| \leqslant |||V_1 - V_2|||,$$

which is to say that the family $\{P_\Omega\}$ of functions from $\mathfrak{B}$ to $\mathbb{R}$ is equicontinuous. The convexity of P_Ω follows from Peierl's theorem, according to which for any self-adjoint operator A on $\mathcal{H}_n$ and any orthonormal basis $\{\Psi_i\}$ in $\mathcal{H}_n$

$$\sum_i \exp \{-(\Psi_i, A\Psi_i)\} \leqslant \mathrm{Tr}\, e^{-A}$$

(when $\mathcal{H}$ is infinite-dimensional, Peierl's theorem still holds, provided that Ψ_i are in the domain of A). We have indeed

$$\begin{aligned}
&P_\Omega(\alpha V_1 + (1-\alpha)V_2) \\
&\quad = N(\Omega)^{-1} \ln \mathrm{Tr}_\Omega \exp \{-\alpha H_1 - (1-\alpha)H_2\} \\
&\quad = N(\Omega)^{-1} \ln \sup_{\{\Psi_i\}} \sum_i \{\exp (\Psi_i, [-\alpha H_1 - (1-\alpha)H_2]\Psi_i)\} \\
&\quad = N(\Omega)^{-1} \ln \sup_{\{\Psi_i\}} \sum_i \{\exp (\Psi_i, -H_1\Psi_i)\}^\alpha \{\exp (\Psi_i, -H_2\Psi_i)\}^{1-\alpha} \\
&\quad \leqslant N(\Omega)^{-1} \ln \sup_{\{\Psi_i\}} \left\{\sum_i \exp (\Psi_i, -H_1\Psi_i)\right\}^\alpha \left\{\sum_i \exp (\Psi_i, -H_2\Psi_i)\right\}^{1-\alpha} \\
&\quad \leqslant N(\Omega)^{-1} \ln [\{\mathrm{Tr} \exp (-H_1)\}^\alpha \cdot \{\mathrm{Tr} \exp (-H_2)\}^{1-\alpha}] \\
&\quad = \alpha P_\Omega(V_1) + (1-\alpha) P_\Omega(V_2).
\end{aligned}$$

† Notice that in the theorem we can write $\beta = 1$ without loss of generality, since $V \in \mathfrak{B}$ implies $\beta V \in \mathfrak{B}$.

‡ Robinson [1967]; Ruelle [1969a].

We have then that (ii) and (iii) are valid for all Ω in $\mathcal{F}$. We then prove in the traditional manner (adaptation of van Hove [1949]) that $P(V)$ exists for all V in $\mathfrak{B}_0$. The rest of the theorem follows from the fact that $\mathfrak{B}_0$ is dense in $\mathfrak{B}$ and the family $\{P_\Omega\}$ is equicontinuous on $\mathfrak{B}$. ▌

Our next problem is to define the time evolution for our infinite system. We first consider, for each Ω_0 in $\mathcal{F}$, each R in $\mathfrak{R}(\Omega_0)$ and each $\Omega \supseteq \Omega_0$:

$$\alpha_t(\Omega)[R] = e^{iH(\Omega)t} R e^{-iH(\Omega)t};$$

we then ask whether the limit of this expression exists as $\Omega \to \infty$ and, if it does, whether it defines a time evolution in the sense of Section II.2. The answer is given by the following proposition:

Theorem 2. *Let* $\mathfrak{R}$ *be the quasi-local algebra of a quantum lattice system; if the interaction* V *satisfies the condition*

$$|||\mathrm{V}|||_1 \equiv \sum_{\Omega \in \mathcal{F}} \|\mathrm{V}(\Omega)\| \; exp\,\{\mathrm{N}(\Omega) - 1\} < \infty,$$

then for each Ω_0 *in* $\mathcal{F}$ *and each* R *in* $\mathfrak{R}(\Omega_0)$

$$\lim_{\substack{\Omega \to \infty \\ \Omega \subseteq \Omega_0}} \alpha_t(\Omega)[\mathrm{R}] \equiv \alpha_t[\mathrm{R}]$$

exists and extends to a continuous, one-parameter group of automorphisms of $\mathfrak{R}$.

Proof. Here again a sketch of the proof will suffice.† Since $H(\Omega) \in \mathfrak{R}(\Omega)$, we can define the derivation $L(\Omega)$ of $\mathfrak{R}$ as $L(\Omega)[R] = [H(\Omega), R]$, which generates $\alpha_t(\Omega)$:

$$\alpha_t(\Omega)[R] = \sum_{n=0}^{\infty} i^n \frac{t^n}{n!} L^n(\Omega)[R].$$

We now recall that

$$H(\Omega) = \sum_{\Omega_1 \subseteq \Omega} V(\Omega_1)$$

and introduce it into the above expression for $\alpha_t(\Omega)$ to get

$$\alpha_t(\Omega)[R] = \sum_{n=0}^{\infty} i^n \frac{t^n}{n!} \sum_{\Omega_1, \ldots \Omega_n \subseteq \Omega} [V(\Omega_n), [\ldots, [V(\Omega_2), [V(\Omega_1), R]] \ldots]].$$

The trick of the proof is to recognize that

$$\|[V(\Omega), R]\| \leqslant \|R\| \, (\exp N(\Omega_0)) n! \, (2|||V|||_1)^n$$

whenever $R \in \mathfrak{R}(\Omega_0)$ and $\Omega_0 \subseteq \Omega$, so that

$$\lim_{\Omega \to \infty} \alpha_t(\Omega)[R] = \sum_{n=0}^{\infty} i^n \frac{t^n}{n!} \sum_{\Omega_1, \ldots, \Omega_n \subset \mathbb{Z}^\nu} [V(\Omega_n), [\ldots, [V(\Omega_2), [V(\Omega_1), R]] \ldots]]$$

† Robinson [1968], Ruelle [1969a]; see also Manuceau and Trottin [1968a].

converges whenever $|t|\, 2\, |||V|||_1 < 1$, hence defines α_t which can immediately be extended by continuity to an automorphism of $\mathfrak{R}$. The norm-continuity at $t = 0$ of $\alpha_t[R]$ for each R separately is then easily proved; we also check that the group property $\alpha_{t_1}\alpha_{t_2} = \alpha_{t_1+t_2}$ holds for t_1, t_2 and $t_1 + t_2$ in the just-defined time interval of convergence of α_t. The last step of the proof consists of extending α_t to all real values of t by the group property; this extension evidently carries with it the continuity and automorphism properties of α_t. ∎

Now that we have the time evolution actually defined as an automorphism of a separable algebra, we are in the best possible circumstances to discuss the KMS condition and, in the cases in which it holds, to use the results of Subsection II.2.f. This has also been considered by Robinson [1968]† who proved that not only does the mean free energy exist in the thermodynamical limit (see Theorem 1) but also that $\lim_{\Omega\to\infty} \langle \phi(\Omega); R\rangle$ exists for all R in $\bigcup_{\Omega\in\mathcal{F}} \mathfrak{R}(\Omega)$, hence provides a proper definition of the Gibbs state ϕ of the infinite system; furthermore, he showed that ϕ satisfies the KMS condition‡ and is extremal $\mathbb{Z}^\nu$-invariant. It should, however, be noted that these last three results are obtained under one further supplementary condition; that is, the graph of the function $P: \mathfrak{B}_1 \to \mathbb{R}$ has a unique tangent plane at $(V, P(V))$; to determine whether the latter condition is satisfied at a particular physical point is, however, not an easy task. Araki [1969] considered a one-dimensional quantum lattice with finite-range interaction and showed, without supplementary conditions, that the Gibbs state ϕ exists, satisfies the uniform clustering property and the KMS condition, and is primary; it is also extremal in $\mathfrak{S}_\beta$ and $\mathfrak{S}_{\mathbb{Z}^\nu}$. Hence ϕ can be interpreted as a pure thermodynamical phase. This interpretation is even strengthened by the fact that Araki also proved that the expectation value of any local observable in this state, as well as the mean free energy, depends analytically on the potential (hence, in particular, on the temperature), thus eliminating the possibility of a phase transition. We should notice in connection with the properties of the Gibbs state derived by Araki that we already know that some of them are consequences of the others; e.g., ϕ primary implies ϕ uniform clustering [Theorem IV.1.7 and $\mathfrak{B}_\phi \subseteq \mathfrak{Z}_\phi(\mathfrak{R})$], ϕ primary and KMS imply ϕ extremal KMS (Corollary 1 to Theorem II.2.11 and $\mathfrak{R}$ separable), ϕ primary and ϕ $\mathbb{Z}^\nu$-invariant imply ϕ extremal $\mathbb{Z}^\nu$-invariant (Theorem II.2.8), the latter condition evidently resulting also from ϕ $\mathbb{Z}^\nu$-invariant and ϕ extremal KMS (Corollary 1 or 3 to Theorem II.2.11) or ϕ uniform clustering (remarks following Theorem IV.1.7). Moreover, Lanford and Ruelle [1969] showed that for a quantum lattice system $\mathfrak{B}_\phi = \mathfrak{Z}_\phi(\mathfrak{R})$, so that conversely (Theorem IV.1.7) ϕ uniform clustering implies ϕ primary. In higher dimensions, however, it is

† See also Ruelle [1969a].

‡ See also Lanford and Robinson [1968a].

possible that $\mathfrak{B}_\phi = \mathfrak{Z}_\phi(\mathfrak{R}) \neq \{\lambda I\}$ for the Gibbs state, the decomposition of which provides a symmetry breaking associated with a phase transition. See in this connection the models discussed by Robinson [1969] and Ginibre [1969b], that is, the anisotropic Heisenberg model, the Ising model with transverse magnetic field, and the quantum lattice gas with hard cores extending over nearest neighbors. Incidentally, the anisotropy in the Heisenberg model in two-dimensions is essential to the occurrence of a phase transition, since Mermin and Wagner [1966] showed that an isotropic Heisenberg model in two dimensions has no phase transition at finite temperature. Concerning the analyticity properties of the anisotropic Heisenberg model, see also the earlier work of Gallavotti, Miracle-Sole, and Robinson [1968].

Since the quasi-local algebra $\mathfrak{R}$ of a quantum lattice system is a norm-separable, simple, and nonabelian algebra on which $\mathbb{Z}^\nu$ acts in a η-abelian manner, we can use Theorem II.2.14 to conclude that every $\mathbb{Z}^\nu$-invariant state ϕ on $\mathfrak{R}$ with ϕ extremal in $\mathfrak{S}_\beta$ for $0 < \beta < \infty$ always generates a type III primary representation of $\mathfrak{R}$. Type III factors actually come in a variety of ways when we deal with this algebra. Størmer [1969] in particular studied the product state on $\mathfrak{R} = \bigotimes_{\gamma\in\mathbb{Z}^+} \mathfrak{R}_\gamma$ (where $\mathfrak{R}_\gamma$ are copies of the same C^*-algebra $\mathfrak{R}_0$), and we quote one of his results here. Let G be the group of all finite permutations of the positive integers in $\mathbb{Z}^+$. For each n in $\mathbb{Z}^+$ let us single out the element g_n in G defined by

$$g_n(\gamma) = \begin{cases} 2^{n-1} + \gamma & \text{if } \quad 1 \leqslant \gamma \leqslant 2^{n-1}, \\ \gamma - 2^{n-1} & \text{if } \quad 2^{n-1} < \gamma \leqslant 2^n, \\ \gamma & \text{if } \quad 2^n < \gamma. \end{cases}$$

Let us now denote by α_g the automorphism of $\mathfrak{R}$ generated by

$$\alpha_g[R_\gamma] = R_{g[\gamma]}.$$

The subgroup of G which was just selected possesses the following remarkable property; for any pair R and S of elements in $\mathfrak{R}$, with $\|R\| \leqslant 1$, $\|S\| \leqslant 1$, and any $\varepsilon > 0$ there exists an integer $m(R, S, \varepsilon)$ and elements R_ε and S_ε in $\mathfrak{R}$ such that $\|R - R_\varepsilon\| < \varepsilon/4$, $\|S - S_\varepsilon\| < \varepsilon/4$ and

$$R_\varepsilon = \sum_\gamma \bigotimes_\delta R_{\gamma\delta},$$

$$S_\varepsilon = \sum_\gamma \bigotimes_\delta R_{\gamma\delta},$$

with $R_{\gamma\delta} = I = S_{\gamma\delta}$ for all $\gamma, \delta > m(R, S, \varepsilon)$. For every n such that $2^{n-1} > m$ we then have $\|[\alpha_{g_n}[R], S]\| < \varepsilon$ from which we conclude

$$\lim_{n\to\infty} \|[\alpha_{g_n}[R], S]\| = 0 \;\forall\; R, S \in \mathfrak{R}.$$

In analogy with what we have seen in Section 1 for the translation group we say here that a G-invariant state ϕ on $\mathfrak{R}$ is strongly clustering if for all R and S in $\mathfrak{R}$:

$$\lim_{n\to\infty} \langle \phi; \alpha_{g_n}[R]S\rangle = \langle \phi; R\rangle \langle \phi; S\rangle.$$

Størmer proves then that this condition is equivalent to each of the following two conditions: (a) ϕ is extremal G-invariant and (b) $\phi = \otimes\, \phi_\gamma$, with $\phi_\gamma = \phi_0$. He also give a complete characterization, in this general case, of the type of primary representation associated with such states when ϕ_0 is a factor state on $\mathfrak{R}_0$: π_ϕ is type I_n, I_∞, or II_1 when ϕ_0 is respectively an homomorphism, is pure and not an homomorphism, or is a trace and not an homomorphism; π_ϕ is type II_∞ when ϕ_0 is neither pure nor a trace and, in addition, the vector state on $\pi_\phi(\mathfrak{R}_0)'$, generated by Φ_0 in $\mathcal{H}_{\phi_0}$, is a trace. Finally π_ϕ is type III when the state just defined on $\pi_\phi(\mathfrak{R}_0)'$ is not a trace. On using this result, we can immediately construct factors of type I_∞, II_1, and III from our algebra of quasi-local observables on a quantum lattice system. Indeed let ϕ_0 be the state on $\mathfrak{M}_2$ considered in the first examples of Subsection II.1.c and II.2.e. if $\beta = \infty$, ϕ_0 is pure and not an homomorphism; hence ϕ is a primary state of type I_∞ (physically it is the ground state of our free system interacting only with the magnetic field). If $\beta = 0$, ϕ_0 is a trace, but not an homomorphism; hence ϕ is a primary state of type II_1 (physically it is the infinite temperature state). If $0 < \beta < \infty$, we can easily see, since we have explicitly $\pi_{\phi_0}(\mathfrak{R}_0)'$, that ϕ_0 belongs to the last class of states, so that ϕ is a primary state of type III. This, by the way, illustrates the fact that Theorem II.2.14 applies precisely to the domain of validity we recognized for it. Furthermore, we can see (Powers [1967a, b]) that the type III factors just obtained are mutually nonisomorphic if they correspond to different temperatures.

The simplicity of quantum lattice systems can also be exploited for the purpose of nonequilibrium statistical mechanics.

On considering the extremely simple case of a generalized Ising model (in the sense described in the beginning of this subsection) Radin [1970] analyzed the behavior in time of $\langle \phi_t; R\rangle$ for a wide class of initial conditions and of local observables; we can easily show that in this case the time evolution does *not* act in a G-abelian manner. More importantly for physical applications, the traditional wisdom, according to which the rate of approach to equilibrium should be linked, in the thermodynamical limit, to the degree of continuity of the spectrum of the effective Hamiltonian, can be given a precise mathematical meaning. It should be emphasized at this point that we are now discussing the time evolution of a local observable embedded in an infinite system, so that the Hamiltonian we are talking about is that which implements locally the time evolution of the infinite system. This Hamiltonian depends, as an operator, on the Hilbert space in which it is made to act by the GNS construction and the degree of continuity of its spectrum is a representation dependent property. Once the initial state ϕ_0 is chosen the degree of continuity of the spectrum of this Hamiltonian can in turn be linked to the space dependence of the function $\varepsilon(|\omega - \omega'|) \equiv \varepsilon_{\omega\omega'}$. It might also be pointed out that Radin's method extends to interactions which are more general than the simple Ising model described above.

Finally the time development of local perturbations away from a KMS state has been the object of recent papers by Abraham et al. [1970], Radin [1971a, b], and Emch and Radin [1971].

Although it is true that the time evolution of quantum spin systems is well defined in the thermodynamical limit (provided, evidently, that the assumptions of Theorem 2 are satisfied!), it might be remarked, in closing this subsection, that the Weiss limit presents a singularity in this otherwise smooth class of physical systems: the time evolution of the finite systems does not converge, as the Weiss limit is taken, to an automorphism of the quasi-local algebra $\mathfrak{R}$. For the representations of physical interest, however (e.g. the Gibbs state and its pure thermodynamical phases components), the time evolution does converge,† in the Weiss limit, to an automorphism of $\pi_\phi(\mathfrak{R})''$. For the pure thermodynamical phases this automorphism is generated locally by the Weiss molecular field Hamiltonian $H = -\sum B_m(\omega)\sigma_\omega^z$, where $B_m(\omega)$ is a c-number that satisfies the usual self-consistency equations. This Hamiltonian clearly generates an automorphism of $\pi(\mathfrak{R})$, and since $\mathfrak{R}$ is simple this automorphism can be uniquely lifted to an automorphism of $\mathfrak{R}$. Since, on the other hand, $B_m(\omega)$ depends on the pure thermodynamical phase to which it is associated, so does the automorphism of $\mathfrak{R}$ which it generates. Consequently $\alpha_t(\Omega)$ cannot converge in the algebraic sense, as $\Omega \to \infty$, to an automorphism of $\mathfrak{R}$: the time evolution must be defined separately for all states of physical interest. In this connection it is nice to know that the time evolution defined for the Gibbs-state representation is consistent with the time evolution that we can define for its pure thermodynamical components; however, the time evolutions defined for different temperatures (below the critical point) are definitely different.

The BCS-model is another example of a physical system in which the molecular field method becomes exact in the thermodynamical limit.‡ We might therefore expect the same phenomena to occur in this case also; this is indeed so, as shown by Dubin and Sewell [1970]. These authors make this statement as an illustration of an ingenious theory which they devised as a substitute to Haag, Hugenholtz, and Winnink's proof that the Gibbs state

† Emch and Knops [1970].

‡ See Subsection I.1.f for a brief description and references.

satisfies the properties characteristic of a KMS state. We discuss another application of their theory in Subsection b.

b. Free Quantum Gases

In this subsection we construct the quasi-local algebras appropriate to the description of nonrelativistic Fermi and Bose gases and we study their time evolution.

Let $\mathcal{A}$ be the algebra of the canonical anticommutation relations defined in Chapter 3 but with $\mathcal{C} = \mathfrak{L}^2(\mathbb{R}^3)$. Clearly

$$\mathcal{A} = \overline{\{a^*(f), a(g) \mid f, g \in \mathfrak{L}^2(\mathbb{R}^3)\}}$$
$$= \overline{\{a^*(f), a(g) \mid f, g \in \mathfrak{D}(\mathbb{R}^3)\}}\,,$$

where $\mathfrak{D}(\mathbb{R}^3)$ is the space of all infinitely differentiable functions in $\mathbb{R}^3$ with compact support; the second of the above equalities is an immediate consequence of the corollary to Theorem III.2.3 and of the fact that $\mathfrak{D}(\mathbb{R}^3)$ is dense in $\mathfrak{L}^2(\mathbb{R}^3)$. The local algebras $\mathcal{A}(\Omega)$ are defined similarly with $\mathfrak{L}^2(\Omega)$ and $\mathfrak{D}(\Omega)$ substituted for $\mathfrak{L}^2(\mathbb{R}^3)$ and $\mathfrak{D}(\mathbb{R}^3)$ in the above equalities. The isotony and space covariance of the theory are readily established. Because of the presence of anticommutators in the CAR, the condition of local commutativity does not hold for $\mathcal{A}$. This circumstance evidently does not violate causality, since the latter is a requirement that we can physically impose only on observable quantities, whereas $a(f)$ and $a^*(f)$ are *not* observables. The local observables for the region Ω are defined as uniform limits of even polynomials in the elements $a^*(f), a(g)$ of $\mathcal{A}(\Omega)$. Let $\mathfrak{R}(\Omega)$ be the C^*-algebra generated by the local observables associated with Ω. These algebras satisfy the isotony condition and we can therefore define the C^*-subalgebra $\mathfrak{R}$ of $\mathcal{A}$ as the C^*-inductive limit of the algebras $\mathfrak{R}(\Omega)$ when Ω runs over the directed set of all bounded open regions of $\mathbb{R}^3$. $\mathfrak{R}$ clearly inherits the space covariance of $\mathcal{A}$ and now satisfies the local commutative condition. $\mathfrak{R}$ is therefore a quasi-local algebra in the sense of Section 1. An alternative characterization of $\mathfrak{R}$ can be obtained as follows. Let us denote by α the automorphism of $\mathcal{A}$ defined by $\alpha[a(f)] = -a(f)$ for all f in $\mathcal{C}$. We then have

$$\mathfrak{R} = \{R \in \mathcal{A} \mid \alpha[R] = R\}.$$

A natural question to ask at this point is whether $\mathfrak{R}$ inherits the nice algebraic properties of $\mathcal{A}$ such as the fact that $\mathcal{A}$ is simple. Doplicher and Powers [1968] answered the latter question affirmatively and Størmer [1970] sharpened the result into the following proposition:

Theorem 3. *$\mathfrak{R}$ is *-isomorphic to $\mathcal{A}$.*

Proof. Let $\{e_\gamma \mid \gamma \in \mathbb{Z}^+\}$ be an orthonormal basis in $\mathfrak{T}$. We recall that $\mathcal{A}$ is defined as $\mathcal{A} = \bigotimes \mathcal{A}_\gamma$, where $\mathcal{A}_\gamma$ is, for each γ in $\mathbb{Z}^+$, a copy of the algebra $\mathfrak{M}_2$ of all 2×2 matrices with complex entries. Let $\mathfrak{T}_n$ be the finite dimensional subspace of $\mathfrak{T}$ spanned by $\{e_i \mid i = 1, 2, \ldots, n\}$. We now use the elements σ_i^x, σ_i^y, and σ_i^z defined on p. 272 and notice that $\alpha[\sigma_i^z] = \sigma_i^z$, $\alpha[\sigma_i^x] = -\sigma_i^x$ and $\alpha[\sigma_i^y] = -\sigma_i^y$ so that the automorphism α is implemented on each $\mathcal{A}(\mathfrak{T}_n)$ by the unitary operator $V_n = \sigma_1^z\sigma_2^z \cdots \sigma_n^z$. The eigenvalues of these operators are evidently ± 1 and appear with the same multiplicity, that is, 2^{n-1}. Let P_n and Q_n denote the two eigenprojectors of V_n: $V_n = P_n - Q_n$. For any R in $\mathcal{A}(\mathfrak{T}_n)$ the condition $\alpha[R] = R$ then becomes

$$\begin{aligned} R = V_nRV_n &= P_nRP_n + Q_nRQ_n - P_nRQ_n - Q_nRP_n \\ &= P_nRP_n + Q_nRQ_n. \end{aligned}$$

Since $\mathcal{A}(\mathfrak{T}_n)$ is isomorphic to the algebra of all bounded operators on the 2^n-dimensional Hilbert space $\mathcal{H}_n$, we have

$$\begin{aligned} \mathfrak{R}(\mathfrak{T}_n) &\equiv \{R \in \mathcal{A}(\mathfrak{T}_n) \mid \alpha[R] = R\} \\ &\simeq \mathfrak{B}(P_n\mathcal{H}_n) \oplus \mathfrak{B}(Q_n\mathcal{H}_n) \\ &\simeq \mathcal{A}(\mathfrak{T}_{n-1}) \oplus \mathcal{A}(\mathfrak{T}_{n-1}). \end{aligned}$$

Hence $\mathfrak{R}$ is the norm-closure of an increasing sequence $\{\mathfrak{R}(\mathfrak{T}_n) \mid n = 1, 2, \ldots\}$ of type I_{2^n}-factors, which is to say that $\mathfrak{R}$ is a UHF algebra of type $\{2^n\}$; so is $\mathcal{A}$. Consequently $\mathfrak{R}$ is *-isomorphic to $\mathcal{A}$. ∎

REMARK. The last assertion of the proof follows from Lemma 1.2 in Glimm [1960], which asserts that two UHF algebras $\overline{\bigcup_i \mathfrak{M}_i}$ and $\overline{\bigcup_i \mathfrak{N}_i}$ are *-isomorphic if they are the same type. The proof of this result rests on the existence, *in general*, of an *-isomorphism from $\bigcup_i \mathfrak{M}_i$ onto $\bigcup_i \mathfrak{N}_i$. In our case the existence of such an isomorphism is an immediate consequence of the fact that (see above):

$$\mathfrak{R}(\mathfrak{T}_n) \subset \mathcal{A}(\mathfrak{T}_n) \subset \mathfrak{R}(\mathfrak{T}_{n+1}).$$

We could have then bypassed any reference to UHF algebras in the last part of the proof.

The discussion carried out at the end of Subsection I.1.c can trivially be adapted to the (antisymmetric) Fock space constructed on $\mathfrak{T} = \mathfrak{L}^2(\mathbb{R}^3)$. In this manner we define the time evolution α_t of $\mathcal{A}$ (hence of $\mathfrak{R}$) from the one-particle-space time-evolution. In this case we take for H the minimal closed extension of the operator $-\Delta/2m$ defined on $\mathfrak{D}(\mathbb{R}^3)$. We note that $-\Delta/2m$ is essentially self-adjoint on $\mathfrak{D}(\mathbb{R}^3)$, that H is its Friedrichs extension, and hence is a positive self-adjoint operator. The time evolution is defined without ambiguity and determined by its action on the creation and annihilation operators (see Theorem I.1.1);

$$\alpha_t[a^\natural(f)] = a^\natural(f_t).$$

This discussion can also be carried out for $\mathcal{H} = \mathfrak{L}^2(\Omega)$ and leads to a time evolution $\alpha_t(\Omega)$ of $\mathcal{A}(\Omega)$ [and hence of $\mathfrak{R}(\Omega)$]. In this case it is convenient to take for Ω a cube of edge L centered at the origin and for the one-particle Hamiltonian $H(\Omega)$ that particular extension of $-\Delta/2m$ [the latter defined on $\mathfrak{D}(\Omega)$] which corresponds to periodic boundary conditions.

The thermodynamical limit can then be taken without any problem and its result offers no surprise.

To this effect we chose a sequence $\{\Omega_n\}$ of cubes of edge L^n centered at the origin. Clearly $\{\Omega_n\}$ is absorbing in $\mathbb{R}^3$; i.e., every bounded open region Ω of $\mathbb{R}^3$ is included in some Ω_n and all subsequent cubes. With the help of Theorem I.1.1 and Theorem III.2.1 we check that, for every f in $\mathfrak{D}(\mathbb{R}^3)$, $\alpha_t(\Omega_n)[a(f)]$ converges to $\alpha_t[a(f)]$ as $n \to \infty$. The "second quantized" Cook-Fock Hamiltonian (see Subsection I.1.c) allows us to compute the grand canonical equilibrium state $\phi(\Omega_n)$ for $\mathcal{A}(\Omega_n)$ and we check again that, for every R in $\bigcup \mathfrak{R}(\Omega)$, $\langle\phi(\Omega_n); R\rangle$ converges as $n \to \infty$, thus defining a state on $\mathfrak{R}$ which we identify as the Gibbs state of the free Fermi gas for the natural temperature β and the chemical potential μ.† The KMS analysis of Haag, Hugenholtz, and Winnink (see Subsection II.2.e) can then be performed for this model without any further difficulties.

As far as the equilibrium thermodynamical limit is concerned, the free Fermi and Bose gases can be treated in rather close parallel; for the case of the free Bose gas see the summary given in Subsection II.1.c.

The time evolution of the free Bose gas, however, presents some new phenomenon. Let f be any function in $\mathfrak{D}(\mathbb{R}^3)$ and f_t its time evolution under the free Hamiltonian defined above for the one-particle space $\mathfrak{L}^2(\mathbb{R}^3)$. Its Fourier transform $(f_t)^\sim$ is then related to the Fourier transform $\tilde{f}$ of f by

$$(f_t)^\sim(k) = e^{i(k^2/2m)t}\tilde{f}(k)$$

so that f_t does not belong to $\mathfrak{D}(\mathbb{R}^3)$ (as can be seen, for instance, from the Paley-Wiener theorem). As a consequence of this simple remark, and of Theorem I.1.1, we see‡ that if we define the local algebra $\mathfrak{W}(\Omega)$ as the norm-closure (or as the weak-operator closure) in $\mathcal{H}_F(\Omega)$ of the algebra generated by $\{U(f), V(g) \mid f, g \in \mathfrak{D}(\Omega)\}$ the time evolution $W(f) \to W(f)_t = W(f_t)$ will carry local observables outside the algebra of quasi-local observables.

† For further details see Araki and Wyss [1964].

‡ To our knowledge the forthcoming argument was first published by Dubin and Sewell [1970]; see also Wieringa [1970].

The time evolution therefore cannot be defined in this case as an automorphism of the algebra of quasi-local observables: $\mathfrak{W} = \overline{\{U(f), V(g) \mid f, g \in \mathfrak{D}\}}$.

Incidentally, the reason why the difficulty did not appear in the case of Fermi gas is to be looked for in the continuity property of the Fermi field obtained as a consequence of Theorem III.2.1. In the case of Bose gas we could bypass this problem by considering the CCR algebra over another test function space; for instance, $\mathcal{S}(\mathbb{R}^3)$ would do. The inconvenience of following this alternate procedure, however, is that we might lose much of the quasi-local structure of the theory unless we were able to find some guiding principle to govern the choice of this test function space.

Another way to bypass this problem has been devised by Dubin and Sewell [1970]. Define the quasi-local algebra $\mathfrak{W}$ as the C^*-inductive limit of

$$\{\mathfrak{W}(\Omega) \equiv \mathfrak{B}(\mathcal{H}_F(\Omega)) \mid \Omega \in \mathcal{F}\},$$

where $\mathcal{F}$ is the collection of all bounded open sets of $\mathbb{R}^3$ and $\mathfrak{B}(\mathcal{H}_F(\Omega))$ is the algebra of all bounded observables on the Fock space constructed over $\mathfrak{L}^2(\Omega)$. Then pick up in $\mathcal{F}$ an increasing sequence $\{\Omega_n\}$ which is absorbing for $\mathbb{R}^3$ and such that the local Hamiltonian and the local (grand-canonical) equilibrium density matrices can be properly defined for each Ω_n. Let $\alpha_t(\Omega_n)$ and $\phi(\Omega_n)$ denote respectively the time-evolution and the grand-canonical equilibrium state defined on $\mathfrak{W}(\Omega_n)$ from these operators. Instead of assuming that $\alpha_t(\Omega_n)$ converges (as $n \to \infty$) to an automorphism of $\mathfrak{W}$, assume that

$$\lim_{m\to\infty} \lim_{n\to\infty} \langle \phi(\Omega_n); \alpha_{t_1}(\Omega_n)[R_1] \cdots \alpha_{t_k}(\Omega_n)[R_n] \times \alpha_{t_{k+1}}(\Omega_m)[R_{k+1}] \cdots \alpha_{t_{k+l}}(\Omega_m)[R_{k+l}]\rangle$$

exists for all $R_1 \cdots R_{k+l}$ in $\bigcup \mathfrak{W}(\Omega)$ and all $t_1 \cdots t_{k+l}$ in $\mathbb{R}$ with $k, l < \infty$. Notice in particular ($t_i = 0$ for $i = 1, 2, \ldots, k + l$) that this condition implies the existence of the Gibbs state ϕ. Under those assumptions we can prove that the time evolution can be defined in the limit as an automorphism of $\pi_\phi(\mathfrak{W})''$ and is implemented by a continuous one-parameter group $\{U_\phi(t)\}$ of unitary operators on $\mathcal{H}_\phi$ such that $U_\phi(t)\,\Phi = \Phi$. Furthermore, the following consequence of the KMS condition persists in the thermodynamical limit of the system satisfying these assumptions; there exists a conjugation operator C acting on $\mathcal{H}_\phi$ and such that $C\Phi = \Phi$, $[C, U_\phi(t)] = 0$ for all t in $\mathbb{R}$ and $C\,\pi_\phi(\mathfrak{W})''C = \pi_\phi(\mathfrak{W})'$.

The interest in this result is that it extends the domain of applicability of the KMS condition beyond the range of Subsection II.2.f. In particular, these assumptions (and not the stronger assumptions of Haag, Hugenholtz, and Winnink) are satisfied for the free Bose gas as well as for the BCS model (compare with the last remarks in the preceding subsection).

Bibliography

Abraham, D. B., E. Barouch, G. Gallavotti, and A. Martin-Löf [1970], Thermalisation of a magnetic Impurity in the Isotropic XY Model, *Phys. Rev. Lett.* **25**, 1449–1450.

Akhiezer, N. I., and I. M. Glazman [1961], *Theory of Linear Operators in Hilbert Space*, Vols. I and II (translated by M. Nestell), Ungar, New York.

Albert, A. A. [1934], On a Certain Algebra of Quantum Mechanics, *Ann. Math.* **35**, 65–73.

Albert, A. A. [1946], On Jordan Algebras of Linear Transformations, *Trans. Amer. Soc.* **59**, 524–555.

Albert, A. A. [1950], Structure Theory of Rings and Algebras. Power-Associative Algebras, *Proc. Int. Congr. Math.*, Cambridge, Mass., Vol. II, 25–32, American Mathematical Society, Providence, R.I.

Albert, A. A. [1950], A Theory of Power-Associative Commutative Algebras, *Trans. Amer. Math. Soc.* **69**, 503–527.

Albert, A. A. [1958], A Construction of Exceptional Jordan Division Algebras, *Ann. Math.* **67**, 1–28.

Ambrose, W. [1944], Spectral Resolution of Groups of Unitary Operators, *Duke Math. J.* **11**, 589–595.

Araki, H. [1960], Hamiltonian Formalism and the Canonical Commutation Relations in Quantum Field Theory, *J. Math. Phys.* **1**, 492–504.

Araki, H. [1963], A Lattice of von Neumann Algebras Associated with the Quantum Theory of a Free Bose Field, *J. Math. Phys.* **4**, 1343–1362.

Araki, H. [1964a], Von Neumann Algebras of Local Observables for Free Scalar Field, *J. Math. Phys.* **5**, 1–13.

Araki, H. [1964b], On the Algebra of all Local Observables, *Progr. Theoret. Phys. (Kyoto)* **32**, 844–854.

Araki, H. [1964c], Type of von Neumann Algebra Associated with Free Field, *Progr. Theoret. Phys. (Kyoto)* **32**, 956–965.

Araki, H. [1969], Gibbs States of a One-Dimensional Quantum Lattice, *Commun. Math. Phys.* **14**, 120–157.

Araki, H. [1969b], Local Quantum Theory I, in *Local Quantum Theory*, R. Jost, Ed., Academic Press, New York.

Araki, H. [1970a], *On Quasifree States of CAR and Bogoliubov Automorphisms*, Preprint, Kyoto.

Araki, H. [1970b], *Local Quantum Theory*, Benjamin, New York.

Araki, H. [1971], On Representations of the Canonical Commutation Relations, *Commun. Math. Phys.* **20**, 9–25.

Araki, H., and E. J. Woods [1963], Representations of the Canonical Commutation Relations Describing a Nonrelativistic Infinite Free Bose Gas, *J. Math. Phys.* **4,** 637–662.

Araki, H., and E. J. Woods [1971], Topologies on Test Function Spaces Induced by Representations of the Canonical Commutation Relations (to appear).

Araki, H., and W. Wyss [1964], Representations of Canonical Anticommutation Relations, *Helv. Phys. Acta* **37,** 136–159.

Arnold, V. I., and A. Avez [1968], *Ergodic Problems of Classical Mechanics*, Benjamin, New York.

Balsev, E., J. Manuceau, and A. Verbeure [1968], Representations of Anticommutation Relations and Bogoliubov Transformations, *Commun. Math. Phys.* **8,** 315–326.

Balsev, E., and A. Verbeure [1968], States on Clifford Algebras, *Commun. Math. Phys.* **7,** 55–76.

Bardeen, J., L. N. Cooper, and J. R. Schrieffer [1957], Theory of Superconductivity, *Phys. Rev.* **108,** 1175–1204.

Bargmann, V. [1962], Remarks on a Hilbert Space of Analytic Functions, *Proc. Natl. Acad. Sci.* **48,** 199–204.

Bargmann, V. [1964], Note on Wigner's Theorem on Symmetry Operations, *J. Math. Phys.* **5,** 862–868.

Bargmann, V., and E. P. Wigner [1948], Group Theoretical Discussion of Relativistic Wave Equations, *Proc. Natl. Acad. Sci.* **34,** 211–223.

Barton, G. [1963], *Introduction to Advanced Field Theory*, Wiley-Interscience, New York.

Behncke, H. [1969], A Remark on C^*-Algebras, *Commun. Math. Phys.* **12,** 142–144.

Berberian, S. K. [1965], *Measure and Integration*, Macmillan, New York.

Birkhoff, G., and J. von Neumann [1936], The Logic of Quantum Mechanics, *Ann. Math.* **37,** 823–843.

Bishop E., and K. de Leeuw [1959], The Representations of Linear Functionals by Measures on Sets of Extreme Points, *Ann. Inst. Fourier* (*Grenoble*), **9,** 305–331.

Bloch F., and A. Nordsieck [1937]. Note on the Radiation Field of an Electron, *Phys. Rev.* **52,** 54–59.

Borchers, H. J. [1960], Über die Mannigfaltigkeit der Interpolierenden Felder zu einer Kausalen S-Matrix, *Nuovo Cimento*, **15,** 784–794.

Borchers, H. J. [1961], Über die Vollständingkeit Lorentz invarienten Felder in einer zeitartigen Röhre, *Nuovo Cimento* **19,** 787–793.

Borchers, H. J. [1962], On Structure of the Algebra of Field Operators, *Nuovo Cimento* **24,** 214–236.

Borchers, H. J. [1965a], On the Vacuum State in Quantum Field Theory II, *Commun. Math. Phys.* **1,** 57–79.

Borchers, H. J. [1965b], Local Rings and the Connection of Spin with Statistics, *Commun. Math. Phys.* **1,** 281–307.

Borchers, H. J. [1967], On the Theory of Local Observables, in *Cargèse Lectures in Theoretical Physics*, F. Lurçat, Ed., Gordon and Breach, New York.

Borchers, H. J. [1969], On Groups of Automorphisms with Semi-bounded Spectrum, in *Systèmes à un nombre infini de degrés de liberté*, CNRS, Paris.

Borchers, H. J., R. Haag, and B. Schroer [1963], The Vacuum State in Quantum Field Theory, *Nuovo Cimento* **29,** 148–162.

Bourbaki, N. [1940], *Eléments de mathématiques*, *topologie générale*, Hermann, Paris.

Bourbaki, N. [1948], *Eléments de mathématiques*, *algèbre*, Chapter III, Hermann, Paris.

Bourbaki, N. [1959], *Eléments de mathématiques*, *algèbre*, Chapter IX, Hermann, Paris.

Bourbaki, N. [1964], *Eléments de mathématiques*, *espaces vectoriels topologiques*, Chapter IV, Hermann, Paris.

Brascamp, H. J. [1970], Equilibrium States for a Classical Lattice Gas, *Commun. Math. Phys.* **18,** 82–96.

Braun, H., and M. Koecher [1966], Jordan-Algebren, *Grundlehren der Mathematischen Wissenschaften*, **128,** Springer, Berlin.

Bures, D. J. C. [1963], Certain Factors Constructed as Infinite Tensor Products, *Comp. Math.* **15,** 169–191.

Cannon, J. T., and A. M. Jaffe [1970], Lorentz Covariance of the $\lambda(\varphi^4)_2$ Quantum Field Theory, *Commun. Math. Phys.* **17,** 261–321.

Chaiken, J. M. [1966], MIT Thesis (unpublished).

Chaiken, J. M. [1967], Finite-Particle Representations and States of the Canonical Commutation Relations, *Ann. Phys.* **42,** 23–80.

Chaiken, J. M. [1968], Number Operators for Representations of the Canonical Commutation Relations, *Commun. Math. Phys.* **8,** 164–184.

Choquet, G. [1969], *Lectures on Analysis*, Vol. I, II, and III, Benjamin, New York.

Choquet, G., and P. A. Meyer [1963], Existence et unicité des représentations intégrales dans les convexes compacts quelconques, *Ann. Inst. Fourier* (*Grenoble*), **13,** 139–154.

Cohn, P. M. [1954], On Homomorphic Images of Special Jordan Algebras, *Can. J. Math.* **6,** 253–264.

Combes, F. [1967], Sur les états factoriels d'une C^*-algèbre, *C.R. Acad. Sc. Paris*, **265,** 736–739.

Cook, J. M. [1953], The Mathematics of Second Quantization, *Trans. Am. Math. Soc.* **74,** 222–245.

Cook, J. M. [1961], Asymptotic Properties of a Boson Field with Given Sources, *J. Math. Phys.* **2,** 33–45.

Courbage, M., S. Miracle-Sole, and D. Robinson [1971], Normal States and Representations of the Canonical Commutation Relations, *Ann. Inst. Henri Poincaré* **A14,** 171–178.

Dähn, G. [1968], Attempt of an Axiomatic Foundation of Quantum Mechanics and More General Theories IV, *Commun. Math. Phys.* **9,** 192–211.

Davies, E. B. [1968], On the Borel Structure of C^*-algebras, *Commun. Math. Phys.* **8,** 147–163.

Davies, E. B. [1969], The Structure of C^*-Algebras, *Quart. J. Math. Oxford* (2) **20,** 351–366.

Davies, E. B. [1969b; 1970; 1971], Quantum Stochastic Processes I, *Commun. Math. Phys.* **15,** 277–304; II, *ibid.*, **19,** 83–105; III, *ibid.*, **22,** 51–70.

Davies, E. B. and J. T. Lewis, [1970], An Operational Approach to Quantum Probability, *Commun. Math. Phys.* **17,** 239–260.

Day, M. M. [1957], Amenable Semi-Groups, *Illinois J. Math.* **1,** 509–544.

Day, M. M. [1962], *Normed Linear Spaces*, Springer, Berlin.

Dell'Antonio, G. F. [1966], On Some Groups of Automorphims of Physical Observables, *Commun. Math. Phys.* **2,** 384–397.

Dell'Antonio, G. F. [1967], On the limits of Sequences of Normal States, *Comm. Pure Appl. Math.* **20,** 413–429.

Dell'Antonio, G. F. [1968], Structure of the Algebras of Some Free Systems, *Commun. Math. Phys.* **9,** 81–117.

Dell'Antonio, G. F. [1971], Can Local Gauge Transformations Be Implemented? *J. Math. Phys.* **12,** 148–156.

Dell'Antonio, G. F., and S. Doplicher [1967], Total Number of Particles and Fock Representation, *J. Math. Phys.* **8,** 663–666.

Dell'Antonio, S. Doplicher, and D. Ruelle [1966], A Theorem on Canonical Commutation and Anticommutation Relations, *Commun. Math. Phys.* **2,** 223–230.

Dirac, P. A. M. [1930], *The Principles of Quantum Mechanics*, Clarendon, Oxford.

Dixmier, J. [1950], Les moyennes invariantes dans les semi-groupes et leurs applications, *Acta Sci. Math. (Szeged)* **12,** 213–227.

Dixmier, J. [1952], Algèbres quasi-unitaires, *Comment. Math. Helv.* **26,** 275–322.

Dixmier, J. [1957], *Les algèbres d'opérateurs dans l'espace hilbertien (Algèbres de von Neumann)*, Gauthier-Villars, Paris.

Dixmier, J. [1958], Sur la relation $i(PQ - QP) = 1$, *Comp. Math.* **13,** 263–270.

Dixmier, J. [1964], *Les C*-algèbres et leurs représentations*, Gauthier-Villars, Paris.

Dobrushin, R. L. [1968a], Gibbsian Random Fields for Lattice Systems with Pairwise Interactions, *J. Funct. Anal. Appl.* **2,** 31–43.

Dobrushin, R. L. [1968b], The Problem of Uniqueness of a Gibbsian Random Field and the Problem of phase transitions, *J. Funct. Anal. Appl.* **2,** 44–57.

Doplicher, S. [1965], An Algebraic Spectrum Condition, *Commun. Math. Phys.* **1,** 1–5.

Doplicher, S., R. Haag, and J. E. Roberts [1969a, b], Fields, Observables and Gauge Transformations I, *Commun. Math. Phys.* **13,** 1–23; II, *ibid.* **15,** 173–200.

Doplicher, S., R. V. Kadison, D. Kastler, and D. W. Robinson [1967], Asymptotically Abelien Systems, *Commun. Math. Phys.* **6,** 101–120.

Doplicher, S., and D. Kastler [1968], Ergodic States in a NonCommutative Ergodic Theory, *Commun. Math. Phys.* **7,** 1–20.

Doplicher, S., D. Kastler, and D. W. Robinson [1966], Covariance Algebras in Field Theory and Statistical Mechanics, *Commun. Math. Phys.* **3,** 1–28.

Doplicher, S., D. Kastler, and E. Størmer [1969], Invariant States and Asymptotic Abelianess, *J. Funct. Anal.* **3,** 419–433.

Doplicher, S., and R. T. Powers [1968], On the Simplicity of the Even CAR Algebra and Free Field Models, *Commun. Math. Phys.* **7,** 77–79.

Doplicher, S., T. Regge, and I. M. Singer [1968], A Geometrical Model Showing the Independence of Locality and Positivity of the Energy, *Commun. Math. Phys.* **7,** 51–54.

Dubin, D. A., and G. Sewell [1970], Time-Translations in the Algebraic Formulation of Statistical Mechanics, *J. Math. Phys.* **11,** 2990–2998.

Dunford, N., and J. T. Schwartz [1957], *Linear Operators, Part I: General Theory; Part II: Spectral Theory*, Wiley-Interscience, New York.

Dyson, F. J. [1962], The Threefold Way. Algebraic Structure of Symmetry Groups and Ensembles in Quantum Mechanics, *J. Math. Phys.* **3,** 1199–1215.

Edwards, C. M. [1970], The Operational Approach to Algebraic Quantum Theory I. *Commun. Math. Phys.* **16,** 207–230.

Edwards, C. M. [1971], Classes of Operations in Quantum Theory, *Commun. Math. Phys.* **20,** 26–56.

Effros, E. G., and F. Hahn [1967], Locally Compact Transformation Groups and C^*-Algebras, *Memoirs Ann. Math. Soc.* **75,** 1–92.

Emch, G. G. [1961], *Representations du Groupe de Lorentz*, Seminaire de l'Institut de Physique Théorique, Genève.

Emch, G. G. [1963a, b], Mécanique quantique quaternionienne et relativité restreinte, I, *Helv. Phys. Acta* **36,** 739–769; II, *ibid.*, **36,** 770–788.

Emch, G. G. [1965], Representations of the Lorentz Group in Quaternionic Quantum Mechanics, *Lectures in Theoretical Physics*, **VIIa,** W. E. Brittin Ed., University of Colorado Press.

Emch, G. G. [1966a], The Definition of States in Quantum Statistical Mechanics, *J. Math. Phys.* **7,** 1413–1420.

Emch, G. G. [1966b], Rigourous Results in Non-Equilibrium Statistical Mechanics, *Lectures in Theoretical Physics*, **VIIIa,** W. E. Brittin, Ed., University of Colorado Press.

Emch, G. G. [1968], *Projective Group Representations and Quaternionic Hibert Spaces*, talk presented at the Conference on Orthomodular Lattices, Amherst, Mass.

Emch, G. G. [1971], Theory and Results about Phase Transitions from the C^*-algebraic Approach, in *Phase Transitions and Critical Phenomena*, C. Domb and M. S. Green, Eds., Academic Press, London.

Emch, G. G., and M. Guenin [1966], Gauge Invariant Formulation of the BCS-Model, *J. Math. Phys.* **7**, 915–921.

Emch, G. G., and H. J. F. Knops [1970], Pure Thermodynamical Phases as Extremal KMS States, *J. Math. Phys.* **11,** 3008–3018.

Emch, G. G., H. J. F. Knops, and E. J. Verboven [1968a], On the Extension of Invariant Partial States in Statistical Mechanics, *Commun. Math. Phys.* **7,** 164–172.

Emch, G. G., H. J. F. Knops, and E. J. Verboven [1968b], On Partial Weakly Clustering States with Application to the Ising Model, *Commun. Math. Phys.* **8,** 300–314.

Emch, G. G., H. J. F. Knops, and E. J. Verboven [1970], The Breaking of Euclidian Symmetry with an Application to the Theory of Crystallization, *J. Math. Phys.* **11,** 1655–1668.

Emch, G. G., and C. Piron [1963], Symmetry in Quantum Theory, *J. Math. Phys.* **4,** 469–473.

Emch, G. G., and C. Radin [1971], Relaxation of Local Thermal Deviations from Equilibrium, *J. Math. Phys.* **12,** 2043–2046.

Ezawa, H. [1964], The Representation of Canonical Variables as the Limit of Infinite Space Volume: the case of the BCS Model, *J. Math. Phys.* **5,** 1078–1090.

Ezawa, H. [1965], Quantum Mechanics of a Many-Boson System and the Representation of Canonical Variables, *J. Math. Phys.* **6,** 380–404.

Ezawa, H., and J. A. Swieca [1967], Spontaneous Breakdown of Symmetries and Zero-Mass States, *Commun. Math. Phys.* **5,** 330–336.

Fabrey, J. D. [1970], Exponential Representations of the Canonical Commutation Relations, *Commun. Math. Phys.* **19,** 1–30.

Fano, G., and G. Loupias [1971], On the Thermodynamical Limit of the B.C.S. State, *Commun. Math. Phys.* **20,** 143–166.

Feldman, J., J. M. G. Fell [1957], Separable Representations of Rings of Operators, *Ann. Math.* **65,** 241–249.

Fell, J. M. G. [1960], The Dual Spaces of C^*-Algebras, *Trans. Am. Math. Soc.* **94,** 365–403.

Fillmore, P. A., and D. M. Topping [1967], Operator Algebras Generated by Projections, *Duke Math. J.* **34,** 333–336.

Finkelstein, D., J. M. Jauch, S. Schiminovitch, and D. Speiser [1962], Foundations of Quaternion Quantum Mechanics, *J. Math. Phys*, **3,** 207–220.

Finkelstein, D., J. M. Jauch, S. Schiminovitch, and D. Speiser [1963], Principle of General Q-covariance, *J. Math. Phys.* **4,** 788–796.

Finkelstein, D., J. M. Jauch, and D. Speiser [1959], Notes on Quaternion Quantum Mechanics, Preprints, CERN.

Finkelstein, D., J. M. Jauch, and D. Speiser [1963], Quaternionic Representations of Compact Groups, *J. Math. Phys.* **4,** 136–140.

Fock, V. [1932], Konfigurationsraum und zweite Quantelung, *Z. Phys.* **75,** 622–647.

Foias, C., L. Gehér, and B. Sz-Nagy [1960], On the Permutability Condition of Quantum Mechanics, *Acta. Sci. Math.* (*Szeged*) **21,** 78–89.

Friedrichs, K. O. [1953], *Mathematical Aspects of the Quantum Theory of Fields*, Wiley-Interscience, New York, Parts I to V of this book were published in *Commun. Pure Appl. Math.* **4** [1951], 161–224; **5** [1952], 1–56 and 349–411; and **6** [1953], 1–72.

Gallavotti, G., O. E. Lanford, III, and J. L. Lebowitz [1970], Thermodynamic Limit of Time-Dependent Correlation Functions for One-Dimensional Systems, *J. Math. Phys.* **11,** 2898–2905.

Gallavotti, G., S. Miracle-Sole, and D. W. Robinson [1968], Analyticity Properties of the Anisotropic Heisenberg Model, *Commun. Math Phys.* **10,** 311–324.

Gårding, L. [1947], Note on Continuous Representations of Lie Groups, *Proc. Nat. Acad. Sc.* **33,** 331–332.

Gårding, L., and A. S. Wightman [1954a], Representations of the Anticommutation Relations, *Proc. Nat. Acad. Sc.* **40,** 617–621.

Gårding, L., and A. S. Wightman [1954b], Representations of the Commutation Relations, *Proc. Nat. Acad. Sci.* **40,** 622–626.

Gelfand, I., and M. A. Naimark [1943], On the Imbedding of Normed Rings into the Ring of Operators in Hilbert Space, *Mat. Sborn., N. S.* **12 [54],** 197–217.

Gelfand, I. M., and N. Y. Vilenkin [1964], *Generalized Functions*, Academic Press, New York.

Giles, R. [1970], Foundations for Quantum Mechanics, *J. Math. Phys.* **11,** 2139–2160.

Ginibre, J. [1968], Reduced Density Matrices of the Anisotropic Heisenberg Model, *Commun. Math. Phys.* **10,** 140–154.

Ginibre, J. [1969a], On Some Recent Work of Dobrushin, in *Systèmes à un nombre infini de degrés de liberté*, CNRS, Paris.

Ginibre, J. [1969b], Existence of Phase Transitions for Quantum Lattice Systems, *Commun. Math. Phys.* **14,** 205–234.

Glimm, J. [1960], On a Certain Class of Operator Algebras, *Trans. Am. Math. Soc.* **95,** 318–340.

Glimm, J. [1961], Type I C^*-Algebras, *Ann. Math.* **73,** 572–612.

Glimm, J. [1969], Models for Quantum Field Theory, in *Local Quantum Theory*, R. Jost, Ed., Academic, New York.

Glimm, J. C., and R. V. Kadison [1960], Unitary Operators in C^*-Algebras, *Pacific J. Math.* **10,** 547–548.

Godement, R. [1944], Sur une généralisation d'un théorème de Stone, *C.R. Acad. Sci. Paris*, **218,** 901–903.

Goldstine, H. H., and L. P. Horwitz [1962], On a Hilbert Space with Non-Associative Scalars, *Proc. Nat. Acad. Sci.* **48,** 1134–1142.

Goldstine, H. H., and L. P. Horwitz [1964], Hilbert Space with Non-Associative Scalars I, *Math. Ann.* **154,** 1–27.

Greenberg, W. [1969], Correlation Functionals of Infinite Volume Quantum Spin Systems, *Commun. Math. Phys.* **11,** 314–320.

Greenberg, O. W., and S. S. Schweber [1958], Clothed Particle Operators in Simple Models of Quantum Field Theory, *Nuovo Cimento* **8,** 378–405.

Greenleaf, F. P. [1969], *Invariant Means on Topological Groups*, Van Nostrand-Reinhold, New York.

Grothendieck, A. [1957], Un résultat sur le dual d'une C^*-algèbre, *J. Math. pures et appl.* **36,** 97–108.

Gudder, S. P. [1970], *Representations of Groups as Automorphisms on Orthomodular Lattices and Posets*, Preprint, Denver, 1970.

Gudder, S. [1970b], Axiomatic Quantum Mechanics and Generalized Probability Theory, *Probabilistic Methods in Applied Mathematics*, Vol. 2, 53–129.

Guenin, M. [1966], On the Interaction Picture, *Commun. Math. Phys.* **3,** 120–132.

Guenin, M. [1969], Remarks on the Operator Solution of the Scalar-Field Model, in *Local Quantum Theory*, R. Jost, Ed., Academic Press, New York.

Guenin, M., and B. Misra [1963], On the von Neumann Algebras Generated by Field Operators, *Nuovo Cimento* **30,** 1272–1290.

Guenin, M., and G. Velo [1967], Automorphisms and Broken Symmetries in Algebraic Quantum Field Theories, *Nuovo Cimento* **47A,** 36–48.

Guenin, M., and G. Velo [1968], Mass Renormalization as an Automorphism of the Algebra of Field Operators, *Helv. Phys. Acta* **41**, 362–366.

Guichardet, M. A. [1966], Produits tensoriels infinis et représentations des relations d'anticommutation, *Ann. Scient. Éc. Norm. Sup.* **83**, 1–52.

Guichardet, A., and D. Kastler [1970], *Désintegration des états quasi-invariants des C*-algèbres*, *J. Math. Pures et Appl.* **49**, 349–380.

Gunson, J. [1967], On the Algebraic Structure of Quantum Mechanics, *Commun. Math. Phys.* **6**, 262–285.

Haag, R. [1955], On Quantum Field Theory, *Dan. Mat. Fys. Medd.* **29**, No. 12.

Haag, R. [1959], Discussion des "axiomes" et des propriétés asymptotiques d'une théorie des champs locale avec particules composées, in *Les problèmes mathématiques de la théorie quantique des champs*, CNRS.

Haag, R. [1962], The Mathematical Structure of the Bardeen-Cooper-Schrieffer Model, *Nuovo Cimento* **25**, 287–298.

Haag, R., N. Hugenholtz, and M. Winnink [1967], On the Equilibrium States in Quantum Statistical Mechanics, *Commun. Math. Phys.* **5**, 215–236.

Haag, R., R. V. Kadison, and D. Kastler [1970], Nets of C^*-Algebras and Classification of States, *Commun. Math. Phys.* **16**, 81–104.

Haag, R., and D. Kastler [1964], An Algebraic Approach to Quantum Field Theory, *J. Math. Phys.* **5**, 848–861.

Haag, R., and B. Schroer [1962], Postulates of Quantum Field Theory, *J. Math. Phys.* **3**, 248–256.

Hagen, C. R., G. Guralnik, and V. S. Mathur [1967], *Proceedings of the 1967 International Conference on Particles and Fields*, Wiley-Interscience, New York.

Halmos, P. R. [1950], *Measure Theory*, Van Nostrand, Princeton, N.J.

Halmos, P. R. [1958], *Lectures on Ergodic Theory*, Chelsea, New York.

Halperin, H. [1967], Finite Sums of Irreducible Functionals on C^*-Algebras, *Proc. Amer. Math. Soc.* **18**, 352–358.

Hegerfeldt, G. C. [1971], Equivalence of Basis-Dependent and Basis-Independent Approaches to Canonical Field Operators, *J. Math. Phys.* **12**, 167–172.

Hegerfeldt, G., and J. R. Klauder [1970], Metric on Test Function Spaces for Canonical Field Operators, *Commun. Math. Phys.* **16**, 329–346.

Hegerfeldt, G. C., and O. Melsheimer [1969], The Form of Representations of the Canonical Commutation Relations for Bose Fields and Connection with Finitely Many Degrees of Freedom, *Commun. Math. Phys.* **12**, 304–323.

Henle, M. [1970], Spatial Representation of Groups of Automorphisms of von Neumann Algebras with Properly Infinite Commutant, *Commun. Math. Phys.* **19**, 273–275.

Herman, R. H., and M. Takesaki [1970], States and Automorphism Groups of Operator Algebras, *Commun. Math. Phys.* **19**, 142–160.

Hewitt, E., and K. A. Ross [1963], *Abstract Harmonic Analysis*, Vol. I and II, Springer, Berlin.

Hille, E., and R. S. Phillips [1957], *Functional Analysis and Semi-Groups*, American Mathematical Society, Providence, R.I.

Holland, S. S., Jr. [1968], *Why the Interest in Orthomodular Lattices?* Talk presented at the Conference on Orthomodular Lattices, Amherst, Mass.

Horwitz, L. P. [1966], Gauge Fields of an Algebraic Hilbert Space, *Helv. Phys. Acta* **39**, 144–154.

Horwitz, L. P., and L. C. Biedenharn [1965], Intrinsic Superselection Rules of Algebraic Hilbert Space, *Helv. Phys. Acta* **38**, 385–408.

Hove, L. Van, *see* van Hove, L.

Hugenholtz, N. M. [1967], On the Factor Type of Equilibrium States in Quantum Statistical Mechanics, *Commun. Math. Phys.* **6,** 189–193.

Hugenholtz, N. M., and J. D. Wieringa [1969], On Locally Normal States in Quantum Statistical Mechanics, *Commun. Math. Phys.* **11,** 183–197.

Jacobson, N. [1951], General Representation Theory of Jordan Algebras, *Trans. Amer. Math. Soc.* **70,** 509–530.

Jacobson, N., and C. E. Richart [1950], Jordan Homomorphisms of Rings, *Trans. Amer. Math. Soc.* **69,** 479–502.

Jadczyk, A. Z. [1969a], On the Spectrum of Internal Symmetries in the Algebraic Quantum Field Theory, *Commun. Math. Phys.* **12,** 58–63.

Jadczyk, A. Z. [1969b], On Some Groups of Automorphisms of von Neumann Algebras with Cyclic and Separating Vector, *Commun. Math. Phys.* **13,** 142–153.

Jaffe, A. M. [1969], Constructing the $\lambda(\phi^4)_2$ Theory, in *Local Quantum Theory*, R. Jost, Ed., Academic Press, New York.

Jauch, J. M. [1958], Theory of the Scattering Operator, I, Simple Systems, *Helv. Phys. Acta* **31,** 127–158; II, Multichannel Scattering, *ibid.*, **31,** 661–684.

Jauch, J. M. [1959], Seminar notes, CERN.

Jauch, J. M. [1968a], *Foundations of Quantum Mechanics*, Addison-Wesley, Reading, Mass.

Jauch, J. M. [1968b], Projective Representations of the Poincaré Group in a Quaternionic Hilbert Space, in *Group Theory and Its Applications*, E. Loebl, Ed., Academic Press, New York.

Jauch, J. M. [1970], *Foundations of Quantum Mechanics*, Lecture notes, Varenna.

Jauch, J. M., B. Misra, and A. G. Gibson [1968], On the Asymptotic Condition of Scattering Theory, *Helv. Phys. Acta* **41,** 513–527.

Jauch, J. M., and C. Piron [1963], Can Hidden Variables be Excluded in Quantum Mechanics? *Helv. Phys. Acta* **36,** 827–837.

Jelinek, F. [1968], BCS-Spin Model, Its Thermodynamic Representations and Automorphisms, *Commun. Math. Phys.* **9,** 169–175.

Jordan, P. [1932], Über eine Klasse nicht assoziativer-hyperkomplexen Algebren, *Gött. Nachr.*, 569–575.

Jordan, P. [1933a], Über Verallgemeinerungsmöglichkeiten des Formalismus der Quantenmechanik, *Gött. Nachr.* 209–217.

Jordan, P. [1933b], Über die Multiplikation quantenmechanischen Grössen, *Z. Phys.* **80,** 285–291.

Jordan, P. [1968], Über das Verhältnis der Theorie der Elementarlänge zur Quantentheorie, *Commun. Math. Phys.* **9,** 279–292.

Jordan, P. [1969], Über das Verhältnis der Theorie der Elementarlänge zur Quantentheorie II, *Commun. Math. Phys.* **11,** 293–296.

Jordan, P., J. von Neumann, and E. Wigner [1934], On an Algebraic Generalization of the Quantum Mechanical Formalism, *Ann. Math.* **35,** 29–64.

Jordan, P., and E. P. Wigner [1928], Über das Paulische Äquivalenzverbot, *Z. Phys.* **47,** 631–651.

Jost, R. [1965], *The General Theory of Quantized Fields*, American Mathematical Society, Providence, R.I.

Jost, R. [1969], *Local Quantum Theory, Proceedings of the International School of Physics "Enrico Fermi,"* Academic Press, New York.

Kadison, R. V. [1951], Isometries of Operator Algebras, *Ann. Math.* **54,** 325–338.

Kadison, R. V. [1952], A Generalized Schwartz Inequality and Algebraic Invariants for Operator Algebras, *Ann. Math.* **56,** 494–503.

Kadison, R. V. [1957a], Unitary Invariants for Representations of Operator Algebras, *Ann. Math.* **66,** 304–379.

Kadison, R. V. [1957b], Irreducible Operator Algebras, *Proc. Natl. Acad. Sci.* **43,** 273–276.

Kadison, R. V. [1962], States and Representations, *Trans. Amer. Math. Soc.* **103,** 304–319.

Kadison, R. V. [1963], Remarks on the Type of von Neumann Algebras of Local Observables in Quantum Field Theory, *J. Math. Phys.* **4,** 1511–1516.

Kadison, R. V. [1965], Transformations of States in Operator Theory and Dynamics, *Topology* **3,** Suppl. 2, 177–198.

Kadison, R. V. [1967a], Lectures on Operator Algebras, in *Cargèze Lectures in Theoretical Physics*, F. Lurgat, Ed., Gordon and Breach, New York.

Kadison, R. V. [1967b], The Energy-Momentum Spectrum of Quantum Fields, *Commun. Math. Phys.* **4,** 258–260.

Kadison, R. V. [1968], Appendix to a paper by Davies [1968], *Commun. Math. Phys.* **8,** 161–163.

Kadison, R. V. [1969], Rings of Operators, in *Encyclopedia Britannica.*

Kadison, R. V., and J. R. Ringrose [1967], Derivations and Automorphisms of Operator Algebras, *Commun. Math. Phys.* **4,** 32–63.

Kadison, R. V., and I. Singer [1959], Extensions of Pure States, *Amer. J. Math.* **81,** 383–400.

Kaplansky, I. [1951], A Theorem on Rings of Operators, *Pacific J. Math.* **53,** 227–232.

Kaplansky, I. [1952], Algebras of Type I, *Ann. Math.* **56,** 460–472.

Kaplansky, I. [1968], *Rings of Operators*, Benjamin, New York.

Kastler, D. [1964], A C^*-Algebra Approach to Field Theory, in *Analysis in Function Space*, T. Martin and I. Segal, Eds., MIT Press, Cambridge, Mass.

Kastler, D. [1965], The C^*-Algebra of a Free Boson Field, *Commun. Math. Phys.* **1,** 14–48.

Kastler, D. [1967a], Topics in the Algebraic Approach to Field Theory, in *Cargèse Lectures in Theoretical Physics*, F. Lurcat, Ed., Gordon and Breach, New York.

Kastler, D. [1967b], Broken Symmetries and the Goldstone Theorem in Axiomatic Field Theory, in *Proceedings of the 1967 International Conference on Particles and Fields*, C. R. Hagen, G. Guralnik, and V. A. Mathur, Eds., Wiley-Interscience, New York.

Kastler, D. [1969], Quasi-Free States of Fermion Systems, in *Systèmes à un nombre infini de degrés de liberté*, CNRS, Paris.

Kastler, D., R. Haag, and L. Michel [1968], *Central Decomposition of Ergodic States*, Séminaire de Physique Théorique I, Marseille.

Kastler, D., J. C. T. Pool, and E. T. Poulsen [1969], Quasi-Unitary Algebras Attached to Temperature States in Statistical Mechanics. A comment on the Work of Haag, Hugenholtz, and Winnink, *Commun. Math. Phys.* **12,** 175–192.

Kastler, D., and D. W. Robinson [1966], Invariant States in Statistical Mechanics, *Commun. Math. Phys.* **3,** 151–180.

Kastler, D., D. W. Robinson, and A. Swieca [1966], Conserved Currents and Associated Symmetries; Goldstone Theorem, *Commun. Math. Phys.* **2,** 108–120.

Kato, T. [1951], Fundamental Properties of Hamiltonian Operators of Schrödinger Type, *Trans. Amer. Math. Soc.* **70,** 195–211.

Kato, T. [1953], Perturbation Theory of Semi-Bounded Operators, *Math. Ann.* **125,** 435–447.

Kato, T. [1966], *Perturbation Theory for Linear Operators*, Springer, Berlin.

Kato, Y. [1961], Some Converging Examples of the Perturbation Series in the Quantum Field Theory, *Progr. Theoret. Phys. (Kyoto)* **26,** 99–122.

Kehlet, E. T. [1969], On the Monotone Sequential Closure of a C^*-Algebra, *Math. Scand.* **25,** 59–70.

Kelley, J. L. [1955], *General Topology*, Van Nostrand, New-York.

Klauder, J. R. [1963, 1964], Continuous-Representation Theory: I, Postulates of Continuous-Representation Theory, *J. Math. Phys.* **4,** 1055–1057; II, Generalized Relation

Between Quantum and Classical Mechanics, *J. Math. Phys.* **4,** 1058–1072; III, On Functional Quantization of Classical Systems, *J. Math. Phys.* **5,** 177–187; (IV, *see* McKenna and Klauder [1964]); (V, *see* Klauder and McKenna [1965]).

Klauder, J. R. [1965], Rotationally—Symmetric Model Field Theories, *J. Math. Phys.* **6,** 1666–1679.

Klauder, J. R. [1967], Weak Correspondence Principle, *J. Math. Phys.* **8,** 2392–2399.

Klauder, J. R. [1969], Hamiltonian Approach to Quantum Field Theory, Lecture given at the VIII Internationalen Universitätswochen für Kernphysik, Schladming.

Klauder, J. R., and J. McKenna [1965], Continuous-Representation Theory: V, Construction of a Class of Scalars Boson Field Continuous Representations, *J. Math. Phys.* **6,** 68–87.

Klauder, J. R., J. McKenna, and E. J. Woods [1966], Direct-Product Representations of the Canonical Commutation Relations, *J. Math. Phys.* **7,** 822–828.

Klauder, J. R., and L. Streit [1969], Properties of "Quadratic" Canonical Commutation Relation Representations, *J. Math. Phys.* **10,** 1661–1669.

Knops, H. J. F. [1969], Properties of Crystal States, *Commun. Math. Phys.* **12,** 32–42.

Koopman, B. O. [1931], Hamiltonian Systems and Transformations in Hilbert Spaces *Proc. Natl. Acad. Sci.* **17,** 315–318.

Kovács, I., and J. Szücs [1966], Ergodic Type Theorems in von Neumann Algebras, *Acta, Sc. Math.* (Szeged) **27,** 233–246.

Kraus, K. [1968], Algebras of Observables with Continuous Representations of Symmetry Groups, *Commun. Math. Phys.* **9,** 99–111.

Kraus, K. [1970], An Algebraic Spectrum Condition, *Commun. Math. Phys.* **16,** 138–141.

Kubo, R. [1957], Statistical Mechanical Theory of Irreversible Processes. I, *J. Phys. Soc. Japan* **12,** 570–586.

Kuratowski, C. [1948], *Topologie* (2 vols.) *Monografie Matematyczne,* Warszawa-Wroclaw; English translation, *Topology* (2 vols.), Academic Press, New York.

Lanford, O. E. [1968, 1969], The Classical Mechanics of One-Dimensional Systems of Infinitely Many-Particles: I, Existence Theorem; II, Kinetic Theory, *Commun. Math. Phys.* **9,** 179–191; **11,** 257–292.

Lanford, O. E., and D. W. Robinson [1968a], Statistical Mechanics of Quantum Spin Systems III, *Commun. Math. Phys.* **9,** 327–338.

Lanford, O. E., and D. W. Robinson [1968b], Mean Entropy of States in Quantum Statistical Mechanics, *J. Math Phys.* **9,** 1120–1125.

Lanford, O. E., and D. Ruelle [1967], Integral Representations of Invariant States on B^*-Algebras, *J. Math. Phys.* **8,** 1460–1463.

Lanford, O. E., and D. Ruelle [1969], Observables at Infinity and States with Short Range Correlations in Statistical Mechanics, *Commun. Math. Phys.* **13,** 194–215.

Lebowitz, J. L. [1968], Statistical Mechanics—A Review of Selected Rigorous Results, *Ann. Rev. Phys. Chem.* **19,** 389–418.

Lee, T. D., and C. N. Yang [1952], Statistical Theory of Equations of State and Phase Transitions: II, Lattice Gas and Ising Model, *Phys. Rev.* **87,** 410–419, (for I, see Yang and Lee [1952]).

Loupias, G., and S. Miracle-Sole [1966], C^*-Algèbres des systèmes canoniques: I, *Commun. Math. Phys.* **2,** 31–48.

Lowdenslager, D. B. [1957], On Postulates for General Quantum Mechanics, *Proc. Amer. Math. Soc.* **8,** 88–91.

Ludwig, G. [1964], Versuch einer axiomatischen Grundlegung der Quantenmechanik und allgemeinerer physikalischer Theorien, *Z. Phys.* **181,** 233–260.

Ludwig, G. [1967], Attempt of an Axiomatic Foundation of Quantum Mechanics and More General Theories: II, *Commun. Math. Phys.* **4,** 331–348.

Ludwig, G. [1968], Attempt of an Axiomatic Foundation of Quantum Mechanics and More General Theories: III, *Commun. Math. Phys.* **9**, 1–12.

Lurçat, F. [1967], *Cargèse Lectures in Theoretical Physics. Applications of Mathematics to Problem in Theoretical Physics*, Gordon and Breach, New York.

Mackey, G. W. [1949], Imprimitivity for Representations of Locally Compact Groups. I, *Proc. Nat. Acad. Sci.* **35**, 537–545.

Mackey, G. W. [1951], Induced Representations of Groups, *Amer. J. Math.* **73**, 576–592.

Mackey, G. W. [1952], Induced Representations of Locally Compact Groups. I, *Ann. Math.* **55**, 101–139.

Mackey, G. W. [1955], *The Theory of Group Representations*, Lecture notes, Department of Mathematics, University of Chicago.

Mackey, G. W. [1957], Borel Structures in Groups and Their Duals, *Trans. Amer. Math. Soc.* **85**, 134–165.

Mackey, G. W. [1963], *Mathematical Foundations of Quantum Mechanics*, Benjamin, New York.

Mackey, G. W. [1968], *Induced Representations and Quantum Mechanics*, Benjamin, New York.

Mallarmé, S. [1885], *Le "Ten o'clock" de M. Whistler*, Conférence faite à Londres.

Manuceau, J. [1968a], Etude de quelques automorphismes de la C^*-algèbre du champ de bosons libres, *Ann. Inst. Henri Poincaré* **8**, 117–138.

Manuceau, J. [1968b], C^*-algèbre de relations de commutation, *Ann. Inst. Henri Poincaré*, **8**, 139–161.

Manuceau, J., F. Rocca, and D. Testard [1969], On the Product Form of Quasi-Free States, *Commun. Math. Phys.* **12**, 43–57.

Manuceau, J., and J. C. Trottin [1969], On Lattice Spin Systems, *Ann. Inst. Henri Poincaré* **A10**, 359–380.

Manuceau, J., and A. Verbeure [1968], Quasi-Free States of the C.C.R.-Algebra and Bogoliubov Transformations, *Commun. Math. Phys.* **8**, 293–302.

Manuceau, J., and A. Verbeure [1970], Non-factor Quasi-free States of the CAR Algebra, *Commun. math. Phys.* **18**, 319–326.

Martin, P. C., and J. Schwinger [1959], Theory of Many-Particle Systems: I, *Phys. Rev.* **115**, 1342–1373.

McKenna, J., and J. R. Klauder [1964], Continuous-Representation Theory IV. Structure of a Class of Function Spaces Arising from Quantum Mechanics, *J. Math. Phys.* **5**, 878–896.

Mermin, N. D. [1967], Absence of Ordering in Certain Classical Systems, *J. Math. Phys.* **8**, 1061–1064.

Mermin, N. D., and H. Wagner [1966], Absence of Ferromagnetism or Antiferromagnetism in One- or Two-Dimensional Isotropic Heisenberg Models, *Phys. Rev. Letters* **17**, 1133–1136.

Michel, L. [1966], Relativistic Invariance and Internal Symmetries, in *Axiomatic Field Theory*, M. Chretien and S. Deser, Eds., Gordon and Breach, New York.

Michel, L., and A. S. Wightman, [undated], *On the Representations of the Complete Poincaré Group*, Seminar notes, Princeton, N.J.

Misra, B. [1965], On the Representations of Haag Fields, *Helv. Phys. Acta* **38**, 189–206.

Montvay, J. [1965], The Representations of Symmetry Groups on the Algebra of Quasi-Local Observables, *Nuovo Cimento* **40A**, 121–131.

Murray, F. J., and J. von Neumann [1936], On Rings of Operators, *Ann. Math.* **37**, 116–129.

Nachbin, L. [1964], *Lectures on the Theory of Distributions*, Instituto de Física e Matemática Universidade do Recife.

Naimark, M. A. [1943], Positive-Definite Operator Functions on a Commutative Group, *Izv. Akad. Nauk SSSR. Ser. Mat.* **7,** 237–244.

Naimark, M. A. [1964], *Normed Rings* (translated by Leo F. Boron), Noordhoff, Groningen, The Netherlands.

Narnhofer, H. [1970], *On Fermi Lattice Systems with Quadratic Hamiltonians, Acta Phys. Austriaca* **31,** 349–353.

Naudts, J. [1970], *Observables at Infinity*, Preprint, Leuven.

Nelson, E. [1959], Analytic Vectors, *Ann. Math.*, **70,** 572–615.

Nelson, E. [1964], Interaction of Non-relativistic Particles with a Quantized Scalar Field, *J. Math. Phys.* **5,** 1190–1197.

Neumann, J. von, *see* von Neumann, J.

Paige, L. J. [1963], Jordan Algebras, in *Studies in Modern Algebra*, A. A. Albert, Ed., Prentice-Hall, Englewood Cliffs, N.J.

Pedersen, G. K. [1966, 1968, 1969a, b], Measure Theory for C^*-Algebras: I, *Math. Scand.* **19,** 131–145; II, *Math. Scand.* **22,** 63–74; III, *Math. Scand.* **25,** 71–93; IV, *Math. Scand.* **25,** 121–127.

Pedersen, G. K. [1969c], On Weak and Monotone σ-Closures of C^*-Algebras, *Commun. Math. Phys.* **11,** 221–226.

Phelps, R. [1966], *Lectures on Choquet's Theorem*, Van Nostrand, Princeton, N.J.

Pier, J. P. [1965], *Sur une classe de groupes localement compacts remarquables du point de vue de l'analyse harmonique*, Thèse, Nancy.

Piron, C. [1964], Axiomatique quantique, *Helv. Phys. Acta* **37,** 439–468.

Piron, C. [1969], Les règles de supersélection continues, *Helv. Phys. Acta* **42,** 330–338.

Plymen, R. J. [1968a], C^*-Algebras and Mackey's Axioms, *Commun. Math. Phys.* **8,** 132–146.

Plymen, R. J. [1968b], A Modification of Piron's Axioms, *Helv. Phys. Acta* **41,** 69–74.

Pohlmeyer, K. [1968], Eine scheinbare Abschwächung der Lokalitätsbedingung, *Commun. Math. Phys.* **7,** 80–92.

Pool, J. C. T. [1968a], Baer *-Semigroups and the Logic of Quantum Mechanics, *Commun. Math. Phys.* **9,** 118–141.

Pool, J. C. T. [1968b], Semi-Modularity and the Logic of Quantum Mechanics, *Commun. Math. Phys.* **9,** 212–228.

Porta, H., and J. T. Schwarz [1967], Representations of the Algebra of all Operators in Hilbert Space, and Related Analytic Function Algebras, *Commun. Pure Appl. Math.* **20,** 457–492.

Powers, R. T. [1967a, b], Representations of Uniformly Hyperfinite Algebras and their Associated von Neumann Rings, *Bull. Am. Math. Soc.* **73,** 572–575; *Ann. Math.* **86,** 138–171.

Powers, R. T. [1967c], Absence of Interaction as a Consequence of Good Ultraviolet Behavior in the Case of a Local Fermi Field, *Commun. Math. Phys.* **4,** 145–156.

Powers, R. T. [1967d], *Representations of the Canonical Anti-Commutation Relations*, Thesis, Princeton.

Powers, R. T. [1971], Self-adjoint Algebras of unbounded Operators, *Commun. Math. Phys.* **21,** 85–124.

Powers, R. T., and E. Størmer [1970], Free States of the Canonical Anticommutation Relations, *Commun. Math. Phys.* **16,** 1–33.

Putnam, C. R. [1967], *Commutation Properties of Hilbert Space Operators and Related Topics*, Springer, Berlin.

Radin, C. [1970], Approach to Equilibrium in a Simple Model, *J. Math. Phys.* **11,** 2945–2955.

Radin, C. [1971a], Noncommutative Mean Ergodic Theory, *Commun. Math. Phys.* **21,** 291–302.

Radin, C. [1971b], Gentle Perturbations, *Commun. Math. Phys.* (to appear).

Reed, M. C. [1968], *On the Self-Adjointness of Quantum Fields and Hamiltonians*, Stanford University thesis.

Reed, M. C. [1969a], A Gårding Domain for Quantum Fields, *Commun. Math. Phys.* **14,** 336–346.

Reed, M. C. [1969b], The Damped Self-Interaction, *Commun. Math. Phys.* **11,** 346–357.

Reeh, H., and S. Schlieder [1961], Bemerkungen zur Unitäräquivalenz von Lorentzinvarianten Feldern, *Nuovo Cimento* **22,** 1051–1068.

Reeh, H., and S. Schlieder [1962], Über den Zerfall der Feldoperator algebra im Falle einer Vakuumentartung, *Nuovo Cimento* **26,** 32–42.

Rellich, F. [1946], Der Eindeutigkeitssatz für die Lösungen der quantenmechanischen Vertauschungsrelation, *Gött, Nach.*, 107.

Rickart, C. E. [1960], *General Theory of Banach Algebra*, Van Nostrand, Princeton, N.J.

Rideau, G. [1968], On Some Representations of the Anticommutation Relations, *Commun. Math. Phys.* **9,** 229–241.

Riesz, F., and B. Sz-Nagy [1955], *Leçons d'analyse functionnelle*, Gauthier-Villars, Paris, Akadémiai Kiadú, Budapest.

Roberts, J. E., and G. Roepstorff [1969], Some Basic Concepts of Algebraic Quantum Theory, *Commun. Math. Phys.* **11,** 321–338.

Robinson, D. W. [1965a], A Theorem Concerning the Positive Metric, *Commun. Math. Phys.* **1,** 89–94.

Robinson, D. W. [1965b], The Ground State of the Bose Gas, *Commun. Math. Phys.* **1,** 159–174.

Robinson, D. W. [1966a], Algebraic Aspects of Relativistic Quantum Field Theory, in *Axiomatic Field Theory*, M. Chretien and S. Deser, Eds., Gordon and Breach, New York.

Robinson, D. W. [1966b], *Symmetries, Broken Symmetries, Current and Charges*, Lecture notes, Istanbul.

Robinson, D. W. [1967], Statistical Mechanics of Quantum Spin Systems. I, *Commun. Math. Phys.* **6,** 151–160.

Robinson, D. W. [1968], Statistical Mechanics of Quantum Spin Systems. II, *Commun. Math. Phys.* **7,** 337–348.

Robinson, D. W. [1969], A Proof of the Existence of Phase Transitions in the Anisotropic Heisenberg Model, *Commun. Math. Phys.* **14,** 195–204.

Robinson, D. W. [1970], Normal and Locally Normal States, *Commun. Math. Phys.* **19,** 219–234.

Robinson, D. W., and D. Ruelle [1967a], Extremal Invariant States, *Ann. Inst. Henri Poincaré* **6,** 299–310.

Robinson, D. W., and D. Ruelle [1967b], Mean Entropy of States in Classical Mechanics, *Commun. Math. Phys.* **5,** 288–300.

Rocca, F., M. Sirugue, and D. Testard [1969a], Translation Invariant Quasi-Free States and Bogoliubov Transformations, *Ann. Inst. Henri Poincaré* **A10,** 247–258.

Rocca, F., M. Sirugue, and D. Testard [1969b], Quasi-Free States as Equilibrium States under the Kubo-Martin-Schwinger Boundary Condition, *Commun. Math. Phys.* **13,** 317–334.

Rocca, F., M. Sirugue, and D. Testard [1970], On a Class of Equilibrium States Under the KMS Condition. II. Bosons, *Commun. Math. Phys.* **16,** 119–141.

Roos, H. [1970], Independence of Local Algebras in Quantum Field Theory, *Commun. Math. Phys.* **16,** 238–246.

Ruelle, D. [1962], On the Asymptotic Condition in Quantum Field Theory, *Helv. Phys. Acta* **35,** 147–163.

Ruelle, D. [1966], States of Physical Systems, *Commun. Math. Phys.* **3,** 133–150.

Ruelle, D. [1967a], The States of Classical Statistical Mechanics, *J. Math. Phys.* **8,** 1657–1668.

Ruelle, D. [1967b], A Variational Formulation of Equilibrium Statistical Mechanics and the Gibbs Phase Rule, *Commun. Math. Phys.* **5,** 324–329.

Ruelle, D. [1967c], Quantum Statistical Mechanics and Canonical Commutation Relations in *Cargèse Lectures in Theoretical Physics*, F. Lurcat, Ed., Gordon and Breach, New York.

Ruelle, D. [1968], Statistical Mechanics of a One-Dimensional Lattice Gas, *Commun. Math. Phys.* **9,** 267–278.

Ruelle, D. [1969a], *Statistical Mechanics*, Benjamin, New York.

Ruelle, D. [1969b], Symmetry Breakdown in Statistical Mechanics, *Cargèse Lecture*, Vol. 4, D. Kastler, Ed., Gordon and Breach, New York.

Ruelle, D. [1970], Integral Representation of States on a C*-Algebra, *J. Funct. Anal.* **6,** 116–151.

Sakai, S. [1956], A Characterization of *W**-Algebra, *Pacific J. Math.* **6,** 763–773.

Sakai, S. [1957], On Topological Properties of *W**-Algebras, *Proc. Japan. Acad.* **33,** 439–444.

Sakai, S. [1962], *The Theory of W*-Algebras*, Lecture notes, Yale University.

Sakai, S. [1965], On the Central Decomposition for Positive Functionals on *C**-Algebras, *Trans. Amer. Math. Soc.* **118,** 406–419.

Sakai, S. [1971], *C*-Algebras and W*-Algebras*, Springer, Berlin.

Schatten, R. [1950], A Theory of Cross-Spaces, *Ann. Math. Studies No. 26*, Princeton University Press.

Schrieffer, J. R. [1964], *Theory of Superconductivity*, Benjamin, New York.

Schwartz, L. [1951], *Théorie des distributions*, Hermann, Paris.

Schwartz, J. T. [1967], *W*-Algebras*, Gordon and Breach, New York.

Schweber, S. S. [1961], *An Introduction to Relativistic Quantum Field Theory*, Row, and Peterson, Evanston, Ill.

Schwinger, J. [1951], On the Theory of Quantized Fields. I, *Phys. Rev.* **82,** 914–927.

Segal, I. E. [1947], Postulates for General Quantum Mechanics, *Ann. Math.* **48,** 930–948.

Segal, I. E. [1951], A Class of Operator Algebras Which Are Determined by Groups, *Duke Math. J.* **18,** 221–265.

Segal, I. E. [1956], Tensor Algebras Over Hilbert Spaces: I, *Trans. Amer. Math. Soc.* **81,** 106–134; II, *Ann. Math.* **63,** 160–175.

Segal, I. E. [1959, 1961, 1962], Foundations of the Theory of Dynamical Systems of Infinitely Many Degrees of Freedom: I, *Mat. Fys. Medd. Dan. Vid. Selsk.* **31,** No. 2; II, *Can. J. Math.* **13,** 1–18; III, *Ill. J. Math.* **6,** 500–523.

Segal, I. E. [1963], Mathematical Problems of Relativistic Physics, *American Mathematical Society*, Providence, R.I.

Sewell, G. L. [1970], Unbounded Local Observables in Quantum Statistical Mechanics, *J. Math. Phys.* **11,** 1868–1884.

Shale, D., and W. Stinespring [1964], States of the Clifford Algebra, *Ann. Math.* **80,** 365–381.

Shelupsky, D. [1966], Translations of the Discrete Bose-Einstein Operators, *J. Math. Phys.* **7,** 163–166.

Sherman, S. [1950], The Second Adjoint of a *C**-Algebra, *Proc. Intern. Congr. Math., Cambridge, Mass.*, Vol. I, 470. American Mathematical Society, Providence, R.I.

Sherman, S. [1951], Non-Negative Observables are Squares, *Proc. Amer. Math. Soc.* **2,** 31–33.

Sherman, S. [1956], On Segal's Postulates for General Quantum Mechanics, *Ann. Math.* **64,** 593–601.

Sinha, K. [1969], *A Non-Interaction Theorem for a Class of Bose Fields*, Thesis, Rochester.

Sinha, K., and G. G. Emch [1969], Adaptation of Powers' No-Interaction Theorem to Bose Field, *Bull. Am. Phys. Soc.* **14,** 86.

Sirugue, M., and M. Winnink [1970], Constraints Imposed upon a State of a System that Satisfies the KMS Condition, *Commun. Math. Phys.* **19,** 161–168.

Stone, M. H. [1932a], *Linear Transformations in Hilbert Space and Their Applications to Analysis*, American Mathematical Society, Providence, R.I.

Stone, M. H. [1932b]. On One-Parameter Unitary Groups in Hilbert Space, *Ann. Math.* **33,** 643–648.

Stone, M. H. [1949], Boundedness Properties in Function-Lattices, *Can. J. Math.* **1,** 176–186.

Størmer, E. [1965], On the Jordan Structure of C^*-Algebras, *Trans. Amer. Math. Soc.* **120,** 438–447.

Størmer, E. [1966], Jordan Algebras of Type I, *Acta Mathematica* **115,** 165–184.

Størmer, E. [1967a], Large Groups of Automorphisms of C^*-Algebras, *Commun. Math. Phys.* **5,** 1–22.

Størmer, E. [1967b], Types of von Neumann Algebras Associated with Extremal Invariant States, *Commun. Math. Phys.* **6,** 194–204.

Størmer, E. [1968a], On Partially Ordered Vector Spaces and their Duals, with Applications to Simplexes and C^*-Algebras, *Proc. London Math. Soc.* (3) **18,** 245–265.

Størmer, E. [1968b]. Irreducible Jordan Algebras of Self-Adjoint Operators, *Trans. Amer. Math. Soc.* **130,** 153–166.

Størmer, E. [1969a], Symmetric States of Infinite Tensor Products of C^*-Algebras, *J. Funct. Anal.* **3,** 48–69.

Størmer, E. [1969b], Asymptotically Abelian Systems, in *Cargèse Lectures*, Vol. 4., D. Kastler, Ed., Gordon and Breach, New York.

Størmer, E. [1970], The Even CAR-Algebra, *Commun. Math. Phys.* **16,** 136–137.

Streater, R. F. [1964], Intensive Observables in Quantum Theory, *J. Math. Phys.* **5,** 581–590.

Streater, R. F. [1965], Spontaneous Breakdown of Symmetry in Axiomatic Theory, *Proc. Roy. Soc.* (*London*) **287A,** 510–518.

Streater, R. F. [1967], The Heisenberg Ferromagnet as a Quantum Field Theory, *Commun. Math. Phys.* **6,** 233–247.

Streater, R. F., and A. S. Wightman [1964], *PCT, Spin and Statistics, and All That*, Benjamin, New York.

Streater, R. F., and I. F. Wilde [1970], The Time Evolution of Quantized Fields with Bounded Quasi-Local Interaction Density, *Commun. Math. Phys.* **17,** 21–32.

Streit, L. [1967], Test Function Spaces for Direct Product Representations of the Canonical Commutation Relations, *Commun. Math. Phys.* **4,** 22–31.

Streit, L. [1968], A Generalization of Haag's Theorem, Preprint, Pittsburgh.

Streit, L. [1970], Generalized Free Fields as Cyclic Representations of the Canonical Commutation Relations, *Acta Phys. Austriaca* **32,** 107–112.

Stueckelberg, E. C. G. [1960], Quantum Theory in Real Hilbert Space, *Helv. Phys. Acta* **33,** 727–752.

Stueckelberg, E. C. G., and M. Guenin [1961a], Quantum Theory in Real Hilbert Space, *Helv. Phys. Acta* **34,** 621–628.

Stueckelberg, E. C. G., and M. Guenin [1962], Quantum Theory in Real Hilbert Space, *Helv. Phys. Acta* **35,** 673–695.

Stueckelberg, E. C. G., M. Guenin, C. Piron, and H. Ruegg [1961b] Quantum Theory in Real Hilbert Space, *Helv. Phys. Acta* **34,** 675–698.

Swieca, J. A. [1967], Range of Forces and Broken Symmetries in Many-Body Systems, *Commun. Math. Phys.* **4,** 1–7.

Swieca, J. A. [1969], Goldstone's Theorem and Related Topics, in *Cargèse Lectures,* Vol. 4., D. Kastler, Ed., Gordon and Breach, New York.

Symanzik, K. [1967], Euclidean Proof of the Goldstone Theorem, *Commun. Math. Phys.* **6,** 228–232.

Takeda, Z. [1954a], Conjugate Spaces of Operator Algebras, *Proc. Japan. Acad.* **30,** 90–95.

Takeda, Z. [1954b, c], On the Representations of Operator Algebras: I, *Proc. Japan Acad.* **30,** 299–304; II, *Tohoku Math. J.* **6,** 212–219.

Takeda, Z. [1955], Inductive Limit and Infinite Direct Product of Operator Algebras, *Tohoku Math. J.* **7,** 68–86.

Takesaki, M. [1958], On the Conjugate Space of an Operator Algebra, *Tohoku Math. J.* **10,** 194–203.

Takesaki, M. [1960], On the nonseparability of Singular Representations of Operator Algebras, *Kodai Math. Sem. Reports,* **12,** 102–108.

Takesaki, M. [1970a], Disjointness of the KMS-States of Different Temperatures, *Commun. Math. Phys.* **17,** 33–41.

Takesaki, M. [1970b], *Tomita's Theory of Modular Hilbert Algebras and its Applications,* Springer, Berlin.

Takesaki, M. and M. Winnink [1971], Local Normality of the KMS states (to appear).

Tavel, M. [1964], *Weak and Electromagnetic Interactions in Quaternion Quantum Mechanics,* Preprint, Yeshiva.

Teichmuller, O. von [1935]; see von Teichmuller, O.

Testard, D. [1970], Types des représentations quasi-libres de l'algèbre de Clifford, *Ann. Inst. Henri Poincaré* **A12,** 329–341.

Thirring, W. [1968], On the Mathematical Structure of the B.C.S. Model, *Commun. Math. Phys.* **7,** 181–189.

Thirring, W. [1969], The Mathematical Structure of the BCS Model and Related Models, in *The Many-Body Problem,* L. M. Garrido, A. Cruz, and T. W. Preist, Eds., Plenum Press London.

Thirring, W., and A. Wehrl [1967], On the Mathematical Structure of the B.C.S. Model, *Commun. Math. Phys.* **4,** 303–314.

Tilgner, H. [1970], A Class of Solvable Lie Groups and Their Relation to the Canonical Formalism, *Ann. Inst. Henri Poincaré,* **A13,** 103–107.

Tilgner, H. [1971], A Class of Lie and Jordan Algebras Realized by Means of the Canonical commutation Relations, *Ann. Inst. Henri Poincaré,* **A14,** 179–188.

Tillman, H. G. [1963, 1964], Zur Eindeutigkeit der Lösungen der quantenmechanischen Vertauschungs Relationen: I, *Acta Sci. Math.* (Szeged) **24,** 258–270; II, *Arch. Math.* **15,** 332–334.

Topping, D. M. [1965a], Vector Lattices of Self-Adjoint Operators, *Trans. Amer. Math. Soc.* **115,** 14–30.

Topping, D. M. [1965b], Jordan Algebras of Self-Adjoint Operators, *Mem. Amer. Math. Soc.* **53,** 48 pp.

Trotter, H. F. [1958], Approximation of Semi-Groups of Operators, *Pacific J. Math.* **8,** 887–919.

Trotter, H. F. [1959], On the Product of Semi-Groups of Operators, *Proc. Amer. Math. Soc.* **10,** 545–551.

Turumaru, T. [1952, 1953, 1954, 1956], *On the Direct Product of Operator Algebras:* I, *Tohoku Math. J.* **4,** 242–251; II, *ibid.* **5,** 1–7; III, *ibid.* **6,** 208–211; IV, *ibid.* **8,** 281–285.

Uhlhorn, U. [1963], Representation of Symmetry Transformations in Quantum Mechanics, *Arkiv Fys.* **23,** 307–340.

Van Daele, A. [1971], Quasi-Equivalence of Quasi-Free States on the Weyl Algebra, *Commun. Math. Phys.* **22,** 171–191.

Van Daele, A., and A. Verbeure [1971], Unitary Equivalence of Fock Representations of the Weyl Algebra, *Commun. Math. Phys.* **20,** 268–278.

Van Hove, L. [1949], Quelques propriétés générales de l'intégrale de configuration d'un système de particules avec interaction, *Physica* **15,** 951–961.

Van Hove, L. [1951], Sur l'opérateur Hamiltonien de deux champs quantifiés en interaction, *Acad. Roy. Belg. Bull. Classe Sci.* **37,** 1055–1072.

Van Hove, L. [1952], Les difficultés de divergences pour un modèle particulier de champ quantifié, *Physica* **18,** 145–159.

Varadarajan, V. S. [1962], Probability in Physics and a Theorem on Simultaneous Probability, *Commun. Pure Appl. Math.* **15,** 189–217.

Varadarajan, V. S. [1968, 1970], *Geometry of Quantum Theory,* I and II, Van Nostrand, Princeton, N.J.

Verbeure, A. F., and E. J. Verboven [1966], Quantum States of an Infinite System of Harmonic Oscillators with Linear Interaction Terms, *Phys. Letters* **23,** 672–673.

Verbeure, A. F., and E. J. Verboven [1967a], States of Infinitely Many Oscillators, *Physica* **37,** 1–22.

Verbeure, A. F., and E. J. Verboven [1967b], States of Infinitely Many Oscillators with Linear Interaction Terms, *Physica* **37,** 23–31.

Verboven, E. J. [1966a, b], Quantum Thermodynamics of an Infinite System of Harmonic Oscillators; I, *Phys. Letters* **21,** 391–393; II, *Physica* **32,** 2081–2101.

Vigier, J. P. [1946], *Etude sur les suites infinies d'opérateurs hermitiens*, Thèse, Genève.

Viswanath, K. [1968], *Contribution to Linear Quaternionic Analysis*, Indian Statistical Institute, Calcutta, 1968.

Von Neumann, J. [1927a], Mathematische Begründung der Quantenmechanik, *Gött. Nachr.*, 1–57; Collected Works, Vol. 1, No. 9.

Von Neumann, J. [1927b], Wahrscheinlichkeitstheoretischer Aufbau der Quantenmechanik, *Gött. Nachr.*, 245–272; Collected Works, Vol. 1, No. 10.

Von Neumann, J. [1927c], Thermodynamik quantenmechanischer Gesamheiten, *Gött. Nachr.* 273–291; Collected Works, Vol. 1, No 11.

Von Neumann, J. [1931], Die Eindeutigkeit der Schrödingerschen Operatoren, *Math. Ann.* **104,** 570–578; Collected Works, Vol. 2, No. 7.

Von Neumann, J. [1932a], *Grundlagen der Quantenmechanik*, Springer, Berlin; English translation by R. T. Beyer, Princeton University Press, Princeton, N.J.

Von Neumann, J. [1932b], Zur Operatorenmethode in der klassischen Mechanik, *Ann. Math.* **33,** 587–642; Collected Works, Vol. II, No. 17.

Von Neumann, J. [1936], On an Algebraic Generalization of the Quantum Mechanical Formalism (Part I), *Mat. Sborn.* **1,** 415–484; Collected Works, Vol. 3, No. 9.

Von Neumann, J. [1938], On Infinite Direct Product, *Compos. Math.* **6,** 1–77; Collected Works, Vol. 3, No. 6.

Von Neumann, J., and R. Schatten [1946], The Cross-Space of Linear Transformations. *II, Ann. Math.* **47,** 608–630; Collected Works, Vol. 4, No. 29.

Von Teichmuller, O. [1935], Operatoren im Wachschen Raum, *J. Reine Angew. Math.* **174** 73–124.

Wagner, H. [1966], Long-Wavelength Excitations and the Goldstone Theorem in Many-Particle Systems with "Broken Symmetries," *Z. Phys.* **195,** 273–299.

Wentzel, G. [1943], *Einführung in die Quantentheorie der Wellenfelder*, Franz Deuticke, Wien, English translation, *Quantum Theory of Fields*, Interscience, New York, 1949.

Wick, G. C., A. S. Wightman, and E. P. Wigner [1952], Intrinsic Parity of Elementary Particles, *Phys. Rev.* **88,** 101–105.

Wielandt, H. [1949], Über die Unbeschränktkeit der Schrödingerschen Operatoren der Quantenmechanik, *Math. Ann.* **121,** 21.

Wieringa, J. D. [1970], *Thermodynamic Limit and KMS States in Quantum Statistical Mechanics: a C*-Algebraic Approach*, Thesis, Groningen.

Wightman, A. S. [1959a], Quelques Problèmes mathématiques de la theorie quantique relativiste, in *Les problèmes mathématiques de la théorie quantique des champs*, CNRS, Paris.

Wightman, A. S. [1959b], Relativistic Invariance and Quantum Mechanics, *Nuovo Cimento*, Suppl. XIV, 81–94.

Wightman, A. S. [1960], L'Invariance dans la Mécanique Quantique Relativiste, in *Relations de Dispersion et Particules Elémentaires*, C. de Witt and R. Omnes, Eds., Hermann, Paris.

Wightman, A. S. [1962], Localizability of Quantum Mechanical Systems, *Rev. Mod. Phys.* **34,** 845–872.

Wightman, A. S. [1964], La théorie quantique locale et la théoric quantique des champs, *Ann. Inst. Henri Poincaré*, **I,** No. 4, Section A, 403–420.

Wightman, A. S. [1967], Progress in the Foundations of Quantum Field Theory, in *Proceedings of the 1967 International Conference on Particles and Fields*, C. R. Hagen, G. Guralnik, and V. A. Mathur, Eds., Wiley-Interscience, New York.

Wightman, A. S., and S. S. Schweber [1955], Configuration Space Methods in Relativistic Quantum Field Theory. I, *Phys. Rev.* **98,** 812–837.

Wigner, E. P. [1931], *Gruppentheorie und ihre Anwendung*, Vieweg, Braunschweig; English translation by J. J. Griffin, Academic Press, New York, 1959.

Wigner, E. P. [1939], On Unitary Representations of the Inhomogeneous Lorentz Group, *Ann. Math.* **40,** 149–204.

Wigner, E. P. [1962], Representations of the Poincaré Group, in Group Theoretical Concepts and Methods in Elementary Particle Physics, *Proc. Nato Summer School, Istanbul*, F. Gursey, Ed., Gordon and Breach, New York.

Wils, W. [1968], Desintégration centrale des formes positives sur les C^*-algèbres, *C.R. Acad. Sc. Paris*, **267,** 810–812.

Wils, W. [1969], *Central Decomposition of C^*-Algebras*, Lecture notes, Aarhus.

Winnink, M. [1968], *An Application of C^*-Algebras to Quantum Statistical Mechanics of Systems in Equilibrium*, Thesis, Groningen.

Wintner, A. [1947], The Unboundedness of Quantum-Mechanical Matrices, *Phys. Rev.* **71,** 738–739.

Woods, E. J. [1970], Continuity Properties of the Representations of the Canonical Commutations Relations, *Commun. Math. Phys.* **17,** 1–20.

Wulfsohn, A. [1963], Produit tensoriel de C^*-algebres, *Bull. Sc. Math.* 2ᵉ serie, **87,** 13–27.

Wyss, W. [1969], On Wightman's Theory of Quantized Fields, in *Lectures in Theoretical Physics*, Vol. XI–D, K. T. Mahanthappa and W. E. Britten, Eds., Gordon and Breach, New York.

Yang, C. N., and T. D. Lee [1952], Statistical Theory of Equations of State and Phase Transitions. I, Theory of Condensation. *Phys. Rev.* **87,** 404–409; II, *see* Lee and Yang [1952].

Yukawa [1935], On the Interaction of Elementary Particles. I, *Proc. Phys. Math. Soc. Japan* **17,** 48–57; reprinted in *Foundations of Nuclear Physics*, R. T. Beyer, Ed., Dover, New York, 1949.

Index